PESTS OF FRUIT CROPS

NIPA GENX ELECTRONIC RESOURCES & SOLUTIONS P. LTD.
New Delhi-110 034

About the Editors

Dr. Nripendra Laskar, Professor of Agricultural Entomology completed his B. Sc. (Ag.) and M. Sc. (Ag.) in Agricultural Entomology degree with specialisation in Economic Entomology from Bidhan Chandra Krishi Visvavidyalaya Mohanpur, Nadia, West Bengal and qualified NET organised by ASRB, ICAR, New Delhi in 2000. He obtained his Ph. D. Degree from Visva-Bharati, Santiniketan, West Bengal. Dr. Laskar has 20 years of research, teaching and extension experience in his concerned field of specialisation. He has successfully guided a number of students in both M. Sc. and Ph.D. level and presently working on Tephritids of economic importance and Beekeeping. He has published more than 40 research papers in different journals of National and International repute, 14 book chapters, 05 books and several popular articles. He is the member of a number of scientific societies in India and member of editorial board of the "Journal of Agriculture and Technology" and Editor of a half yearly bengali magazine of agriculture, folk culture and literature, "Ujaan". In addition, Dr. Laskar is presently acting as the Principal Investigator of 05 research projects. He has attended and presented scientific findings on Agricultural Entomology in different National and International Seminar, Symposium and Conferences.

Dr. Victor Phani, completed his B. Sc. (Ag.) from Bidhan Chandra Krishi Viswavidyalaya, and M. Sc. and Ph. D. from ICAR-Indian Agricultural Research Institute, New Delhi, India and qualified NET organised by ASRB, ICAR, New Delhi in 2016. Currently he is serving as Assistant Professor, Departmentof Agricultural Entomology, College of Agriculture, Uttar Banga Krishi Viswavidyalaya, Majhian, Dakshin Dinajpur, WB, India. Dr. Phani is a young, energetic and dynamic worker and his area of research is concerned with the understanding of host-parasite relationships, biocontrol and taxonomic studies of the plant-parasitic nematodes and insect pests. He has published several research papers in different journals of national and international repute.

PESTS OF FRUIT CROPS

Cheif Editor

Nripendra Laskar

Professor

Department of Agricultural Entomology

Uttar Banga Krishi Viswavidyalaya

Pundibari, Cooch Behar, West Bengal

Co- Editor

Victor Phani

Assistant Professor

Department of Agricultural Entomology

College of Agriculture, Uttar Banga Krishi Viswavidyalaya

Majhian, Dakshin Dinajpur, West Bengal

NIPA GENX ELECTRONIC RESOURCES & SOLUTIONS P. LTD.

New Delhi-110 034

NIPA GENX ELECTRONIC RESOURCES & SOLUTIONS P. LTD.

101,103, Vikas Surya Plaza, CU Block
L.S.C. Market, Pitam Pura, New Delhi-110 034
Ph. +91 11 27341616, 27341717, 27341718
E-mail: newindiapublishingagency@gmail.com
www: www.nipabooks.com

For customer assistance, please contact

Phone: + 91-11-27 34 17 17
Fax: + 91-11-27 34 16 16
E-Mail: feedbacks@nipabooks.com

ISBN: 978-93-94490-70-3

Composed and Designed by NIPA.

Preface

Fruits play an important role in providing nutritional security and several other advantages in the society. Fruits are also valuable from religious, mythological and artistic aspects. These are not only delicious but also have many nutrients which are very much essential for human health. Man cannot live on cereals alone and importance of fruits in balanced diet is well recognized. These are good sources of vitamins and minerals without which human body cannot maintain proper health and develop resistance to diseases. In the modern era, the Global conception of transformation in the food consumption pattern is away from rice, wheat and pulses and this health consciousness are causing major shift in the focus towards the fruit crops that resulted in greater demand for fruits in domestic as well as international market. The fruit crops produce good yield and generate higher return even ona small piece of land.

Due to prevalence of diverse edaphic and climatic conditions almost all kinds of fruits can be grown successfully in our country. Even the dry land can also beutilised in cultivating the hardy fruit crops like ber, custard apple, cashew nut, amla etc. Increased transport and cold storage facilities and development of Agro-Industries are paving the way smooth for making fruit production an attractive and remunerative employment opportunity. By growing fruit, farmers get more income from the field as compared toother cash crops. Production of fruits thus, play a pivotal role in strengthening the agrarian Indian economy by way of creating greater employment opportunity vis-a-vis providing raw materials to the Agro-Industries.

India is the second largest producer of fruits in the world.Fruit production of India is 88.98 million tonnes from an area of 7.21 million hectares that is rated second in the world after China. Fruits are grown throughout the country, but the state of Maharashtra is contributing significantly to the extent of 15.12% toward fruit basket of India followed by A. P. (11.81%), Gujarat (8.99%), T. N. (8.28%), U. P. (7.74%) and Karnataka (7.47%). Maharashtra is not only major fruit growing state of India but also variety of fruits like mango, grapes, citrus, sapota, pomegranate, banana, papaya etc. are grown here.

In the era of climate change and chemicalised cultivation, successful production of fruits is hampered badly due to the incidence of several insect pests, mites and nematode parasites. Sustainable management of the pests of fruit crops is really a complex issue in view of the intensive cultivation system. Lack of adequate knowledge with regard to plant nutrition, protection and economic constraints added to the complexity of fruit production. In the changing climatic condition the pest scenario is changing day by day. The invasive alien species (IAS) of pests are not uncommon under free trade and frequent travelling of peoples from one country to another. A considerable portion of both pre-harvest and post-harvest crop losses have been witnessed every year. To cope up with this dreaded problem, toxic synthetic pesticides have offered a simple solution to the growers as they are cheap, easy to apply and readily available. Farmers are using this tool indiscriminately even when it is not at all required.

Rampant use of this ultimate weapon for managing the pests in orchard ecosystem creating a number of incurable problems like resistance, resurgence, secondary pest outbreak and human health hazards. These necessitated seeking scientific information on management of pests in fruit production and re-evaluating the role of pesticides. A good number of sustainable IPM pest management strategies have also been developed using traps, bioagents, botanicals and safer pesticides. The information on the insect pests, mites and nematode parasites of fruit crops and their sustainable management are scattered. There is scarcity of books having exhaustive and exclusive information on pests of fruit crops and their eco-friendly management.

This book deals comprehensively on the pests of fruit crops along with their identification, biology, mode of feeding and symptoms of infestation, economic threshold level (ETL) and integrated management. Good quality photographs of some of the important pests have been accommodated. The book will be useful for the students, researchers, plant protection specialists, extension workers and the growers as well. Suggestions to improve the contents of the book are most welcome.

The editors profusely thank Professor Biswanath Bandyopadhyay, Ex-Vice-Chancellor, Uttar Banga Krishi Viswavidyalaya and also Retired Professor of Agricultural entomology, Bidhan Chandra Krishi Viswavidyalaya for his valuable suggestions and continuous encouragement. We express heartfelt gratitude to the contributors who have kindly cooperated by providing their valued articles in spite of their busy schedule in respective profession. The publisher, NIPA, New Delhi deserves the commendation for their professional contribution.

Editors

Contents

List of Contributors

Adrish Dey, Department of Entomology, Uttar Banga Krishi Viswavidyalaya, Pundibari Cooch Behar, West Bengal-736165

Amitava Banerjee, Bidhan Chandra Krishi Viswavidyalaya, Mohanpur, Nadia West Bengal-741221

Anil Kumar, Department of Entomology, Faculty of Agriculture, Dr. Rajendra Prasad Central Agricultural University, Pusa, Samstipur, Bihar-848 125

Atanu Maji, Department of Entomology, Uttar Banga Krishi Viswavidyalaya, Pundibari Cooch Behar, West Bengal-736165

Atanu Seni, Regional Research and Technology Transfer Station, Odisha University of Agriculture & Technology, Chiplima, Sambalpur, Odisha-768025, India

Biswajit Patra, Regional Research Station (Hill Zone), Uttar Banga Krishi Viswavidyalaya Kalimpong, West Bengal-734301

Biswanath Bandyopadhaya, Bidhan Chandra Krishi Viswavidyalaya, Mohanpur, Nadia, West Bengal-741221

Biwash Gurung, Department of Entomology, Uttar Banga Krishi Viswavidyalaya, Pundibari Cooch Behar, West Bengal-736165

Debashis Roy, Bidhan Chandra Krishi Viswavidyalaya, Mohanpur, Nadia, West Bengal-741221

G. Mahendiran, ICAR-Central Institute of Temperate Horticulture, K. D. Farm, Old Air Field Rangreth, Srinagar, Jammu and Kashmir-190007

Gobinda Roy, Department of Entomology, Uttar Banga Krishi Viswavidyalaya, Pundibari Cooch Behar, West Bengal-736165

Jagdish Jaba, The International Crops Research Institute For The Semi-Arid Tropics (ICRISAT), Patancheru, Hyderabad, Telangana

Jaydeep Halder, Indian Institute of Vegetable Research (IIVR), Varanasi, Uttar Pardesh-221305

Kalmesh Manganavi, Department of Entomology, Bihar Agricultural University, Sabour Bhagalpur, Bihar

Mudasir Ahmad Dar, ICAR-Central Institute of Temperate Horticulture, K.D. Farm, Old Air Field, Rangreth, Srinagar, Jammu and Kashmir-190007

Nagendra Kumar, Department of Entomology, Faculty of Agriculture, Dr. Rajendra Prasad Central, Agricultural University, Pusa, Samstipur, Bihar-848 125

Narendra Prakash, ICAR Research Complex For Neh Region, Manipur Centre, Lamphelpat Manipur-795004

Nithya Chandran, Division of Entomology, Indian Agricultural Research Institute (IARI) New Delhi-110012

Nripendra Laskar, Department of Agricultural Entomology, Uttar Banga Krishi Viswavidyalaya, Pundibari Cooch Behar, West Bengal, India-736165

P. Venkata Rao, Department of Entomology, Uttar Banga Krishi Viswavidyalaya, Pundibari, Cooch Behar, West Bengal-736165

Pijush Kanti Sarkar (Retd. Professor), Bidhan Chandra Krishi Viswavidyalaya, Mohanpur, Nadia, West Bengal-741221

Puran Pokhrel, School of Agriculture, Swami Vivekananda University, Malir Math, Kolkata, West Bengal-700121

Rachana R. R., Department of Insect Ecology, National Bureau of Agricultural Insect Resources (NBAIR), P.bag No:2491, H.a. Farm Post, Bellary Road, Bengaluru - 560 024 Karnataka

Rajesh Kumar, Divyayan Krishi Vigyan Kendra, Ramakrishna Mission Vivekananda University, Ranchi, Jharkhand - 834008, India

Richa Varshney, Department Of Insect Systematics, National Bureau Of Agricultural Insect Resources (NBAIR), P. Bag No:2491, H.A. Farm Post, Bellary Road, Bengaluru - 560 024, Karnataka

Riju Nath, Department Of Entomology, Uttar Banga Krishi Viswavidyalaya, Pundibari, Cooch Behar, West Bengal-736165

Romila Akoijam, ICAR Research Complex for NEH Region, Manipur Centre, Lamphelpat-795004, Manipur

Roshna Gazmer, Department of Agricultural Entomology, Uttar Banga Krishi Viswavidyalaya, Pundibari, Cooch Behar - 736165, West Bengal

S.J. Prasanthi, Department of Agriculture, Jharkhand Rai University, Ranchi-834008 Jharkhand

Samrat Saha, Department of Entomology, Uttar Banga Krishi Viswavidyalaya, Pundibari Cooch Behar, West Bengal-736165

Sandip Patra, ICAR Research Complex for NEH Region, Umroi Road, Umiam Meghalaya-793103

Shahid Ali Akbar, ICAR-Central Institute of Temperate Horticulture, K.D. Farm, Old Air Field, Rangreth, Srinagar, Jammu and Kashmir-190007

Shyamal Kumar Sahoo, Department of Entomology, Uttar Banga Krishi Viswavidyalaya, Pundibari, Cooch Behar, West Bengal-736165

Siva Kumar Golla, The International Crops Research Institute For The Semi-Arid Tropics (ICRISAT), Patancheru, Hyderabad, Telangana

Suprakash Pal, Department of Entomology, Uttar Banga Krishi Viswavidyalaya, Pundibari, Cooch Behar, West Bengal-736165

Suraj Sarkar, Cooch Behar Krishi Vigyan Kendra, Uttar Banga Krishi Viswavidyalaya Pundibari, Cooch Behar, West Bengal-736165

T. Bharathimeena, Indian Institute of Vegetable Research (IIVR), Varanasi, Uttar Pradesh- 221 305, India

Tamoghna Saha, Department of Entomology, Bihar Agricultural University, Sabour, Bhagalpur-813210, Bihar

Tarak Nath Goswami, Department of Entomology, Bihar Agricultural University, Sabour, Bhagalpur-813210, Bihar.

Tushar Kanti Dutta, Division of Nematology, Indian Agricultural Research Institute (IARI) New Delhi-110012

Umashankar Nayak, Regional Research Station, Ranital, Odisha University Of Agriculture & Technology, Bhadrak, Orissa, India-756100

Victor Phani, Department of Entomology, College Of Agriculture, Uttar Banga Krishi Viswavidyalaya, Majhian, Dakshin Dinajpur, West Bengal- 733141

1

Pests of Mango Ecosystem and Their Integrated Management

Shyamal Kumar Sahoo, Uma Shankar Nayak and Atanu Maji

Mango (*Mangifera indica* Linn.) (Family-Anacardiaceae) is one of the most popular tropical and sub-tropical fruit crops in India and is well known as the "King of Fruits". Mango is the national fruit of India and Pakistan as well as national tree of Bangladesh. It is cultivated worldwide in 49.46 lakh hectares with a production of 371.20 lakh tonnes. In India, the crop is cultivated in the largest area of 23.12 lakh hectares among all the countries in the world and production is around 150.30 lakh tonnes, contributing 40.48% of the total world production. India exports 59.22 thousand tonnes of mangoes to 40 different countries worldwide, valuing Rs. 162.92 crores (Anonymous, 2019).

Mango is grown in wide range of soil and climatic conditions of India in almost all the States right from Kashmir to Kanyakumari. Mango fruit is rich in dietary fibre, vitamins (vitamin-A, vitamin-B6, vitamin-C and vitamin-E), minerals and antioxidant compounds, which plays a significant role in improving human health and nutrition.

Average productivity of mango in our country is just 8 tonnes/ha compared to 15 tonnes/ha of Israel and Australia which have introduced mango recently. Several factors are responsible for lower productivity such as unfavourable weather in this sub-continent during flowering and fruiting stage, improper pollination, lack of pollinators, percentage of hermaphrodite flowers, improper cultural and other management practices, insect-pests and diseases etc. Among them losses due to different insect-pests is most significant sometimes upto the tune of 90%. Mango tree is infested by more than dozen major insects round the year because of evergreen perennial nature of the tree and rich nutritional properties of the mango fruit. Tandon and Verghese (1985) have listed more than 400 insect-pests associated with mango during different growth stages in India. Among pestiferous arthropods, majority (about 45% of total species) are foliage feeders followed by fruit feeders (32%) (Reddy and Sridevi, 2016).

Mango hoppers, mealy bug, bark eating caterpillar, leaf cutting weevil, shoot borer, stem borer and fruit fly occurs in almost all the mango growing areas of India and cause significant damage in every year. Occurrence of stone weevil, shoot gall psylla, defoliator, gall causing insects and termite is region specific in India.

Table 1: Pest complex in mango ecosystem

Sl. No.	Common Name	Scientific Name	Family	Order
1.	Mango hopper	*Amritodus atkinsoni* (Lethiery) *Idioscopus clypealis* (Lethiery) *Idioscopus niveosparsus* (Lethiery) *Idioscopus nitidulus* (Lethiery) *Amrasca splendens* Ghauri	Cicadellidae	Hemiptera
2.	Mealy bug	*Drosicha mangiferae* (Green) *Drosicha stebbingi* (Green) *Perissopneuman ferox* Newst	Margarodidae	Hemiptera
3.	Fruit fly	*Bactrocear dorsalis* (Hendal) *Bactrocear zonata* Sanuders *Bactrocera diversa* Coq. *Bactrocera tau* Walker *Bactrocera correcta* Bezzi	Tephritidae	Diptera
4.	a. Flower gall midge	*Erosomyia indica* Grover *Dasineura amaramanjarae* Grover *Procystiphora mangiferae* (Felt) and *P. indica* Grover	Cecidomyiidae	Diptera
	b. Leaf gall midge	*Amradiplosis allahabadensis* Grover *Procontarinia mangifoliae* Grover *Procontarinia matteiana* Kieffer and Cecconi	Cecidomyiidae	Diptera
	c. Branch gall midge	*Rhabdophaga mangiferae* Mani	Cecidomyiidae	Diptera
5.	Stone weevil	*Sternochetus mangiferae* (Fab.)	Curculionidae	Coleoptera
6.	Shoot gall psylla	*Apsylla cistellata* Buckton	Psyllidae	Hemiptera
7.	Stem borer	*Batocera rufomaculata* de Geer *Batocera rubus* (Linn.) *Batocera titana* (Thompson)	Cerambycidae	Coleoptera
8.	Bark eating caterpillar	*Indarbela quadrinotata* (Walker) *Indarbela tetraonis* (Moore) *Indarbela theivora* (Hampson)	Metarbelidae	Lepidoptera

Sl. No.	Common Name	Scientific Name	Family	Order
9.	Leaf cutting weevil	*Deporaus* (*Eugnamptus*) *marginatus* Pascoe	Attelabidae	Coleoptera
10.	Fruit borer	*Autocharis* (*Noorda*) *albizonalis* Hampson *Dichocrocis (Conogethes) punctiferalis* (Guenee)	Pyralidae	Lepidoptera
11.	Shoot borer	*Chlumetia transversa* Walker	Noctuidae	Lepidoptera
12.	Scale insect	*Aspidiotus destructer* Sign *Rastrococcus iceryoides* (Green)	Diaspidae	Hemiptera
13.	Leaf webber or nesting caterpillar	*Orthaga exvinacea* Hampson *Orthaga euadrusalis* Walker *Orthaga mangiferae* Mishra	Pyraustidae	Lepidoptera
14.	Mite	*Aceria mangiferae* Sayed	Eriophyidae	Acarina
		Oligonychus mangiferus (Rahman and Sapra)	Tetranychidae	Acarina
15.	Pulp weevil	*Sternochetus frigidus* (Fab.). *Sternochetus gravis* (Fab.).	Curculionidae	Coleoptera
16.	Flower webber	*Eublemma versicolor* Walk.	Noctuidae	Lepidoptera
17.	Thrips	*Rhipiphorothrips cruentatus* Hood *Scirtothrips dorsalis* Hood *Frankliniella occidentalis* (Pergande)	Thripidae	Thysanoptera

Mango tree harbours different arthropod pests in different phenological stages because of the perennial and evergreen nature of the plant. Proper knowledge about the time of infestation of the insect pests is necessary, so that suitable management strategy could be implemented timely. Therefore, among the different arthropod pests recorded in mango, the more severe and common pests causing considerable economic loss to the crop along with their period and stage of damage are discussed below.

1. Mango hopper, *Amritodus atkinsoni* Lethiery, *Idioscopus clypealis* Lethiery, *Idioscopus niveosparsus* Lethiery, *Idioscopus nitidulus* Walker, *Amrasca splendens* Ghauri (Cicadellidae: Hemiptera)

Totally 18 species of mango hopper have been reported throughout the world (Pena and Mohyuddin, 1997). However, generally three species of hoppers are reported from different mango growing areas of India *i.e.*, *Amritodus atkinsoni* (Lethierry), *Idioscopus clypealis* (Lethierry) and *I. niveosparsus* (Lethierry). Das *et al.* (1969) recorded a new hopper, *Amrasca splendens* Ghauri from Kerala causing severe damage to mango plantation. Viraktamath and Viraktamath (1985) reported three new species of mango hoppers namely *Busoniominus manjunathi, Idioscopus anasnyal* and *Idioscopus jayshriae* on mango in Karnataka. *A. atkinsoni* is more common in North India. *I. clypealis* is found all over India.

Distribution

Mango hoppers are the most serious major pests of mango and widespread throughout the mango growing areas of our country (Verghese, 2000). Apart from India it is distributed in Pakistan, Bangladesh, Sri Lanka, Indonesia, Philippines, Taiwan, Vietnam and Burma. In India it is prevalent in almost all the mango growing areas like of West Bengal, UP, Bihar, Orissa, Punjab, Gujarat, Maharashtra, T.N., A.P., H.P. and Rajasthan.

Hosts of commercial importance

According to Pruthi and Batra (1960) it is a monophagous pest of mango while another record from Udaipur, India that mango hopper also infests fig (*Ficus carica*) and cesiman (*Thrunstira deliciose*). However, almost all the research workers have reported this hopper as a monophagous pest of mango.

Identifying features

Nymph

Freshly hatched nymphs are wedge-shaped and whitish in colour with two small red eyes. Gradually with each moulting, colour changes to yellow, yellowish green, green and ultimately greenish brown. They are very active moving rapidly around the inflorescence and shoots. The nymphs are wingless, unable to fly or jump but having side movement. To notice their presence our hand has to be placed at the opposite side of the trunk or inflorescence then they will move in front of us.

Adult

All the species of mango hopper have a wedge-shaped body with a broad head and narrow abdomen towards the anal end. The hind pair of legs are adapted for hopping. Different identifying features of the commonly found hoppers are mentioned below.

A. atkinsoni	*I. niveosparsus*	*I. clypealis*
i) Two spots on the scutellum and abdomen.	i) Three spots on scutellum and white cross band across the forewing.	i) Two spots on the scutellum, dark spot on the vertex.
ii) Dark grey colour	ii) Dark colour	ii) Light brown colour
iii) Largest in size (4.2 - 5mm length)	iii) Medium in size (4 mm)	iii) Smallest in size (3.5 mm)

Biology

The female hopper inserts the minute egg singly into the plant tissue through slits make on the shoots, flower buds and inflorescence stalk. The eggs hatch out in

4 - 7 days. The nymphs undergo 4- 5 moulting in 10-13 days and become adults. The period from egg to adult takes about 12 - 17 days and during a flowering season two or more broods of the pest may occur (Paul, 2007). According to Atwal (1963), the incubation period ranges from 4-7 days, nymphal period from 8-13 days with three instars and the total life cycle is completed in 15-19 days. Hoppers hibernate in cracks and crevices of the barks on the tree.

Seasonal incidence

The hoppers are active during flowering period and in the remaining period, they remain confined to the under surface of leaves, situated in dark and moist areas of the tree (Ramesh Babu and Singh, 2014). Warm, humid and cloudy climate is most congenial for their population build up (Chowdhury, 2015). They remain active throughout the year but the incidence is severe during February to April synchronizing with the flowering and fruit setting of mango trees. Hopper population are also noticed during June-August on vegetative flush.

Mode of feeding and symptoms of infestation

The damage is done by two ways, (i) Primary damage is done in the inflorescences and in other tender parts of the mango plant by physical injury due to insertion of the ovipositor for egg laying led to drying and withering of such plant parts. (ii) Secondary damage is caused by both the nymph and adult hoppers which puncture and suck the cell sap by their piercing and sucking type of mouthparts from tender leaves, shoots, inflorescences, flowers and tender fruits. Due to heavy drain out of vital plant sap withering and shedding of flower buds as well as wilting and drying of shoots and leaves occur. In addition to these, hoppers also secrete honey dew on the different plant parts after heavy sucking of the plant sap which encourage the growth of black sooty mould fungus (*Capnodium mangiferae* Cooke) hampering the normal photosynthesis of the tree thereby affect the growth and development of the plant.

In case of severe infestation, the entire garden shows a sickly appearance. The trees are deprived of buds and blossom and the leaves appear shiny and covered with sooty mould and thousands of cast skins of the hoppers are found on the shoots and leaves. This damage caused by hoppers is known as "Honey Dew Disease" (Theni Manzu) in many parts of Andhra Pradesh. Due to continuous injury by this pest, the trees gradually lose their vigour and affecting yield capacity in the long run. Srivastava (1997) found that the orchard having closer spacing and tall plants attracted higher hopper population which facilitated better breeding of these hoppers and cause lot of damage. During higher infestation, characteristic clicking sounds of leaf hoppers can be heard. Old, neglected and closely planted orchards that are shady and with high humid conditions favour the multiplication of mango hopper (Chowdhury, 2015).

Economic threshold level (ETL)

Corey *et al*. (1989) determined the ETL for *Idioscopus clypealis* as 4 - 4.5 hoppers/ panicle.

Integrated management

a. Cultural approaches

i. Avoiding close planting as the hopper incidence is severe in overcrowded orchards.

ii. Pruning of dense overlapping branches after rainy season for better light interception to check the pest population build up.

iii. Keeping the orchard clean by regular ploughing and removal of weeds as well as alternate hosts like custard apple, hibiscus, guava etc.

iv. Water logging and swampy conditions in the orchard should be avoided through proper drainage.

v. Avoiding excess application of nitrogenous fertilizers.

b. Mechanical approaches

i. Installations of light trap @ 1/ acre and operate between 6 pm to 10 pm followed by collection and disposal of the trapped hopper.

c. Biological approaches

i. Neuropteran insects like *Mallada boninensis* and *Chrysopa lacciperda*, some unidentified mantids, twelve species of spider, Coccinellid beetle like *Coccinella septempunctata*, *C. transversalis* and *Menochilus sexmaculatus* predate upon mango hopper (Srivastava, 1997). Mymarid like *Gonatocerus* sp. and *Polynema* sp., Eulophid like *Aprostocetus* sp. and *Tetrastichus* sp. parasitize the eggs of *A. atkinsoni* and *I. nitidulus*. Dipteran parasites *Pipunculus annulifemer*, Lepidopteran parasite *Epipyrops fuliginosa* and Strepsiptera parasite are also recorded against mango hopper. Entomopathogenic fungus (EPF), *Lecanicillium lecanii* and *Beauveria bassiana* are pathogenic to *I. clypealis* (Srivastava and Tandon, 1986).

ii. Application of neem-based formulations like neem oil @ 3% and NSKE @ 5% to keep the pest population below ETL.

iii. Application of biopesticides, *Metarhizium anisopliae* or *Beauveria bassiana* or *Lecanicillium lecanii* @ 6 ml/lt. (1 x 10^8 cfu /ml) on tree trunk once during off season and twice at 7 days interval during flowering season to keep the pest population below ETL (Anonymous, 2014).

d. Chemical approaches

i. Need based spraying of insecticides like Imidacloprid 17.8 SL @ 150 ml/ ha or Thiomethoxam 25 WG @ 150 g/ ha or Ethofenprox 10 EC @ 500 ml/ ha or Lambda cyhalothrin 5 EC @ 500 ml/ ha on rotational basis to prevent insecticide resistance development. First spray should be done at early stage of panicle formation if hopper population is more than 5-10/ panicle, second spray at full length stage of panicle and the third spray after fruit setting at pea size (Prakash, 2012).

ii. Following insecticides may be used against mango hopper as per the Department of Agriculture and Cooperation, Govt. of India (2014):

Buprofezin 25% SC @ 1.25ml/ l of water; Deltamethrin 2.8% EC @ 0.33 to 0.5 ml/lit; Dimethoate30% EC @ 1.65ml /lit.; Imidacloprid17.8% SL @ 0.33ml /l; Lambda-cyhalothrin 5% EC @ 0.5-1.0 ml/ l of water; Malathion 50% EC @ 1.5 ml / l of water; Oxydemeton methyl 25% EC@ 1 ml/ l of water. Insecticide to be mixed properly with 10-15 lit. of water for spraying in each 20-25 years old mango tree.

iii During flowering spraying of insecticides must be done in the afternoon hours to protect the pollinators.

2. Mango mealy bug, *Drosicha mangiferae* Green, *Drosicha stebbingi* Green, *Rastrococcus iceryoides* Green (Margarodidae: Hemiptera)

Mango mealy bug, *Drosicha mangiferae* Green is another serious pest of mango that causes severe damage and considerable yield loss to the crop. Based on the extent of yield loss, it is ranked 2nd important insect pest after mango hopper and the yield loss may extend up to 50-80 percent in some years (Atwal, 1976 and Karar *et al*., 2012). More than 20 species of mealy bug were reported in mango. Among them *Drosicha mangiferae* Green, *Drosicha stebbingi* Green and *Rastrococcus iceryoides* Green are reported as the major pest of mango in India, Bangladesh, Nepal, Bhutan, China and Pakisthan.

Distribution

According to Tandon and Lal (1976) the pest is observed in all the mango growing states of our country except Kerala and Karnataka. The level of infestation is serious in Punjab, UP, Delhi and Bihar while moderate in West Bengal and low in Orissa, Gujarat, HP, MP, TN, AP, and Maharashtra.

Hosts of commercial importance

The pest being polyphagous in nature has a wide host range which includes fruit plants, ornamental plants, vegetables, weeds etc. So far, it has been recorded in 71 plants species viz, apple, apricot, ber, cherry, citrus, jamun, loquat, papaya, litchi, guava, pear, peach, plum, phalsa, pomegranate, sapota, tamarind, sunflower, arhar, arjun, brinjal etc. Although the pest feed on different plant species mentioned above but it prefers mostly mango crop.

Identifying features

Freshly laid eggs are oval in shape and pink in colour and later on transform into yellow colour. First instar nymphs are chestnuts brown in colour, very active and are negatively geotropic in nature. Second instar nymph resembles first instar and body colour is dull, grey brown with black colour antennae and proboscis. Third instar female nymph is more densely covered with waxy material secreted by dermal glands. Third instar male nymph is known as prepupa and it resemble third instar female nymph but body size is smaller. In case of male, pupal stage is found in cracks and crevices of the bark in form of cottony cocoon. A well-defined sexual dimorphism is seen in the adult stage with the females are wingless and large-bodied having mouthparts and the males, on the other hand, are winged with only one pair of wings and atrophied.

Biology

The adult gravid females after fertilization crawl down along the tree-trunk to the ground where they lay eggs in clusters of 300 - 500 at the depths of about 2-6 inches round the base of the plant in cottony ovisac. This depth of soil depends upon the soil texture and soil moisture. The egg remains in the diapause stage for about 7 months (June- December). Migration of female insects from the tree downwards to the ground and oviposition in the soil generally happen during the months from April to June (Kumar *et al.*, 2009). Eggs can hatch as early as December of the same year or as late as March of the succeeding year. Late monsoon and winter rains have been reported to delay hatching of the insect. The young nymphs soon after hatching crawl about in search of suitable sites for colonization. Thereafter, they move up in the tree-trunks and this upward migration lasts for several weeks. About 80 % of the newly hatched nymphs immediately climb on the mango tree and settled on suitable tender portion of the crop. On reaching the fresh growths, the nymphs congregate there and start sap sucking. They moult thrice during their nymphal period which lasts about three months or more, depending on the environmental factors.

Mango mealy bug passes through three nymphal instars and the duration of first, second and female third instar varies from 45 to 71 days, 18 to 38 days and 15 to 26 days, respectively. Total nymphal duration of female is 77-135 days and male is 67-119 days. The adult female lives for 18-51 days whereas adult male lives for 4-6 days only.

Mode of feeding and symptoms of infestation

Both the adult females and nymphs suck sap from tender leaves, shoots, inflorescence and fruit peduncle thereby reduce the plant growth, vigour and causes flower and fruit drop. Under severe infestation, excessive draining of plant sap causes drying of infested plant parts and the young plants in the newly developed orchards are very much vulnerable. Like other Hemipteran insects, mealy bugs also excrete honey dew, which encourages the development of sooty mould which interferes with the normal photosynthetic activity of the plant. Interestingly, adult female and nymphs of both sexes cause damage while adult male survive only for mating. The presence of white cottony cushioned nymphs and adults is a conspicuous symptom of infestation (Mani, 2016).

Seasonal incidence

First instar nymphs are found during December to February, second instars during February to mid-March and third instars from March to April and then adults.

Integrated management

a. Cultural approaches

i. Ploughing of the orchards in the month of November exposes the eggs to Sun and predators.

ii. Removal of alternate weed host like *Hibiscus*, okra, croton, guava, *Clerodendrum infortunatum* etc. to be done.

iii. Raking of the soil around tree trunk followed by application of Chlorpyriphos 1.5 % D @ 250 g per tree preferably during last week of December or 1st week of January to control the newly hatched out nymphs.

b. Mechanical approaches

i. The first instar nymph having the crawling habit which could be restricted to manage the pest in efficient manner by fixing various banding materials like grease, coal tar, grease + coal tar, coal tar + glue + resin and alkathene sheet around the tree trunk as suggested by various worker (Srivastava, 1980). It was found that alkathene sheet is most effective among all the banding materials.

ii. Banding with alkathene sheet (400 gauge) of 25 cm width around the tree trunk at 30 cm above the ground level have been found effective to check the upward movement of nymphs to the trees. The band should be placed well in advance before the hatching of eggs *i.e.*, during November- December. After wrapping both the upper and lower ends of sheet to be tied with jute rope which act as deterrent to the crawler nymphs.

iii. Manual picking of mealy bug can be done in small plant or spraying of strong jet of water in the initial stage of infestation to remove the bugs.

iv. Destruction of ant colonies which act as the carrier of mealy bug.

c. Biological approaches

i. After banding with alkathene sheet application of biopesticides like *Beauveria bassiana* @ 5g/l or NSKE 5% in second week of December around the tree trunk can reduce the pest infestation (Srivastava, 1980).

ii. Natural biocontrol agents like Coccinellids (11 species), Spiders (6 species), predatory mites (2 species) and parasitoids (3 species) have been observed to predate upon mealy bug in mango. So, conservation of these natural biocontrol agents to be done through ecological engineering (Tandon, 1995).

d. Chemical approaches

i. If nymphs have already started ascending the trees, spraying of Carbosulfan 25 EC @ 2 ml/liter of water or Dimethoate 30 EC @ 2 ml/liter to be done during January- March as per the requirements (Ravishankar, 2011).

ii. The crawlers can be easily managed with the application of systemic insecticides like Imidacloprid 17.8 SL @ 0.5 ml/lit or Thiamethoxam 25 WG @ 0.4 g/ lit of water.

3. Fruit fly, *Bactrocera dorsalis* Hendel, *Bactrocera zonata* Sanuders, *Bactrocera diversa* Coq., *Bactrocera tau* Walker, *Bactrocera correcta* Bezzi (Tephritidae: Diptera)

There are several species of fruit flies reported to feed on mango across the world. Among them six to seven species are usually detected to infest mango as mentioned here. *B. dorsalis* is considered to be the dominant and most common which is known as oriental fruit fly (Tandon, 1995).

Distribution

Fruit flies are widely distributed in the Asia-Pacific region, ranging from India to Hawaii and encompassing all of South-East Asia. In India almost all of them have been detected in different mango growing states viz., Uttar Pradesh, Bihar, Madhya Pradesh, Maharashtra, Punjab, West Bengal, Assam etc.

Hosts of commercial importance

The pests have wide host range and climate tolerance and high dispersal capacity (Peterson and Denno, 1998). Mango is one of the important hosts of these fruit flies. In addition to mango, they also found to infest on guava, loquat, peach, apple, fig, quince, banana, plum, pomegranate, citrus, apricot, and a variety of vegetables.

Identifying features (*B. dorsalis*)

Eggs are white in colour and banana shaped. The young larvae *i.e.*, the maggots are white translucent in colour become yellowish in later stage and are acephalous. The fully developed larva has a jumping habit. It is cylindrical and dull deep reddish yellow. The adult fly is brown or dark brown in colour with hyaline wing and yellow legs.

The other flies are superficially similar in appearance with the *B. dorsalis*.

Biology

Female adult fly puncture the matured fruit and deposit the eggs in cluster (2-15 eggs). The incubation period lasts for about 1- 3 days, larval period 6- 29 days, pre-pupal period 1.5- 2 days and pupal period lasts for 6- 44 days in different seasons of the year. When the third instar larva finishes feeding, they leave the fruits, fall on the ground, crawl away to a sheltered spot and pupate in soil at about 5-10 cm depth from soil surface. Adult flies start emerging from April onwards with maximum population during May to July which coincides with fruit maturity. The pest has many generations per year depending on host fruit availability.

Mode of feeding and symptoms of infestation

Female fruit flies deposit their eggs in the fruit mesocarp by puncturing with their long and sharp ovipositor at the physiologically ripe stage of fruit. Bacteria enter through the puncture of the fruit and infected fruit starts decaying. On hatching, maggots bore inside the fruit by making tunnels and feed on the decaying fruit pulp. Area fed by the maggot is discolored due to rotting of the fruit and the fruit drops prematurely. Infested fruits become unfit for human consumption and market acceptance is significantly reduced.

Seasonal incidence

Sarada *et al.* (2001) reported the seasonal incidence and population fluctuation of fruit flies. The population gradually increased from the first week of February (34 fruit flies/ trap) to last week of June (235.0 fruit flies/ trap). The peak population was observed from May to June coinciding with fruit maturity period. Similarly, Patel *et al.* (2013) observed peak infestation of fruit fly with 36.67 % of fruit damage during 22nd SW (28 May - 3 June). Sahoo (2006) also observed the peak population of mango fruit fly (383.33 flies/ trap) during May- June in Malda district of West Bengal.

Integrated management

Being polyphagous pests with high reproductive potential, wide host range, greater adaptability to varied climatic situations and overlapping generations, their management is rather difficult. The conventional management strategies mostly focus on chemical insecticides and due to concealed nature of feeding, the maggots of fruit flies mostly remain unaffected by such insecticides. So, the under mentioned integrated management strategy developed by IIHR, Bengaluru may be followed against this pest.

i. Regularly collect the fallen infested fruits and dispose them to minimize the further spread of infestation.

ii. Ploughing the interspaces in the orchard during November- December will expose fruit fly puparia to hot sun rays and predators.

iii. **Post-harvest treatment of fruits**: Within 24 hours of harvest, fruits to be immerged in hot water at $48^0C \pm 1\ ^0C$ for 1 hour to prevent further infestation in the storage. This process also provides 100% mortality of the fruit fly eggs inside the fruit without affecting the quality. For export to Europe or Japan, vapour heat treatment (VHT) and to the USA, gamma radiations (400 Gy) are to be done mandatorily (Venkata Rami Reddy *et al.*, 2018).

iv. Adoption of Male Annihilation Technique (MAT) through setting up of Methyl eugenol sex lure traps in the orchard @ 15/h throughout the fruiting season. The traps are used to monitor and to mass trap the male flies which ultimately reduce the population density of fruit flies in the orchard.

v. Bait application technique (BAT) comprising of both bait placement and bait spray is also found to be effective in suppressing the fruit fly infestation. Poison bait may be prepared by adding 100g of jaggery and 2ml of Deltamethrin 2.8 EC or in 1 lit of water. BAT can be initiated at 45 days prior to fruit maturity.

vi. The bait solution should be sprayed as spot application at four directions of the tree canopy or on the tree trunks at weekly interval. Spray in 200 spots @ 40 ml per spot (approximately 8 lit/ha). The bait could be sprayed on the nearby hedges and vegetation to prevent pest population builds up in the alternate hosts.

4. Mango midges

a. Flower gall midge, *Erosomyia indica* Grover, *Dasineura amaramanjarae* Grover, *Procystiphora mangiferae* Felt and *P. indica* Grover (Cecidomyiidae: Diptera)

b. Leaf gall midge, *Amradiplosis allhabadensis* Grover, *Procontarinia mangifoliae* Grover, *Procontarinia matteiana* Kieffer and Cecconi (Cecidomyiidae: Diptera)

c. Branch gall midge, *Rhabdophaga mangiferae* Mani (Cecidomyiidae: Diptera)

Gall midges (Cecidomyiidae) are gaining importance throughout the mango growing areas of the world due to changing agro-ecological condition. Among the insect pests associated with mango there are nearly about 26 of them produce galls on various parts of mango tree (Raman *et al.*, 2009). Among the fourteen types of galls reported to occur in mango through insect infestation, eleven are on the leaf, one in the leaf axil, one on the inflorescence and one on the stem have been found. In India, 12 species of midges representing three genera are known to produce different types of galls on mango leaves while five species including *Erosomyia indica* Grover and Prasad and *E. mangiferae* Felt are reported to attack mango flowers (Verghese *et al.*, 1988).

Distribution

Mango inflorescence midge (*Erosomyia indica*) has gained much attention in the recent past in all mango growing areas of the world. Similarly, the leaf gall midge, *Procontarinia matteiana* Kieffer & Cecconi is also reported to be a serious pest of mango in India, Indonesia, Kenya, Mauritius, Oman, South Africa and Unite Arab Emirates (Venkata Rami Reddy *et al.*, 2018). Mango shoot gall insect, *Rhabdophaga mangiferae* is another serious pest in UP, Punjab, Bihar and West Bengal (Mani, 1935).

Biology

The adult midges lay egg singly by inserting inside the floral parts, tender leaves and other parts. Incubation period 2-3 days, larval period varies from 7-10 days and pupal period 5-7 days. The mature larvae slip down into the soil for pupation through the small exit holes in the axis of the inflorescence. Second attacks start at fruit setting as young maggots bore into these tender fruits which slowly turn yellow and finally drop (Prakash, 2012).

The light orange coloured fly of *Procystiphora mangiferae* lays eggs inside the immature blossoms and the maggots upon hatching feed on stalks of stamens, anthers and ovary of the flowers. Only one maggot is found in each bud and it pupates inside the bud itself. However, the adult flies of *Dasineura amaramanjarae* insert the eggs into unopened flower buds. The maggots on hatching feed inside the buds and the infested buds fail to open and drop down. The maggots hibernate in the soil and thus carry-over of the pest to the next year and under favourable conditions they pupate and emerge as adults.

Seasonal incidence

Among the gall midge species recorded from mango, *Procontarinia matteiana* Kieffer & Cecconi, commonly known as leaf gall midge. In India, it infests mango throughout the year particularly during vegetative and fruit maturity period of the crop (Kaushik *et al.*, 2012). The pests are multivoltime and remain active almost throughout the year with highest populations in April and September.

Mode of feeding and symptoms of infestation

Erosomyia indica: It affects the crop at 3 stages, *i.e.*, at floral bud burst, fruit set and tender leaves particularly encircling the inflorescence. The first phase is more damaging as the entire inflorescence destroyed before flowering and fruit set. The larvae tunnel the axis of inflorescence and destroy it completely and causes bending and drying of the inflorescences before fruit set. Its attack on inflorescence could be recognized by presence of tiny black spots and infested panicles have characteristic right-angle bend with small exit hole.

Procystiphora mangiferae: The infested flower buds fail to open, dry and drop down.

Procontarinia matteiana: It forms solitary or grouped galls on the upper and lower surfaces of the leaves. In case of severe attack, leaves become curled, ultimately dry and fall down and finally reduce fruit yield. Younger trees may die while older trees fail to recover normal growth after repeated attacks of this pest.

Integrated management

i. Periodic Collection and disposal of infested panicles, leaves and twigs.

ii. Deep ploughing of orchard in October- November to expose pupae and diapausing larvae to the Sun and natural predation.

iii. Monitoring of larval population on white paper in April-May and apply Chlopyriphos (1.5%) based on population.

iv. The soil below the plant canopy to be covered with plastic sheet. It prevents the emergence of adults from soil and also prevents the dropping larvae to go into soil for pupation.

v. Spray systemic insecticides like Dimethoate 30 EC @ 2ml/ lit or Imidacloprid 17.8 SL @ 0.5 ml/l at bud burst stage. Insecticide application may be repeated with the interval of 10 days in the flowering season up to pin head stage of the fruit based on pest severity.

vi. In a survey of parasitoids of Cecidomyiid pests of mango in India, Grover (1986) found that *Platygaster* sp., *Systasis* sp. and *Eupelmus* sp. were associated with *Dasineura* sp. and *Tetrastychus* sp. was associated with

Eryosomyia indica. Conservation of these bioagent to be done through habitat manipulation.

5. Stone weevil, *Sternochetus mangiferae* Fab. (Curculionidae: Coleoptera)

Distribution

Mango stone weevil is cosmopolitan in distribution and is found in India, Australia, Malaysia, Burma, Africa, Philippines, Pakistan, Bangladesh, Sri Lanka, Thialand, Indonesia, Japan, Uganda and Queensland. In India the pest is serious in humid areas of South Indian Coastal region on the late maturing varieties whereas in North India it is rarely found because the pest is very sensitive to temperature and low humidity.

Hosts of commercial importance

Mango stone weevil, *Sternochetus mangiferae* (Fab.) is the most dominant and important monophagous pest of mango.

Identifying features

Eggs are minute, whitish and cigar shaped. The grubs are whitish, apodous, cylindrical and fleshy with dark head. Pupae are exarate, whitish when newly formed, but change to a very pale red colour just before eclosion. The abdominal apex has paired urogomphi. Adult weevils are greyish brown in colour resembles the bark of the mango tree and 5-8 mm long, stout and 4 mm in breadth, poor flier. The colour varies and they may be covered with black, greyish, reddish or yellowish scales. When disturbed, adults press their legs to the body, and fit their beak into a ventral groove. Females have an elevated ridge at the pygidial apex while it is rounded in the male.

Biology

The female scoops a small boat shaped cavity on the skin (epicarp) of fruit and lays egg singly and cover the cavity with a small quantity of brown exudate. The wound creates a sap flow, which solidifies and covers the egg with a protective opaque covering (Venkata Rami Reddy *et al.*, 2018). As many as 36 eggs may be laid on a single fruit but according to Butani (1993) only 6 or 7 may mature and become adult weevils. Oviposition starts 3-4 days after mating and the insect prefer marble size fruits. Incubation period is 5-7 days. On hatching grubs bore through the pulp (mesocarp) and soon reach to the region where seed coat (endocarp) is still soft. Then the grub crosses the outer coat of the seed and reaches near the seed endosperm where it spends the rest period of the life. Grub makes zigzag tunnels in the pulp and eats unripe tissue and bore into cotyledons.

Grub undergoes five instars and pupation takes place inside the seed. If the seed coat has become harden then the larva fails to bore the seed endosperm, therefore age of the fruit is an important limiting factors for the survival of the stone weevil. Larval period is about 40-50 days and pupal period is 6-8 days.

Mode of feeding and symptoms of infestation

The grub tunnels in a zig-zag manner through the pulp (endocarp), seed coat and finally reaches the cotyledon and complete stone is destroyed only leaving behind black mass. This pest causes damage to the crop in four ways *i.e.*, premature fruit drop, spoilage of fruit pulp during post-harvest period, hindrance in export of fresh fruits due to quarantine issues and germination failure of seeds leading to the unavailability of quality rootstock. Infestation of *S. mangiferae* on mango can be noticed by the appearance of reddish-brown spots and water-soaked areas in the pulp of immature fruit (Maheswari and Purushotham, 1999). This pest spreads to other areas through the transport of fruit, seeds, seedlings and/or cuttings containing larvae, pupae or adults. Infested fruits become unfit for human consumption and market acceptance is significantly reduced.

Seasonal incidence

During the non-fruiting period, the weevil diapause under loose bark and cracks on mango tree trunks or under the fallen leaves. The adult weevil emerged from the fruit during June-July and remains in diapause stage till March of the next season (Shukla *et al.*, 1985).

Management

i. Collection and destruction of all fallen fruits at weekly intervals till harvest.

ii. Cleaning by using old brooms at junctions of branches on the trunk prior to flowering (October).

iii. Spraying Acephate 75 SP @1.5 g/l when the fruits are of lime size (2.5-4 cm diameter). This spray should be followed by Dewltamethrin 2.8 EC @ 1ml/l after two to three weeks.

iv. Destruction of all left over seeds in the orchard after harvest.

6. Shoot gall psylla, *Apsylla cistellata* Buckton (Psyllidae: Hemiptera)

Distribution

Mango shoot gall psylla is distributed in North India (U.P., H.P., Punjab and Uttaranchal), Bihar, Jharkhand, West Bengal, Nepal and Bangladesh (Tandon and Verghese, 1985) and it has a localized and definite distribution in those regions.

Hosts of commercial importance

Mango shoots gall psylla, *Apsylla cistellata* Buckton (Hemiptera: Psyllidae) is a monophagous and univoltime serious pest of mango. No other host plant of this pest has so far been recorded.

Identifying features

Freshly laid eggs look like a rectangular block with rounded corner. It is almost transparent (Singh, 1954). Newly hatched nymphs are minute, yellowish with red marking all over the body. As many as 80 nymphs may remain within a gall. The adult psyllid is 3- 4 mm in length, head and thorax are black and abdomen is light brown. Wings cover the abdomen entirely, hind legs are longer than other legs. Sit on the substratum in a 45° angle.

Biology

Adults soon after emergence, rolled down on the earth to clean their waxy materials from the body and start mating. Then mated female lays up to 150 eggs in the midrib of the young leaf on the underside by inserting singly with its ovipositor. They only lay eggs in the young leaf of matured fruiting tree not in the nursery plant and adult are capable of sensing the chemical constitution of the leaf. The most important peculiarity in their life cycle is that the embryo inside the egg chorion is capable of sucking plant sap in the ovipositional site after the formation of stylet through penetration of the plant tissue before hatching. This stage is arrived generally in July end.

While sucking sap in the midrib, the embryo releases certain chemical (probably phenyl amino acids) with their saliva into the plant tissue which induces axillary and apical bud to grow and develop into conical shaped gall. The egg stage passes from March - April to September *i.e.*, almost 6 - 7 months. After their entry into the gall, nymphs settle and attach themselves to the central axis of the bud gall where they insert their mouth parts and suck sap. Then they undergo moulting 5 times *i.e.*, six nymphal instars within 5 - 6 months from October to February.

Mode of feeding and symptoms of infestation

The nymphs and adults of psyllids attack the terminal shoots and the damage is manifested by the formation of hard conical galls resulted from the swelling of the auxiliary and apical buds. The galls provide shelter to the newly emerged nymphs. The affected shoots contain three to five numbers of galls at the terminal end or even below it. The gall formation directly interferes with the flowering in mango and thereby yield is adversely affected. In due course of time infested twig dries up and shows die back symptom (Singh *et al.*, 1975).

Seasonal incidence

Psyllid adult emerges from the galled shoot during March- April which coincides with the emergence of new flush in mango. The peak adult emergence happens in the first week of March (Singh, 2003).

Integrated management

i. Pruning of gall bearing shoots to kill the nymphal stages which could check further spread of the pest (Singh, 2000).

ii. Conservation and augmentation of parasitoids, *Inostemma apsyllae* and predators like purplish pirate bug and lacewings to facilitate biological control of the pest.

iii. Need based spraying of insecticides like imidacloprid 17.8 SL @ 200 ml/ ha or thiamethoxam 25 WG @ 200 g/ ha coinciding with the period of the emergence of nymphs during August and September.

iv. Srivastava *et al.* (1982) recommended Quinalphos (0.05%) and Dimethoate (0.06%) for the control of the pest.

7. Mango stem borer, *Batocera rufomaculata* de Geer, *Batocera rubus* Linn., (Cerambycidae: Coleoptera)

The species recorded in India are *Batocera rufomaculata* de Geer, *B. rubus* L., *B. roylei* Hope, *B. numitor* Newman and *B. titana* Thompson. Among them the first two are comparatively prevalent and destructive in mango orchard.

Distribution

The pest is distributed in India, Sri Lanka, Mauritius, Trinidad, Bornea and Virgin Island (Ballou, 1916). In India the pest is more serious in south India (Ayyar, 1963).

Hosts of commercial importance

Besides mango the insect can also infest jackfruit, fig, mulberry, apple, avocado, rubber, pomegranate, walnut, forest plant etc. (Sundra Babu, 1975).

Identifying features

The freshly laid eggs are white and shiny with oval, narrowly rounded ends. The colour then changes to dirty white or yellowish white. Full grown larva is fleshy, yellowish creamy, shining, tapering towards eighth segment and cylindrical. Pupae are yellowish brown to dark brown. The adults are large, greyish brown beetle with pubescence, thorax has a pair of lateral horn like structure, colour in general resembles with that of mango bark, antennae is longer than the body that's why it is known as longicorn beetle.

Biology

The mated adult selects a suitable site on the bark of old mango tree and makes an incision with the powerful mandible and lays egg singly. Then the eggs are protected by the exudation from the bark and also by a viscous fluid secreted by the beetle. More number of eggs are laid on the trunk and primary branches than in secondary branches and adult beetles lay eggs continuously till their death. A single beetle lays upto 200 eggs, which hatch in 7- 13 days. After hatching the neonate larvae feed initially under the bark. The larva tunnel through the sapwood and may go deep into the core of the tree trunk. Larva passes through seven instars and they pass the winter in inactive stage. According to Butani (1993) larval stage remains active for 140- 160 days. Pupation takes place in the tunnel and lasts for 20- 25 days. Total life cycle takes 170- 190 days and adult longevity is 60- 100 days. The adults are nocturnal, may live for several months and can fly for long distances. This is a univoltime pest which complete one generation per year.

Mode of feeding and symptoms of infestation

The damaging stage is only the larva and the adult beetle feed on the foliage of different orchard plants. The grubs after hatching feed on bark and make irregular cavities and later they bore into the sap wood, make irregular tunnels and feed the vascular tissues, resulting in the interruption of nutrient and water transport in the plant. Frass comprising of the chewed material and excreta comes out from the hole and sometimes sap oozes out in the holes also. Normally their attack in the early stage remains unnoticed till a branch or two starts shedding leaves and drying up. The damage results in yellowing, drying and shedding of leaves followed by dieback of shoots and drying of affected branches which ultimately leading to death of the whole plant.

Seasonal incidence

Most severe incidence of mango stem borer was noticed during summer to rainy season (25.33 to 74.66%) while, the least during winter season (4.17 to 12.00%) (Jadhav *et al.*, 2015).

Integrated management

i. Orchard should be kept clean. Cut and destroy affected branches with grubs and pupae. Extraction of grubs using hooks or wires and destruction of the same.

ii. Stem wrapping with a nylon mesh during May- August helps in capturing freshly emerging adult beetles (Reddy *et al.*, 2014).

iii. Apply Coal tar + Kerosene (1: 2) on the basal part of the trunk upto 3 feet height after scraping the loose bark in order to deter the females from laying eggs.

iv. Clean the bore hole, remove the frass and insert cotton wool soaked in emulsion of Dichlorvos (0.05%) or Kerosine oil in each hole and plug them with mud.

v. A formulation developed by IIHR, Bangaluru known as "sealer cum healer", when applied on the stem along with an insecticide (Dichlorovos) and a fungicide (Copper Oxychloride) helps to protect the trunk from egg laying by the beetle (Shivananda *et al.*, 2012).

vi. Spray trunk portion with Chlorpyriphos 20 EC @ 3 ml/l or Imidacloprid 17.8 SL @ 3 ml/ 10 lit or Thiamethoxam @ 1 g/lit five times at weekly interval by changing the chemical after the onset of monsoon protect from their infestation (Upadhyay *et al.*, 2013).

8. Bark-eating caterpillar, *Indarbela quadrinotata* Walker, *Indarbela tetraonis* Moore, *Indarbela theivora* Hampson (Metarbelidae: Lepidoptera)

Distribution

This pest is reported from all jackfruit growing areas of the world including Burma, Sri Lanka, Pakistan, Bangladesh and India also.

Hosts of commercial importance

This pest has wide range of host plants including many fruit trees plants like guava, ber, litchi, orange, pomegranate, mulberry and Eugenia, ornamentals and forest trees.

Identifying features

Eggs are spherical and dirty white in colour. Full grown caterpillar is about 3.5- 4 cm long, smooth skinned with dark chitinised patches on the segments. The pupa having rows of teeth or hooks on the abdominal segments by means of which it climbs out to release the moth. The adult moth is pale brown in colour, fore wings with rows of dark rusty red spots.

Biology

The adult female lays eggs in clusters under the loose bark of the trees during summer months (May). Incubation period lasts for 10 days. The larval stage of the pest continues for about 9- 10 months from June to April. The caterpillars become fully grown by December but continue feeding till March-April. With the rise of temperature, the fully developed caterpillar undergoes pupation inside the bore hole at tree trunk and the pupal period lasts for 21-31 days. The life cycle completes one generation throughout the year in India.

Mode of feeding and Symptom of damage

Caterpillars bore into the trunk or junction of branches and feed on the tree bark by making zigzag galleries on the wood and later bore inside the stem and branches. Caterpillars remain hidden in the tunnel during day time and come out at night, feed on the bark. Affected plant shows dried galleries of brownish ribbon-like masses composed of frass and excreta mixed with adhesive material secreted by the caterpillar on the stem and shoots. Heavy infestations retard the growth of tree, resulting in drying of the branches and finally of the tree itself.

Seasonal incidence

Adults emerges during in May-July and larval feeding period during June-April (Srivastava, 1997).

Integrated management

i. The population of this pest can be reduced by keeping the orchard clean with proper management practices.

ii. The caterpillars can be killed by inserting an iron spike into the tunnels.

iii. Periodic removal of webs from tree trunks and filling the insect holes with cotton swab soaked with DDVP (0.05%) and plugging with mud.

iv. The pest can be successfully managed by injecting ethylene glycol and kerosene oil (1:3 ratio) into the tunnel followed by sealing with mud.

9. Mango leaf cutting weevil, *Deporaus* (*Eugnamptus*) *marginatus* Pascoe (Attelabidae: Coleoptera)

The mango leaf-cutting weevil is an emerging and serious pest particularly in mango nurseries and young plantations. The vegetative growth of the grafts in the nursery is arrested due to their attack (Bhole and Dumbre, 1989). The export of mango saplings from India to other countries is embargoed due to the leaf cutting weevil (Srivastava, 1997).

Distribution

This is found all over India wherever mango is grown. It is also reported in Bangladesh, Pakistan, Myanmar, Srilanka, China and Malaysia (Srivastava, 1997).

Hosts of commercial importance

The infestation of the pest is generally observed in mango, however their presence as well as attack is also noticed in litchi (Gupta and Singh, 1986) and in *Butea frondosa* (Beesan, 1941). Tigvattnanont (1988) reported this pest in *Mangifera caloneura*, *M. foetida* and *Bouea burmanica* from Thialand.

Identifying features

Freshly laid eggs are white, very soft, cylindrical, smooth and both ends rounded. Later on, it becomes light yellowish before hatching. Full grown larva is dull yellow or dirty green in colour and more active than the earlier instar. Larva passes through three instars. Pupa is exarate, whitish yellow when freshly formed and then transformed into greyish yellow. The body is densely hairy, soft with numerous setae. In adult, the elytra is reddish brown with dark longitudinal stripes, pygidium exposed. Female weevil are slighter bigger than the male with the longer snout. Female possess distinctly curved genitalia while in case of male it is tubular. Adults are active flier, highly mobile having the habit of feigning death.

Biology

Mating occurs throughout the day repeatedly even after the oviposition. The adult female surveys the whole leaf before oviposition and then make an initial cut basally up to the midrib and give punctures along the midrib by their sharp snout. Then the weevil inserts the egg singly in each puncture which is covered by the oozed-out sap. The insect has the habit of depositing its' egg on right or left alternately alongside the midrib starting from the base of the leaf towards tip. Immediately after completion of egg laying, the female weevil gives the final cut and the cut leaf fall below in the soil. The incubation period lasts for 2.5 days and a female can lay 134-309 eggs in her lifetime. Hatched out larva feed as leaf miner in between the two epidermal layers of the leaf. The larva completes its development within 6 -9 days in three instars (Khanna, 1952 and Butani, 1993). Full grown larva pupates in the soil by forming the hemispherical earthen shell for their protection. Pupal period varies from 5- 15 days depending upon weather and location. The total life cycle is completed within 17- 65 days in different locations (Sahoo and Jha, 2006).

Mode of feeding and symptoms of infestation

The weevil cut away young mango leaves as if it cut by a sharp blade or scissor and also cause damage by scrapping epidermis of young leaves as a result curling and drying of the leaves occur. They cut the leaves of young plants leaving about only 1/4 of it on the tree. The fallen cut leaf pieces seen in the plant as well as under the plant indicate the infestation of insect. Ultimately photosynthesis of the plant affected and growth of the sapling arrested which reduces the consumer acceptability and market value.

Seasonal incidence

Pest incidence remains low during winter and gradually increase from March and the highest infestation is observed during June to September (Kumawat and Mamocha, 2013). Sahoo and Jha (2008) reported maximum damage by the leaf cutting weevil in West Bengal during August- September (23- 29 % cut leaf) and minimum damage during January.

Integrated management

i. Cut leaves from the soil to be collected and burned regularly to destroy the egg and larval stages.

ii. Ploughing of the soil in the nursery as well in the orchard to kill the pupa.

iii. Application granular insecticides viz., Cartap hydrochloride 4G in the mango nursery reduce their damage.

iv. Need based applications of insecticides like, Deltamethrin 2.8% EC or Lambda cyhalothrin 5% EC in rotation will protect the young shoots from weevil attack.

10. Mango fruit borer (Red banded caterpillar), *Autocharis* (*Noorda*) *albizonalis* Hampson, *Dichocrocis (Conogethes) punctiferalis* Guenee (Pyralidae: Lepidoptera)

Mango red banded caterpillar (*Autocharis albizonalis* Hampson) is a new and emerging pest of mango and sometimes cause extensive damage to the crop in many mango growing areas of the country. Earlier it was a rare and minor pest of mango and gained the status of a major pest recently.

Distribution

Autocharis (*Noorda*) *albizonalis* (Hampson) was reported first time from India by Sengupta and Behura (1955). Now, is recorded throughout the Asian countries like India, Indonesia, Philippines, Java and Thialand but information on its bio-ecology and varietal preference is scanty (Waterhouse, 1993).

Hosts of commercial importance

Mango is the primary host of this fruit borer but also have wild hosts like *Cyperus rotundus* (coco grass) and *Mangifera odorata*. While, Golez (1991) reported this as the monophagus pest in mango.

Identifying features

Eggs are milky white, oval in shape. Larva is red in colour with white inter segmental streaks and are about 25 mm long when fully grown. Pupa is obtect, dark brownish. Tip portion of the pupal cage slightly pointed. Male adult moths are bold and healthier (23 mm in wing span) than female moths (21 mm in wing span). Fore wings of the adult moth are ashy wood in colour and mixed with bluish pink metallic shine and hind wings are lighter in colour than fore wings and without any black spots. Sexual dimorphism is the presence of brush like dark brown hairs on lower side of tibia and tarsus only in male.

Biology

Female moths lay eggs *en masse* near the distal end of the fruit or fruit apex and concealing it from natural enemies after a pre-oviposition period of 1.5 -2.5 days. The egg stage lasts for 2-4 days. Immediately upon hatching, the larvae enter in the tender fruits at the early stage via distal end of the fruit and feed by making tunnels. Entrance hole is plugged with excreta. The first and second instar larva remains in the flesh while the later instars feed and complete their development inside the seed. Early instar larvae generally prefer small tender fruits but later instar larvae prefer large size fruits. After completing five larval instars in 11 -16 days, pupation occurred preferably inside the fruits or in soil inside a brownish cocoon and sometimes in the dry twigs of the braches or even in the cracks and crevices. (Bhattacharya, 2014). Pre-pupal and pupal stage lasts for 2- 6 and 9 -12 days, respectively. Life cycle is completed within 30- 41 days in West Bengal (Sahoo and Jha, 2009). Adults are generally nocturnal and during the day they spent more of their time resting under mango leaves.

Mode of feeding and symptoms of infestation

Larvae enter the fruit by boring through the distal end or narrow tip of the fruit and make tunnel and feed inside the flesh and seed. The first sign of infestation is the presence of a sap stain running from the entry hole and collecting on the drip point of the fruit apex. The saps darken over time and become very noticeable (Fenner, 1997). Typical damage symptom is the appearance of small hole at the distal end of the fruit. Sometimes cracking also take place in the infested fruits. Fruit borer damaged fruits become unsuitable for human consumption and subsequent fruit rotting and necrosis of the fruit due to secondary microbial infection further aggravates the problem and therefore threatens the commercial mango production.

Seasonal incidence

April-May is the congenial period for the fruit borer infestation when the fruits are medium in size.

Integrated management

This is an internal borer pest so its management is very difficult only by using insecticides. So, the following integrated approaches to be adopted for managing the pest.

i. Periodic collection and destruction of the infested fruits to prevent further damage and completion of life cycle.

ii. Orchard sanitation through removal of alternate hosts, removal and destruction of pupae inside the dry branches of mango.

iii. Larval parasitoids viz., *Apanteles* sp., *Angitia trochanterata* and *Bracon brevicornis* have been reported (Venkata Rami Reddy *et al.*, 2018).

iv. Golez (1991) observed *Trichogramma chilonis* and *T. chilotraeae* as egg parasitoid and vespid wasp, *Rhychium attrisimum* as larval predator of mango fruit borer in Philippines.

v. Sujatha *et al.* (2002) recorded different natural enemies of the pest viz., Red ant (*Oecophylla smaragdina*), Earwig (*Acrania picta*) and Mantid (*Mantis religiosa*). Entomopathogenic fungus viz., *Metarrhizum anisopliae* and *Beauveria bassiana* on prepupa of mango fruit borer were also observed by them.

vi. Infestation can be checked by spraying insecticides like Indoxacarb 14.5 SC (0.5 ml/lit.) at fruit set and Alphamethrin 10 EC or Deltamethrin or Lambda cyhalothrin or Cypermethrin 10 EC @ 1 lit/ ha at late stage depending upon the severity will check the pest incidence.

11. Mango shoot borer, *Chlumetia transversa* Walker (Noctuidae: Lepidoptera)

Distribution

Mango shoots are infested by several species of borer viz., *Chlumetia transversa* Walker in India, Bangladesh, Taiwan, Java, Philippines and China; *C. alternans* Moore and *C. brevisegna* Hollaway in Japan.

Hosts of commercial importance

Besides mango, it also infests on litchi.

Identifying features

Eggs are oval, pale yellowish. Full grown caterpillar is pink with dirty white spots dorsally and is pale white ventrally. Fore wings are dark grey, beautifully patterned with wary design, hind wings are fuscous.

Biology

Eggs are laid singly on tender leaves, shoots or flower panicles which hatch in 2-3 days. Freshly hatched out caterpillars bore into midrib of tender leaves and come out after a couple of days to bore into tender shoots near growing point and make downward tunnels. Full grown caterpillar after completing five instars in 10- 12 days, enters into cracks in the bark of the trees or dried malformed inflorescence or cracks and crevices in the soil for pupation. Pupal period lasts for 15- 18 days. The severely affected trees during winter show leafless shoots with green and dark black tops. There are four overlapping generation of this pest in a year.

Mode of feeding and symptoms of infestation

Leaves of infested shoots wither and droop down and finally the entire new growth dries up resulting in the stunted growth of plant with terminal bunchy appearance (Singh, 1957). The infestation has also been noticed on inflorescence stalk. The damage can be noticed by presence of excreta at the entry hole and the drying and withering of affected shoots. Young grafted seedlings in the nursery are severely affected and may even be killed.

Seasonal incidence

The pest remains active during August to November with peak period of activity during the second fortnight of September (Handa, 2006).

Integrated management

i. The attacked shoots may be clipped off and destroyed to check further infestation.

ii. Plough orchard to expose the pupae for natural enemies and to the Sun.

iii. Installation of light traps for monitoring as well as for moth trapping.

iv. Need based spraying of Triazophos 40 EC or Quinalphos 25 EC or Profenophos 50EC @ 2 ml/l of water at fortnightly intervals from the commencement of new flush.

v. Conservation of parasitoid, *Bracon greeni* in the mango eco-system can reduce the pest population naturally.

12. Scale insect, *Aspidiotus destructor* Sign, *Rastrococcus iceryoides* Green (Diaspidae: Hemiptera)

Scale insect is a regular pest of mango that damages the plant through the continuous sucking of sap by both the nymphs and adults. This insect is becoming serious pests of mango in regions of Western U.P.

Distribution

Over 70 species of scale insects belonging to Coccidae, Diaspidae and Lacciferidae are attacking the mango in India but out of them only few are serious. Among all of them, *Aspidiotus destructor* Sign. is serious in mango nursery and *Rastrococcus iceryoides* Green is common on mango tree. Besides India these insects are distributed in Tanzania, Pakistan, Sri Lanka, China, Taiwan, Fiji Island, Iran, California, Australia and Mexico.

Hosts of commercial importance

Several species are reported in mango as well as these species feed on other hosts like Mango, citrus, coconut, banana, avocado, cocoa, ginger, guava, yam, grapevine, papaya, sapota, peach, tea, coffee etc. (Srivastava, 1997).

Identification and biology

Adult females are wingless, while males are winged and atrophied. The female leads a sedentary life on the plant parts and feed the plant sap. The mature female produces numerous minute eggs beneath their body in a distinct cottony sac or pouch. When the eggs inside the mother become mature and about to hatch, the female dies away. Hatched out tiny larvae are known as crawler, which possess 3 pairs of legs, a pair of feelers, a well-developed feeding tube and minute eyes. Incubation period lasts for 7- 8 days in *Aspidiotus destructor*. The newly hatched out very active larvae move to tender parts of the plants and shortly attach themselves at a suitable spot. The larva loses its original form and become legless covered under the scale like covering after periodic moulting.

The female fix themselves with the place for entire life and lose their legs, feelers, etc. and grow by feeding the plant sap and secrete waxy covering. Scale formed by the last moulting skin with some waxy material. In most of the case males are rare, when present develop through a pupa-like stage, sheltered beneath a scale like covering, termed as test. The life cycle is completed in 31-35 days in *Aspidiotus destructor* and there are about 3-4 generations in a year.

Mode of feeding and symptoms of infestation

The nymphs and adult female scale insect suck the sap from the leaves and other tender parts of branches, flowers and also fruits and reduce the growth and vigour of the plants. They also secrete honeydew, subsequently leading to the development of sooty mould on leaves and other tender parts of the tree that interfere with photosynthetic activity of the plants. Initially the attacked leaves show discoloration resulting in dropping of younger leaves and under severe infestation, growth and fruit bearing capacity of the tree is affected adversely.

Seasonal incidence

High temperature and high humidity prevailing during July- August enhances the infestation of scale insect.

Integrated management

i. Pruning of infested branches and burning them can be done to reduce scale infestation.

ii. Use of planting material free from scales to minimise the problem.

iii. Due to presence of thick waxy layer on the body of the scale, it becomes very difficult to control with insecticides but at the same time they are found heavily attacked by different natural enemies. So, conservation of the natural enemies, parasitiods viz., *Aneristus ceroplastae* (Aphelinidae; Hymenoptera), *Anagyrus pseudococci* (Encyrtidae; Hymenoptera), and *Tetrastichus perpureus* (Eulophidae). Among predators, *Cryptolaemus montrouzieri, Chilocorus nigritus, Scymnus* spp., (Cocconellidae) and *Spalgis epius* (Lycanidae; Lepidoptera) feed voraciously on the coccids (Mani *et al.*, 1995).

iv. Need based spraying of systemic pesticides, Imidacloprid 17.8 SL @ 3 ml/ 10 lit or Thiomethoxam 25% WG @ 1 g/lit.

13. Mango leaf webber, *Orthaga exvinacea* Hampson, *Orthaga euadrusalis* Walker, *Orthaga mangiferae* Mishra (Pyraustidae: Lepidoptera)

Distribution

The species are important pests of mango in all the mango growing areas of the country including U.P., Bihar, West Bengal and North India.

Hosts of commercial importance

Mango is the most important host of the pest.

Identifying features

Eggs are dull greenish in colour. Full grown caterpillar measures about 2.5- 3 cm, brownish blue colour with whitish striation dorsally. The larvae when disturbed fall with a sudden jerk. The adult moths are medium size and amber in colour.

Biology

Eggs are laid on mango leaves either in single or in clusters and after 4-5 days the larvae upon hatching feed on chlorophyll content of the leaves. Generally, the first instar larvae feed on leaves by scrapping the chlorophyll of the epidermal surface. From second instar onwards the larvae start webbing the leaves and feed on entire leaves leaving behind the midrib and veins. There are five larval instars and larval period varies between 15 - 30 days depending upon weather conditions. Pupation takes place within the webbed foliage or in soil.

Mode of feeding and symptoms of infestation

Initially caterpillars feed on leaf surface gregariously by scrapping and later they make web of tender shoots and leaves together and feed within giving a burnt appearance to leaves. Several caterpillars may be found in a single webbed up cluster of leaves (Anant, 2016). As a result, photosynthesis of the crop hampers, which ultimately affect the growth and productivity of the tree. The infestation varied between 25 to 100 per cent (Kannan and Rao, 2006).

Seasonal incidence

The leaf webber remains active throughout year, except during the month of February to May (*i.e.*, flowering, fruit setting, fruit maturity and ripening stages) with the peak during first fortnight of November.

Integrated management

i. Pruning of infested shoots, removal of leaf webs and burning them to check further infestation.

ii. Proper canopy management of the old senile orchards.

iii. Plough orchard to expose the pupae for natural enemies and to the Sun.

iv. Installation of light traps for monitoring as well as for moth trapping.

v. Need based spraying of lambda cyhalothrin 5 EC or quinolphos 25 EC (2 ml/l of water) to manage the pest.

14. Mango mite, *Aceria mangiferae* Sayed (Eriophyidae: Acarina), *Oligonychus mangiferae* (Rahman and Sapra) (Tetranychidae: Acarina)

Among the non-insect pests, mango mite (*Aceria mangiferae*) has significant economic importance due to its association with mango malformation. It is commonly known as mango bud or gall mite. Besides this, spider mite (*Oligonychus mangiferae*) is also found to feed on the upper surface of mango foliage.

Distribution

In India, mango bud mite is a major pest in Punjab, Delhi and U.P. (Singh and Mukherjee, 1989). It is also distributed in Egypt, Israel, South Africa, Southern Asia, Florida and Mexico.

Hosts of commercial importance

They are monophagous pests of mango.

Identification (*Aceria mangiferae*)

Adult mite is whitish, tiny about 0.2 mm in length, cylindrical, worm like with two pairs of anteriorly placed legs, the prodorsum bears a pair of backward-pointed setae set on tubercles with three strong ridges located between them.

Biology

The mango bud mite lives throughout the year within closed adventive buds and moves into the terminal buds when population increases. It reproduces by arrhenotoky and its entire life cycle completes within 2-3 weeks in summer. The pest shows 8 generations in Egypt with distinct spring and autumn peaks (Gerson and Applebaum, 2015).

Mode of feeding and symptoms of infestation

Mite sucks sap from internal and auxiliary buds resulting in the cessation of growth and development of close lateral buds, leading to crowded and malformed buds. This results in proliferation of terminal shoot showing "witches' broom" like symptom. Ochoa *et al.* (1994) reported that this mite usually occurs along with the pathogenic fungus, *Fusarium mangiferae* results in vegetative and flower malformation resembles witches' broom. In Florida, this mite is associated with malformed mango flower (Pena, 1993), reflected that it may be the vector of the pathogen that could be real cause of malformation. Gerson and Applebaum (2015) noted that the pathogen is transmitted between trees and probably between tree parts by the mite, which enhances fungal penetration into the host via the feeding wounds. Under severe infestation dropping of flowers and buds occurs.

Seasonal incidence

The infestation starts from April and gradually reaches a peak in June (Venkata Rami Reddy *et al*., 2018).

Integrated management

i. Periodic pruning and destruction of mite infested plant parts and malformed vegetative and flower buds to restrict further damage.

ii. Planting of resistant cultivars reduce the loss.

iii. Conservation of predatory mites, *Amblyseius swirskii*, Coccinellids and *Chrysoperla* to encourage natural control of mites.

iv. Need based spraying of acaricides like Propergite @ 2 ml/ liter of water.

15. Mango pulp weevil, *Sternochetus frigidus* Fab., *Sternochetus gravis* Fab. (Curculionidae: Coleoptera)

Pulp weevil has the potential to cause high economic loss to cultivated mango (*Mangifera indica*) as well as it can attack the wild species of *Mangifera sylvatica* grown in hilly areas. The insects usually infest the pulp (mesocarp) of the mango fruit but not the stone or seed. The pest is known to exist in North East India. It is also reported from Assam, West Bengal, Manipur, Meghalaya and Tripura (Dhang, 2015).

Adults of the pest feed on the developing fruits by making very small punctures on the peel. However, the larvae are the most destructive because they feed and develop on the pulp. The weevils feed exclusively on mango regardless of the variety (De Jesus *et al*., 2005). The infestation by the pest is difficult to detect since the damage they cause cannot be seen from the outside. By the time the fruits are harvested, the tiny entry holes created by the young larvae, is not detectable anymore.

As a result of their attack fruits become unfit for consumption and quality of the fruit affected. No external symptom of damage is noticed except only the exit hole on the fruit due to emergence of the adult weevil is the sign of infestation.

The pest can be kept under suppression through orchard sanitation, conservation of natural enemies and need based spraying of Acephate 75 SP @ 1.5 g/lit of water when fruits are of marble followed by Deltamethrin 2.8 EC @ 1ml/lit after two or three weeks based on the pest severity.

16. Flower webber, *Eublemma versicolor* Walk. (Noctuidae: Lepidoptera)

Mango flower webber is a minor pest of mango distributed in all mango growing areas of the country. The larvae web the flowers in the inflorescence together and start feeding inside a silken gallery and cause significant damage. The larvae also bore into the inflorescence stalk.

The female moth lays eggs on sepals and the larva hatches after an incubation period of 3-4 days. The larval period is 18-20 days and pupation take place inside the inflorescence and adult emergence takes place in 8-9 days.

The pest can be managed by collection and destruction of webbed flowers along with the larvae and pupae, conservation and augmentation of natural enemies and need based application of Lamda Cyhalothrin @ 750 ml/ ha at fortnightly interval at flowering stage.

17. Thrips (Thripidae: Thysanoptera)

Thrips are the emerging problem to the mango growers that infest leaves, flowers and fruits of mango. They are generally polyphagous and infest large number of horticultural (mango, guava, custard apple, cashew nut, almond and pomegranate) and agricultural crops. Primarily they have recognised a pest and also act as the vector of different viral, bacterial and fungal diseases of crops (Srivastava, 1997).

Commonly occurring thrips on mango plants are *Rhipiphorothrips cruentatus* (feed on leaves) and *Scirtothrips dorsalis* (feed on inflorescence and young fruits). Patel *et al.* (1997) reported *Pantachaetothrips* spp., *Selenothrips rubrocinctus* Giard, *Caliothrips impurus* Priesner and *Scirtothrips dorsalis* on mango rootstock. Krishnamoorthy and Ganga Visalaksh (2012) reported *Frankliniella schultzai* and *Thrips subnudula* feeding on inflorescence of mango in India for first time.

Damaging stage is both the adult and nymph. They colonize and feed on plant sap by puncturing, lacerating and rasping the surface of growing leaves, buds and flowers. The infested leaf tips blacken, curl and drop off and sometimes defoliation occurs. Lewis (1973) noted that apart from weakening of the inflorescence and reduction in fruit set, thrips cause serious bronzing of fruit surface due to presence of air in emptied cell cavities of the mature fruits.

Need based spraying of Fipronil 80 WG @ 40 g/ ha or Imidacloprid 17.8 SL @ 0.3 ml/ lit or Spinosad @ 0.25 ml/l may suppress the pest population.

In addition to the aforementioned insect and non-insect pests of mango some other minor pests have also been reported to feed on mango in various mango growing regions of the world. They are as follows:

a. Fruit sucking moth, *Othreis fullonica* Clerck, *Othreis materna* Linn. (Noctuidae: Lepidoptera)

b. Leaf eating caterpillar, *Cricula trifenestrata* Helf. (Saturniidae: Lepidoptera)

c. Leaf twisting weevil, *Apoderus transquebaricus* (Curculionidae: Coleoptera)

d. Grey weevil, *Myllocerus discolor* Boheman (Curculionidae: Coleoptera)

e. Termite, *Odontotermes obesus* Rambur, *Odontotermes wallonensis* (Termitidae: Isoptera).

References

Anant, A. K. 2016. Studies on insect-pests of mango with special reference to seasonal incidence and management of mango leaf hoppers. M.Sc. (Ag) Thesis submitted to IGKV, Raipur, 1-89 pp

Anonymous, 2014. AESA based IPM in Mango. Published by Department of Agriculture and Cooperation, Ministry of Agriculture, Government of India, 1-84 pp.

Anonymous, 2019. Product profile of mango. APEDA website, http://agriexchange.apeda.gov.in/ Market Profile/ one/ Mango.

Atwal, A. S. 1963. Insect pests of mango and their control. *Punjab Hort. J.*, 3(2-4): 234-258.

Atwal, A. S. 1976. Agricultural Pests of India and South East Asia. Kalyani, 226-233.

Ayyar, T. V. 1963. Hand book of economic entomology for South India, Govt. Press, Madras, pp-516.

Bagle, B. G. and Prasad, V. G. 1984. Varietal incidence and control of stone weevil, *Sternochetus mangiferae* Fab. *Ind. J. Ent.*, 46(4): 389-392.

Ballou, H. A. 1916. Insects in the virgin island. *Agric. News West Indies*.

Balock, T. W. and Kozuma, T. T. 1964. Note on the biology and economic importance of mango weevil, *Sternochetus mangiferae* in Hawaii. *Proc. Hawaiian Ent. Soc.*, 17: 353-364.

Bana, J. K., Kumar, S. and Sharma, H. 2018. Diversity and nature of damage of mango insect-pest in south Gujarat ecosystem. *Journal of Entomology and Zoology Studies*, 6(2):274-278.

Beesan, C. F. C. 1941. The ecology and control of the forest insects of India and the neighbouring countries. Forest Research Institute and College, Dehradun.

Bhattacharyya, M. 2014. Impact of ecological factors on the infestation of mango red banded caterpillar. *Journal of Entomology and Zoology Studies*, 2 (4):68-71.

Bhole, S. D. and Dumbre, R. B. 1989. Bionomics and chemical control of mango leaf cutting weevil, *Deporaus marginatus* Pascoe (Coleoptera; Curculionidae). Indian J of Ent., 51(3):234-237.

Bhumannavar, B. S. 1991. Record of *Citripestis eutraphera* Meyrick (Pyralidae; Lepidoptera) on *Mangifera indica* in India. *J. Bombay Nat. His. Soc.*, 88(2):299.

Bindra, O. S., Verma, G. C. and Sandhu, G. S. 1970. Studies on the relative efficacy of the banding materials for control of mango mealy bug, Drosicha stebbingi Green. J. Res. Punjab agric. Univ., 8(4):491-493.

Buckton, G. B. 1982. The mango shoot psylla, *Psylla cistellata*. Indian Mus. Notes, 3(2):91-92.

Butani, D. K. 1979. Insects and fruits. Periodical Expert Book Agency, Delhi.

Butani, D. K. 1993. Mango pests problems. Periodical Expert Book Agency, New Delhi, 290 pp.

Chowdhury, S. K. 2015. Diversity and nature of damage of mango insect pests at Kaliachak-II Block of Malda, West Bengal, India. J Ent. Zoo. Std, 3(4):307-311.

Corey, F. M. J., Cadapan, E.P. and Sanchez, F. F. 1989. Attractivity of different light trap colours in various sizes of landwick to mango hopper, *Idioscopus* spp. at different phases of the moon. C.M.Y.J. of science, 2(1): 2-12, Philippines.

Das, N. M., Ramamony, K. S. and Nair, M. R. G. K. 1969. Biology of a new jassids of mango, *Amrasca splendens* Ghauri, Indian J. Ent., 31(3): 288-290.

De Jesus, L. R. A., Calumpang, S. M. F., Median, J. R. and Ohsawa, K. 2005. Feeding behaviour of mango pulp weevil, *Sternochetus frigidus* Fab. At different physiological stages of mango, *Mangifera indica* L. Philippines Agricultural Scientist, 86: 282-289.

Dhang, P. 2015. Mango Pulp Weevil, *Sternochaetus frigidus*- an invasive species with economic importance in Philippines. International Pest Control.

Fenner, T. 1997. Red-banded mango caterpillar: biology and control prospect. Northern Territory Department of Primary Industry and Fisheries, Darwin, pp-2.

Gerson, U. and Applebaum, S. 2015. Plant pests of Middle East. Aceria mangiferae Sayed.

Golez, H. G. 1991. Bionomics and control of mango seed borer, *Noorda albizonalis* Hampson (Pyralidae; Lepidoptera). Acta Horticulturae, 291:418-424.

Grover 1986. Integrated control of midge pests. *Cecidol. Int.* 7:1-28.

Gupta, B. P. and Singh, Y. P. 1986. New record of *Deporaus marginatus* as a pest of *Litchi chinensis*. Prog. Hort. 18(3-4):295-296.

Handa, S. 2006. Seasonal incidence of mango shoot borer in different cultivars of mango and efficacy of different insecticides and neem oil for its control. Agric. Sci. Digest, 26 (4):306-308.

Jadhav, S. B., Patil, R. S.; Dhumal, S. S. and Nale, V. N. 2015. Seasonal impact and influence of mango stem borer on rejuvenation of old mango orchard. *I*nternational Journal of Agriculture Innovations and Research, 4(1):91-96.

Kannan, M. and Venugopala Rao, N. 2006. Ecological studies on mango leaf webber (*Orthaga exvinacea* Hamp.) in Andhra Pradesh as a basis for IPM. International Journal of Agriculture Science., 2(2):308-311.

Karar, H., Arif, M. J., Ali, A., Hameed, A., Abbas, G. and Abbas, Q. 2012. Assessment of yield losses and impact of morphological markers of various mango genotypes on mango mealy bug (*Drochisa mangiferae*; Margarodidae; Homoptera). Pak. J. Zool., 44(6):1643-1651.

Kaushik, D. K, Baraiha, U., Thakur, B. S. and Parganiha, O. P. 2012. Pest complex and their succession on mango (*Mangifera indica*) in Chhattisgarh, India. Plant Arch., 12: 303-306.

Khanna, S. S. 1952. Biology of *Deporaus marginatus* Pascal (Curculionidae; Coleoptera). Proc. Nat. Aca. Sci. India B., 22(1-4):72-78.

Krishnamoorthy, A. and Ganga, V. P. N. 2012. Record of thrips on mango. J. Hort. Sci., 7(1):110-111.

Kumar, A., Pandey, S. K. and Kumar, R. 2009. Population dynamics of mango mealy bug, *Drosicha mangiferae* Green from Jhansi, U.P. Biological Forum, 1(2):66-68.

Kumawat, M. M. and Singh, Mamocha, K. 2013. Population dynamics and management of mango leaf cutting weevil, *Deporaus marginatus* Pascoe in Arunachal Pradesh. Indian Journal of Entomology, 75(1):62-67.

Lewis, T. 1973. Thrips, their biology, ecology and economic importance. Academic, New York.

Maheswari T._U., Purushotham, K. 1999. A simple method to diagnose nut weevil incidence in mango. Insect Environment, 5(3):101.

Mani, M. S. 1935. Studies on Indian Itonidae (Cecidomyiidae; Diptera). Description of new midges and galls. Rec. Indian Mus., 37(4):424-454.

Mani, M. S, Krishnamoorthy, A. and Pattar, G. L. 1995. Biological control of mango mealy bug, *Rastrococcus iceryoides* Green (Homoptera; Pseudococcidae). Pest Management in Horticultural Ecosystem, 1(1):15-20.

Mani, M. S. 2016. Fruit crops: mango. In: Mani, M.S. and Shivaraju, C. (eds) Mealy bug and their management in agricultural and horticultural crops. Springer, New Delhi, pp-385.

Ochoa, R., Aguilar, H. and Vargas, C. 1995. Phytophagous mite of Central America: an illustrated guide. Centro Agronomico Tropical de Investigacion Ensenanza, Technical series 6. Turrialba, Costa Rica.

Patel, K. B., Saxena, S. P. and Patel, K. M. 2013. Fluctuation of fruit fly oriented damage in mango in relation to major abiotic factors. Hort. Flora Research Spectrum, 2(3): 197-201.

Patel, J. R., Patel, M. B., Radadia, G. C., Shah, A. A. and Pandya, H. V. 1997. Insect pest management in mango nursery. J. Appl. Hort. (Navsari), 3:125-128.

Paul, A. V. N. 2007. Insect Pest and their management. A Manual. Division of Entomology Indian Agricultural Research Institute, New Delhi. pp-42-47.

Pena, J. E. and Mohyuddin, A. I. 1997. Insect pests. The mango-Botany, Production and Uses. Edited by: R.E. Litz. CAB International.

Pena, J. E. 1993. Pests of mango in Florida. Acta Horticulture, (341):395-406.

Peterson, M. A. and Denno, R. F. 1998. The influence of dispersal and diet breadth on patterns of genetic isolation by distance in phytophagous insects. Amr. Nat. 152: 428-446.

Prakash, O. 2012. IPM schedule for mango pests. Extension Bulletin No. 1. National Horticulture Mission, Ministry of Agriculture Department of Agriculture & Cooperation, Krishi Bhawan, New Delhi, pp-48.

Pruthi, H. S. and Batra, H. N. 1960. Important fruit pests of North West India. Bull. No. 80, 113 pp. ICAR, New Delhi.

Raman, A., Burckhardt, D. And Harris, K. M. 2009. Biology and adaptive radiation in the gall inducing Cecidomyiidae (Insecta: Diptera) and Calophyidae (Insecta: Hemiptera) on *Mangifera indica* (Anacardiaceae) in the Indian subcontinent. Trop. Zool., 22:27-56.

Ramesh Babu, S. and Singh, V. 2014. Bio-efficacy of Tolfenpyrad 15 EC against hopper complex in mango. Pest Management in Horticultural Ecosystems, 20(1):22-25.

Ravishankar, H. 2011. Management of Mango Mealy Bug. Press Release / Advisory. Central Institute for Subtropical Horticulture, Rehmankhera, PO Kakori, Lucknow.

Reddy, P. V. R., Chakravarty, A. K., Sudhagar, S. and Kurian, R. 2014. A simple technique to capture, contain and monitor the fresh emerging beetles of tree borer. Current Biotica, 8(2):191-194.

Reddy, P. V. R and Sreedevi, K. 2016. Arthropod Communities Associated with Mango (*Mangifera indica* L.): Diversity and Interactions. Book Chapter in *Economic and Ecological Significance of Arthropods in Diversified Ecosystems*. Springer Publication, Edited by Chakravarti and Sridhara, pp-271-298.

Sahoo, S. K. 2006. Insect-pests complex in mango with special reference to bioecology and management of mango fruit borer and leaf cutting weevil. Ph. D. thesis submitted to Bidhan Chandra Krishi Viswavidyalaya, Mohanpur, Nadia, West Bengal, India.

Sahoo, S. K., and Jha, S. 2006. Behaviour and development of Mango leaf cutting weevil, *Deporaus marginatus* (Pascoe) (Attelabidae; Coleoptera). Journal of Plant Protection and Environment, 3(1):68-75.

Sahoo, S. K. and Jha, S. 2008. Screening of mango varieties against leaf cutting weevil in the nursery. Crop Research, 35(3):226-232.

Sahoo, S. K. and Jha, S. 2009. Bioecology of Mango Fruit borer, *Autocharis (=Noorda) albizonalis* Hampson (Pyralidae, Lepidoptera)-A Recent Threat to Mango Growers in West Bengal, India. (edt: S. A. Oosthuyse, Proc.VIII Int. Mango Symposium Acta Hort.820, ISHS 2009) Acta Horticulture, 820:1345-1425.

Sarada, G., Maheswari, T. U. and Purushotham, K. 2001, Effect of trap colour, height and placement around trees in capture of mango fruit flies. Journal of Applied Zoological Researches, 12:108-110.

Sengupta, G. C. and Behura, B. K. 1955. Some new records of crop pests from India. Indian J. Ent., 17:283-285.

Shivananda, T. N., Verghese, A., Hegde, R. M., Kumar, N. K. K. and Singh, A. S. 2012. Management of mango fruit borer using sealer cum healer. Technical bulletin no. 72/12, IIHR, Bangalore.

Singh, G. 2000. Physiology of shoot gall formation and its relationship with juvenility and flowering in mango. VI Int. Sym. Mango. 509: 803-810.

Singh, G. 2003. Mango shoot gall: its causal organism and control measures. ICAR, New Delhi, 94 pp.

Singh, H. S. 2014. Management of fresh leaf-cutting weevil, *Deporaus marginatus* (Pascoe) in mango. Insect Environment, 20(1):19-21.

Singh, S. P. 1989. Relative incidence of mango nut weevil, *Sternochetus mangiferae* Fab. on different mango varieties. Acta Hort., 231:575-580.

Singh, S. M. 1954. Studies on the *Apsylla cistellata* Buckton causing mango shoot galls in India. Indian J. Econ. Ent., 47(4):563-564.

Singh, S. M. 1957. A serious damage to mango shoots by the borer, *Chlumetia transversa* Walker in UP. Indian J. Hort., 14:236-238.

Singh, G., Kumar, A. and Everrett. T. R. 1975. Biological observations and control of *Apsylla cistellata* Buckton (Hemiptera; Psyllidae). Indian J. Ent., 37:46-50.

Singh, J. and Mukherjee, I. N. 1989. Pest status of phytophagous mite in some northern states of India. In: Proc. of the first Asia-Pacific conference of entomology. Chiang mai, Thailand. 192-203 pp.

Srivastava, R. P. 1980. Relative effectiveness of different mechanical bands in preventing ascent of mango mealy bug. Indian J. Ent., 42(1):110-115.

Srivastava, R. P. 1997. Mango insect pest management. Published by International Book Distributing Co., Lucknow, U.P. 159-162.

Srivastava, R. P. and Tandon, P. L. 1986. Natural occurrence of two entomogenous fungi pathogenic to mango hopper, *Idioscopus clypealis*. *Ind. J. Plant Patho*., 4(2):121-123.

Srivastava, R. P., Tandon, P. L. and Verghese, A. 1982. Evaluation of insecticides for the control of mango shoot gall psylla, *Apsylla cistellata* Buckton (Psyllidae; Homoptera). Entomon, 7(3):281-284.

Shukla, R. P. and Tandon, P. L. 1985. Bioecology and management of mango stone weevil, *Sternochetus mangiferae* Fab. Int. J. of Tropical Agri. 3:292-303.

Shukla, R. P., Tandon, P. L. and Suman, C. L. 1985. Spatial distribution of different stages of mango stone weevil, *Sternochetus mangiferae* Fab. Agric. Ecosystem and Environment. 12(2):135-140.

Sujatha, A. and Zaheeruddin, S. M. 2002. Biology of pyralid fruit borer, *Deanolis albizonalis*: a new pest of mango. J. Applied Zoological Res., 13:1-5.

Sujatha, A., Zaheeruddin, S.M. and Arjuna Rao, P. 2002. Record of natural enemy fauna on the mango fruit borer, *Deanolis albizonalis* Hampson. Insect Env., 8(3):140-141.

Sundra Bahu P. C. 1975. Annual report on the scheme bionomics and control of mango stem borer, *Batocera rufomaculata* De Geer in Tamil Nadu.

Tandon, P. L. 1995. Integrated pest management in mango. In: Proceedings National Symposium on "Integrated pest management and Environment" organized by the Plant Protection Association of India and FAO at Madras, 2-4th February, 1995.

Tandon, P. L. and Lal, B. 1976. New record of predatory mites on mango mealy bug, *Drosicha mangiferae* Green (Margarodidae: Hemiptera). Current Science, 45(15): 566-567.

Tandon, P. L. and Verghese, A. 1985. World list of insect, mite and other pests of mango, Technical document, vol 5. Indian Institute of Horticultural Research, Bangalore.

Tigvattnanont, S. 1988. Biological and autecological studies of the mango leaf cutting weevil, *Deporaus marginatus* (Attelabidae; Coleoptera). Khonkaan Agri. J., 16(1):51-62.

Upadhyay, S. K., Chaudhury, B. and Sapkota, B. 2013. Integrated management of mango stem borer (*Batocera rufomaculata* Dejan) in Nepal. Glob J Biol Agric. Health Sci., 2(4):132-135.

Venkata Rami Reddy, P., Gundappa, b. and Chakravarty, A. K. 2018. Pests of mango. In Omkar (ed.), Pests and Their Management. Published by Springer Nature Singapore Pte Ltd.

Verghese, A., Tandon, P. L. and Rao, G. S. P. 1988. Spatial distribution pattern and sampling plan for the blister midge, *Erosomyia indica* Grover (*Cecidomyiidae: Diptera*) in India. *Insect Science and Application,* 9(4):515-518.

Verghese, A. 2000. Effect of Imidacloprid on mango hopper, *Idioscopus niveosparsus* (Leth.) (Homoptera: Cicadellidae). Acta Horticulturae, 509(2):733-736.

Viraktamath, S. and Viraktamath, C. A. 1985. New Species of *Busoniomimum* and *Idioscopus* (Homoptera; Cicadellidae; Idiocerinae) breeding on mango in South India. Entomon., 10(4):305-311.

2

Insect Pests of Citrus and Their Management

Biswajit Patra, Rajesh Kumar and S. J. Prasanthi

Citrus genus belongs to the family Rutaceae that comes in various forms and sizes (from round to oblong) that commonly known as oranges, mandarins, limes, lemons and citrons. Besides being consumed as raw/fresh fruits, they are also used in the industries of beverages, cosmetic and pharmaceuticals. Citrus fruits have got abundant quantity of vitamin-C and other micronutrients like sugars, dietary fiber, potassium, folate, calcium, thiamin, niacin, vitamin B6, phosphorus, magnesium, copper, riboflavin and pantothenic acids etc. (Economos, 1999). Secondary metabolites like flavonoids, limonoids, coumarins, alkaloids, phenolic acids and other essential oils of this fruit also have anti-oxidative, anti-inflammatory, anti-cancerous as well as cardiovascular protective effects (Lu *et al.*, 2015).

According to statistics given by FAO, citrus fruits are grown worldwide with a global production of about 124.246 million tons (FAO, 2017). In India, Citrus comprises the third largest fruit industry after banana and mango. During 2015-16, it was grown in an area of about 9.35 lakh hectares attaining a yield of 11.5 million tons (Anonymous, 2017). Over the years, India's citrus industry is suffering heavy economic losses due to several abiotic and biotic stresses. The biotic stresses include insect pests, plant pathogenic fungi, bacteria, virus, nematodes and weeds.

More than 875 species of insect pests and mites have been reported to infest on various species of citrus plants throughout the world. The distribution of different citrus pest species in various countries also varies according to the ecological variation. Among insect pests, aphid, psylla, leaf miner, mealy bug, fruit sucking moth, and lemon butterfly etc. are of national significance (NIPHM, 2014).

The major insect pests of citrus are as follows.

Table 1: Insect and mite pests of citrus.

Sl. No.	Common Name	Scientific name	Family	Order
1.	Citrus trunk borer	*Anoplophora versteegi* (Ritsemai)	Cerambycidae	Coleoptera
2.	Lime or orange tree borer	*Cheledonium cinctum* Guer-Mene.	Cerambycidae	Coleoptera
3.	Citrus fruit fly	*Bactrocera minax* (Enderlein)	Tephritidae	Diptera
4.	Citrus leaf miner	*Phyllocnistes citrella* Stainton	Gracillaridae	Lepidoptera
5.	Citrus psylla	*Diaphorina citri* Kuwayama	Psyllidae	Hemiptera
6.	Fruit sucking moth	*Eudocima (Othreis) fullonica* Clerck	Noctuidae	Lepidoptera
7.	Lemon butterfly	*Papilio polytes* *P. demoleus* Linn.	Papilionidae	Lepidoptera
8.	Brown citrus aphid	*Toxoptera citricidus* Kirkaldy	Aphididae	Hemiptera
9.	Black citrus aphid	*Toxocoptera aurantia* Boyer de Fons.	Aphididae	Hemiptera
10.	Mealy bug	*Planococcus citri* (Risso)	Pseudococcidae	Hemiptera
11.	Citrus blackfly	*Aleurocanthus woglumi* Ashby	Aleurodidae	Hemiptera
12.	Citrus whitefly	*Dialeurodes citri* (Ashmead)	Aleurodidae	Hemiptera
13.	Citrus thrips	Thrips nilgiriensis Rama.	Thripidae	Thysanoptera
14.	Cottony cushion scale	*Icerya purchasi* Mask.	Margarodidae	Hemiptera
14.	Citrus rust mite or Silver mite	*Phyllocoptruta oleivora* Ashmaed	Eriophyidae	Acarina
15.	Citrus red mite	*Panonychus citri* (McGregor)	Tetranychidae	Acarina

1. Citrus trunk borer, *Anoplophora versteegi* Ritsema (Cerambycidae: Coleoptera)

Citrus trunk borer, *Anoplophora versteegi* Ritsema is one of the severe pests of citrus in the north eastern states of India (Sachan and Gangwar, 1982). The species was originally described as *Monohammus versteegi,* and named after Mr. W. F. Versteeg, a member of the Committee for the Scientific Sumatra-Expedition.

Distribution

Indian sub-continent including Assam, West Bengal, Sikkim etc. It is a severe pest of citrus in North-East Indian states as because it can causes 41.4 to 62.3% damage in case of severe infestation.

Hosts of economic importance

The plants under the family Rutaceae, including crops orange-group like *Citrus reticulata*, *C. limon*, *C. sinensis*, *C. grandis*, *C. jambhiri*, *C. aurantifolia* and *C. medico* (Singh *et al*, 2012).

Identification and biology

The adult cerambycid beetle measures about 25-31 mm and grey in colour. It has a number of black dots on the hard elytra. The antennae are larger than the. Full grown larva is about 40-50 mm long and creamy in colour. Its body consists of a number of well-marked segments that are slightly broader at the base of the head.

The adult females after copulation, lay eggs during May-June on the cracks of trunks just above the soil level in a slit dug by the females. Oviposition takes place after dusk. A female lay about 8-15 eggs in the vicinity of collar region of *khasi* mandarin plant (Thakur and Shylesha, 1996). Upon hatching, the newly emerged grubs bore into the trunk through egg laying site and feed. Larval stage lasts for about 282 days and the adult emerges from pupa in about 53 days. There is only one generation in a year.

Mode of feeding and symptoms of infestation

It has been reported that about 15-60% orchards are getting infested every year by this pest in Assam (Shylesha *et al*., 1996) and in "Khasi" mandarin orchards, damage is as high as 68.15%. This pest also causes citrus decline in Darjeeling district of West Bengal.

The early larval stages are passed underneath the bark. The grubs live by forming galleries in the sapwood. The grubs first bore horizontally in the sapwood and later they bore the heartwood and gradually reach the central portion of the trunk from where they move upward by making a vertical tunnel. Before pupation the grubs tunnel at right angle to the vertical gallery and make a bigger and circular exit hole which is plugged with chewed fibrous material. Presence of resinous exudation and saw dust like powder in the affected tree trunk is the symptom of attack by this pest. Under severe infestation the fruiting tree may die.

The affected branches gradually dry up and the leaves also wither away. The grub bore into the xylem and phloem tissues during development resulting in deterioration of the tree that may lead to death. Both the larval and pupal periods are passed in the tree trunk and the adult emerges from the trunk. The adult beetles feed on leaf lamina along the mid-rib leaving the leaf margin intact.

Management

i. Pruning and burning the dried and withered branches kills the overwintering stages and the larvae or pupae of trunk borer.

ii. Collection of the adult by shaking the tree during the emergence period *i.e.* April to August and destruction thereafter.

iii. Killing of the grub by hooking with a curved wire is also effective to manage this pest.

iv. Injecting 5 ml of 0.1% Dichlorvos and then sealing the treated holes by mud.

v. Injection of kerosene @5 ml/hole is more effective as compared to petrol.

2. Lime or orange tree borer, *Cheledonium cinctum* Guer-Mene. (Cerambycidae: Coleoptera)

Distribution

Majorly destructive in the southern parts of India particularly in the citrus growing areas of Bangalore and Mysore.

Hosts of commercial importance

Orange and other crops like lime under the family Rutaceae.

Identification and biology

Adults are large sized longicorn beetle, dark violet to dull metallic green in colour with conspicuous yellow band in the middle of elytra. Female lays about 30-50 eggs on the twigs (also on thorns) that remains covered by a resinous secretion. The egg period varies from 11-72 days depending upon environmental condition. Creamy white grubs have a flat head. Larval period is about 10 months. The full-grown grubs pupate in tunnel inside the stem and lasts for three weeks. Adult emergence takes place during the month of April-May and life cycle gets completed in one year.

Mode of feeding and symptoms of infestation

The grubs generally bore inside stem and feed on the internal tissues from within. Injury to the crop is caused by the feeding of the grub on the woody portion of the plant. Feeding of the grubs within the stem results in withering of the terminal twigs in early stages that eventually spreads to the thicker branches. Gum exudation accompanied with powdery materials are seen on the ground below the affected area. When the plant is in three to four years old, a single grub is sufficient to kill it. But in case of older, it takes several years to destroy the plant. Infestation by this pest also results in qualitative and quantitative loss in yield.

Management

i. Pruning of affected twigs or branches along with the grubs and destruction thereof is effective. This operation may be done during June and end of July in south Indian condition.

ii. Plugging of fresh holes on stem and twigs with cotton soaked in insecticidal solution having fumigating action is helpful in killing the grub within the tunnel.

iii. Stem padding with Organophosphate insecticide like Chlorpyriphos or DDVP @ 2.5 ml+2.5 ml of water may also be practised to reduce pest population, reproduction and multiplication.

3. Citrus fruit fly or Chinese fruit fly, *Bactrocera minax* (Enderlein) (Tephritidae: Diptera)

The Citrus fruit fly which is also known as Chinese citrus fly, *Bactrocera minax* (Enderlein), is a destructive pest of citrus. In severe case of infestation, it can cause up to 100% fruit damage (Huang *et al.*, 2007).

Hosts of commercial importance

The pest feed on several plants of the family Rutaceae.

Distribution

The pest has been found to infest citrus crops in South-Central China. The pest also found to occur in limited regions of Bhutan and North-West India bordering with China (Wang and Luo, 1995). Interest in this pest outside of China has risen in recent years primarily due to the concern that *B. minax* might spread to other major citrus producing areas through fresh fruit trade and travelers.

Identification

The insect is large in size, adults can reach as long as 24 mm (Chen and Xie, 1955). It is an extremely large species of fruit fly. Face of the adult fly is fulvous with narrow elongate facial spots reaching oral margin. Scutum is red-brown without dark patterns, scutellum yellow with a narrow red-brown basal band. Legs with all segments mostly fulvous; wings with cells bc and c fuscous, microtrichia found in outer corner of cell bc and outer 1/2 of cell c, a broad fuscous costal band overlapping R(4+5) and becoming darker towards the apex but not expanding into a spot. A narrow fuscous cubital streak is there but not reaching margin of wing, supernumerary lobe weak.

Abdomen is elongated oval and petiolated (similar to many other *Dacus* species). Abdominal terga III-V are orange-brown in colour with a moderately broad transverse fuscous band across anterior margin of tergum III and a medium width medial longitudinal pale fuscous band over all three segments (Adhikar and Joshi, 2016).

Biology

B. minax has several biological characteristics that separate this species from the vast majority of species in the genus *Bactrocera*. The species is a univoltine one and is oligophagous that exclusively feeding on citrus plants. It is among the most cold tolerant species in the genus, with larvae able to survive freezing temperatures for days (Luo and Chen, 1987).The adults, post emergence from the pupae, fly to nearby woods feeding on such items as honeydew secreted by aphids (Yang, 1989). By the second to third week, adults reach sexual maturity and starts mating. Females lay fertilised eggs about 15 days after mating (Lu *et al.*, 1997). The insect prefers laying eggs in young fruits with diameters of about 2 to 4 cm. Fecundity varies from 50 to 200. It appears that adults prefer laying eggs on sweet orange. Incubation period is about 30 days. The larvae feed on flesh of the fruit from within the fruit. There remain three larval instars in their life cycle.

The mature larvae fall from the infested hanging fruits or exit from affected fallen fruits into soil for overwintering. The larval stage lasts roughly for about two months. The pupal stage is the longest stage of the species, lasting from 150 to almost 200 days. Most pupae overwinter in soil at an approximate depth of 4 - 6 cm (Wang and Luo, 1995). Pupal diapause in *B. minax* is obligatory and cannot be terminated if directly exposed to high temperature or if organic chemicals are topically applied.

Mode of feeding and symptoms of infestation

Fruits infested with the fruit fly drop off from the tree. Damaging stage of the pest is maggot. The larvae cause damage to the fruit by tunneling within the fruits and feeding on pulp differs with the type and maturity of fruit, number of larvae and prevailing environmental conditions. By the time maggot gets matured and reaches third instar, a large portion of fruit gets destroyed. Heavily infested fruits often drop prematurely causing a great loss to citrus grower (Drew, 2006). As high as 100% yield loss may take place due to heavy infestation by this dreaded pest.

Management

Various management strategies for this pest have been recommended from time to time that include use of para-pheromones followed by sanitation, chemicals, botanicals, cultural methods, pest exclusion measures and food lures etc. (Adhikari, 2019). Some of the strategies are being elaborated hereunder-

i. Field sanitation

The fallen and hanging infested fruits are to be collected in large bags. These bags with infested fruits are subjected to various treatments like heat treatment (exposing the bags under the sun), boiling, deep bury (at least 50 cm below the soil surface), burning or putting the infested fruits in shallow pond may be practiced. It has been observed that good sanitation practices can significantly decrease *B. minax* infestation (Wang and Zhang, 2009). As *B. minax* pupae overwinter in soil, raking or shallow ploughing in is also helpful in exposing the pupae to outer temperature, soil insecticides and natural enemies like birds.

ii. Bagging

Fruit bagging is an effective mechanical method to protect fruits from the pest. The practice not only has the potentiality to protect fruit reaching (or near) 100% but also has small effect on fruit quality. Bagging should be conducted during young fruit period or before colour changing. Application of pesticides, if required, should be done about 3-5 days before bagging.

iii. Foliar pesticide spray

It is a common management option followed for *B. minax*. Commonly used insecticides are organophosphates such as Trichlorfon and/or Pyrethroids such as Deltamethrin (Shu and Xiao, 2006). Organophosphate pesticides found to have strongest influence on pupation and emergence of the pest respectively but the pesticide use should follow careful optimisation due to high toxicity and residual effects (Liu, 2015).

iv. Sterile Insect Technique (SIT)

Sterile Insect Technique have been found successful in reducing *B. minax* infestation on oranges considerably.

v. Food based lures (bait spray)

The most common food-based lure for the pest is hydrolysed protein, waste brewer's yeasts, sugar + vinegar mix etc. Malathion can effectively control the fruit fly and is usually combined with protein hydrolysate to make as bait spray. It works on the principle that both male and female flies are attracted to the protein source, as protein is essential for the reproduction and vitellogenesis. Bait sprays also offer an advantage over the cover sprays for spot treatment where the toxicity on attracted flies is maximum with minimal impact on natural enemies (Wang, 1995).

3. Brown citrus aphid, *Toxoptera citricida* Kirkaldy (Aphididae: Hemiptera)

Brown citrus aphid (BrCA) is now one of the world's destructive pests of citrus. Besides causing direct damage to the crop, it also serves as a vector of citrus tristeza virus (CTV). This aphid is cosmopolitan in nature and thought to be originated in the Southeast Asia (Rocha-Pena *et al.,* 1905).

Distribution

Africa, Australia, eastern Asia, India, Central and South America, the Mediterranean region, and the Pacific Islands (Martin *et al.,* 2012).

Hosts of commercial importance

Host range brown citrus aphid is restricted to citrus and some non-rutaceous species, including *Pyracantha* (Rosaceae), *Cudramia* (Moraceae) and *Maclura* (Moraceae).

Identification and biology

Globally, around 16-20 aphid species are reported from different citrus growing areas, out of which five species are most commonly encountered. The species are *T. citricida, Aphis spiraecola*; *Aphis gossypii*; *Toxoptera aurantia* and *Aphis craccivora*. Adults of *T. citricida* are shiny black in colour and the nymphs are grey to reddish brown. However, colour variation may be observed in some species.

Winged adult female (alata) are 1.1-2.6 mm in length, antennae six segmented with I, II, and III heavily black and other segments banded at joints. Secondary rhinaria 7-20 on III and 0-4 on IV, setae on antennae III subequal. Siphunculi are black and elongated; cauda elongated black with 25-40 setae. Stridulatory apparatus present on abdomen, forewing with pterostigmata light brown and media usually twice branched.

Wingless adult female (aptera) are 1.5-2.8 mm in length and oval in shape. Antennae six segmented with no secondary rhinaria and the segments are not banded. Segments I and II are black, III and IV are pale and slightly swollen. Segments V and VI are dark at least at joints, setae on antennae III at least as long as the diameter of the segment; siphunculi black, elongate, and only slightly longer than cauda. Cauda black and elongate with about 30 setae; 'knees' of all three pairs of legs very dark and stridulatory apparatus are present.

The life cycle of brown citrus aphid is much simple when compared to those of other aphids. Generally, brown citrus aphid exhibits a holocyclic type of life clycle, where the females give birth to live birth clones of themselves without the help of males. Hence there is no sexual cycle in the autumn *i.e.*, no males, no oviparae, and no eggs. All individuals throughout the year are viviparous parthenogenetic females (Halbert and Brown, 1996). Nymphs attain maturity within 6-8 days at a favorable temperature of 20°C or higher. Populations of the aphids increase very quickly compared to other species because of their simple life cycle. A single aphid could produce several thousand aphids within 3 weeks, in absence of natural enemies (Komazaki, 1988).

Mode of feeding and symptoms of infestation

Aphids cause losses in citrus orchards by following three ways.

- Firstly, the nymphs and adults suck plant sap from the developing shoot tips, leaves, and buds using their piercing and sucking type of mouthparts. Their feeding results in stunted plant growth, curled and distorted leaves, and premature dropping of buds, respectively (Martin *et al.*, 2012).
- Secondly, *T. citricida* is the important and efficient vectors of *Citrus tristeza closterovirus*, a phloem-limited virus, when compared to other aphids (Halbert and Brown, 1996). CTV induces quick decline in sour oranges and stem pitting on root stocks of grape fruit and sweet orange.
- Thirdly, a sugary sweet fluid or honey dew secreted by these sucking insects not only attracts the ants but also encourages sooty mold fungus on leaves and fruits (Martin *et al.*, 2012).

Economic Threshold Level (ETL)

About 25% infested shoots.

Management

i. Only CTV-tolerant or resistant rootstock should be used for planting.
ii. When CTV problems are anticipated, closer plant spacing should be considered to optimize land use during the grove's early years.

iii. Trees that decline or become stunted can either be replaced or simply removed and neighboring trees allowed to fill in.

iv. Under natural conditions, a hymenopteran wasp *Lysiphlebus testaceipes* seems to lay eggs parasitises the adult aphids. Besides, release of coccinellid predator, *Menochilus sexmaculatus* @ 50 per tree is also recommended.

v. Spraying the plants with Dimethoate during March and again in September is effective to manage the pest.

4. Citrus leaf miner, *Phyllocnistis citrella* Stainton (Gracillaridae: Lepidoptera)

Phyllocnistis citrella is one of the serious pests of citrus orchards almost worldwide. This pest was first described from Kolkata, India (Stainton, 1856) was confirmed by Don Davis, a specialist in the Gracillariidae family, at the Smithsonian Institution (USNM).

Distribution

Leaf mines have infiltrated most of the citrus – generating world areas including the Mediterranean Basin and North, Central, and South America over the past 20 years (Sarada *et al.*, 2014).

Hosts of commercial importance

It is most frequently found on leaves of Rutaceae including citrus, orange, lemon, lime, tangerine, etc. A few representatives of Oleaceae, Loranthaceae, Leguminaceae, Salicaceae, and Tiliaceae have also been recorded to have infested this species (Sarada *et al.*, 2014).

Identification and biology

Citrus leaf miner is a very small, light-colored moth, less than 1/4 inch long and about 4 mm wingspread. The forewings contain white to silvery iridescent scales with black and tan markings. Black spot is also seen on tip of the wings. The hind body portion and wings are white with long fringed scales extending from wing margins. In resting pose the wings remain folded. The head is white and smoothly scaled (Sarada *et al.,* 2014).

The eggs are singly laid on the underside of tender leaves and hatch within 2-10 days. Larvae are minute (up to 3 mm), translucent and greenish yellow in colour. Upon hatching, larvae feed inside the leaf and get protected within the epidermis. Caterpillar moults for three times and enter pupation within 5 to 20 days depending upon environmental condition. CLM is easily detected by the characteristic serpentine larval mine on leaves. The pupa characteristically is in a pupal cell at the leaf margin where the adults emerge within 6 to 22 days during the dawn and are crepuscular in nature (Heppner, 1996).

Mode of feeding and symptoms of infestation

Upon hatching, the larvae mine on the upper surface during feeding and form serpentine galleries. These mines are filled with a central line of frass. Under heavy infestation, both the surfaces of the leaves are infested and occasionally the fruits are also get infested. These results in the following symptoms (Heppner, 1996).

- Leaf curl, leaf deformation, chlorosis, necrosis and leaf dropping.
- Epidermis appearing as a silvery film over leaf mines
- Pupation chamber near leaf margin, the edge of which is rolled over, and exposed portion of chamber with a distinct orange color; and
- Succulent branches of green shoots may also be attacked
- Mines may provide access for the citrus canker bacterium, *Xanthomonas axonopodis.*

Generally, the infestation of Citrus leaf miner (CLM) is severe in younger and flushing foliage. Mature citrus trees of more than 4 years old could withstand minor leaf damage without any effect on tree growth or fruit yield. CLM is rarely reported to infest the fruit. CLM is reported to cause a leaf damage of 12-85% (Lara *et al.*, 1998).

Management

i. Pest populations may be partially attenuated by altering the flushing pattern of the tree so that long intervals occur without flushes.
ii. Pruning of CLM damaged leaves should be avoided because this causes off-season new flush growth for CLM oviposition.
iii. Nitrogenous fertiliser should not be applied at times of the year when leaf miner populations are high and flush growth will be severely damaged, such as in the summer and fall (Grafton-Cardwell *et al.*, 2010).
iv. Vigorous shoots known as water sprouts often develop on branches and above the graft union on the trunk of mature trees which grow rapidly and produce new leaves for a prolonged period of time. Where citrus leaf miner is a problem, remove water sprouts that might act as a site for the moths to lay eggs (oviposition). Suckers, the vigorous shoots which grow from the trunk below the graft union, should always be removed since they originate from the root stock and do not usually produce desirable fruit.
v. Traps baited with a pheromone (insect sex attractant) is a useful tool for detecting leaf miners, determining when moths are flying and depositing eggs, and timing insecticide applications. However, they do not catch enough of the population to be used for control. Pheromone traps to be installed at shoulder height on a citrus tree.

vi. Under natural conditions, these larvae are parasitized by minute wasps such as *Cirrospilus* sp. (Grafton-Cardwell *et al.,* 2010).

vii. Spraying NSKE 5% (50 g/l) or neem cake extract 5% or neem oil 3 % or Imidacloprid 17.8 SL 125 ml per ha is effective in managing the pest. To protect bees, avoid applying Imidacloprid during the period 1 month prior to or during bloom.

5. Citrus Psylla, *Diaphorina citri* Kuwayama (Psyllidae: Hemiptera)

It is one of the most devastating insect pests of citrus in northern part of India, especially in Punjab, Haryana, Himachal Pradesh and Maharashtra but is of minor importance in south India (Randhawa and Srivastava, 1986).

Distribution

D. citri is widespread throughout the southern parts of Asia, from the southern islands of Japan in the east, through Southern China, southeast Asia to India and Pakistan in the west (Aubert 1987), eastern Iran (Bove *et al.,* 2000), Saudi Arabia near the Red Sea (Bove, 1986), the Indian Ocean islands of Reunion and Mauritius (Catling, 1973), Brazil and St. Helena (Aubert, 1987), Florida (Hardy, 1995) etc.

Hosts of commercial importance

Mainly *Citrus* spp., at least two species of *Murraya* and at least three other genera, all in the family Rutaceae (French *et al.,* 2001).

Identification and biology

Nymphs are flattened, oval in shape with orange colour. Adults are minute insect, shiny black with grey dusting on the body, wings are extending beyond the tip of the abdomen. The adults are 3 to 4 mm long with a mottled brown body. The head is light brown, whereas *Trioza erytreae* has black head. The forewing is broadest in the apical half, mottled, and with a brown band extending around the periphery of the outer half of the wing. This band is slightly interrupted near the apex (in *Trioza erytreae*, band is broadest at middle, unspotted and transparent). The antennae have black tips with two small, light brown spots on the middle segments (in *Trioza erytreae*, antennae are nearly all black). A living *Diaphorina citri* is covered with whitish, waxy secretion, making it appear dusty. The adult psyllid feeds with head down, and keeping the rest of its body raised from the leaf surface at an angle of 45-degree with its tail end which is unique for psylla (Grafton-Cardwell *et al.,* 2006).

Females live for 1-2 days and lay several hundreds of eggs depending on the temperature and host plant (Tsai and Liu, 2000). The egg of *D. citri* is anchored on a slender, stock-like process arising from the plant tissue. It is elongate with a broad basal end and tapering towards its distal and curved end. The average size of the egg measures 0.31 mm long and 0.14 mm wide. Freshly deposited eggs are light yellow, and turn bright orange with two distinct red eye spots at maturity (Grafton-Cardwell *et al.*, 2006). The eggs are placed on plant tissue with the long axis vertical to surface. *Trioza erytreae* eggs are laid with the long axis horizontal to surface (Maed, 1998).

The eggs hatch into wingless nymphs. The nymphs are flattened, orange to yellow to brownish in colour and 1/100 to 1/14 inch long. They are 0.25 mm long during the 1st instar, and 1.5 to 1.7 mm in last (5th) instar. Their color is generally yellowish-orange. There are no abdominal spots, whereas in *Trioza erytreae*, advanced nymphs have two basal dark abdominal spots. The wing pads in *Diaphorina citri* are large, while *Trioza erytreae* has small pads. In *Diaphorina citri*, large filaments are confined to the apical plate of the abdomen (in *Trioza erytreae*, there is a fringe of fine white filaments around the whole body, including head) (Maed, 1998). Each nymph also produces a waxy tubule from its posterior end to help removing the extra sugary waste product away from its body. For Asian citrus psyllid, the tubule is curly in shape with a bulb at the end, which is an unique taxonomic character (Grafton-Cardwell *et al.*, 2006). The total life cycle requires from 15 to 47 days, depending upon the season.

Mode of feeding and symptoms of infestation

The nymphs, while feeding on new flushes damage the citrus plants. They remove plant sap from the tissue and also inject salivary toxins during feeding. This results in deformity of new leaves like twisting and curling and inhibits or kills new shoots by burning them back. Secretion of honeydew promotes the sooty mold growth, which is unsightly but not harmful. Most important, the Asian citrus psyllid can kill citrus trees through its feeding activity if the insect infects the tree with the bacterium causing huanglongbing disease.

Management

- Only clean and healthy plants should be used.
- Prune the affected trees and dried shoots.
- Conservation of natural enemies like parasitoids, *Tetrastichus radiates* and predators like *Coccinella septumpunctata, C. rependa, Chilomenes sexmaculata, Chilocorus nigritus, Brumus suturalis, Chrysoperla carnea.*
- Spraying any of the insecticides *viz.*, NSKE 5 %, neem oil 10 L, Imidacloprid 200 SL 250 ml in 1500-2000 l of water per ha during March and again in September.

6. Fruit sucking moth, *Eudocima (Othreis) fullonica* Clerck., *E. maternal*, *E. ancilla* (Noctuidae: Lepidoptera)

Distribution

In 1869, a French botanist Thozet had observed for the first time that *Eudocima (Othreis) fullonica* (Clerck) moths sucking juice from ripe orange fruits at Rockhampton in Australia (Baptist, 1944). Presenly, the pest is distributed in China, Australia, Korea, Japan, Philippines, Hawai, Thailand etc. In India, the moth has been recorded in 1903 by Stebbing. *Eudocima fullonica* (Clerck), *Eudocima maternal* (Linnaeus), *Eudocima homaena* Hubner and *Eudocima cajeta* (Cramer) and others are also known to occur in India. In many parts of India, the pest has been found active during the months of rainy season after which there was decline in their activity (Bhumannavar and Viraktamath, 2012).

Hosts of commerial importance

These moth feeds on a wide range of trees/crops e.g., banana, citrus, fig, guava, kiwifruit, litchi, mango, stone fruit, persimmon, ripening papaya etc. Larval hosts include native vines of the family Menispermaceae and the preferred species are *Tinospora smilacina* and *Stephania* spp (Anonymous, 2012).

Identification and biology

E. materna: The moth has brownish black forewings with a white stripe and yellowish hind wings with a circular black spot in the middle.

E. fullonica: The moth has brownish forewings and yellowish black hind wings with a half moon or kidney shaped black spot.

E. ancilla: The forewings of this moth are dark brown with a green band in the middle; hind wings are yellowish with a kidney shaped black spot (Raghavaiah, 2011).

Adults lay round and translucent eggs singly on wild foliage of plants and weeds like *Tinospora cordifolia*, *Cocculus pendulus*, *C. hirsutus* in and around citrus orchard. Eggs hatch into larvae after completion of 3-4 days of incubation period. Larvae are brightly colored semi-loopers bearing a dorsal hump on the last segment of the body. They have distinct eye spots on the head and two large spots (mainly white with dark centre) on either side of the body just before the first pair of prolegs. Their velvety dark-brown back ground makes them cryptic.

Full-grown larva assumes a characteristic snake like posture when they get disturbed. Larva moults for four times during feeding on the native vines for about 4 weeks and after attaining maturity enters pupation. It is dark brown pupa in a delicate silk cocoon between webbed leaves. After about two weeks, adult emergence takes place from the pupa.

Mode offeeding and symptoms of infestion

Adults of the genus *Eudocima (Othreis)* are the prominent fruit piercing moths, causing extensive damage in pomegranate in the south and oranges in central India (Bhumannavar and Viraktamath, 2012). The moths are nocturnal in habit. During daytime, they hide in fallen leaves and in weeds and become active at dusk and swarm in large numbers when citrus fruits are about to ripen.Besides causing direct damage, various micro-organisms gain entry through the hole pierced by these moths leading to secondary damage.

In contrast to other moths, only adults of fruit sucking moths are destructive to fruit crops. During nights, moths pierce the skin of the fruit using their proboscis and suck the juice. The moths continue feeding throughout the night and cause colossal damage. The pin-hole made during piercing serves as an entry for various micro-organisms causing secondary rots. Secondary moth-feeders also drill these fruit taking the advantage of pin-holes. The main symptoms are:

- Premature dropping of the fruit
- Rotting of the fruits
- Internal injury consisting of bruised, dry area beneath the skin.

Achaea janata is also reported to suck the juice from unripe fruits (Anonymous, 2012 and Raghavaiah, 2011).

Management

- Destruction of all weeds host of larvae before June-July (Susainathan, 1924).
- Collection and destruction of rotten and dropped fruits.
- Smoking of orchard at wind entering edge of the orchard at least half an hour before dusk and continued for two or three hours after dusk, so that the moths fail to detect fruits and do not enter the orchard (Baptist,1944)
- Installation of light traps to attract moths
- Bagging of fruits with polythene or paper covers is effective, but expensive.
- Adults are highly phototropic. One fluorescent light/ha one month before fruit maturation between 7.00 PM to 6.00 AM below which poison baits with sugar solution 1% + fruit pulp + Malathion 1 ml should be placed.
- Trap crop – growing tomato crop in orchards to attract the adult moth.

7. Citrus butterfly, *Papilio demoleus* Linnaeus *P. polytes, P. Helenus* (Papilionidae: Lepidoptera)

The lemon butterfly is an important destructive pest whose larval forms cause severe damage by heavily devouring large quantity of foliage. The larvae are a serious pest of citrus nursery stock (trees 1-2 ft. in height) and other young citrus trees and are capable of defoliating entire nursery groves (Sarada *et al.,* 2014).

The genus *Papilio* is a widespread member of the family Papilionidae worldwide. Among various species of this pest *P. demoleus* is the most commonly found and is known as the lime, citrus or chequered swallowtail. Unlike most of the Swallowtail butterflies, it does not contain a prominent tail.

Distribution

It was widely distributed from Farmosa to Arabia including Bangladesh, Burma, Ceylon, Pakistan and India (Butani, 1973). This species is found throughout tropical and subtropical including Saudi Arabia, Iran and the Middle East to India, Nepal, southern China, Taiwan, and Japan. It is also found in Malaysia, Indonesia and Australia. In recent years, *Papilio demoleus* has been recorded in the Dominican Republic, Puerto Rico, and Jamaica also.

Hosts of commercial importance

It can feed and develop on all varieties of cultivated or wild citrus and other species of family Rutaceae. Besides citrus, it also attacks ber, wood apple and curry leaf (Raghavaiah, 2011). The larval population density remains high during October to December and July to December. Citrus butterfly is also able to survive during the cool winter when the temperatures drop below 0°C (Sarada *et al.,* 2014).

Papilio demoleus has been observed ovipositing on *Citrus aurantium* (bitter orange) and *Citrus aurantifolia* (Mexican lime, West Indian lime) (Common and Waterhouse, 1982, Braby, 2000; Rafi *et al*., 1999a; Rafi *et al*., 1999b).

Identification and biology

It is a black yellow coloured butterfly with wingspan range of 80-100 mm and hind wing has no tail. Head and thorax are black with creamy yellow streaks on each side. The legs and abdomen are dusky black with creamy yellow colouration on the ventral side of abdomen and the body was covered with black and yellow hairs. The upper portion of the forewing is largely black and the outer wing margin has a series of irregular yellow spots. Two yellow spots are present at the upper end of the discal cell with several scattered yellow spots in the apical region. The upper hind wing has a red tornal spot and the discal black band is dusted with yellow scales. The underside is paler yellow with the black areas more heavily dusted with yellow antennae are black and club shaped. In females the tip of the abdomen was flat and in males it is pointed. The adults fly in every month but are more abundant after monsoons (Sarada *et al.*, 2014). *P. polytes* males are black and females vary in form. *P. Helenus* has black wings with three white distal spots (Raghavaiah, 2011).

The female butterfly visits several plants, laying yellowish white, smooth, round eggs in groups of 2-3. Eclosion period is 3-8 days. Newly emerged dark brown caterpillars stay in the middle of the upper side of the leaf and soon develop irregular white markings on their body resembling bird's dropping. Early instars bear two rows of sub-dorsal fleshy spines. The fully grown caterpillar turns cylindrical in form, uniformly pale green, measures about 40-50 mm in length with a white sub-spiracular band. The osmeterium, normally hidden but can be everted to emit smelly compounds that deter some predators. Larval period ranges from 8 to 16 days in summer and four weeks during November to December. Larvae eat their own exuviae after each moult. Full grown larva spins a girdle around its body and pupates on a twig, pupal period 8-11 days.

Mode of feeding and symptoms of infestion

The damaging stage of the pest is larva which have cutting and chewing type mouthparts. The caterpillars feed gregariously on tender leaves keeping off the mid ribs and defoliate the entire seedlings or plant. Larvae being the voracious feeder cause severe defoliation of plants. In case of severe infestation, entire tree is defoliated.

Seasonal incidence

The pest attacks acid lime and mandarin plantations almost year-round but is serious during July-August.

Management

i. In small orchards and nurseries with mild infestation, early stage of the larvae (which look like bird dropping) may be hand-picked and destroyed to reduce the pest population.

ii. Natural enemies of the larvae are yellow wasp (*Polistes hebreus* F.), egg parasitoids: *Trichogramma evanescens; Telenomus* sp., larval parasitoids: *Apanteles (=Ooencyrtus) papilionis, Bracon hebetor, Distatrix papilionis* etc. The pupa has also been found to parasitised by *Pteromalus puparium.* Population of these natural enemies may be augmented through controlled use of toxic chemicals in the orchards.

iii. Spraying of *Bacillus thuringenesis* formulation is effective in managing the pest as it is exposed feeder.

iv. Entomopathogens like bacterium *Serratia marcesscens* and fungus *Fusarium* sp. also kill the pest population substantially.

v. The pest can be managed chemically by foliar spraying of any of the contact or systemic insecticides viz., Dimethoate @ 1.5 ml, Fenitrothion 1ml, or Quinalphas 2ml/l at an early larval stage

9. Citrus aphids, a. Brown citrus aphid and b. Black citrus aphid

a. Brown citrus aphid, *Toxoptera citricida* (Kirkaldy) (Aphididae: Hemiptera)

The brown citrus aphid, *Toxoptera citricida* (Kirkaldy), is one of the world's most serious pests of citrus. Although brown citrus aphid alone can cause serious damage to citrus.

Distribution

The distribution of brown citrus aphid includes Southeast Asia (Carver, 1978), Africa south of the Sahara, Australia, New Zealand, the Pacific Islands, South America, the Caribbean and Florida. So far, the remainder of U.S. citrus-producing areas and the Mediterranean, except (since 1994) for the island of Madeira (Aguiar *et al.*, 1994) have remained free of the pest (Blackman and Eastop, 1994).

Hosts of commercial importance

The preferred host range of brown citrus aphid is limited to citrus and a few close relatives (Aguiar, 1994 and Yokomi, 1994). However, brown citrus aphid has been reported to form large colonies on the new growth of other plants in several families. It is not known which, if any, of these reports are the result of misidentifications or collections of incidental specimens. Non-rutaceous plants are not normally preferred by *T. citricidus*, but may be colonised in absence of young and tender citrus foliage.

Identification

The genus *Toxoptera* closely resembles the genus *Aphis*, but are easily distinguished from the latter by the presence of a stridulatory apparatus with latero-ventral ridges on the abdomen and peg-like hairs on the hind tibiae. Both winged and wingless forms of adults are present in an aphid population.

T. citricidus are medium-sized aphids, 1.5–2.8 mm long. They are shiny, black to dark brown. They need to be examined microscopically to observe the very long, fine and erect hairs on the legs and body margins. Siphunculi are relatively shorter. The cauda is thick and bluntly rounded at the apex. Full grown aphids are 1.5–2.8 mm long. The median vein of the forewings is normally forked twice. Siphunculi are about 1/6 body length and strongly sculptured, while the cauda is rather bulbously rounded at the apex.

However, field identification of brown citrus aphid can be difficult because four of the five regularly collected species can be dark in color, and all five species colonize new growth. Additionally, mixed colonies of two or more species are common.

Biology

T. citricida prefer warm climates. It can, however, tolerate colder areas also by developing a holocyclic stage and overwintering as eggs. Development period of its life is temperature dependent. At 20°C, *T. citricida* has a nymphal development time is 6-8 days with an average pre-reproductive period of 8.1 days, longevity is 28.4 days. Fecundity is about 58.5 offspring/female with an intrinsic rate of natural increase. Mean generation time is about 11.2 days. Its thermal threshold is 8.4°C and required 125-degree days for development.

Takanashi (1989) reported slightly longer generation time under similar conditions and differentiated between alata and aptera development time. Winged morphs develop when populations become crowded and/ or food source declines in quality and disperse in search of new hosts to begin new colonies.

Mode offeeding and symptoms of infestion

New, tender shoots are vulnerable to *T. citricida* colonization. The aphid prefers those plant parts that help in support rapid population buildup. The pest is an external feeder and suck plant sap from the host by penetrating their stylets into phloem. Excess plant sap is excreted out from the body as honeydew which supports sooty mold growth. Heavy infestation by *T. citricida* is noted when growing points of citrus are covered by the dark-colored aphid and the flush bends under the physical weight of the colony. On the colony, aphid-tending ants are often present that collect honeydew. When disturbed, *T. citricida* populations sway rapidly *en masse*, making stridulatory movements with their hind legs probably to fend off their enemies. Flowers are not a preferred host tissues of this aphid. On mature leaves, stems, and fruit *T. citricida* population can notable to sustain.

High population of aphids during bloom periods can cause severe direct damage to citrus plants by sucking plant sap. In addition to that the major damage associated with *T. citricida* is the transmission and spread of severe strains of CTV. Such strains cause rapid decline and death of citrus plants regardless of tree age. The most virulent strains of CTV cause stem pitting in twigs, branches, and trunks of citrus trees. Stem pitting CTV weakens a tree and reduces fruit size, quality, and quantity. This occurs over a period of 6 to 25 years depending on the virulence of CTV. Grapefruit cultivars are most sensitive to stem pitting but sweet orange varieties are also susceptible. However, mandarins are most tolerant.

Management

i. Cultural management

a. Efforts to manage virus inoculum are the most important control strategy (Garnsey *et al*., 1996) because spread of severe strains of CTV is the major problem associated with *T. citricida*. If virulent stem pitting strains and *T. citricida* are endemic, citrus scion varieties tolerant to CTV should be planted. These include mandarins, pummelos, tagelos and tangor.

b. When CTV problems are anticipated, closer plant spacing should be considered to maximize land use during the grove's early years.

c. Trees that decline or become stunted can either be replaced or simply removed and neighboring trees allowed to fill in.

c. Close plant spacing is becoming a common practice in the United States in new groves.

d. Several areas have managed CTV by eradication of infected trees. This program is cost effective if virus incidence is low and spread is slow.

ii. Biological management

Although natural enemies are important in regulating aphid populations, they alone may not be satisfactory for controlling plant virus diseases. Given that alternate prey are available, natural enemies could reduce *T. citricida* populations to mitigate secondary spread of CTV (tree to tree within a field), especially if conservation and augmentation efforts are used. A classical biological control effort has been undertaken using this strategy in Florida with the release of *L. japonica* and *Aphelinus spiraecolae*.

iii. Chemical management

It is not known whether controlling aphids will reduce spread of CTV in production situations, but insecticides may be beneficial in protecting nursery stock and valuable budwood sources. As the pest is a sap sucking one hence, use of systemic insecticides help suppressing the aphid population.

b. Black citrus aphid, *Toxocoptera aurantia* Boyer de Fons. (Aphididae: Hemiptera)

Distribution

The aphid is found wherever coffee is grown throughout the tropics and subtropics. It is present in South America, Africa, India, eastern Asia and Australia, as well as the Mediterranean region, central America and southern U.S.A. (Carver, 1978).

Hosts of commercial importance

This aphid has over 120 hosts that include camellia, cocoa, coffee, *Hibiscus*, kamani, lime, mango, pomelo and *Vanda* orchid etc.

Identification and biology

Eggs are not produced by this species as they are parthenogenetic viviparous in nature. Females give birth to living young. There are four nymphal stages of this aphid. The first stage is approximately 1/36 inch in length and the last about 1/17 inch. The nymphs are without wings and brownish in colour. Only females are found in a population. They are oval in shape, shiny black, brownish-black or reddish brown in colour and either with or without wings. They are measuring about 1/25 to 1/12 inch in body length and having short black-and-white banded antennae. Winged individuals tend to have darker abdomens and are slightly thinner. The incidence of winged individuals is dependent on the population density and age of the plant part.

Reproduction is parthenogenetic. Females start reproducing immediately after attaining sexual maturity. They produce 5 to 7 live young per day, up to a total of about 50 young per female.

Mode of feeding and symptoms of infestion

Aphids feed by sucking sap from their hosts. This often causes the plants to become deformed, the leaves curled and shrivelled. In some cases, galls are formed on the leaves (Metcalf, 1962). This pest congregates on the tender young shoots, flower buds and the undersides of young and tender leaves. They are not known to feed on the older and tougher plant tissues (Carver, 1978).

Like other soft bodied insects, aphids also produce honeydew. This sweet and watery excrement is fed on by bees, wasps, ants and other insects. In addition to that the honeydew serves as a medium on which a sooty fungus, sooty mold fungus grows. This fungus blackens the leaf, interfere badly with the photosynthesis activity and decreases vigor and causes disfigurement of the host. When the sooty mold occurs on fruit, it often becomes unmarketable or of a lower grade as the fungus is difficult to wash off (Elmer and Brawner, 1975). Secondary damage caused by the aphid infestation is transmission of viral diseases from infected to healthy plants. It causes substantial losses than by direct feeding injury and is often the most damaging feature of an aphid infestation. On Citrus it is a vector of Citrus tristeza virus, citrus infectious mottling virus and little leaf and lemon-ribbing virus of lemon.

Management

i. Several natural enemies of the black citrus aphid are there that keep this pest under control. There are many other predators and parasites to this pest throughout the world. Predators in include *Allograpta obliqua* Say, *Chrysopa basalis*, *Chrysopa microphya* McLachlan, *Coccinella inaequalis* Fab., *Coelophora inaequalis*, *Platyomuslividi gaster* Mulsant and *Scymnodeslivi dgaster*. The parasites in Hawaii include *Aphelinus semiflavus* Howard and *Lysiphlebus testaceipes* (Cresson).

ii. This pest can also be kept under suppression by the entomogenous fungus *Acrostalagmus albus*.

iii. If chemical control becomes necessary, particularly when heavy infestation found, either insecticidal oil, or a synthetic insecticide may be used. Chemical control should only be applied at the first signs of damage during periods of flush growth.

10. Citrus mealybug, *Planococcus citri* Risso (Pseudococcidae: Hemiptera)

Planococcus citri is a polyphagous pest, which is adapted to feed and cause damage on wide range of host plants particularly citrus and glass house tomatoes. It is known to transmit some plant virus diseases like Cacaoa swollen shoot virus (Watson, 2016).

Distribution

This pest has a pan tropical distribution that sometimes extends into subtropical regions. It is native to Asia and found all over the globe.

Hosts of commercial importance

The citrus mealy bug, *P. citri* were observed infesting 65 plant species belonging to 56 genera in 36 families in Egypt (Ahmed and Shaaban, 2010). It is a minor pest of annona, Arabica and Robusta coffee (young trees are occasionally killed), cotton and various other vines. Other crops found to infest are banana, carambola (starfruit), cocoa, flowering ginger, macadamia, mango, and plants belonging to the Citrus genus.

Identification and biology

The wingless female deposits eggs as white-cottony masses, called ovisacs, on the trunk and stems of citrus plants, giving the appearance of cotton spread on the plant. A female can lay from 300 to 600 eggs in her lifetime, which are deposited in groups of five to 20 and they may hatch after six to 10 days or several weeks (Gill *et al.*, 2012).

On hatching, nymphs (crawlers) settle along the midribs, underside leaves, twigs, and at the places where fruits and leaves touch each other. Wax and honey dew secreted by these nymphs serve as indicators of infestation. The nymphs are yellow, oval-shaped with red eyes, and covered with white, waxy particles (Griffiths and Thompson, 1957) and they take six to 10 weeks to reach maturity. In contrast to female nymphs, fourth instar of males produce cottony-appearing cocoon and pupate (Gill *et al.*, 2012). On emergence, the length of the adult varies up to 3 mm. The females are wingless, white to light brown in color, with brown legs and antennae. The body of adult females is coated with white wax and bears a characteristic faint gray stripe along their dorsal side.

Mode of feeding and symptoms of infestation

The feeding action by both the nymphs and adults by sucking plant sap. This pest has two forms, a root form that attacks the roots of its host and an aerial form that attacks the foliage and fruit. Honeydew secreted by mealy bugs coats the surface of fruits and leaves favoring the development of sooty mould, inhibits photosynthesis eventually weakening the plant. Leaves of plants infested by the root form that wilt and turn yellow as if affected by drought. Roots are sometimes stunted and encrusted with greenish white fungal tissue (*Polyporus* sp.). Citrus mealybugs are visible beneath the fungus when it is peeled away. When the root form is associated with fungal tissue, it has the capability to kill the plant. The aerial form of this insect is found on leaves, twigs, and at the base of fruits. Premature dropping of leaf, fruit and flower takes place in case of heavy infestation by the pest (Matin *et al.*, 2012)

This pest is a vector of Swollen Shoot Disease in cocoa plant.

Management

i. **Cultural management:** Plants with brown leaves should be uprooted and replaced. High pressure water sprays are moderately effective.

ii. **Biological management:** The coccinellid, *Cryptolaemus montrouzieri* Mulsant (Coleoptera: Coccinellidae) an effective predator. It offers good results in reduction the population of adults of *P. citri*, one month after releasing the predator. Parasitic wasp *Leptomastix dactylopiiat* is also a good parasite of the pest.

iii. **Chemical management:** Chemical control is often inefficient management strategy due to their habit of hiding in crevices between foliage. However, the infested plants with green or yellow leaves may be saved with careful treatment. Malathion and Dimethoate may be effective in controlling this pest, especially if white oil is mixed with the parathion and malathion preparations. In addition to that Imidachloprid, Thiomethoxam are also effective in reducing the pest load from the crop.

11. Citrus blackfly, *Aleurocanthus woglumi* Ashby. (Aleyrodidae: Hemiptera)

Distribution

Citrus blackfly was first detected in the Western Hemisphere in 1913. Then it spread to Cuba in 1916, Mexico in 1935 (Smith *et al.*, 1964). Now it is distributed in almost all over the World including India.

Hosts of commercial importance

The citrus blackfly is found on over 300 host plant species. However, citrus trees such as lemon, orange and pomelo are most heavily infested by this pest.

Identification and biology

Eggs

Eggs are laid by the adult female in a spiral pattern on the underside of the leaf. Each female lays two to three egg-spirals during her life span. Incubation period is about 7 to 10 days (Dowell *et al.*, 1981).

Nymph

The first instar is elongate-oval, brown coloured, measuring about 0.30 mm long × 0.15 mm wide with two glassy filaments curving over the body. The first instar lasts 7 to 16 days. The second instar is more ovate and convex than the first instar and is dark brown in color. Duration of the second instar is about 7 to 30 days. The third instar is more convex and much longer than the second. The body is glossy black with stouter spines.

Adult

On emergence, the head of adult is pale yellow, legs are whitish and eyes are reddish-brown in colour. But within 24 hours after emergence, the insect is covered with a fine wax powder, which gives it a slate blue appearance.

The life cycle from egg to adult ranges from 45 to 133 days depending on the temperature and humidity. Six generations per year are produced in south Florida condition (Nguyen *et al.*, 1996).

Mode of feeding and symptoms of infestation

It damages citrus plant by sucking nutrients from foliage which weakens the plants. In case heavy infestation, citrus blackflies excrete honeydew on which sooty molds develop. Sooty molds coat citrus leaves, causing them to appear black that can severely impair leaf respiration and photosynthesis.

Management

i. Biological

Citrus blackfly has several natural enemies. The parasitic wasps, *Encarsia perplexa* (Huang & Polaszek) and *Amitushes peridum* (Silvestri) are important. A female citrus blackfly larva will support two and occasionally three or four parasites while a male citrus blackfly larva will support only one parasite. Some other biocontrol agents of this pest have also been detected.

ii. Chemical

a. Whiteflies also are managed by spraying some systemic insecticides like Acetamiprid, Imidacloprid, Thiomethoxam etc.
b. Spraying Profenophos 2.0 ml /l or Chlorpyriphos 2 ml/l or Imidacloprid 0.5 ml/l or Acephate 1.5 g/l are effective. The entire plant canopy should be drenched with the solution.
c. Spray oil has some insecticidal properties, but is primarily used to remove sooty mold that grows on the fruit and leaves.

12. Citrus whitefly, *Dialeurodes citri* (Ashmead) (Aleyrodidae: Hemiptera)

Distribution

The citrus whitefly is apparently of Asiatic origin. Now it has sporadically invaded several citrus growing areas worldwide. *Dialeurodes citri* first appeared in California in 1907. Citrus whitefly was found in France around 1945, and then spread throughout the Mediterranean Basin. Presently it is also known to occur in the Soviet Union, Turkey as well as in Israel.

Hosts of commercial importance

Citrus is the most important host, but the following are also food plants like, banana, chinaberry, coffee, jasmine, pear, persimmon, pomegranate etc.

Identification and biology

Eggs

The citrus whitefly lays yellowish eggs with smooth surface when freshly laid, but soon turn black and have a surface that is netted with a system of ridges.

Nymph

Nymphs are flat, elliptical, scale-like object that closely fastened to the underside of a leaf. The nymphs, after the first instar, are flattened, oval and are similar in appearance to the early instars of the unarmored scale insects.

Like scale insects, whiteflies lose their normal legs and antennae after the first molt. But unlike scale insects, the females gain them back in the adult stage. The nymphs lack a fringe of conspicuous, white, waxy plates or rods extending out from the margin of the body. Nymphs of the species are translucent, oval in outline and very thin. Pupae are thickened and are somewhat opaque and eye spots of the developing adult may show through the pupal skin.

Adult

The adult is a tiny, mealy white insect with four mealy white wings that expand about 3.2 mm. The adults of both sexes have two pairs of wings covered with a white, powdery wax which gives the insects their common name.

Tiny female adult lays about 150 eggs on the foliage that hatch in 8 to 24 days. The unfertilised eggs develop into males. The larvae after hatching settle on the growing twigs to feed and do not move until reaching adult stage. Nymphal life averages 23 to 30 days. Pupae opaque with visible eye spots. Pupal development requires about 13 to 30 days. Adults live for almost 10 days. The entire life cycle from egg to adult requires about 41 to 333 days. There may be several overlapping broods in a year.

Mode of feeding and symptoms of infestation

Nymphs and adults suck large quantities of sap from tender parts of the plant. Further injury is caused by sooty mold fungus which grows over fruit and foliage in copious amount of honeydew excreted by the whitefly. Heavily infested trees become weak and produce small fruits of insipid taste.

Management

i. Regularly pruning should be done to avoid whitefly population build up in the orchard.

ii. Conserve predators like Coccinellids viz., *Cryptognatha flavescens.*, *Verania cardoni* etc.

iii. Whiteflies are difficult to control with insecticides. Most less toxic products such as insecticidal soaps, neem oil, or petroleum-based oils control only those whiteflies that are directly sprayed. Therefore, plants must be thoroughly covered with the spray solution, and repeated applications may be needed.

iv. Insecticides recommended for the management of Citrus blackfly are also effective in suppression of Citrus whitefly.

13. Citrus rust mite or silver mite, *Phyllocoptruta oleivora* Ashmaed (Eriophyidae: Acarina)

The citrus rust mite, which is also called silver mite, is an important pest of citrus. It was first described in 1879 from Florida (Burditt *et al.*, 1963).

Distribution

The pest is mostly distributed in tropical areas and present in all the humid citrus growing areas (Hoy, 2011). Majorly infested areas include countries of the South Africa, Middle East, southern Asia, Australia, USA and (Japan and China) (Tropea Garzia and De Lillo, 2018).

Identification and biology

The species has four developmental stages in its life cycle, viz., egg, first instar (larva), second instar (nymph) and adult. The adults are cigar shaped, minute wormlike yellow mites and are found on under surface of the leaves and fruits. The female lays one to two spherical transparent eggs per day and as many as 30 eggs are laid during her lifetime. Eggs are spherical, clear and found along leaf midribs or clustered in fruit depressions. Incubation period is about 3 days. The freshly emerged larva resembles the adults and changing in color from clear to lemon yellow. The nymph moults twice to become an adult.

Mode of feeding and symptoms of infestation

Citrus rust mite damages epidermal cells of plant leaves, fruit and green twigs of citrus using piercing and sucking type of mouthparts. As a result of this mite infestation the fruit surface blemishes developed which can lower external grade of fresh fruit, reduce fruit size and increase fruit drop. When a fruit is injured in summer, the injured surface is smooth and dark brown in color, commonly referred to as "bronzing". Feeding destroys rind cells and the surface becomes silvery on lemons, rust brown on mature oranges or black on green oranges. Mite feeding on fruit early in the spring produces a peel which is somewhat rough in texture but lighter in color than damage caused by feeding in the summer and referred to as "shark skin"(Raghavaiah, 2011).

Seasonal incidence

These mites can be found any time during the year with peak populations usually occurring during June and July and are known to avoid most sun exposed portion of the fruit.

Management

Foliar sprays with wettable Sulphur@ 3g/l or Propargite 1ml/l or Spiromesifen 240 SC 8-12 ml/20 l of water once in a month can effectively manages mite population development in citrus.

Other pests of citrus

Citrus thrips, *Thrips nilgiriensis* Rama., Citrus red mite, *Panonychus citri* (McGregor), Cottony cushion scale, *Icerya purchasi* Mask. etc. that sometimes cause damage to the citrus plants in some citrus growing areas.

References

Adhikari, D. 2019. "Status and Management of Fruit Fly in Nepal (Power point Presentation)". National Plant Protection Workshop, Lazimpat, Kathmandu (2019).

Adhikari, D. and Joshi, S. L. 2016. Occurrences and field identities of fruit flies in sweet orange (*Citrus sinensis*) orchards in Sindhuli, Nepal, J. Nat. Hist. Mus., 30:47-54.

Aguiar, A. M. F., Fernandes A, Ilharco F. A. 1994. On the sudden appearance and spread of the black citrus aphid Toxoptera citricidus (Kirkaldy), (*Homoptera: Aphidoidea*) on the island of Madeira. BocagianaMuseu Municipal do Funchal (Historia Natural),168: 1-7.

Ahmed, N. H., Abd-Rabou, S. M. 2010. Host plants, geographical distribution, natural enemies and biological studies of the citrus mealybug, *Planococcus citri* (Risso) (*Hemiptera: Pseudococcidae)*. Egyptian Academic Journal of Biological Sciences, 3(1): 39-47.

Anonymous, 2012. Retrieved from https://www.daf.qld.gov.au/plants/fruit-and-vegetables/a-z-list-of-horticultural-insect-pests/fruit-piercing-moth (Accessed date: 08.01.2018).

Anonymous, 2017. Area, Production and yield of horticulture and plantation crops. In: Agricultural statistics at a glance 2017. Government of India. pp. 218.

Aubert, B. 1987. Trioza erytreae Del Guercio and Diaphorina citri Kuwayama (*Homoptera: Psyllidae*), the two vectors of citrus greening disease: Biological aspects and possible control. Fruits, 42:149-162.

Baptist, B. A. 1944. The fruit-piercing moth (*Othreis fullonica* L.) with special reference to its economic importance. Indian Journal of Entomology, 6:1-13.

Bhumannavar, B. S., Viraktamath, C. A. 2012. Review article: Biology, ecology and management of fruit piercing moths (*Lepidoptera: Noctuidae*). Pest Management in Horticultural Ecosystems. 18(1):1-18.

Blackman, R. L. and Eastop, V. F. 2000. Aphids on the World's Crops: an Identification and Information Guide, 2nd edn. i–x, 1–466. Wiley, Chichester (GB).

Blackman, R. L, Eastop, V. F. 1994. Aphids on the World's Trees. Commonwealth Agricultural Bureau International, Wallingford, UK, pp: 987.

Bove, J. M. 1986. Greening in the Arab Peninsula: Towards new techniques for detection and control.FAO Plant Prot. Bul, 34: 7-14.

Bove, J. M., Danet, J. L., Bananej, K., Hassanzadeh, N., Taghizadeh, M., Salehi, M, Garnier, M. 2000. Witches' broom disease of lime (WBDL) in Iran. Proc. 14th Conf. Int. Org. Citrus Virol:207-212.

Braby, M. F. 2000. Butterflies of Australia: Their identification, biology and distribution. Volume 1. CSIRO Publishing.

Burditt, A. K., Reed, D. K., Crittenden, C. R. 1963. Observations on the mites Phyllocoptruta oleivora (Ashmead) and Aculus pelekassi Keifer under laboratory conditions. Florida Entomologist 46:1-5.

Butani, D. K. 1973. Insect pests of fruit crops- citrus. Pesticides, 7(12):23-26.

Carver, M. 1978. The Black Citrus Aphids, Toxoptera citricidus (Kirkaldy) and T. auranti (Boyer de Fons.) (*Homptera: Aphidiae*). J. Aust. Entomol. Soc., 17:263-270.

Catling, H. D. 1973. Results of a survey for psyllid vectors of citrus greening disease in Reunion. FAO Plant Prot. Bull, 21:78-82.

Chen, S. X. and Xie, Y. Z. 1955. Taxonomic Notes on the Chinese Citrus Fly, *Tetradacus citri Chen*. Acta Entomologica Sinica, 5, 123-126. [In Chinese]

Common, I. F. B, and Waterhouse, D. F. 1982. Butterflies of Australia. Angus and Robertson Publishers.

Dowell, R. V., Cherry, R. H., Fitzpatrick, G. E., Reinert, J. A., Knapp, J. L. 1981. Biology, plant-insect relations, and control of the citrus blackfly. Florida Agricultural Experiment Station Bulletin, 818:1-48.

Drew, R. A. I. 2006. "Attractiveness of various combinations of colors and shapes to females and males of Bactrocera minax (*Diptera: Tephritidae*) in a commercial mandarin grove in Bhutan". Journal of Economic Entomology 99(5):1651-1656.

Economos, C. 1999. Nutritional and health benefits of citrus fruits. Food Nutr Agric, 24:11–18.

Elmer, H. S. and Brawner, O. L. 1975. Control of Brown Soft Scale in Central Valley. Citrograph. 60(11):402-403.

FAO 2017. Citrus fruit-Fresh and processed statistical bulletin, pp:4.

French, J. V., Kahlke, C. J., da Graca, J. V. 2001. First Record of the Asian Citrus Psylla, Diaphorina citri Kuwayama (*Homoptera: Psyllidae*), in Texas. Subtropical Plant Science, 53:14-15.

Gill, H. K., Goyal, G. and Gillett, K. 2012. Featured creatures. Retrieved from http://entnemdept.ufl.edu/creatures/CITRUS/Planococcus_citri.htm (Accessed date: 22.01.2018)

Grafton-Cardwell, E. E., Godfrey, K. E., Rogers, M. E., Childers, C. C. and Stansly, P. A. 2006. Asian Citrus Psyllid (PDF). Oakland: UC ANR Publication 8205.

Grafton-Cardwell, E. E., Headrick, D. H., Godfre K. E., Kabashima 2010. Citrus leaf miner: Integrated Pest Management for Home Gardeners and Landscape Professionals. Pest notes. Retrieved from http://ipm.ucanr.edu/PMG/PESTNOTES/pn74137.html (Accessed on: 02.01.2018).

Griffiths, J. T., Thompson, W. L. 1957. Insects and mites found on Florida citrus. University of Florida Agricultural Experiment Station Bulletin, 591:30-33.

Halbert, S. E. and Brown, L. G. 1996. DPI Entomology Curcular374. Retrieved from http://entnemdept.ufl.edu/creatures/citrus/bc_aphid.htm (Accessed date: 31.12.2017).

Heppner, J. B. 1996. Feature creatures. http://entnemdept.ufl.edu/creatures/citrus/citrus_leafminer.htm (Accessed on: 02.1.2018).

Hoy, M. A., 2011. Agricultural Acarology: Introduction to Integrated Mite Management. CRC Press, Boca Raton, Florida, 430.

Huang, D. S., Xiao, G. Z., Yang, Z. A. and Wen, J. Z. 2007. The Damage and Control Technology of Bactrocera (Tetradacus) minax Enderlein in Taoyuan County. Plant Quarantine, 4, 233-235. [In Chinese]

Komazaki, S. 1988. Growth and reproduction in the first two and summer generations of two citrus aphids, Aphis citricola van der Goot and Toxoptera citricidus (Kirkaldy) (*Homoptera: Aphididae*), under different thermal conditions. Applied Entomology and Zoology, 23: 220-227.

Lara, G. J., Quiroz, M. H., Sanchez, J. A., Badii, M. H., Rodriguez, C. V. A. 1998. Citrus leaf miner Phyllocnistis citrella Stainton, incidence, damage and natural enemies in Montemorelos, Nuervo Leon, Mexico. South Western. Entomol, 23(1):93-94.

Liu, H. 2015. "Effect of Six Insecticides on Three Populations of Bactrocera (*Tetradacus*) minax (*Diptera: Tephritidae*)". Current Pharmaceutical Biotechnology 16(1):77-83.

Lu, H. X., He, K. P., Ruan, H. F. and Mu, B. Z. 1997. The Biological Features of Chinese Citrus Fly Dacus citri (Chen). Journal of Hubei Agricultural College, 17:169-173. [In Chinese]

Luo, Y. L. and Chen, C. F. 1987. Pupal Biological Characteristics of *Tetradacus citri Chen.* China Citrus, 4, 9-10. [In Chinese]

Martin, K. W., Hodges, A. C. and N. C. Leppla 2012. Citrus pests. Retrieved from idtools.org/id/citrus/pests/factsheet.php?name=Citrus%20mealybug (Accessed date: 23.01.2018).

Metcalf, R. L. 1962. Destructive and Useful Insects Their Habits and Control. McGraw-Hill Book Company; New York, San Francisco, Toronto, London. Pp-1087.

Nguyen, R. 1996. Featured creatures. Retrieved from http://entnemdept.ufl.edu/creatures/citrus/citrus_blackfly.htm (Accessed date: 17.01.2018).

NIPHM, 2014. AESA based IPM packages for citrus. GoI, pp-2.

Rafi, M. A., Khan, M. R. and Ilyas, M. 1999b. Host preference of lemon butterfly Papilio demoleus Linn. in the northern Barani areas of Pakistan. Pakistan Journal of Science 51(3-4):93-94.

Rafi, M. A., Matin, M. A. and Khan, M. R. 1999a. Number of generations and their duration of the lemon butterfly, (Papilio demoleus Linn.) in the rain fed ecology of Pakistan. Pakistan Journal of Scientific Research 51(3-4):131-136.

Raghavaiah, G. 2011. Pests of citrus. In: Pests of crops and their Management. Dept.of Entomology, Agriculture College, Bapatla, pp:107-111.

Rocha-Pena, M. A., Lee, R. F., Lastra, R., Niblett, C. L., Ochoa-Corona, F. M., Garnsey, S. M., Yokomi, R. K. 1995. Citrus tristeza virus and its aphid vector Toxoptera citricida: Threats to citrus production in the Caribbean and Central and North America. Plant. Dis, 79:437-443.

Sachan, J. N., Gangwar, S. K., 1982. Insect pests of citrus. In: Technical Bulletin No. 16 on Mandarin orange decline in North Eastern Hill Region and its control. ICAR Research Complex, Shillong, pp-33.

Sarada, G., Gopal, K., Sankar, T. G., Laxmi, L., Gopi, V., Nagalaxmi, T., Ramana, K. T. V. 2014. Citrus Leaf Miner (*Phyllocnistis citrella Stainton, Lepidptera: Gracillariidae*): Biolology and Management: A Review. Research & Reviews: Journal of Agriculture and Allied Sciences, 3(3): 39-48.

Sarada, G., Gopal, K., Ramana, K. T. V., Lakshmi, L. M., Nagalakshmi, T. 2014. Citrus Butterfly (*Papilio demoleus Linnaeus*) Biology and Management: A Review. Research & Reviews: Journal of Agriculture and Allied Sciences, 3(1):17-25.

Shu, X. L. and Xiao, Z. J. 2006. "The Occurrence and Control of Bactrocera (Tetradacus) minax Enderlein". Plant Doctor 19:26.

Singh, K. M. and Singh, T. K. 2012. Life cycle and host preference of citrus trunk borer Anoplophora versteegi (Rits.) (*Cerambycidae: Coleoptera*). Indian Journal of Entomology. 74(2):120-124.

Smith, H. D., Maltby, H. L., Jimenez, E. J. 1964. Biological control of the citrus blackfly in Mexico. U.S. Dept. Agric. Tech. Bull, 1311.

Stainton, H. T. 1856. Descriptions of three species of Indian Micro-Lepidoptera. Transactions of the Entomological Society of London,3:301-304.

Susainathan, P. 1924. Fruit-sucking moths of South India.Proceedings of 5th Entomological Meeting, Pusa, pp:23-27.

Takanashi, M. 1989. The reproductive ability of apterous and alate viviparous morphs of the citrus brown aphid, Toxoptera citricida (Kirkaldy) (*Homoptera: Aphididae*). Jpn. J. Appl. Entomol. Zool. 33:266-269.

Thakur, N. S. A. and Shylesha, A. N. 1996. Management of citrus trunk borer - major pest of khasi mandarin in Meghalaya. In: Proceedings of the National Symposium on IPM & Sustainable Agriculture: An Entomological Approach. 22-24 September, 1995, SanatanDharm College, Muzaffarnagar India. pp-206-208.

Tropea, G., G. and De Lillo, E. 2018. Geographic distribution of Phyllocoptruta oleivora in the Mediterranean Basin, with particular emphasis on Italy. Systematic and Applied Acarology 23: 1021-1023.

Tsai, J. H., Liu, Y. H. 2000. Biology of Diaphorina citri (Homoptera: Psyllidae) on four host plants. Journal of Economic Entomology, 93:1721-1725.

Wang, X. L. and Zhang, R. J. 2009. "Review on Biology Ecology and Control of Bactrocera minax Enderlein". Journal of Environmental Entomology, 31:73-79.

Wang, X. J. and Luo, Y. L. 1995. Advanced Study on Bactrocera minax. Chinese Bulletin of Entomology, 32, 310-315. [In Chinese]

Watson, G. 2016. Planococcus citri (citrus mealybug). Retrieved from https://www.cabi.org/isc/datasheet/45082 (Accessed date: 22.01.2018).

Yang, S. G. 1989. Occurrence and Control of Chinese Citrus Fly in Yanhe, Guizhou. Journal of Guizhou Agricultural Science, 6, 23-26. [In Chinese]

Yokomi, R. K., Lastra, R., Stoetzel, M. B., Damsteegt, V. D., Lee, R. F., Garnsey, S. M., Gottwald, T. R., Rocha-Pena, M. A. and Niblett, C. L. 1994. Establishment of the brown citrus aphid (*Homoptera: Aphididae*) in Central America and the Caribbean Basin and transmission of citrus tristeza virus. Journal of Economic Entomology, 87(4):1078-1085.

3

Pests of Jackfruit and Their Management

Gobinda Roy, Nripendra Laskar and Samrat Saha

Jackfruit (*Artocarpus heterophyllus* Lam.) belongs to family Moraceae, is the largest edible fruit in the world (Naik, 1949 and Sturrock, 1977). It is one of the most popular and important fruit crops and is the national fruit of Bangladesh (Haque, 1977). The jackfruit tree is widely cultivated in tropical regions of India, Bangladesh, Nepal, Sri Lanka, Vietnam, Thailand, Malaysia, Indonesia and the Philippines. Jackfruit is alsofound across Africa, e.g., in Cameroon, Uganda, Tanzania, and Mauritius, as well asthroughout Brazil and Caribbean nations such as Jamaica. However, jackfruit originally native to Indian subcontinent, is now widely cultivated in the tropics of both hemispheres (Ochas *et al*., 1981). In our country, the trees are found distributed in southern states like Kerala, Tamil Nadu, Karnataka, Goa, coastal Maharashtra and other states like, Assam, Bihar, Tripura, Uttar Pradesh and foothills of Himalayas. The name originated from its Malayalam name *Chakka.* It is also called *kathhal* (hindi and urdu), *pala* (tamil), *halasinahannu* (kannada) *panasapandu* (telugu) and *phanos* (marathi and Konkani). The fleshy carpel which is botanically the perianth is the edible portion. It is the favourite fruit of many, owing to its sweetness.

The pulp of the ripe fruit is eaten fresh and sometimes it is preserved in syrup or dried. The seed is also cooked and used for cooking. In Greek, 'Arto' means 'bread' and 'carpus' means 'fruit', therefore, jackfruit is also called 'breadfruit'. The rink (skin of fruit) and leaves are excellent cattle feed (Anonymous, 1995). Jackfruit wood is valuable for making furniture, the 'yellow heart wood' produces a yellow dye and the latex is used as sealing material. The tree can also be used to improve the environment of homesteads. So, jackfruit is also called multipurpose fruit tree.

The fruit is rich in several nutrients. It can act as source of complete nutrition to the consumers. The fruit is equivalent to Avocado and olive in terms of the healthier mix of nutrients for human dietary needs, almost having the exact nutrient equivalents of mother's milk. It is rich in vitamin B and C, potassium, calcium, iron, proteins and high level of carbohydrates, affordable and readily available supplement to our staple food. Its seeds are rich in proteins and can be relished as a nutritious nut. The fruit is also the source of chemical "Jacalin" useful in preventing colon cancer, AIDS etc.

Many insect pests are attracted to jack fruit. Insects that cause less than 5% damage are not considered as pests. The insects which cause damage between 5-10% are called minor pests and those that cause damage above 10 are considered as major pests. Jack fruit plants are infested by about thirty-five (35) species of insect pests. Among them, shoot and fruit borer, Spitle bugs, Mealy bugs, Bud weevil, Bark eating caterpillar, Aphid and Leaf weeber *etc.* are very important. Of these, the shoot and fruit borer, *Deaphania caesalis* is considered as the major pest of jackfruit (Tandon, 1998). An average of 27.44% jackfruits is infested by *D. caesalis* in Bangladesh (Khan and Islam, 2004).

Table 1: The arthropod pests recorded to infest jackfruit.

Sl.No.	Common name	Scientific name	Family	Order
1.	Shoot and fruit borer	*Deaphania caesalis* Walker	Pyralidae	Lepidoptera
2.	Bark eating caterpillar	*Indarbela tetraonis* Moore, *I. quadrionotata* Walker.	Metarbelidae	Lepidoptera
3.	Bud weevil	*Ochymera artocarpi* Marshall	Curculionodae	Coloeptera
4.	Aphid	*Greenidia artocarpi* Schout	Aphididae	Hemiptera
5.	Leaf webber	*Diaphiniabi vitralis* Guenee	Lymantriidae	Lepidoptera
6.	Mealy bug	*Drosicha mangiferae* Green	Monophebidae	Hemiptera
7.	Spittle bug	*Cosmocarta relata* Distant	Cercopidae	Hemiptera

1. Fruit and shoot borer, *Deaphani acaesalis* Walker (Pyralidae: Lepidoptera)

Jack fruit borer, *Diaphania caesalis*, is a destructive lepidoteran insect pest. It was originally described by Walker in 1859 as *Glyphodes caesalis* sp. nova. Subsequently, it was transferred to genus *Margaronia*, then to genus *Palpita* and finally Wang (1963) placed it under genus *Diaphania*.

It is considered as the one of the major pests of Jackfruit, which reduces both the quality and quantity of fruit. It is a very common pest in India and Bangladesh. The incidence of this pest occurs sporadically or in epidemic form every year throughout India. In the favorable weather severe infestation may occur and maximum fruits may be infested.

Distribution

Jack fruit borer was found almost all the jack fruit growing countries. In India it is found in several states. Fletcher (1914) reported *G. caesalis* as a pest of jack fruit in Karnataka and Maharashtra. It is also recorded from Assam, Sikkim, Bihar, Uttar Pradesh, Andhra Pradesh and Tamil Nadu of India (Chowdhury and Majid, 1954). Outside India, it has been reported from Borneo, Burma and Sri Lanka (Hampson, 1986) and also in Bangladesh (Alam *et al.*, 1964).

Hosts of commercial importance

Plants affected by this pest is mainly jack fruit. Other hosts recorded include fruit trees such as Mango, citrus, guava, litchi, etc. from the Asian countries.

Identification and biology

The adult moth is about half-an inch long with black edge bands and pale-yellow patches on wings. The adult female moth lays her eggs on leaf, shoot and flower bud.

The moth lays eggs singly on the surface of the spikes or on the surfaces of the spathe and on the tip of the tender shoots of jack fruits. Eggs are spindle shaped with smooth and shiny surface and these were laid in batches. The eggs are pale brown to brown in colour. The average incubation period is about 4-5 days. However, the difference of incubation period between two generation depends on environmental factor specially temperature and humidity.

The first instar larva is very tiny. The full-grown caterpillar is pinkish, each segment banded with numerous black flattened horny warts from which arise single short bristly hairs. Head and prothoracic shield become yellow. Pupation takes place in a silken cocoon made by the caterpillar inside the fruit tunnel several inches long, twisted dried of leaf or on the surface of nearest two more fruits. Pupa is red brown in colour and pupal period is about a week.

The adult moth comes out from the pupa through the opening of cocoon. Newly emerge moth is pale brown in colour which after a few hours changes to cream colour. The adults after emergence remain quite silent and they do not take any food and rarely fly. The female is slightly larger than male. The body is covered with cream colour scales. It bears compound eyes. Antennae are segmented and covered with scales. The wings colour is combination of cream and brown. Semi-circular brown spots present on the surface of the wings. Both the front and hind pairs of wings are covered with scales and fringed with hairs at the edge. The legs are walking type and more or less equal in size and shape. Average longevity of the adult was 4-6 days.

Mode of feeding and symptoms of infestation

Jack fruit borer moth lay eggs on the surface of both male and female spikes or on the surface of the spathe and on the tip of the tender shoots of jackfruit (Khan *et al*., 2003). On hatching the caterpillar bore into shoots, flowers buds and fruits of all developing stages (Soepadmo, 1991). Early infestation of jackfruit borer results in deformation of fruits and sometimes dropping off the immature fruits. The larvae bore into the mature fruit and cause damage to the edible part. Later infested fruits frequently get rotten due to entrance of rain water into the fruits. The entrance hole is easily visible and associated with mass of excreta (Khan and Islam, 2004).

Jack fruit borer also attacks in nursery. Caterpillar bore into the tip of jackfruit sapling and proceeds towards the base by making tunnel. As a result, the affected parts wilt and dry resulting lateral branching of sapling. Caterpillar also attacks the tip of tender shoot of mature plant and collapses the growing tip.

As jack fruit borer can attack different stage of fruits as well as the tender shoots, so it is considered as a major constraint of jackfruit production in India. On the other hand, its damage severity is increasing day by day. But unfortunately, very little research works have so far been done on its biology, pest status and management.

Management

As it is an internal feeder, so it's effective control measure is very much difficult. However, following management strategies may be undertaken so as to reduce its infestation severity.

i. Removal and destruction of the affected shoots, flower buds and fruits in the initial stage of attack.

ii. To protect the fruits from ravages of this pest covering the fruits with alkathene bags may be practiced (Madhava Rao, 1965 and Tendon, 1998).

iii. Fallen, overripe and damaged fruits should be collected and buried under ground.

iv. Hayes (1996) reported that this pest might be controlled by hand picking.

v. Spraying of Carbaryl during flowering may be recommended.

vi. Trees showing the infestation sign at flowering should be sprayed with Cypermethrin 10EC or Fenitrothion 50EC. A second spray should be given after 15 days of first application where necessary (Karim, 1995).

2. Bark eating caterpillar, *Indarbela quadrinotata* Walker, *I. tetraonis* Moore, (Metarbelidae: Lepidoptera)

Bark eating caterpillar is a polyphagous pest infesting a large number of plant species. There are 13 species of the genus of which, *I. quadrinotata* (Walker), *I. tetraonis* (Moore) and *I. dea* (Swinhoe) are of economic importance. *I. tetraonis* is the only lepidopteran borer found boring into trunk and main stems of jack fruit trees in India.

Distribution

This pest is reported from all jack fruit growing areas of the world including Burma, Sri Lanka, Pakistan, Bangladesh and India also.

Hosts of commercial importance

Plants affected by this caterpillar include fruit trees such as mango, citrus, guava, litchi, and cashews and forest trees including *Casuarina* spp., *Acacia* spp., and *Albizia* spp. (Patil *et al,* 1990, Sharma and Kumar 1986; Sharma and Verma, 1987).

Identification

The full-grown caterpillar is smooth, with sparse hairs and measures about 3.5 to 4 cm in length. Light brown sclerotized patches are present dorso-laterally. The thoracic legs are simple with the last segment ending in a curved claw. Abdominal legs are present on segments 6-9. The pupa is light brownish in colour measuring about 1.5 cm in length and with 2 short pointed cephalic processes. Abdominal segments bear transverse rows of spine-like processes, of which there are two rows each on segments 5 to 8. Segments 3, 4 and 8 have only one row and only two spines on segment 10. The caudal end (cremaster) bears several small spines – like processes. These spines and teeth - like processes help the pupa to orient itself towards the tunnel mouth prior to eclosion. Pupal period lasts for 3 weeks.

The moth is light brown in colour and measures 15-18 mm across the wings. Fore wing has a sub-apical brown spot and with several transverse rows of brown scales. Hindwing is light black in colour. Head is depressed and antennae strongly pectinate, with the pectination uniform throughout.

Biology

The life cycle of this insect has been worked out on various hosts by different workers. Adults start emerging during May to July. They are 35-40 mm in size, pale brown or grey in colour. Female moths deposit eggs in cluster under loose bark, in forks in the older wood, or where twigs have broken off or been badly pruned in early June. Eggs hatch 15-25 days after being laid. Larvae are 50-60 mm and have pale brown bodies with dark brown heads. Larvae that hatch out initially feed on the bark and subsequently bore into the trunk. The tunnel entry remains closed with a frass covering which is drawn out into a sleeve through which the larva moves. The tunnel is used as shelter by the larva and is kept clean of the faecal pellets and frass which are added on to the distal end of the sleeve. The larval period lasts for 9- 10 months. Several caterpillars may attack the same tree at different locations with serious injury to the bark and the death of small branches.

Pupation occurs within the larval tunnel, with the cephalic end of the pupa slightly protruding outside. The pupal period lasts for about 15-25 days (Sharma and Kumar, 1986). Life cycle is annual with one generation per year.

Mode of feeding and symptoms of infestation

Although moths were available in the field from March onwards, there is no evidence for the beginning of a fresh generation. However, with the onset of rains in late May, new generation starts to appear. During this period, eggs as well as the newly hatched larvae could be seen on the bark of trees. Usually, it takes 2 to 3 months for the larva to make the tunnel in the wood and to construct the sleeve. The sleeve extending from the borer hole is a clear indication of the existence of the larva inside. As there is only one generation in a year, this insect is univoltine.

Seasonal incidence

Fresh attack continued until October, after which the population remained more or less steady.

Management

Management strategies reported against this pest are either mechanical or chemical. It is necessary during all the life cycle stages but the best management time is when eggs are hatching and caterpillars are small.

i. Pruning of affected branches and destruction thereof.

ii. Keeping the orchards clean and avoiding overcrowding of trees will help to minimize attacks of the pest.

iii. The mechanical method involved killing the larvae within the tunnels by inserting a sharp metallic probe and sealing the tunnel entrance using mud.

iv. In the chemical method, application of a toxic substance either by injection or by inserting a cotton swab soaked in the chemical was the most widely used method. Spot application either by brushing or spraying was also tried in certain cases. Diclorvos (0.08%), Carbaryl (1%), Quinalphos (0.05%) any one among them are most effective against this pest.

3. Bud weevil, *Ochyromera artocarpi* Marshall (Curculionidae: Coleoptera)

Distribution

The pest has narrow distribution and has been found in only 21 countries (Anonymous, 1973). It is widespread in Asia and some tropical and sub-tropical countries. It is found all over India and has been recorded as a major pest from

Assam and Western coast of Karnataka. Two more curculionid weevils, viz., *Onychocnemis careyae* Marshall and *Teluropus ballardi* Marshall have been reported feeding on leaves of jack fruit trees in South India (Nair, 1995), but not causing any appreciable loss.

Hosts of commercial importance

Plants affected by this insect pest include fruit trees such as Mango, citrus, guava, litchi, potato and sweet potato are recorded from the Asian countries mainly.

Identification

Adult weevils are reddish brown to blackish gray, and are covered with short, stiff, erect bristles and scales. They resemble soil particles and are hard to detect in the soil. Eggs are grayish yellow to yellow. Larvae are white. Pupae are whitish and sedentary.

Biology

Adult females lay yellow greyish eggs singly in cavity excavated in the buds/ stems. Roots are the preferred site for oviposition. Larvae tunnel, feed and grow within buds/stems. Fully grown larvae enter a resting stage (pupa), thereafter, adults emerge. Adults resemble soil particle and are hardly possible to detect in soil. The body of the adult has short erect bristles and scales are reddish brown to grayish black. The plant produces terpenoids as a result of the feeding and the damage result in discoloration. Adults feed on buds, inflorescence and leaves of the jackfruit tree.

Mode of feeding and symptoms of infestation

The small whitish grubs bore into tender flower buds and fruits and induce premature drop. The affected shoots, leaves, flower buds and fruits should be removed and destroyed. It is also reported to bore into the tender buds and shoots. Adults are grayish brown weevils found nibbling the leaves.

Management

i. Removal of volunteer plants, wild hosts and crop debris (sanitation).

ii. Planting away from weevil infested field.

iii. Removal and destruction of affected shoots as well as infested and fallen flower-buds and fruits.

iv. Collection and destruction of grubs and adults in the initial stage of infestation.

v. Ants, spider, carabids and earwings are important generalist predators that predate upon the pest.

vi. Spraying with 0.2% Carbaryl followed by 0.1-0.15% Malathion at 10 days interval is quite effective in controlling the weevils (Singh, 1970).

4. Jackfruit aphid, *Greenidia artocarpi* Schout (Aphididae: Hemiptera)

This aphid is belongs to the genus *Greenidia* but from the original description and the figure it is very difficult to decide to which genus it belongs. Among several species of aphids, *G. artocarpi* play a vital role to infest the jack fruit tree.

Distribution

From paleontological data it is evident that at one time the tribe occurred in Europe. This, however, remains doubtful and at present the tribe would seem to be restricted to Eastern Asia. The western boundary seems to be in India, the northern boundary somewhere in Siberia, the eastern boundary in Japan, while in the south several species are known from Java.

Hosts of commercial importance

The species is highly polyphagous in nature. In addition to that it is found to feed on those plants belonging to families Annonaceae, Anacardiaceae, Apocynaceae, Euphorbiaceae, Fagaceae, Guttiferae, Hamamelidaceae, Loranthaceae, Moraceae, Myrtaceae, etc. They also breed on some weeds when the main host plants are absent or when conditions in the field are unfavorable.

Identification

This aphid is light to dark green, sometimes changing to other colour forms like yellowish and soft bodied insect. Eyes, cornicles and the tips of leg joints are black. The cornicles taper towards apex. Cauda is black with 6-12 hairs. Both nymphs and adults are found colonizing on tender parts of plants.

Biology

The winged and wingless forms reproduce parthenogenetically and are oviparous. It lives throughout year without producing sexual forms. One single female can produce 8-25 young ones parthenogenetically in life span of 10-12 days. Young nymphs moult four times and become adults in 5-8 days. The adults also lay eggs in overwinter. The apterous females start producing offspring after 24 hours of getting maturity. Several overlapping generations are completed during a year. Breeding occurs almost throughout the year and both apterous and alate are present.

Mode of feeding and symptoms of infestation

Damage is caused by both nymphs and adults who congregate in large numbers at the growing tips, flowers, fruits and all succulent plant parts guarded by ants. The nymphs and adults suck the sap from the tender parts and leaves of jack fruit and cause them to roll up curl and yellowing and stunted growth of the plant is occurs. The younger plants suffer more than the older plants. Cool and humid weather favours multiplication of the host. Hundreds and thousands of these tiny aphids may be seen on a single leaf or tender shoots. In wild cases the shoot wilts and in severe cases it just dies. The honey dew secreted by such a large number of aphids cover practically the whole surface of the tender leaves or shoots.

Management

i. Removal and destruction of affected plants along with the aphids.

ii. Early maturating varieties should be grown to escape the damage.

iii. Keeping the field weed free to reducing alternate host plants of aphids.

iv. Killing above ground plant parts using weedicide (Paraquat formulation @ 2.5ml/lit of water) on sunny day to prevent expose of jack fruit foliage to aphids.

v. At the time of appearance of the pest on growing points, spray any one of the following insecticides in 400 lit. of water per ha like, Malathion or Dimethoate or Imidacloprid or Thiomethoxam.

5. Jack fruit leaf webber, *Perina nuda* Fabricius, *Diaphania bivitralis* Guenee (Lymantriidae: Lepidoptera)

These are the commonly occurring leaf webbers or leaf eating caterpillars found to damage jack fruit trees in India. Both are sporadic pests usually of minor importance.

Distribution

Perina nuda has been reported from China, India and Sri Lanka. *D. bivitralis* cause considerable damage to jack fruit foliage especially in Tamil Nadu (Muthukrishnan *et al.*, 1958). It has also been recorded from Sikkim and Assam and outside India from Taiwan, Borno, Burma and Sri Lanka.

Host of commercial importance

Besides jack fruit, it has also been recorded as a minor pest of fig and mango in India.

Identification and biology

Perina nuda

The male moth has half ocherous and half transparent fore wings. The female is dull ocherous-white in colour. The full-grown caterpillars are about 22-25 mm long which has short erect tufts of husky grey to brownish hairs on their body.

Eggs are laid in clusters on the leaves and they hatch in 4 to 6 days. A female lays 57 to 90 eggs. Larva has six instars. The larval development is completed in 16-20 days. It pupates in leaf folds and the pupal period is 5 to 9 days. The moth emerges in October and lives for 3 to 11 days. Total life cycle is completed in 27 to 39 days.

Diaphania bivitralis

Full grown caterpillars are about 30 mm long, olive-brown in colour with conspicuous white markings. These pupate within the leaf folds in thin white silken cocoons. Pupae are red and 20 to 2 mm long. Moths are chestnut brown in colour dorsally and white ventrally. Forewings are chestnut brown having two semi hyaline white blotches with a small black discocellar spot between the two. Hind wings are iridescent hyaline white having a broad chestnut marginal band with a black line on its inner edge. The eggs hatch in 5 to 6 days; caterpillars pupate after 14 to16 days and pupae become adults in 8 to 10 days.

Mode of feeding and symptoms of infestation

The newly hatched caterpillars feed gregariously on leaves. They web together the leaves and feed from within by scrapping. The webbed leaves further spoiled by the accumulation of excreta of the caterpillars. Pupation takes place within the webbing and sometimes in soil. The caterpillars roll, fold or web together the leaves and feed within.

Seasonal incidence

The pest is active during rainy season.

Management

i. Collection and destruction of webbed leaves as soon as pest infestation detected along with the caterpillar.

ii. Use of light trap @ 1/acre to attract and kill the adults.

iii. In nature, the caterpillar is parasitised by *Microbracon mellus* Ram. and *Apanteles crocidolomiae* Ahmed.

iv. In case of severe infestation chemical insecticides like Carbaryl 0.2% or Malathion 0.15% may be applied. Two fortnightly applications are sufficient to check the caterpillar population.

6. Jack fruit mealy bug, *Drosicha mangiferae* Green (Monophlebidae: Hemiptera)

Distribution

The pest is widely distributed in different countries of the world including India, Pakistan, Bangladesh, Mayanmar, Sri Lanka, Philippines, New Guinea and some other tropical and sub-tropical countries like, Brazil, Colombia, Costa Rica, Cuba, Panama, Peru, Uganda etc.

Hosts of commercial importance

Main host of the pest is Mango. It has also been recorded to infest tender shoots, leaves and sucking the cell sap on ber, citrus, fig, grapevine, guava, mulberry, tamarind, mango and jack fruit etc.

Identification

Eggs are roundish cylindrical, flattened at both ends, chestnut brown in colour and fully covered with fine cretaceous material forming the ovisac. Nymphs are deep chocolate in colour, having their dorsum covered thinly with a whitish mealy material. The adult females are apterous, about 4-5 mm long. They are dark castaneous covered completely with sticky cretaceous white ovisac. If removed from this ovisac the females are seen white ovisac. If removed from this ovisac, the females are seen covered with whitish mealy substance. Adult males are pink, especially during the pre-pupal and pupal stages but appear yellow in the first and second instars and it is approximately 1.0 mm long, with an elongate oval body that is widest at the thorax (0.3 mm).

Biology

Reproduction is oviparous as well as parthenogenetic and the later is very common. Females mate only once during their life time. The sexually mature female lays about 400 to 700 eggs during May to June in the loose soil within radius of 2-3 meter around the infested trees and the eggs remain there for the period of six to seven months. The eggs hatch in about 7 to 10 days and the hatching is influenced by temperature and precipitation (Ashfaq *et al,* 2005). The nymphal development takes about 15 and 20 days in case of males and females respectively. The nymphs crawl to succulent shoots and base of fruiting parts (Birat, 1964, Atwal, 1976) for searching of food and settle at a suitable tender spot.

Mode of feeding and symptoms of infestation

The mealybug infestation appears on above ground parts on leaves, stem and fruits as clusters of cotton-like masses. The nymphs and females of this bug suck sap inserting its stylets into the epidermis from inflorescence, tender leaves, shoots and fruit peduncle. Affected panicles shrivel and become died. Infested plants are affected by the shooty mould on the leaves; photosynthetic activity is affected (Pruthi and Batra, 1960). Sooty mould fungus growth on the honey dew (Smith *et al,* 1997) renders the fruit unmarketable, reduce photosynthetic efficiency of leaves and causes leaf drop (CAB International, 2005). Severe infestation affects the fruit set and causes fruit drop. It causes immense damage and deprive the trees from its nutrient, ultimately quality and quantity of the fruit is severe reduced (Herren, 1981). Mealy bug became a serious pest of mango fruit 50.90% and pest caused social and cultural problem.

Management

Mealybug management often involves the management of attendant ants that are important for the proper development of mealybugs. Without the ants, mealybug populations are small and slow to invade new areas and the field would be free of a serious mealybug infestation.

Again, plant protection products are of limited effectiveness against mealybugs because of the presence of waxy covering of its body.

However, the following management strategies may be undertaken to manage the mealybug:

i. Healthy and pest free planting materials to be used.

ii. Monitoring and scouting to detect early presence of the mealy bug.

iii. Pruning of infested branches and burning them.

iv. Removal and burning of crop residues.

v. Avoiding the movement of planting material from infested areas to other areas.

vi. Application of sticky bands or alkathene sheet or a band of insecticide on arms or on main stem to prevent movement of crawlers.

v. Nitrogenous fertilizer does should be minimised.

vi. Natural enemies of the papaya mealybug include the commercially available mealy bug destroyer *Cryptolaemus montrouzieri*, ladybird beetles, lacewings, hover flies, *Scymnus* sp. and certain hymenopteran and dipteran parasitoids. Conservation of these natural enemies in nature plays important role in reducing the mealy bug population.

vii. Locate ant colonies and destroy them with drenching of Chlorpyriphos 20 EC @ 2 ml/l of water.

viii. If the activities of natural enemies are not observed, use of botanical insecticides such as neem oil (1 to 2%), NSKE (5%), or Fish Oil Rosin Soap (25g/l of water) should be the first choice.

ix. Apply recommended chemical insecticides as the last resort such as Profenophos 50 EC (2ml/l), Chlorpyriphos 20 EC (2ml/l), Buprofezin 25 EC (2 ml/l), Dimethoate 30 EC (2 ml/litre), Thiomethoxam 25 WG (0.6 g/l), Imidacloprid 17.8 SL (0.6 ml/l).

x. Drenching soil with Chlorpyriphos around the collar region of the plant to prevent movement of crawlers of mealy bug and ant activity is useful.

xi. Insecticide resistance and non-target effects on natural enemies make chemical control a less desirable control option.

xii. Avoid repeating the use of the same chemical insecticide as there are chances for development of resistance in the pest.

7. Jackfruit spittle bug, *Cosmocarta relata* Distant (Cercopidae: Hemiptera)

Members of the family Cercopidae in Hemiptera are commonly called the spittle bugs or froghoppers.

Distribution

The species occur worldwide but relatively few in the Holarctic region.

Hosts of commercial importance

It is a polyphagous pest with a wide range of hosts (Weaver and King, 1954, Fagan and Kuitert, 1969). It can survive on almost any host providing a sufficient amount of fluid to meet its feeding demands (Pass and Reed, 1965).

Identification and biology

Spittlebugs are about 3 to 27 mm in length (including the wing tips). Most species resemble leafhoppers, but the hind tibiae have 1 or 2 strong lateral spines rather than longitudinal rows of enlarged setae. Also, the hind tibiae are shorter than those of leafhoppers and the tibial apex has a crown of enlarged radiating spines. Some adult Machaerotidae have a posteriorly directed spine like scutellar process, superficially resembling the pronotal process of treehoppers.

As an apparent adaptation for life within spittle masses, Cercopid nymphs have the abdominal tergites extended downward as flaps, forming a completely or partially closed "breathing" that functions in respiration and the production of the spittle mass.

Eggs are inserted in dead twigs or slits cut into the bark of living stems. The eggs hatch in early May and the young nymphs migrate to the tender one-year-old growth. The pest overwinters in egg stage. The nymphs begin to produce the frothy spittle from their anus. The spittle apparently protects the nymphs from predators, parasites and dry weather. As the spittle drops onto lower branches, black sooty mould may cover the needles. Several nymphs may join together in one large spittle mass and the nymphs constantly abandon old masses to make new ones. The nymphs mature by July and soon leave the spittle in order to moult into the winged adult. The adults do not form spittle masses but quickly jump and fly if disturbed.

Mode of feeding and symptoms of infestation

Both the nymphs and adults of the pest are xylem feeder. Thompson (1994) reported that the family Cercopidae shows a preference for nitrogen-fixing grasses. The insect feeds by piercing intercellularly with its stylets to the xylem vessels and ingesting sap (Weigert, 1964, Horsfield, 1978). Adult feeding results in phytotoxemia or "froghopper burn" (Byers and Wells, 1966 and Taliafferro *et al.,* 1969). Age and sex were shown to be unimportant in the ability of adults to produce phytotoxemia (Byers and Taliaferro, 1967). A plant growth promoter in the salivary gland may be responsible for the damage (Cutler and Stimmann, 1971). However, the identity of this substance remains unknown.

The toxin is injected into the xylem as the insects feed, and damage begins within 24 hours. Symptoms include stippling, streaking and browning. Necrosis of the host can also result. Recovery from spittlebug damage can occur if the plant is not severely affected, but the recovery time is much longer than would be expected, indicating a residual effect of the phytotoxin (Meyer, 1993, Meyer and Root, 1993; Meyer and Whitlow, 1992; Karban and Strauss, 1993).

Management

i. Keeping the orchard clean and healthy.

ii. Collection and destruction of the bug is early stage of infestation. Light, accessible spittlebug infestations can be removed by hand or by a strong water spray.

iii. Pipunculid fly, *Verrallia virginica* causes about 50-60% parasitism of adult spittlebugs. Hence, natural enemy population should be encouraged in the orchard.

iv. Application of systemic insecticide on the basis of severity of infestation is helpful to reduce pest population as well to protect the plant.

In addition to the aforementioned insect pests, some other pests have also been recorded jack fruit in various jack fruit growing parts of India. The pests are of regional importance and listed as follows:

a. Stem borer, *Batocera rufomaculata* De Deer (Cerambycidae: Coleoptera)

b. Pink waxy scale, *Ceroplastes rubens* Maskell (Coccidae: Hemiptera)

c. Aphid, *Toxoptera aurantii* (Boyer de Fons.) (Aphididae: Hemiptera)

d. Thrips, *Pseudodendrothrips* sp. (Thripidae: Thysanoptera)

e. Castor capsule borer, *Dichcrosis punctiferalis* (Guenee) (Crambidae: Lepidoptera)

References

Alam, M, Z., Ahmad, A.; Alam, S. and Islam, M. A. 1964. A Review of Research, Division of Entomology (1947-1964), Department of agriculture, East Pakistan, Dacca., pp-272.

Anonymous, 1973. Distribution map of pests. No. 309. Common Wealth Institute of Entomology, 56, Queen's Gate, London S.W.F.

Anonymous, 1995. Fruit production manual. Horticulture Research and Development Project with DAE & BADC, Dhaka., Pp-286

Ashfaq, M., Khan, R. A.; Khan, M.; Rasheed, A. F. and Haffez, S. 2005. Complete control of mango mealy bug using funnel type slippery trap. *Pak. Ent.*, 27:45-48.

Atwal, A. S. 1976. *Agriculture pests of India and South East Asia*. Kalyani Publishers, Ludhiana, India, pp-224-227.

Birat, R. B. S. 1964. Mango's mealy bug menace. *Ind. Fmg.*, 14:14-15.

Byers, R. A. and Taliaferro, C. M. 1967. Effects of age on the ability of the adult two-lined spittlebug, *Prosapiabi cincta*, to produce phytotoxemia of coastal bermudagrass. *J. Econ. Entomol.*60:1760-1761.

Byers, R. A. and Wells, H. D. 1996. Phytotoxemia of Coastal bermudagrass caused by the two-lined spittlebug, *Prosapiabi cincta* (*Homoptera: Cercopidae*). *Ann. Entomol. Soc. Amer.* 59:1067-1071.

CAB International, 2005.Crop Protection Compendium (2005 edition). Wallingford, UK: CAB International. site:http://.cabicompendium.org/cpc/aclogin.asp?/cpc/fi ndadatasheet.asp.

Chowdhury, S. and S. Majid 1954. Handbook of plant Protection. Department of Agriculture, Assam, Shillong, pp-117.

Cutler, H.G. and Stimmann, M. W. 1971. The presence of a plant growth promoter in isolated salivary glands of *Prosapiabi cincta* (*Homoptera: Cercopidae*). *J. Georgia Entomol. Soc.*, **6:**69-71.

Fagan, E. B., and Kuitert L. C. 1969. Biology of the two-lined spittlebug, *Prosapiabicincta*, (Homoptera: Cercopidae) on Florida pastures. *Fla. Entomol.*, 52:199-206.

Fletcher, T. B. 1914. Some South Indian Insects. Superintendent government Press, Madras. Pp-565.

Hampson, G, F. 1896. The Fauna of British India including Ceylon and Burma Moths, 4: 70,77,356, Taylor & Francis, London.

Haque, M. A. 1977. Notes on variability in fruit characteristics of selected jackfruit plants (*Artocapus heterophyllus* Lam) from some localities of Mymensingh District. Bangladesh J. Agril. Sic., 4**(1):**119-20.

Hayes, W. B. 1996. Fruit Growing in India. Third Revised Edition.The Indian Universities Press, Allahabad., pp-385-389.

Herren, H. R. 1981. Current biological control research at IITA, with special emphasis on the cassava mealy bug (*Phenacoccus manihoti* Mat-Fer), Dakar, Senegal, USAID., pp-92-97

Horsfield, D. 1978. Relationship between feeding of *Philaenus spumarius* (L.) and the amino acid concentration in the xylem sap. Ecol. Entomol., 2:259-266.

Karban, R. and Strauss, S. Y. 1993. Effects of herbivores on growth and reproduction of their perennial hosts, *Erigeron glaucus*. Ecol., 74**:**39-46.

Karim, M. A. 1995. Insect pests of Fruits and their control in Bangladesh. Pp.113. In: Fruit Production Manual. Horticulture Research and Development Project in collaboration with Department of Corporation, Dhaka.

Khan, M. A. M. and Islam, K. S. 2004. Nature and Extent of Damage of Jackfruit Borer, *Diaphania caesalis* Walker in Bangldesh. Journal of Biological Science, 4**(3):**327-330.

Khan, M. A. M., Islam K. S. and Haque, M. A. 2003. Biology of Jackfruit Borer, *Diaphania caesalis* Walker in Bangladesh. Bangladesh Journal of Environment Science, 9(2)**:**417-421.

Madhava Rao, V. N. 1965. The Jackfruit in India Farm. Bull. (New Series), No. 34, 18 pp., Indian Council of Agricultural Research, New Delhi.

Meyer, G. A. 1993. A comparison of the impacts of leaf and sap-feeding insects on growth and allocation of goldenrod. Ecology 74:1101-1116.

Meyer, G. A. and Root, R. B. 1993. Effects of herbivores insects and soil fertility on reproduction of goldenrod. Ecology, 74:1117-1128.

Meyer, G. A. and Whitlow, T. H. 1992. Effects of leaf and sap-feeding insects on photosynthetic rates of goldenrod. Oecologia., 92:480-489.

Muthukrishnan, T. S., Nagaraja, K. R., Subramanian, T. R., Janaki, I. P., Abraham, E. V. 1958. Brief notes on a few crop pests noted for the first lime in Madras. Madras Agric. J., 45(9):363-364.

Naik, K. C. 1949. South Indian Fruits and Their Culture. P. Varadachery and Co. Madras., pp-300-302.

Nair, V. S. C. 1995: Cultivation of tropical and sub-tropical fruit crops. *J. Econ. Entomol.* 67: 321-343.

Ochas, J. J., Soule, M. J., Dijkman, M. J. and Welburg, C. 1981. Tropical and Subtropical Agriculture. MacMillan Co, New York, pp-625-630.

Pass, B. C., and Reed, J. K. 1965. Biology and control of the spittlebug *Prosapiabicincta*in coastal Bermuda grass. J. Econ. Entomol., 58:275-278.

Patil, P. V., Patil, V. K. and Deshpande, A. D. 1990. Incidence and control of bark eating caterpillar, *Indarbela* sp. on guava trees. Journal of Maharashtra Agricultural University, 15(2):230.

Pruthi, H. S. and Batra, H. N. 1960. Some important fruit pests of North West India. ICAR Bull., 80:1-113.

Sharma, D. D. and Kumar, H. 1986. How to control bark eating caterpillars? Indian Horticulture, 31(1):25.

Sharma, S. K. and Verma, A. K. 1987. Insect pest complex of apricot, *Prunus armeniaca*L. in Himachal Pradesh. Bulletin of Entomology, 28(2):115-123.

Singh, J. P. 1970. Elements of vegetable pests. Vora and Co. Publishers Pvt. Ltd., Bombay, pp-275.

Smith, D., Beattie, G. A. C., and Broadley, R. 1997. Citrus pest and their natural enemies: *Integrated pest management in Australia.* Information series Q197030. Queensland Department of Primary Industries, Brisbane.

Soepadmo, E. 1991. *Artocarpus Heterophyllus* Lam. In: Plant Resources in South East Asia, 2. Edible fruits and nuts., pp-86-91.

Sturrock, D. 1977. Fruits for Southern Florida. South Eastern Printing Co., Stuart, Fla. P. 114.

Taliaferro, C. M., Leuck, D. B. and Stimmann, M. W. 1969. Tolerance of *Cynodonclones* to phytotoxemia caused by the two-lined spittlebug. Crop Sci., 9:765-766.

Tandon, P. L. 1998. Management of Insect pests in Tropical Fruit Crops. pp. 237-244. In: Arora, R.K. and V. Ramanatha Rao eds. "Proceeding of the IPGRI-ICAR-UIFANET Regional Training Course on the conservation and use of Germplasm of tropical fruits in Asia held at Indian Institute of Horticultural Research, 18-31 May 1997, Bangalore, Indian.

Thompson, V. 1994. Spittlebug indicators of nitrogen fixing plant. Ecol. Entomol., 19:391-394.

Wang, P. Y. 1963. Notes on Chinese Pyralid moths of genus *Diapania* Hubner. Acta Ent. Sin., 12(3):358-365, Peking (95,253).

Weaver, C. R. and King, D. R. 1954. Meadow spittlebug. Res. Bull. Ohio Agic. Exp. Station 741:1-99.

Weigert, R. G. 1964. The ingestion of xylem sap by meadow spittlebug, Hilaenus spumarius (L.). Am. Midl. Nat. 71:422-428.

4

Pests of Banana and Their Management

T. Bharathimeena and Nripendra Laskar

Banana is the eighth most important food crop in the world. It is widely cultivated in over 130 countries in the tropics and sub tropics of Asia, Africa and the United States. Plantains and bananas are next important to major cereals, forming a key component in food security and agricultural sustainability in the developing economics (Patel and Shukla, 2009, Shukla, 2010; Deepah and Paranjothi, 2010). Banana ranks fifth in the global market of agricultural commodities and is one of the largest traded fruit crops. It is popularly known as "Apple of the Paradise ". The per capita starch output of banana from unit cultivated area is almost near cassava and sweet potatoes (Choudhary, 2015). It's assumed that an acre of banana crop can give a million calories of energy. Besides being a good source of dietary carbohydrates, it fulfills essential vitamin requirements too, especially Vitamin B6. Banana form a vital part of daily diet due to their year-round availability, varietal range, taste and nutritive values. Ripe banana pulp is the first solid food given to infants. The inflorescence and pseudostem can be chopped, cooked and consumed like vegetables. Specific cultivars are also recommended for treating insomnia, gastric ulcer, dysentery, cancer and heart diseases.

Bananas are richer source of potash, calcium, phosphorus and iron than orange or apple. The plant by virtue of its continuous reproduction is considered as a sacred symbol of fertility and prosperity. They are intricately interwoven in the cultural heritage of several communities and are an integral part of festivals and auspicious celebrations. Processed foods like chips, jam, jelly, canned slices, wine and banana powder have excellent industrial scope and fondly preferred by all classes of consumers. Banana fibre is also used to make pots, bags, ropes, hats, mats, wall hangings etc.

India is the largest producer of banana in the world contributing about 24 per cent of the global output. The crop is grown in about 4.75 million ha with a mean productivity of 13.2 million tonnes most of which is locally traded and consumed. India, Ecuador, Brazil and Peru alone account for half the world's banana production. The crop is best suited for small holders because it can be an ideal intercrop, matures fast and can be continuously harvested. Maharashtra, Tamil Nadu, Gujarat, Assam, Karnataka and Kerala are major banana growing states in India.

The crop is infested by over 450 species of insects and mites worldwide (Ostmark, 1974).In India, around 41 species of insect pests are found associated with banana (Nair, 1975) of which 19 species assume economic significance (Padmanaban *et.al.*, 2016). Weevil borers and aphids are key pests of banana often causing devastating yield losses. Occasionally sporadic out breaks of lepidopteran caterpillars cause considerable damage to the plants. Tingids, thrips, scales, mealybugs and scab moth too have assumed the status of significant pests and more recently the banana scarring beetle is a serious menace in Assam, Bihar and adjoining north eastern states.

Table 1: Insect pests of Banana

Sl. No.	Common Name	Scientific Name	Family	Order
1.	Psueudostem borer	*Odoiporus longicollis* Oliver	Curculionidae	Coleoptera
2.	Banana rhizome weevil	*Cosmopolites sordidus* Germar	Curculionidae	Coleoptera
3.	Rugose spiralling whitefly	*Aleurodicus rugioperculatus* (Martin)	Aleyrodidae	Hemiptera
4.	Banana aphid	*Pentalonia nigronervosa* (Coquerel)	Aphididae	Hemiptera
5.	Banana leaf and fruit scarring beetle	*Basilepta subcostatum* (Jacoby)	Chrysomelidae	Coleoptera
6.	Lacewing bug	*Stephanitis typica* (Distant)	Tingidae	Hemiptera
7.	Banana rust thrips	*Chaetanophothrips signipennis* (Bagnall)	Thripidae	Thysanoptera
8.	Banana scales	*Aspidiotus destructor* Signoret	Diaspididae	Hemiptera
9.	Banana mealy bugs	*Planococcus citri* (Risso)	Pseudococcidae	Hemiptera
10.	Banana skipper	*Erionota thrax* (Linn.)	Hesperiidae	Lepidoptera
11.	Banana fruit scab moth	*Nacoleia octasema* (Meyrick)	Noctuidae	Lepidoptera
12.	Banana coreid bug	*Physomerus grossipes* Fab.	Coreidae	Hemiptera
13.	Cyperus root borer	*Athesapeuta cyperi*	Curculionidae	Coleoptera
14.	Banana mite	*Tetranychus urticae* (Koch)	Tetranychidae	Acarina

1.Psueudostem borer, *Odoiporus longicollis* Oliver (Curculionidae: Coleoptera)

Distribution

The banana pseudostem weevil is believed to have originated from south east Asia which is also the centre of origin of banana. It's a regular pest of banana in India, China, Malaysia, Indonesia and Thailand. It occurs throughout India in all banana growing states, often causing total crop failure when not efficiently addressed to.

Hosts of commercial importance

It is a monophagous pest preferring plantains and highland cultivars particularly the "pome" types.

Identification and biology

Eggs are creamy to yellowish white, cylindrical shaped with rounded ends. There is a prominent area of air space around the anterior end. Yellowish white grubs emerged in 3-8 days. They are fleshy, sub-cylindrical and apodous but not "C" shaped. Body surface is wrinkled and covered with sparse brownish setae of varying lengths. Head is yellowish brown with the margin of cranium is reddish brown. Initially, they start feeding on the tissues around the air chamber of the leaf sheath. Later they burrowed deeply, feeding through the succulent tissues and reached as far as the true stem.

Those grubs cut through the ascending flower bud and peduncle, biting exit holes all the way. About 5-8 grubs have been found in peduncle of Nendran and Myndoli cultivars (Padmanaban *et.al.,* 2001). The inflorescence decayed within the pseudostem and faile to emerge. Larval tunnels are widely distributed and measured about 8-10 cm deep. They reached till the fruit peduncle or even down to the lower most collar region near the rhizome. Larva passed through five instars and complets development in 30-35 days. In colder months this growing period stretched to 51-62 days. The third, fourth and fifth instars are typically more voracious than other stages. The fifth instar is identified by reddish brown head and rounded abdominal tip. Sometimes they drill holes in the pseudostem for better aeration making the plant extremely weak. Grubs ceased to feed towards pupation and entered a very transient prepupal phase. A dark brown coloured, elongate, cylindrical cocoon is constructed by winding bits of fibrous leaf sheath around its body. Pupa is yellowish and exarate.

Adult is a robust reddish brown to black weevil measuring 1.3-2.0 cm. Adults emerged in 17-20, 27-44 days in summer and winter respectively. Total developmental period ranged between 44 to 80 days, with significant weather dependent variations.

Males were differentiated from females by larger rostral punctuations placed on slightly raised areas. They are negatively phototropic and some of them tend to remain inside pseudostem throughout their life feeding on bits of leaf sheath. Adult longevity is about 90-120 days. Some of them live even for 4 years. There were six overlapping generations in a year and the pest did not over winter.

Mode of feeding and symptoms of infestation

Grubs tunnell through the pseudostem and fed voraciously. Adults scrap out leaf sheaths superficially. Small pin head sized holes are formed on the stem initially followed by exudation of gummy substance from leaf sheaths and pseudostem. Fibrous extrusions are pushed out from base of leaf petiole.

Leaves of infested plant turned chlorotic, drooped and withered. Bunch size got reduced and fruits were blackish. When as the grub dug through as far as the peduncle, fruits either become disfigured or ripened immaturely. Extensive feeding by grubs weaken the pseudostem making it prone to wind breakage. The frass filled tunnels permitted the entry of secondary microbes. The plant could not bear the weight of the fruit bunch and often snapped and toppled off. All stages of the pest could be seen in the peripheral region of pseudostem. Dead plants remained succulent for a long time offering food and shelter for the developing life stages. The adults remain in the decomposed part of basal pseudostem and wait for monsoons to resume their activity. About 10-90 per cent yield loss is reported depending on crop stage and management efficiency.

Seasonal incidence

Pest incidence very high during late May to June and from late September to mid October (Padmanaban and Sathiamoorthy, 2001, Shukla, 2010; Thippaiah *et al,* 2012; Priyadarshini *et al,* 2014).

2. Banana rhizome or corm weevil, *Cosmopolites sordidus* Germar (Curculionidae: Coleoptera)

Distribution

The corm weevil, *C. sordidus* is a pest of cosmopolitan importance and occurs throughout the world wherever bananas are grown. As with *O. longicollis,* it is believed to have evolved from South-East Asia.

Hosts of commercial importance

It is a strictly oligopahgous insect feeding only on *Musa* and *Ensete.* Plantains and highland cooking bananas are more susceptible than dessert or brewing bananas. Malbhog, Robusta, Kanthali, Adukkan, Thellackarakeli, Red banana, Poovan and

Nendran were all prone to severe damage. Cavendish varieties were quite tolerant and seldom damaged.

Identification and biology

Adult is a black weevil, free living, nocturnal and highly prone to desiccation. It measured about 10-15 mm long with punctuations all over the body and striations on the elytra. They are found resting on leaf sheaths, crop residues and soil. Females cut out slits in rhizome surface with rostrum and deposit eggs in the outermost layers. They lay one to three eggs per week and the mean fecundity is about 270 eggs.

Eggs are oval, yellowish white and crystalline. Incubation period is about 3-15 days. Larval period has been found to complete in 14-45 days passing through 5-8 instars. The development time varied significantly with cultivar, temperature and plant age and soil nutrition. Pupation occurred in naked cells on the outer surface of rhizome. Adults emerge in 7 days and lived for one to four years. Under tropical conditions, life cycle is completed in 35-42 days (Shukla, 2010). Total developmental period could extend from 70.40 - 81.90 days depending on weather variations. Eggs fail to develop at 12^0C or temperatures below than that (Shukla, 2010, Sahayaraj and Kombiah, 2010 and Priyadarshini *et al,* 2014).

Mode of feeding and symptoms of infestation

The early instar grubs burrow through the rhizome at different locations making a complex network of galleries. They never feed on roots. The tunnels intervene badly with nutrient uptake and the existing roots get killed. Initially yellow lines appeared on leaves and the leaf size become reduced. In due course of development, the apical stem thin out and young plants kill outright. Mature plants fail to flower in time and bunch setting become very poor. Tunneling and further damage in rhizome worsen the condition with increased larval burrowing.The plant became highly vulnerable to other pests and diseases also. They become unable to withstand heavy winds and snapp off. Few grubs may move further to damage psuedostem also but this is very rare.

Yield losses could go up to 85 per cent in the bearing stage of the crop (Ahmad *et al,* 2003). However in several regions of the world, the economic importance of rhizome weevil is still under debate as the disease symptoms are too complex to be quantified. Also, it is very difficult to correlate the trap caught adult numbers and larval tunnels in plants.This pest supposedly implicated in the decline and disappearance of highland banana in traditional growing areas of some banana growing countries.

Adults are gregarious and this habbit is mediated by plant volatiles, aggregation and/ or sex pheromones. On moist substrates, they could thrive without food for several months. Some of them tend to rest in the same mat for months together and only few moved beyond 25 metre in six months. Flowered plants and crop residues are preferred for oviposition.

Management of weevil borers of banana

Weevil borers of banana are highly elusive in nature attributed to their endophytic nature.The adults having long life span and their juveniles being cryptic, they curtail the ease of intervening with conventional pest control methods and hence pose a tough challenge to banana growers. However, the integrated approach as mentioned below are helpful in reducing pest load in the banana orchard.

i. Field sanitation, clean cultivation and periodic pruning of suckers help in lowering field population of weevil borers. Wherever, old suckers are planted, establishment of clean mother garden becomes necessary.

ii. After harvest, corms may be cut at soil level and covering with moist earth to impose a physical barrier against oviposting females.

iii. Removal of harvested plant parts denies major breeding sites for weevil pests. Underground corms should also be uprooted, chopped to small pieces and left to desiccate to kill all pre-adult stages of the pests.

iv. Dipping the suckers in hot water at 49-53^{0}C for 20-30 seconds kill the eggs, grubs and pupae in the outermost layers. It gives durable protection for several crop cycles (Coyne *et al,* 2010).

v. Selection of tolerant cultivars.

vi. Systematic trapping system using cut pieces of pseudostem and rhizome considerably reduced the adult populations. One trap per mat per month could reduce rhizome damage by 43-61 per cent but this was a cumbersome, labour demanding process. The pseudostem bits can be treated with Carbofuran 3G @ 6g for immediate killing of trapped weevils. For trapping psedostem weevils, the disc on stump method was more effective than longitudinal split pseudostem traps owing to more profusely oozing plant fluids.

viii. Traps with aggregation pheromone of BPSW in combination with host plant extracts @ 5 per acre attract more weevils than with the sole use of pheromone. *Cosmolure* traps @ 2 numbers per acre at 10metre away from the border and 20 metre apart could be installed for collecting field populations of rhizome weevils.

ix. Treating the egg laying sites with contact insecticides also give successful results. In plantations with heavy occurrence of BPSW, pseudostem should be sprayed with Carbaryl 50 WP 0.25% or Chlorpyriphos 20 EC 0.1 %

along with neem oil emulsion 0.5 %. This should repeated every 45 days till bunches are produced.

x. Application of Celphos tablet @ 0.5 g x 3 tablets per plant could successfully control all stages of *O. longicollis*. But Celphos being highly phytotoxic could kill the central immature rolled leaf when applied in the vegetative phase and hence banned in recent times.

xi. Pseudostem injection with Chlorpyriphos 20 EC (0.2%) or Quinalphos 25 EC (0.4%) + Cypermethrin 25 EC (200 ppm) immediately reduces damage inflicted by BPSW grubs.

xii. Most botanicals have limited potential in control of banana weevils. Crude extracts of *Melia azadirach* L., *Tagetus* sp., *Ricinus communis* L. showed significant reduction in oviposition rates of pseudostem weevil. Spot application of neem based formulations has been found notably effective against the pest. Stem injection with azadirachtin (0.05 %) @ 4:4 ratio or swabbing with the same prevented egg laying.

xiii. Application of powdered neem seed cake @ 60-100 g per mat in four month intervals significantly reduces rhizome damage by *C. sordidus* and the treatment was on par with Carbofuran 5G @ 60 G per mat applied twice yearly. Also treatment of suckers with commercial neem based pesticides (Azadirachtin-0.15%) deterred the weevils from settling on plants (Sahayaraj and Kombiah, 2010).

xiv. The entomogenous fungi, *Beauveria bassiana* Bals. (Vuill) and *Metarhizium anisopliae* (Metschn.) Sorokin are the most elaborately studied and widely exploited biocontrol agents against banana weevil borers. Their commercial use is restricted owing to lapse in adequate delivery systems for field use.

xv. Tissue cultured propagates were free from pest attack and hence are recommended for planting in endemic areas. Tissue cultured banana plants inoculated by dipping roots in *B. bassina* suspension of 1.5 x 10^7 conidia/ml for 2 hours could resist the attack of *C. sordidus*.

xvii. The entomogenous *Steinernema* sp. and *Heterorhabditis* cause potential epizootics under favourable microclimate.

3. Rugose spiralling whitefly, *Aleurodicus rugioperculatus* (Martin) (Aleyrodidae: Hemiptera)

Distribution

Around the globe a new species of whitefly called rugose spiralling whitefly is now becoming a major problem. Central America is considered as the origin of this pest (Evans, 2008). In the Oriental region, India is the only country where the whitefly has been introduced and it has been reported from Kottayam, Kerala during July-August, 2016. Initially this whitefly was observed from a number of coconut farms of Pollachi area of Coimbatore district, Tamil Nadu (Rao *et al.*, 2018) but presently the pest is available in all parts of the country.

Hosts of commercial importance

Rugose spiralling whitefly is the most important pest of coconut, but it also affects a number of fruit crops.Rugose spiralling whitefly incidence on banana is now becoming a major problem in different parts of the country. Beside banana a number of fruit crops like citrus, custard apple, guava, tapioca subjected to oviposition host of rugose spiralling whitefly and sapota, mango, arecanut in which different life cycle of rugose spiraling whitefly has noticed.

Identification and biology

Females adults lay eggs in a concentric circular or spiral pattern on the underside of leaves and then cover it with white waxy matter. That is why this pest is named as rugose spiralling whitefly. Eggs of this pest has an elliptical shape and colour is creamy white to dark yellow. This pest has 5 developmental stages. The 1st instar nymph known as the crawler, comes out of the egg, and then it searches for a place where it can feed by sucking the plant sap with its needle-like mouth parts. The nymphs has a colour variation of light to golden yellow and produce a dense, cottony wax along with thin waxy filaments (Stocks and Hodges, 2012) which over time get denser. Nymphs having a length of 1.1 - 1.5 mm but may vary in size that depends on instars. For taxonomic identification the puparium of this species is used.Compared tocommon whitefly adults, rugose spiralling whitefly adults are nearly three times larger (approx. 2.5 mm) and they have lethargic nature. They have a pair of irregular light brown bands across the wings (Stocks and Hodges, 2012). This characteristic is used to identify the adults of rugose spiralling whitefly. Males flies at the end of their abdomen bear long pincer-like structures.

Mode of feeding and symptoms of infestation

Rugose spiralling white fly interfere with the normal growth of the host plant without killing its host. The adults suck sap from the plant parts. Besides sucking the sap, whitefly also excretes a sticky substance called honeydew, which allows growth of sooty molds and finally turn the shiny liquid into a black-colored viscous liquid. Once it dries, thick layers are formed by the sooty mold on the host leaves. Photosynthetic activity of the plant is hampered because of the formation of this black layer on the leaf surfaces. Ants and wasps are attracted by the honeydew and protect the whiteflies from different natural enemies (Stocks and Hodges, 2012).

Management

i. Cultural management

a. Avoid closer planting and maintaining the proper spacing of 7.5 x 7.5 m.

b. Application of fertilisers at optimum dose and avoiding excess nitrogenous fertiliser.

c. Intercropping with cocoa and nutmeg which increase the activity of different parasitoids in the infested fields.

ii. Physical management

Placing of 2 yellow light traps per acre during night time between 7.00 pm and 11.00 pm is helpful for the monitoring and trapping of flying RSWF adults.

iii. Mechanical management

a. Yellow sticky traps can be used @ 15/acre and placing 1.5m above the ground level and smear it with engine oil or castor oil also efficiently control RSWF.

b. Burning of heavily infested leaves periodically.

c. To restrict the adult movements from one tree to another tree, water can be sprayed on the lower surface of leaves.

iv. Biological management

a. Different coccinellid beetles such as *Cryptolaemus montrouzieri, Menochilus sexmaculatus* Fabricius, *Curinus coeruleus* (Mulsant), *Chilocorus nigrita* (Fabricius), were recorded as the natural enemies of RSWF in the fields.

b. Release of *Chrysoperla zastrowisillemi* eggs 4000/acre is also effective.

c. An entomopathogenic fungus *Isaria fumosorosea* found effective against RSWF and applied @ 5ml/l of water by mixing with detergent/ Khadi soap @ 5g/l and sprayed at fortnightly intervals.

d. For removal of sooty mould from leaflets, sooty mould scavenging beetle *Leiochrinusnil girianus* Kaszab (Coleoptera: Tenebrionidae) is released.

v. Chemical management

Spraying of Dinotefuran 20SG or Pymetrozine 50WG or Thiamethoxam 25WG can effectively control the pest population in field conditions (Pavithran *et al.*, 2021).

4. Banana aphid, *Pentalonia nigronervosa* (Coquerel) (Aphididae: Hemiptera)

Distribution

The aphid, *P. nigronervosa* is the dreaded vector of Banana Bunchy top virus occuring in most banana growing countries.

Hosts of commercial importance

Preferred hosts of the aphid are the plants under family Musacae. In addition to banana and other species in the genus Musa, it is also found on various plant species in the order Zingiberales and in the family Araceae, including important food and ornamental plants such as cardamom, ginger, taro (Colocasia) etc. (Waterhouse, 1987).

Identification and biology

Like most other aphid species, the banana aphid has four nymphal stages. Newborn nymphs are initially oval in shape and become slightly elongated. They are reddish brown, with four segmented antennae. The second stage nymphs are similar in appearance. The third nymphal stage individuals are light brown, the compound eyes are more noticeable beginning with this stage and the nymphs have five-segmented antennae. The fourth stage nymphs have six-segmented antennae, are light brown in color. The first, second, third, and fourth nymphal stages last for about 2-4, 3-4, 2-4, and 2-4 days respectively (Rajan, 1981).

Adults of *P. nigronervosa* are reddish brown with black veined wings. Nymphs and adults congregated at the base of pseudostem and newest unfurled leaf. The adults have six-segmented antennae that are as long as their body.

Reproduction of the banana aphid is entirely parthenogenetic (without mating). Females give birth to live female young. Males are not known in a population of this species. The life cycle (nymph to adult) is completed in 9 to 16 days. The adult life span ranges from 8 to 26 days. The adult starts giving birth to young one day after reaching maturity. They can give birth to about four aphids per day with an average production of about 14 offsprings per female.There could be as many as 30 generations produced per year depending upon environmental condition.

Mode of feeding and symptoms of infestation

Banana aphid is a phloem feeder and uses its long stylets to pierce plant tissues to suck up the plant sap directly from the vessels. This can cause plants to become weak, leaves become yellowish, curled and shriveled. In some cases, galls are formed on the leaves. Infested young plants may be killed or their normal growth would be checked if there is sufficient feeding by the banana aphid. However, this direct damage by this aphid is generally negligible.

Like any other soft bodied and sap sucking insects, such as leaf hoppers, mealybugs and soft scales, this aphid excretes honeydew. The honeydew serves as a good medium on which sooty fungus grows that decreases photosynthetic activity and decreases vigour of the host.

In addition to causing direct damage to the plant the aphid also causes indirect damage by transmitting banana bunchy top disease (BBTD) from diseased to healthy plants. This substantially causes greater losses than that caused by direct feeding injury. This is often the most damaging feature of an aphid infestation. Both wingless (apterous) and winged (alate) aphids are able to transmit virus. Transmission is usually in a non-persistent manner where the virus is taken up into the aphids "mouth" while feeding on an infected plant and transmitted to a healthy plant during subsequent feedings. In non-persistent transmission, the virus reproduces in the plant and aphids simply aid in transporting the virus. With this type of virus-vector association, the aphid acquires the virus and is only able to transmit the virus temporarily.

Once all the infective charge is reduced by feeding or the passing of time, the aphid is unable to transmit the virus until it feeds on infected tissue again. Symptoms of the disease include dark green streaking of leaves, midrib and petioles. This also results in progressive leaf dwarfing, marginal chlorosis and leaf curling. Fruits of diseased plants become un-marketable because they are small and deformed. In India, banana aphid also acts as a vector of mosaic virus of cardamom which is known as "katte". This viral disease causes considerable losses to cardamom production.

Seasonal incidence

Population of nymphs and adults increases during warm humid months *i.e.*, during July and September.

Management

i. Syrphids, coccinellids and lacewings are important natural enemies of banana aphid. Larvae of *Paragus serratus* (F.) (Syrphidae: Diptera) were found actively preying on nymphs and adults of *P. Nigronervosa* (Padmalatha *et al*, 2003).

ii. Entomopathogenic fungus, *Acremonium* spp. was observed infecting the aphid colonies. Infected individuals showed poor development and reproduction.

iii. Clean cultivation and procurement of aphid/disease free suckers is crucial in endemic areas.
iv. Tolerant varieties like Poovan and Pachanadan varieties may be recommended for planting so as to avoid aphid infestation on the crop.
v. Populations of aphids could be monitored with yellow sticky traps installed @4-5 per acre.
vi. Spray recommendations include Dimethoate, Imidachloprid or Thiomethoxam etc. The nozzle of sprayer should be directed towards crown and pseudostem base up to ground level.

5. Banana leaf and fruit scarring beetle, *Basilepta subcostatum* (Jacoby) (Chrysomelidae: Coleoptera)

Banana scarring beetle, *Basilepta subcostatum* (Jacoby) has been considered as one of the most serious pests in different banana growing region of India.

Distribution

The pest is distributed in different countries viz. India, Bangladesh. Myanmar, Thailand, Laos, Cambodia. The insect is considered as one of the most economically important pests in Eastern India which is also reported to occur in West Bengal, Bihar, Assam and some other parts of India (Paul *et al.,* 2020).

Hosts of commercial importance

Musa spp. are the principal hosts of this pest. Beetles are observed feeding on ginger in Meghalaya. In Assam, North-eastern India, adult beetles are also observed feeding on *Canna indica* Linn. and turmeric (*Curcuma longa*).

Identification and biology

The adults are reddish brown coloured beetle with black head and bluish black elytra. They become active in night. The female lays very tiny eggs in the cavities gnawed on leaves or on the dry scales of young suckers in cluster of about 50 to 60 eggs. Incubation period ranged between 5 to 7 days. Newly emerged grubs feed on leaf sheath or flower bud and larval development completed in about 15 to 21 days and the full fed larva forms cocoons to pupate within the soil. Total life cycle completes in 21-29 days. The males are more active than the females. They live inside the central leaves, flower brackets. Adults hibernate during winter.

Mode of feeding and symptoms of infestation

They feed on leaves and fruits by scrapping the chlorophyll producing several scars on affected parts. The grubs feed on roots and the adultsfeed on the epidermis *i.e.*, the green portion on the ventral and dorsal surface of the leaves. Feeding by the adults makes irregular patches on the leaf surface. These scratches cause rusty brown pustules on effected banana fingers due to necrosis of surrounding tissues. Due to development of scars on banana fingers and deterioration of quality, it ultimately fetch low market price.

Young plants and those about to bear bunch suffered the most. Injury sites permitted the entry of secondary pathogens and the pulp also sometimes spoilt. Due to this damage photosynthetic area is reduced and ultimately growth and yield is affected badly (Ahmed, 1963). The beetle also infests the skin of young, tender fruitcausing heavy damage. Formation of innumerable scars develop on affected fruits and they become blemished and their market value get reduced (Prasad and Singh, 1987).

Batra (1952) reported that the pest prefer central rolled up leaf of the banana plants forming the top whorl at the crown. This central whorl become worse affected than other leaves (Sen and Prasad, 1953). As fruiting commenced, the infestation by the pest remain confined only to very young fruits as evidenced by presence of unblemished matured fruits in the same field under natural condition. Scars on the fruits grew bigger as the fruit matured and fruits become disfigured.

As the beetle population causes serious damage (by scars) on leaves and banana peel, it tremendously influences on both quantity and quality of banana that reduces market acceptability. Extent of damage inflicted upon banana crop by this pest has been reported to be around 80% and in case of severe infestation, the percentage of infested orchards, and intensity of the pest have been recorded up to 100 per cent (Roy and Sharma, 1952).

Seasonal incidence

Population build up was highly correlated with rainfall and reached peaks in the months of June- July in Assam and August-September in Bihar. Misra *et al* (2015) observed a gradual rise in adult numbers with the increase in atmospheric temperature from March onwards.

Management

i. Sanitation or clean cultivation to be maintenance to avoid infestation by the pest.

ii. Application of commercial *B. bassiana* formulation 5g per litre + azadirachtin based neem formulation (0.15%) 5ml per litre could significantly control adult numbers and scarring symptoms (Choudhary *et al,* 2010).

iii. Physical barriers with perforated poly bags (40 μm thick) measuring 100 x 60 cm gave optimal protection against scarring beetles. The lower end of the polybag was tied with a string and opened periodically to remove fallen brackets. After removal of the main inflorescence, the bags ought to be kept closed till harvest (Pathak and Mitra, 2014).

iv. Bunch covering with Cypermethrin impregnated nylon net treatment has found to give 100% protection of banana from infestation by this pest whereas 2.21% has been noted in bunch covering with transparent polybag (Rahman *et al.*, 2004).

v. Soil drenching with *B. bassiana* (1×10^9 CFU ml/1) @ 1500 ml/ha + bunch spray with *B. bassiana* (1×10^9 CFU ml/1) @ 1500 ml/ha at flag leaf emergence followed by bunch covering with white polypropylene bunch sleeve showed very promising result as revealed by Mahalanobish *et al.* (2020).

6.Lacewing bug, *Stephanitis typica* (Distant) (Tingidae: Hemiptera)

Distribution

The species has been documented from Philippines, Laos, Japan (Hill, 2008), China (Wang *et al.*, 2016), India (Poorani *et al.*, 2019), Srilanka, Pakistan etc.

Hosts of commercial importance

In addition to banana, it is an important pest of tropical fruits, arborescent ornamentals and shrubs of the family Arecaceae (Howard *et al.*, 2001). The species has been found infesting 16 other host plants in 8 plant families (Sripriya *et al.*, 2000 and Lin *et al.*, 2009).

Identification and biology

Adults are pale yellowish with transparent, shiny reticulated wings exhibiting a characteristic pattern of lacework, in the forewings. They are about 2-8 mm long and 1-5 mm wide and have two segmented tarsi. Nymphs lack wings and possess characteristic integumentary outgrowths referred to as integumentary processes, spines, tubercles or setae that are discarded in the adult except for the cephalic ones (Mathen, 1990).

Eggs are laid singly on ventral surface of leaves. They pierce the leaf tissues and insert eggs within. A single female lays 25 to 50 eggs. These hatch in 10 to 12 days. Emerging nymphs are blackish and spiny. The nymphal period lasts for 13 to 15 days. Adults live for 5 to 10 days. Total developmental period on banana is about 37.82 days and 9 generations per year could be completed under laboratory conditions. The survival rate of juveniles noted as 69.05 per cent in field but could reach cent per cent in the absence of natural enemies. Larger and denser colonies are located in sheltered plantations than in road side probably because they were unexposed to heavy winds. Adults predominantly occurred in the top most parts of the plant (mostly in second to fourth leaf) where about 67 per cent of egg masses are usually found. Nymphs colonize mostly on middle leaves and 70 per cent of their population were seen between the fifth and seventh leaf.

Mode of feeding and symptoms of infestation

Nymphs and adults congregate on ventral surface of leaf and suck the plant sap. The feeding spots become greyish-yellow and appear as whitish blotches from dorsal side of leaves; ultimately the affected patch become brown. In case of severe infestation, plant growth is retarded, yielding fewer and poor sized fruits. Plants appeared sickly and stunted.

Seasonal incidence

Higher numbers of the pest were reported in the dry months from September-May in China and population dwindled with the advent of rains. All life stages could be noticed in winter months.

Management

i. Collection and destruction of the damaged leaves, flowers and fruits along with life stages.
ii. Use of yellow sticky trap at 15/ha.
iii. Spraying Dimethoate 30 EC @ 17 ml/10 lit. water.
iv. Thatillakunnan, Malakali, Padathi, Agneshwar, Krishnavazhai and Kali cultivars have been reported as tolerant to lacewing bugs.
v. Fenitrothion 50 EC (0.05 per cent) and Carbaryl 50 WP (0.1 per cent) could effectively suppress their population. Those spray recommendations mentioned for aphids could control lacewing bugs as well.

7. Banana rust thrips, *Chaetanophothrips signipennis* (Bagnall) (Thripidae: Thysanoptera)

Distribution

This pest has a diverse distribution in Asia particularly in India, Indonesia, Java, Philippines and Sri Lanka. Inthe Western Hemisphere, it has been reported in Brazil, Costa Rica, French West Indies, Honduras, Mexico, Panama, Puerto Rico and Trinidad and Tobago. In USA, it is present in Hawaii and Florida. In Oceania, the banana thrips is recorded in New South Walesand Queensland (Australia) (Schotman, 1989).

Hosts of commercial importance

Banana rust thrips is a polyphagous pest. In addition to banana, it has also been reported to feed on citrus, anthurium, tomatoes, green beans, calocasia etc. (Hara *et al,* 2002).

Identification and biology

Mean fecundity is about 171 and the female lay about 5.54 eggs per day. Eggs are bean shaped, partially inserted in the peel. After two days, they got covered by encrustation. Incubation period on an average is 9.8 days. The nymph passes through three instars. On maturation, they moved to soil for pupation. There is an intermittent prepupal phase differing from the nymph exclusively by the presence of wing pads.

Adults emerge out of the pupal cell in 6-10 days. The adults are slender creamy yellow to brownish in colour with fringed wings. Dark eye like spots were prominent at the wing base. Total developmental period has been observed as 28 and 90 days in summer and winter months respectively. Feeding and breeding activities are spontaneous with the appearance of new flush in plants. Adults lived for about 24-30 days. They tend to hide underneath flower bracts and hence are difficult to manage.

Mode of feeding and symptoms of infestation

Females pierce the skin of young banana fruits with their ovipositor and lay eggs specifically between the fingers. Nymphs and adults suck sap from immature fruits, flowers and foliages. Feeding symptoms varied with hosts on which they infest. In banana, they feed by lacerating the outer epidermis of fruit. Water soaked spots appeared initially at the injured parts. Later, the latex between the fingers oxidized, turned red and ultimately rusty spots are formed. As the fruit become mature, the spots gradually expand, coaleasce and the peel became rusty textured. The feeding tracks are typically dark, smokey coloured random squiggly or circular type. In extreme cases the skin of injured fruit cracked.

Despite peel damage, the pulp remained intact and edible. Thrips also lacerate and suck sap from pseudostem and leaf sheaths. Continuous feeding resulted in characteristic, dark "V" shaped marks on the outer surface of leaf petioles which turned bronze or rust coloured in due course of time. In India, red rust thripsis becoming a significant pest in many banana growing belts of the country, especially Gujarat, Maharashtra and Tamil Nadu etc. The thrips infestation limit marketability of the produce although quality in terms of taste and nutrition remains unaffected (Bisane *et al*., 2018).

Seasonal incidence

Peak damage symptoms due to red rust thrips appeared from August onwards in South Indian condition at fruiting stage of banana.

Some other thrips of minor importance infesting banana have also been found. They are as follows:

a. Banana corky scab thrips, *Thrips florum* Schm. and *Astrothrips parvilimbus* Stu and Mitri

Adults of *T. florum* are attracted to inflorescence and lay eggs on peel of young fruits. Oviposition injuries are marked by a reddish spot and each spot consisted of a single egg. As and when the eggs are laid, water soaked spots appeared. The injured spot rose in few days, coalesced making the skin roughened and blemished. Nymphs emerged in 3-4 days. They actively crawled and desap the fruits. It resulted in powdery blotching and rusty appearance of peel. Peel eruptions are primarily due to proliferation of mesophyll cells in the superficial layers. *T. florum* causes severe oviposition injuries than feeding marks (Mohanasundaram and Sivakumar, 1971).

b. Banana fruit thrips, *Thrips hawaiiensis* (Morgan)

Jhala *et al* (2004) found that females of *T. hawaiiensis* lay eggs on young fruits beneath the bract. Oviposition sites are marred by black pustules and pimples. The peel are scattered with blemishes and spots degrading the commercial value of fruits. High populations are seen in the second week of July in Gujarat and Maharashtra.

c. Banana leaf thrips, *Helionothrips kadaliphilus* (Ramakrishnan and Margabhandu)

Banana and colacasia are host plants of leaf thrips. Adult females are dark brown in colour. Males are darker to yellowish or reddish brown. They fed exclusively on tender leaves and flowers causing whitish spots and streaking (Varatharajan *et al*, 2010).

Management

The thrips hide inside the flower bud. Therefore, conventional spraying by using insecticides during the fruiting stage wouuld be uneconomical and also harmful to natural enemies. Further, non-judicious use of a pesticide may lead to residue problems and such contaminated food situations can pose a health risk a health and other forms of the life. Considering this perspective, following technologies may be adopted judiciously to lower down the pest population.

i. Application of bud injection technique using Imidacloprid 17.8 SL(0.3 ml/500 ml water) @ 1 ml/bud or with Azadirachtin (1%) (5ml/l water) @ 2 ml/bud.

This technique (bud injection) is safer and competent preventive technique for reducing the percentage of blemished fruits (<10%) caused due to rust thrips infestation on banana. The technology required very less quantity of pesticides and keeps the original superficial fruit peel appearance and significance, which in turn fetch better price on quality fruits in market and provide financial profit to banana growers and traders (Bisane *et al.,* 2017).

8. Banana scales, *Aspidiotus destructor* Signoret (Diaspididae: Hemiptera)

Distribution

The pest has been recorded worldwide in tropical and subtropical areas, including China, Southeast Asia, India, Pakistan, Russia, Brazil, Central America and Caribbean, the Pacific Islands, Africa and North America (Claps *et al.,* 2001). In India, the pest is found to infest the crop in different banana growing regions.

Hosts of commercial importance

Itis a highly polyphagous species reported mostly from perennial crops. Coconut palms, mango, bananas have been recorded as the major hosts.

Identification and biology

Coconut scale resembles other armored scales in that the body is protected by a waxy cover. Infestations may be noted by the formation of closely packed colonies composed of what resemble miniature fried eggs. Eggs are laid under the scale cover around the body of the female. Freshly laid eggs are smooth, elongate and whitish, becoming pale yellow over time. Newly hatched nymphs, also called crawlers are free-moving and recognized by the presence of legs, antennae, and a pair of bristles at the tip of abdomen. Crawlers are light green to yellowish brown, translucent and somewhat oblong in shape. They completed development in 38 and 44 days in summer and winter months respectively.

Adult females have circular or broadly oval body cover which is flat and translucent with a subcentral pale exuviae. Diagnostic characteristics include three pairs of lobes with no indication of a fourth pair, flat and fringed and with sclerotization on the dorsum of the pygidium (Williams and Watson, 1988). Adult male scales are small, two-winged, reddish, gnat-like insects with eyes, antennae, three pairs of legs and long appendages. Mouthparts of the adult males atrophied, they do not able to feed and are short lived.

Mode of feeding and symptoms of infestation

This pest is usually found in densely massed colonies on the lower surfaces of leaves. But in case of extremely heavy infestation, it may be present on both the sides. Both the nymphs and adults suck sap from leaf surface causing yellowish spots and debilated plant growth. It may also be found on petioles, peduncles and fruits. Mature scales are usually found on the older leaves. Infestations are typically associated with yellowing of the leaves in areas where the scales are present.

The yellowing is caused by the removal of sap by the sucking mouth parts and the toxic effects of the saliva that kills the surrounding tissues at the feeding site (Waterhouse and Norris, 1987). This scale is classified as an armored scale and unlike other scales, armored scales do not produce honeydew. Armored scales feed on plant juices. Feeding sites are usually associated with discolorations, depressions and other host tissue distortions.

Management

i. Disinfection of planting materials by irradiation or hot water treatment at 47^0 and 49^0 C for 15 and 10 minutes respectively helps in clearing scales well before planting.

ii. The coccinellid beetle, *Chilocorus nigritus* (F.) is an important predator of hard scales.

iii. Several other natural enemies have been used to successfully in controlling outbreaks of this scale have been found very effective throughout the Tropics, especially *Pseudoscymnus anomalus*, *Cryptognatha nodices*, *C. gemellata*, *Rhyzobius satelles*, *Chilocorus nigritus* and *C. malasiae* (the last four are non-specific predators) (Waterhouse and Norris, 1987).

iv. Chemicals used on scales are usually the same as those used on mealybugs. Chemicals include Dimethoate, Formothion, Malathion and Nicotine (Copland and Ibrahim, 1985). Sprays are only effective on the crawler stage of scales and chemical applications should be adopted only when parasites are not economically effective in managing the pest.

9.Banana mealybug, *Planococcus citri* (Risso) (Pseudococcidae: Hemiptera)

Distribution

The pest is a native of Asia but is also found all over the globe including Americas, Europe and Oceania etc.

Hosts of commercial importance

The mealy bug has been observed infesting 65 plant species belonging to 56 genera in 36 families.

Identification and biology

The wingless female deposits eggs as white-cottony masses, called ovisacs giving the appearance of cotton spread on the plant. A female can lay from 300 to 600 eggs in her lifetime.

On hatching, nymphs (crawlers) settle along the midribs, underside leaves. The nymphs are yellow, oval-shaped with red eyes, and covered with white, waxy materials. In contrast to female nymphs, fourth instar of males produce cottony-appearing cocoon and pupate. The females are wingless, white to light brown in colour, with brown legs and antennae. The body of adult females is coated with white wax and bears a characteristic faint grey stripe along their dorsal side.

Mode of feeding and symptoms of infestation

Both the nymphs and adults feed by sucking plant sap. The affected plants become brittle, yellowish in colour and lose vigour and vitality. Honeydew secreted by mealy bugs coats the surface of fruits and leaves favoring the development of sooty mould, inhibits photosynthesis eventually weaken the plant.

In spite of the above mentioned mealybug banana is also infested by root mealy bug, *Geococcus citrinus* Kuwana and *Geococcus coffeae* Green. Nymphs and adults feed on roots of banana thereby weaken the plant. Both nymphs and adults are whitish and covered with waxy filaments. Males are short lived and live for about 5 days. Females live at least 10 days longer than males. Total development period is completed in about 29-34 days. Many Cyperaceae and Poaceae weeds served as collateral hosts.

Management

i. **Cultural management:** Plants with brown leaves should be uprooted and replaced.

ii. **Biological management:** The coccinellid, *Cryptolaemus montrouzieri* Mulsant is an effective predator. It offer good results in reducing population of the adults. Parasitic wasp *Leptomastix dactylopiiat* is also a good parasitie of the pest.

iii. **Chemical management:** Chemical management is often inefficient strategy due to their habbit of hiding in crevices between foliage. However, the infested plants with green or yellow leaves may be saved with careful treatment by using Malathion and Dimethoate. In addition to that

Imidachloprid, Thiomethoxam are also effective in reducing the pest load from the crop.

10. Banana skipper, *Erionota thrax* (Linn.) (Hesperiidae: Lepidoptera)

Distribution

It is a serious pest of banana in Indonesia, Java and other East Asian countries. Initially this hesperiid had been known only from north eastern states and the Andaman Islands (Veenakumari and Mohanraj, 1991). It never was a major pest on banana in peninsular India. However, Tamil Nadu and northern districts of Kerala, witnessed sporadic outbreak of *E. thrax* in the recent years (Sivakumar *et al*, 2014).

Hosts of commercial importance

The pest found to feed on both cultivated and wild banana and various palms.

Identification and biology

Adults lay eggs on older and greener leaves. The early instars are pale greenish with whitish lines. They scrap and consume green matter from the leaf lamina. Fully grown caterpillars measuring about 4.5 cm and are covered by a whitish waxy material. Total developmental period is on an average 70 days.

Mode of feeding and symptoms of infestation

Larvae made sharp cuts on leaf edges, constructing individual leaf folds. They continue feeding while staying within their respective leaf chambers and later pupate in the last fed leaf roll which measures as much as 29 cm long. A single infested leaf may have as many as 5-12 leaf folds. Those leaf folds with prepupal and pupal stages are plugged with excreta at one end and capped at the other. Severely infested fields appeared grazed and the cuts resemble those made by knife or scissors. Only mid ribs are spared in some plants. Defoliation in mass numbers severely limit the photosynthetic area thereby affecting fruit set.Those growers cultivating leaf purpose banana plants suffered gross decline in economic returns.

Seasonal incidence

Peak abundance of the have been recorded during high rainfall period.

Management

i. Collection and destruction of leaf rolls during outbreak period is effective.

ii. In a survey conducted in Kerala, avian predators were found actively feeding on larva and pupa of *E. thrax*. The common crow, *Corvus splendens* (Vieillot), Crow pheasant, *Centropus sinensis* (Stephens) and the Indian tree pie, *Dendrocitta vagabunda* (Latham) could tear open the leaf fold and consume the larvae and pupae. Following this, larval population declined to about 50 per cent and the top young leaves are usually saved from attack. Babblers too, are active predators of *E. thrax*. With their typical habit of hunting in groups of seven to eight numbers, they could strictly keep the pest population under check, especially in rural areas.

iii. Parasitoids on *E. thrax* are active mostly in the pre-flowering stage of the crop. *Cotesia erionotae* (Wilkinson) and *Brachymeria* spp. were the major parasites on larva and pupa respectively.

iv. Commercial formulation of *Bacillus thuringiensis* may be applied to protect the crop from this pest.

v. Need based application of Cypermethrin, Deltamethrin are effective in managing the pest.

Other minor pests of banana are briefly described as follows:

11. Banana fruit scab moth, *Nacoleia octasema* (Meyrick) (Noctuidae: Lepidoptera)

The banana scab moth is a pest of regional significance in parts of Uttar Pradesh (Satyagopal *et al,* 2014). Larva feeds superficially by scraping on young fruits. These regions formed a black callous and fruits lost their consumer appeal. The curve on the fruit adjacent to the bunch stalk, between the fingers are highly preferred by the pest. The mature larva are seen resting under the bract enclosing the male flower. Damage ceased once the hands lifted. The pest is usually become active in warm and humid months. However in some localities they are a regular occurrence. Adult is a small, light brown to tan coloured moth with small black spots on wings. It lays flattened eggs resembling fish scales in clusters of few to thirty. Scab moth could be controlled by careful selection of equal sized suckers which could ensure uniform, concentrated bunching cycle. Bunch sleeves also could protect the fruits from damage.

12. Banana coreid bug, *Physomerus grossipes* Fab. (Coreidae: Hemiptera)

It is a pest of banana in Maharashtra, recently has assumed economic significance (Padmanaban *et al,* 2016). Nymphs and adults aggregate over stem and bunch. They suck sap from fingers causing sunken black spots on fruit peel. Affected fruits lose their cosmetic value and pulp quality become spoilt. In endemic regions, bunch damage could be prevented tying polypropylene bunch sleeves immediately after the opening of all hands. Nair (1975) reported that the coreid bug puncture and damaged banana fruits.

13. Cyperus root borer, *Athesapeuta cyperi* (Curculionidae: Coleoptera)

It has been recently reported from banana on cv. Malbhog at Guwahati, Assam (Padmanaban *et al,* 2016). Adults are tiny, very active and fast moving weevils living in colonies of at least 14 numbers, sheltered in between the outer and inner leaf sheath, a little below the leaf space. They bit through the leaf sheath making small holes and a jelly like substance oozed out. As many as 36-56 weevils in four leaf sheaths per plant have also been observed.

14. Banana red spider mite, *Tetranychus urticae* (Koch) (Tetranychidae Acarina)

The red spider mite, congregated on the abaxial surface of leaves and suck the leaf sap causing bronzing and drying of leaves. The symptoms typically resemble that of leaf rot (Bhaskar *et al,* 2012). Weed flora within plantations support mite population build up. Potential mite predators are usually present in the orchard but their activity become suppressed due to various factors. Indiscriminate use of pesticides and dry climate also favoured flare up of mites. They could be managed by spraying Spiromesifen 240 SC @0.8 ml/l.

In addition to the aforementioned insect and mite pests some other pests are, hairy caterpillars and limacodids were sporadic pests of banana observed in the rainy season. *Macroplectra naiaria* Moore (Limacodidae: Lepidoptera), originally a pest of coconut is a rare occurrence on banana. Bag worms *Kophene cuprea* (Moore) and *Acanthopsyche minima* Hmp (Psychidae: Lepidoptera) have also recorded to feed on leaves mostly in Kerala and parts of south India (Ostmark, 1974). In Gujarat, the semilooper, *Chrysodeixis* acuta (Wlk.) infested young banana plants which are two to four months old (Tayade *et al,* 2014). The larvae of *Spodoptera litura* (F.) (Noctuidae: Lepidoptera) were known to occasionally swarm up in banana plantations causing severe defoliation.

References

Ahmad, M. A., Singh P. P. and Singh, B. 2003. Efficacy of certain synthetic insecticides and plant products used as foliar and whorl application against the scarring beetle (*Nodostoma subcostatum* Jacoby) on Banana. J. Ent. Res., 27(4):325-328.

Ahmed, A. 1963. Notes on the Biology of Banana Leaf and Fruit Beetle, *Nodostoma viridipennis* Most. A review of research, Division of Entomology, Published by Agricultural Information Service. Dhaka-3. Pp-187-190.

Batra,H.N. 1952. Occurrence of three banana pests of Delhi. Indian Journal of Entomology, 14(1):60.

Bhaskar, H., Binisha, K.V. and Jacob, S. 2012. Outbreak of red spider mite in banana plantations, (*Tetranychidae: Prostigmata*) of Thrissur district, Kerala. *Insect Environment*, 17(4):154.

Bisane, K. D., Patil, N. M., Padmanaban,B., Saxena, S. P. and Patil, P. 2018. Technique for management of banana red rustthrips, *Chaetanophothrips signipennis* (Bagnall). Journal of Entomology and Zoology Studies, 6(5):1964-1967.

Bisane, K. D., Saxena, S. P. and Naik, B. M. 2017. Management of red rust thrips, *Chaetanophothrips signipennis* (Bagnall) in banana, Journal of Applied and Natural Science, 9(1):181-185.

Choudhary, S. K., Mukherjee, U. and Ahmad, M. A. 2010. Efficacy of biopesticides against banana scarring beetle, *Basilepta subcostatum* Jacoby. Pest Management in Horticultural Ecosystems, 16(2):120-123.

Chowdhary, S. K. 2015. Study on Major Insect pests and major diseases of banana of Malda, WB,India. *Ind. J. Appl. Res.* 5(8):607-609.

Claps, L. E., Wolff, V. R. S., Gonzalez, R. H. 2001. Catalogo de las Diaspididae (Hemiptera: Coccoidea) exoticas de la Argentina, Brasil y Chile. Revista de la Socieded Entomologica Argentina, 60:9-34.

Copland, M. J. W. and Ibrahim, A. G. 1985. Chapter 2.10 Biology of Glasshouse Scale Insects and Their Parasitoids. pp. 87-90. In: Biological Pest Control The Glasshouse Experience. Eds. Hussey, N.W. and N. Scopes. Cornell University Press; Ithaca New York.

Coyne, D., Wasukira, A., Dusabe, J., Rotifa, Idowu. and Dubois, T. 2010. Boiling water treatment: A simple, rapid and effective technique for nematode and banana weevil management in banana and plantain (*Musa* spp.) planting material. Crop protection 29(1):1478-1482.

Deepah, R. and Paranjothi, S. 2010. Transmission of cucumber mosaic virus (CMV) infecting banana by aphid and mechanical methods. J. Food Agric. 22 (2):117-129.

Evans, G. A. 2013. The whiteflies (Hemiptera: Aleyrodidae) of the world and their host plants and natural enemies. USDA-APHIS.

Hara, A. H., Jacobsen, C. and Niino-DuPonte, R. 2002. Anthurium thrips damage to ornamentals in Hawaii. University of Hawaii at Manoa, College of Tropical Agriculture and Human Resources, publication IP-9. pp-4.

Hill, D. S. 2008. Major tropical crops and their pest spectra. Pests of Crops in Warmer Climates and Their Control, 511- 658.

Howard, F. W., Giblin-Davis, R., Moore, D. and Abad, R. 2001. Insects on palms. Cabi. Pp-125.

Jhala, R. C., Borad, P. K. and Bharpoda, T. M. 2004. Incidence of Thrips hawaiiensis (Morgan) on Banana in Gujarat, Madras Agric. J. 9(1):21-22.

Lin, M., Liu, F., Peng, Z., Li, W., Xu, W. & Wang, X. 2009. Survey and identification of pest insects on banana crop in Hainan. Southwest China Journal of Agricultural Sciences, 22 (6):1619-1622.

Mahalanobish, D., Hore, J. and Roy, K. 2020. Evaluation of treatment modules for managing scarring beetle, *Basilepta subcostatum* Jacoby infestation in banana, Journal of Crop and Weed, 16(2):159-165.

Mathen, K., Rajan, P., Nair, C. R., Sasikala, M., Gunasekharan, M., Govindankutty, M. P. & Solomon, J. J. 1990. Transmission of root (wilt) disease to coconut seedlings through Stephanitis typica (Distant) (*Heteroptera: Tingidae*). Tropical agriculture, 67(1):69-73.

Mishra, H., Bora, D. K., Das, B. B. D., Baruah, K. 2015. Population dynamics of banana leaf and fruit scarring beetle *(Nodostoma subcostatum* Jacoby) in Assam. Indian J. Ent. 77(3):226-229.

Mohanasundaram, M. and Sivakumar, C. V. 1971. The banana rust thrips in Tamil Nadu, India. Madras Agric. Journal 58(5):365-366.

Nair, M. R. G. K. 1975. Insects and mites of crops in India, ICAR. New Delhi, pp-190.

Ostmark. 1974. Economic Insect pests of bananas. Annu.Rev.Entomol.19: 161-176.

Padmalatha, C., Singh, A. J. A. R. and Jeyapaul, C. 2003. Predatory potential of syrphid predators on banana aphid, *Pentalonia nigronervosa* Coq. J. Appl. Zool. Res 14(2): 140-143.

Padmanaban, B. and Sathiamoorthy, S. 2001. The banana pseudostem weevil *Odoiporus longicollis* . Mont pellier (France). INIBAP, 4.

Padmanaban, B., Sundararaju, P. and Sathiamoorthy, S. 2001.Incidence of banana pseudostem borer, *Odoiporus longicollis* Olivier (Coleoptera: Curculionidae) in banana peduncle. Indian J. Ent., 63(2):65-70.

Padmanabhan, B., Alagesan, A., Kalita, D. N. and Mustaffa, M. M. 2016. Occurrence of Cyprus root borer *Athesepeuta cyperi* Marshall (Curculionidae: Coleoptera: Baridinae) as a minor pest of banana. Entomon 41(1):71-72.

Patel, P.R. and Shukla, A. 2009. Studies on host- pathogen interaction of banana bunchy top. Int. J. Pl. Protection, 2(2):176-177.

Pathak, P. K. and Mitra, S. K. 2014. Assessment of low cost perforated polythene covers as non-chemical approach to control scarring beetle and quality banana production. *Proc. IS on Tropical and Subtropical Fruits, Acta hortic.*, 1024, ISHS, pp-283-284.

Paul, P., Das, S. C., Saha, S. and Uma, S. 2020. Studies on population dynamics and damage potential of banana leaf and fruit scarring beetle, *Basilepta subcostatum* (Jacoby) on local and wild banana genotypes of Tripura. Journal of Entomology and Zoology Studies, 8(4):2020-2026.

Pavithran, S., Ashok, K., Arunkumar, P., and Sanath, R. M. 2021. The rugose spiraling whitefly *Aleurodicus rugioperculatus* Martinand its management practices in India. 07:33-36.

Poorani, J., Padmanaban, B. and Thanigairaj, R. 2019. Natural enemies of banana lacewing bug, *Stephanitis typica* (Distant) in India, including first report of Anagrus sp. (Hymenoptera: Mymaridae) as its egg parasitoid. Munis Entomology & Zoology, 14 (1):83-87.

Prasad, R., Singh, O. I. 1987. Insect pests of banana and their incidence in Manipur. *Indian J Hill Farm* **1**(1):71-73.

Priyadharsini, G. I., Mukherjee, U. and Kumar Nagendra. 2014. Biology and seasonal incidence of pseudostem weevil, *Odoiporus longicollis* (Olivier) (Coleoptera :Curculionidae) in banana. Pest management in Horticultural Ecosystems, 20(1):8-13.

Rahman, M. A., Hossain, M. and Islam, K. S. 2004. Effectiveness of weeding and bunch covering in controlling banana leaf and fruit beetle, J. Bangladesh Agril. Univ. 2(2): 237-243.

Rajan, P. 1981. Biology of *Pentalonia nigronervosa* Fab. caladii van der Goot, vector of 'katte' disease of cardamom. J. Plantation Crops 9:34-41.

Rao, N. B. V. C., Roshan, D. R., Rao, G. K., Ramanandam, G. 2018. A review on rugose spiralling whitefly, *Aleurodicusrugioperculatus*Martin (Hemiptera: Aleyrodidae) in India. Journal of Pharmacognosy and Phytochemistry; 7(**5**):948-953.

Roy, R. S. and Sharma, C. 1952. Disease and pest of banana and their control. Indian Journal of Horticulture, 9(4):39-52.

Sahayaraj, K. and Kombiah, P. 2010. Insecticidal activities of neem gold on banana rhizome weevil (BRW) Cosmopolites *sordidus* (Germar), (Coleoptera: Curculionidae). Journal of Biopesticides 3(1):304-308.

Satyagopal, K. S. N. Sushil, P., Jeyakumar, G., Shankar, O. P., Sharma, S. K. Sain, D. R.;,Boina, B. S., Sunanda, Ram Asre, K. S. , Kapoor, Sanjay Arya , Subhash Kumar, C. S. Patni, C. Chattopadhyay, N. M. Patil, P. K. Ray, C. M. Rafee, B. C. Hanumanthaswamy, K. R. Srinivas, A. Y. Thakare, A. S. Halepyati, M. B. Patil, A. G. Sreenivas, N. Sathyanarayana, S. Latha. 2014. *AESA based IPM package for banana*. Dept. of Agriculture and Cooperation, New Dehi, pp-46.

Schotman, C. Y. L. 1989. Plant Pest of Economic Importance reported in the Region covered by the Caribbean Plant Protection Commission. Proveg (RLAC/90/03), pp-1-190.

Sen, A. C. and Prasad, D. 1953.Pests of banana in Bihar.Indian Journal of Entomology, 15(3):240-244.

Shukla, A. 2010. Insect pests of banana with special reference to weevil borers. Int. J. Plant Protec. 3(2):387-393.

Sivakumar, T., Jiji, T., and Anitha, N. 2014. Field observations on banana skipper *Erionota thrax* L. (Hesperiidae: Lepidoptera) and its avian predators from southern peninsular in India. Current Biotica 8(3):220-227.

Sripriya, C., Padmanaban, B. and Uma, S. 2000. Evaluation of banana (*Musa* sp.) germplasm against insect pests. Indian Journal of Entomology, 62(4):382-390.

Stocks, I., Hodges, G. 2012. Pest Alert- DACS-P-01745. Florida Department of Agriculture and Consumer Services, Division of Plant Industry, 6.

Tayade, S., Patel, Z. P., Singh, S. and Phapale, A. D. 2014. Effect of weather parameters on pest complex of banana under heavy rainfall zone of south Gujarat. J. Agrometeorology 16(2): 222-226.

Thippiah, M., Kumar, C. T. A., Sudhir Kumar, S. and Shivaraju, C. 2012. Seasonal incidence of banana scales (*Aonidiella orientalis* Newstead) and mealy bugs (*Planococcus Citri* Risso) on banana. Environment and ecology, 29(2A):900-902.

Varatharajan, V., Taptamani, H., Singh, H. C. and Singh, O. D. 2010. Diversity and Diagnostic features of Thrips (Thripidae: Terebrantia: Thysanoptera) infesting important Crops. Ann. Pl. Protec. Sci., 18(2):283-292

Veenakumari, K. and Mohanraj, P. 1991. *Erionota thrax* thrax L. (Lepidoptera: Hesperiidae), A new record to Andaman islands. J. Andaman Sci. Association, 7(2):91-92.

Wang, L., Wang, Z., Zeng, L. and Lu, Y. 2016. Red imported fire ant invasion reduces the populations of two banana insect pests in South China. Sociobiology, 63(3):889-893.

Waterhouse, D. F. and K. R. Norris. 1987. Chapter 8: *Aspidiotus destructor* Signoret. pp. 62-71. In: Biological Control Pacific Prospects. Inkata Press, Melbourne, pp-454.

Waterhouse, D. F. 1987. *Pentalonia nigronervosa* Coquerel. In: Waterhouse, D.F. & Norris, K.R. (Eds.).Biological Control: Pacific Prospects. Inkata Press, Melbourne, pp- 42–49.

Williams, D. J. and Watson, G. W. 1988. The scale insects of the tropical South Pacific region. Part 1, the armored scales (Diaspididae). CAB International, Wallingford, UK. pp-290.

5

Pests of Litchi and Their Management

Nithya Chandran and Riju Nath

The litchi or 'lychee' (*Litchi chinensis* Sonnen) is an evergreen subtropical fruit tree of the family Sapindaceae. It is known for its delicious fragrance, quality and juicy fruits contributing significantly to the nation's economy. India is the second largest producer of litchi in the world after China with an area and production of 82,000 ha and 555,000 tonnes, respectively during 2012-13 (DAC, 2013).So far as Eastern India is concerned, Jharkhand, Bihar and West Bengal accounts for 85% of the total litchi production in the country. The state Bihar alone contributes 45% of total litchi production and occupies nearly 40% of the area in India. In India litchi is being grown in an area of 89000 ha with a total production of 556 000 MT annually (http://nhb.gov.in).

Insect pests are one of the most important limitations affecting quantity and quality of litchi. Litchi plants are infested by nearly 42 different types of insect and mite species in different growth stages (Wadhi and Batra, 1964, Hameed *et al.*, 1992).Among them the major insect pests recorded on litchi are leaf mite (*Aceria litchi*), leaf miner (*Conopomorpha cramerella*), fruit borers (*Platypeplus aprobola, Conopomorpha cramerella* and *Dichocrosis* sp.), leaf webber/ roller (*Platypepla aprobola* Meyer), litchi bug (*Tessarotoma javanica* Thunb.), bark eating caterpillar (*Indarbela quadrinotata* and *I. tetraonis*) and shoot borer (*Chlumetia transversa*) etc. are most important. Recently, litchi fruit borer and litchi leaf roller have acquired the status of key pests. Litchi bug, litchi looper and bag worm have also detected as the emerging pests of litchi (Hameed *et al.*, 2001, Kumar *et al.*, 2011; Choudhary *et al.*, 2013 and Kumar *et al.*, 2013). The present chapter will give the basic idea about identification, nature of damage and management of litchi insect pests.

Table 1: Pests of Litchi.

Sl. No.	Common name	Scientific name	Family	Order
1.	Fruit borer	*Conopomorpha sinensis* Brad., *C. cramerella* Snell., *C.litchiella* Brad.	Gracillariidae	Lepidoptera
2.	Bark eating caterpillar	*Indarbela tetraonis* Moore. and *I. quadrinotata* Walker	Metarbelidae	Lepidoptera
3.	Litchi stink bug	*Tessarotoma javanica* Thunberg	Tessaratomidae	Hemiptera
4.	Mealybug	*Planococcus litchi* Cox.	Pseudococcidae	Hemiptera
5.	Litchi Weevil	*Myllocerus discolor* Schoen.	Curculionidae	Coleoptera
6.	Litchi trunk borer	*Aristobia testudo* Voet	Lamiidae	Coleoptera
7.	Red weevil	*Apoderus blandus* Faust	Attelabidae	Coleoptera
8.	Leaf roller	*Dudua aprobola* (Meyrick)	Tortricidae	Lepidoptera
9.	Litchi semilooper	*Anisodes illepidaria* Guen.	Geometridae	Lepidoptera
10.	Fruit piercing moths	*Eudocima fullonia* (Clerck) *Eudocima salaminia* (Cramer) *Eudocima jordani* (Holland)	Noctuidae	Lepidoptera
11.	Nut borers	*Blastobasis* sp. *Gatesclarkeana* spp.	Blastobasidae	Lepidoptera
12.	Oriental fruit fly	*Bactrocera dorsalis* (Hendel)	Tephritidae	Diptera
13.	Litchi mite	*Acerialitchii*	Eriophidae	Acari

1. Fruit borers, *Conopomorpha sinensis, C. cramerella, C. litchiella* (Gracillariidae: Lepidoptera)

Fruit borer complex mainly comprising *Conopomorpha* spp. (Lepidoptera: Cossidae) has become a serious problem in the recent past causing 40–80% yield loss. *C. cramerella* and *C. litchiensis* are abundant in India. However, accurate species identification and their abundance is precisely unresolved and under-studied. Previously, *C. cramerella* was considered to be the dominant species, but subsequent molecular characterization of the specimen revealed that the species was in fact *C. sinensis* (Reddy *et al*, 2016).

Distribution

The pest has been found to occur in India, China, Thailand, Taiwan etc.

Hosts of commercial importance

The litchi fruit borer *Conopomorpha sinensis* Bradley (Lepidoptera: Gracillariidae) is a destructive pest of litchi (*Litchi chinensis* Sonn.) and longan (*Dimocarpus longan* Lour.) that causes significant economic losses. (Li *et al*, 2014).

Identification and biology

C. sinensis lays yellow, scale like eggs (which are about 0.4 × 0.2mm in size) on the fruit any time after fruit setting. Incubation period is about 3-5 days. The larvae immediately after hatching penetrates the fruit, leaf or shoot. One or more eggs may be laid on a fruit. However, only one larva in a fruit finally survives. Mature larvae are 6-10 mm in length and brownish in colour or green if they have fed on leaves. After 8-12 days they leave the feeding site to pupate on or under mature leaves in cream-coloured oval cocoons. The light green pupae change to dark brown just before eclosion, and the moth emerges after 5-7 days.

The moth is very small with long filiform antennae and narrow, fringed forewings measuring 8-11 mm when expanded. The moths leave for about 5-8 days. During the off season, when fruits are not available to the pest, it can survive by feeding on young leaves and shoots similar to *C. litchiella*. (Waite and Hwang, 2002).

The female of *C. litchiella* lays eggs on new shoot. The small, light yellow egg hatches within 3-5 days. Newly hatched larvae are creamy white and it bores in to the shoot. Larvae also mine in the leaf blade. Mature larvae feed on the midrib and veins of young leaves. The larval stage occupies 10-14 days after which larvae pupates on mature leaves. The pupal stage lasts 7-10 days and moths live for about a week (Waite and Hwang, 2002).

Mode of feeding and symptoms of infestation

The caterpillars enter the fruit from the pedicel end and feed on the pulp resulting in rotting and premature dropping of fruits. Besides fruits, they also bore tender shoots (Reddy *et al*., 2016) causing irreparable damage to the plants.

Seasonal incidence

As per Hameed *et al.*(1999), *C. cramerella* cause maximum damage to litchi during May-June. Pest population remains insignificant from October to March but reappeared again during April. Lal and Sharma (1978) recorded maximum population of the pest in the month of September and lowest during December. Almost similar observation was recorded by Sharma (1985). During off season, the borer survives on wildly grown alternate hosts *viz.* Kath Jamun *(Syzygium jambolanum)* and Chota Amaltas (*Casia tora*) in nearby litchi orchards (Lall and Sharma, 1978).

Management

i. Cultural and Mechanical: Pruning of trees after harvest, field sanitation and removal of young fallen fruits has been found effective in reducing pest population. Purbey and Nath (2013) reported that, the fruits can be protected by enclosing in nylon mesh bags. Bagging also improves colour and quality of litchi fruits.

ii. **Biological:** Application of *Trichogramma chilonis* @50000 eggs/ha two times right from flowering to colour break stage. *Mesochorus* sp., *Chelonus* sp., *Apanteles* sp. and *Bracon* sp. are parasitizing this pest. Kumar and Kumar (2007) also reported control of litchi fruit borer by using *Trichogramma chilonis* @50000 eggs/ha at flowering time followed by two spraying of Nimbicidine @3ml/l of water at lentil fruit stage and colour break stage.

iii. **Semiochemicals:** Pheromone trap may be installed for monitoring the pest incidence and mass destruction.

iv. **Botanicals:** Manuring the litchi trees with 4 kg of castor and 1 kg of neem cake per tree in the root zone after the first shower of monsoon reduced the fruit borer infestation (Hameed *et al.,* 2001).

iii. **Chemicals:** Spraying the orchard with Deltamethrin 2.8 EC @ 1ml/l or Fipronil 5 SC @ 2 ml/l or Flubendiamide 39.35 SC @ 1ml/ 5l of water after fruit set and second spray may be given before 15 days of fruit harvest for effective control of fruit borer.

2. Bark Eating Caterpillar, *Indarbela tetraonis* Moore, *I. quadrinotata* Walker (Metarbelidae: Lepidoptera)

Both the species of bark eating caterpillars (*Indarbela tetraonis* Moore and *I. quadrinotata* Walker) are highly polyphagous in nature.

Distribution

The species are distributed in Indian sub-continent especially in the states of M.P., U.P., Rajasthan, Bihar, A.P., Haryana, T.N., Orissa and Maharashtra. Sharma and Kumar (1986) found *I. tetraonis* to be a very serious problem of guava, mango and ber in Bihar.

Hosts of commercial importance

The host plants include mango, mulberry, aonla, ber, citrus, jamun, guava, loquat, jackfruit, pomegranate, phalsa and rose (Verma and Khurana, 1974 and Khurana and Gupta, 1972).

Identification and biology

The adult is a stout yellowish brown moth with brown wavy markings on the forewings. Hind wings are white in colour. Males are smaller than females. Larvae are stout and dirty brown in colour. Adults usually emerge during summer months and female layon average 15-25 eggs in clusters under loose bark of the trees. Eggs hatch in about 8-10 days. The larvae remain hidden in the bore hole during day time. At night, they come out and feed on the bark of the tree. They make webs and feed by making zig-zag galleries on the wood filled with frass and excreta and later bores inside the wooden trunk. Larval period is about 9 -11 months and then pupates inside the stem. Pupal period is about 21- 41 days.

Total life cycle lasts 4-5 months in south India and more than a year in north India. There is only one generation in a year.

Mode of feeding and symptoms of infestation

Litchi is badly affected by the bark eating caterpillars. It mainly attacks the older trees reducing their vigour and vitality by boring into the trunk, main stem and thick branches. The newly emerged caterpillars nibble on the bark. Infestation by the pest is characterized by the presence of thick, blackish or brownish ribbon-like excreta, made up of cut, chewed fibrous materials, in long-winding pattern from the stem (Atwal and Dhaliwal, 2008; Hameed *et al.*, 2001). The caterpillar bore into the affected trunk or branches usually near the forked junction and make tunnels within. Usually in each hole only one larva is found. There may be 2-16 holes under severe infestation in an aged tree.

Larvae make tunnel within the stem by continuous feeding. They remain within the tunnel during day time and normally come out at night. As a result of feeding on the tree trunk, the sap translocation get disrupted and plant growth is arrested under severe infestation (Kuldeep *et al.*, 2015).

Several caterpillars may infest a same tree at different locations with serious injury to the bark and death of small branches. The holes left on the trunk may lead to various pathogenic infections. Affected trees also break at the points of infestation.

Management

i. **Cultural:** Removal and burning of the ribbon like silken webs that may contain the caterpillars hiding under them.

ii. **Mechanical:** Using an iron spoke and inserting them inside the tunnel, the caterpillars can be killed.

iii. **Chemical:**

a. Injection of kerosene oil into the tunnel and sealing with mud can successfully control the pest.

b. Any fumigant may be applied within the tunnel soaked in cotton. Here also the opening of the tunnel to be sealed with clay or mud so as to make it air tight.

c. Cleaning of the tree trunk by removing the frassy webs and putting emulsion of any insecticide having good fumigant toxicity in the holes and plugging them is helpful in managing the pest. Application of Chlorpyrphos @ 2.0-2.5 ml per litre of water inthe pest affected orchard at 15 days interval is effective.

3. Litchi stink bug, *Tessarotoma javanica* Thunberg. (Tessaratomidae: Hemiptera)

The pest is a minor one of litchi in India. However, a few years back, an outbreak of litchi stink bug has been observed in the Chotanagpur plateau of Jharkhand during February–April 2011.The outbreak of the pest may be due to the migration of bugs from wild kusum [*Schleichera oleosa* (Lour.) Oken] plants (a host plant of the litchi stink bug) to the cultivated litchi crop (Jaipal *et al*., 2013).

Distribution

This bug is most destructive in Australia, China, India, Myanmar and Thailand (Han *et al*.1999; Leksawasdi and Kumchu, 1991; Lu *et al*.2006; Menzel, 2002). In India, pest is also noticed in all litchi growing areas particularly, Jharkhand, Bihar, U.P., Punjab, H.P. and Jammu and Kashmir.

Hosts of commercial importance

In addition to litchi, longan, rambutan, kusum, pummelo, castor, pomegranate, eucalyptus, loquat and rose are major host plants of this bug (Singh *et al*., 2009).

Identification and biology

The colonization of adults will start on litchi plants during the first week of February, and during second week of February the first egg masses are usually observed on lower surface of the tender leaves. Normally, the adult females were found to prefer to lay eggs on the underside of young leaves but egg laying was also found on inflorescence, flowers, fruits of litchi and even on the already laid eggs. The mean density of egg masses is about 11.66 egg clusters/m^2 plant canopies (range 8-15 egg clusters) from first week of March to third week of April. A single egg cluster consisted mostly of 14 eggs but ranged between 10 and 16 eggs per cluster (Jaipal *et al*, 2013).

The newly laid eggs are globular in shape, pink in color (sometimes white) but their colour became slightly blackish just before hatching. Under the field cages, a maximum of 16 egg batches has been found to lay by the single gravid female. Incubation period is about 12.80±1.40 days. Number of nymphal instar is four. The mean developmental period of first, second, third and fourth instar nymphs has been recorded as 11.69±0.58, 7.23±0.20,8.63±0.55 day and 13.04±0.55 days respectively. The average longevity of male and female adults are around 43-45 and 47-48 days respectively. Newly emerged nymph is soft bodied and dirty white in colour. However, colour changes to yellow-red after a few days (Jaipal *et al*, 2013).

Seasonal incidence

In the north western part of India, the insect emerges on litchi from last week of April and disappears from the orchard after the last week of August and undergoes hibernation in adult stages (Kumar *et al.*, 2008), however, in eastern India incidence of this bug has been observed from February to September (Jaipal *et al*, 2013).

Mode of feeding and symptoms of infestation

The nymphs and adults feed on the tender plant parts like leaf petioles, buds, fruit stalks and soft branches. They feed by sucking the plant sap. Excessive feeding results in drying of developing buds and soft shoots, ultimately resulting in fruit dropping. When the developing fruits are infested, they drop down after a couple of days.

Management

i. Mechanical

The insects can be collected by shaking the trees during winter and killed in kerosene.

ii. Biological

a. Egg parasitisation by the natural enemies is about 70-90%. The adults are also attacked by several entomopathogenic fungi, predatory birds and ants. These may be encouraged to suppress the pest population.

b. Classical biological control of litchi stink bug utilizing *Anastatus japonicus* (Hymenoptera: Eupelmidae) and *Ooencyrtus phongi* has resulted in effective management of the pest in China, Hong Kong and Thailand (Han *et al.*, 1999; Leksawasdi and Kumchu, 1991).

If synthetic chemicals are used, the time of application and their doses are critical because susceptibility to of the insect to insecticide vary in different times of the year. This depends on fat content of their body (Singh *et al.*, 2009).

iii. Chemical

Any insecticide having good systemic mode of action as well as less residual toxicity may be recommended for its effective management.

4. Mealy bug, *Planococcus litchi* Cox. (Pseudococcidae: Hemiptera)

Distribution

The mealybug distributed in different countries of the world including China, Hong Kong, Japan, Philippines, Singapore, Thailand and Vietnam.

Hosts of commercial importance

The pest is a polyphagous one and found to feed on the crops of family Annonaceae (*Annona squamosa*), Rosaceae (*Eriobotrya japonica*) and Sapindaceae (*Dimocarpus longan, Litchi chinensis*, *Nephelium lappaceum*).

Identification and biology

Egg

Eggs are deposited in ovisacs. The eggs are very small (approximately 0.3 mm long) glossy, light yellow and oval. A female lay about 300 to 600 eggs in her life period which are layed in groups of 5 to 20.

Nymph

After passing the incubation period, the nymphs emerge out from the ovisacs and typically settle along the midribs and veins on the underside of leaves and young twigs. The nymphs are yellow, oval-shaped with red eyes. They are covered with white waxy particles substances. The female nymphs resemble the adult female in appearance, while male nymphs are more elongated. Female nymphs passes four nymphal instars.

Adult

Size (length) of the adult mealy bug ranges from 3 (females) to 4.5 (males) mm. Females are wingless, white to light brown in color with brown legs and antennae. The body of adult females is coated with white wax and bears a characteristic faint gray stripe along their dorsal side. Short waxy filaments can be seen around the margins of their oval body with a slightly longer pair of filaments present at the rear end of their body. Males are winged but short lived. Several overlapping generations occur in a year.

Mode of feeding and symptoms of infestation

The mealybug cause damage of the crop by sucking plant sap by their suctorial mouthparts. Both nymphs and adults are responsibe for causing damage to litchi plant. Young plants susceptible for heavy infestation by the pest. The pest prefers to feed on the tender branches, nodes, leaves, spikes, berries etc. Severe infestation severe infestation of the plant results in chlorotic leaves, aborted flowerbuds and small and underdeveloped berries that ultimately reduces yeild. Severe infestation also results secretion of honey dew on the affected area and leads to development of sooty mould fungus (affects photosynthesis). Mealybugs are most common during the spring and early summer.

Management

Several natural enemies have been identified that are effective at controlling citrus mealybug. Among them, *Leptomastix dactylopii* (Parasitoid), Ladybird beetle, *Cryptolaemus montrouzieri*, spider, reduviid (Predators) etc. are important.

Raking of soil around tree trunk may be helpful to expose the eggs to natural enemies and sun. In case of heavy infestation systemic insecticides are recommended for managing mealybugs. Pre and post-bloom spray applications are recommended for management of mealybugs.

5. Litchi weevil, *Myllocerus discolour* Boheman. (Coleoptera: Curculionidae)

Distribution

Incidence of *M. discolor* Marsh. has been reported from many Indian states including Assam (Ramamurthy and Ghai,1988). Severe infestation to litchi leaves by the pest has been reported from North-East region of India. The pest has also been reported from Maharashtra (Nagpur, Kolhapur), Madhya Pradesh (Indore), Assam, Jammu and Kashmir, Andhra Pradesh, Odisha, Karnataka (Paunikar, 2015).

Hosts of commercial importance

It is a polyphagous pest. In addition to litchi it also infests a number of other commercial crops including maize, sugarcane, sunflower, citrus, mango, jute, brinjal, soyabean, litchi, Mulberry (*Morus alba*) and *Dalbergia sisoo*.

Identification and Biology

Head is prognathus, snout is very short and truncated, antennae capitate, 12 segmented with 10 flagellar segments, flagellum with 3 funicular segments. Frons with brown reclining hairs antennal socket deep and oblong, clypeus with a wide cleft with transparent long setae, body hairs scale like brown, deep brown, whitish and blackish in colour. Elytra with punctuations arranged in parallel lines longitudinally; mid and hind femur with spines; mid femoral spines 1 no; three hind femoral spines (biggest is the proximal), femur enlarged distally. Venter is light yellowish with iridescence; pulvilli well developed; claw highly sclerotized and with two tarsal segments (Mazumder *et al*., 2014).

Mode of feeding and symptoms of infestation

Adult weevils were found to cause severe damage to both matured and immature leaves of litchi. They feed on leaves, nibbling the leaves from the margins and eating away small patches of leaf lamina. Similar nature of leaf damage by the adults of *Myllocerus* spp has been reported (Butani, 1979). Adult beetles per plant in the range of 25-40 were recorded during the morning hours during April-July (Mazumder *et al.*, 2014). Infestation by the pest becomes more severe during the period of new shoot emergence.

Management

i. The immature stages (grubs) feed on soil organic matter. Hence, deep ploughing and thereby exposing the grubs to natural enemies reduces the pest population to some extent.

ii. Hand picking the adult weevils help in curbing down pest population.

iii. Carbaryl 2ml/l may be advocated for application in case heavy crop damage.

6. Litchi trunk borer, *Aristobia testudo* Voet. (Lamiidae: Coleoptera)

Distribution

The pest is distributed in China, India (Sikkim, Assam), Myanmar, Bangladesh, Thailand, Laos and Vietnam etc.

Hosts of commercial importance

Litchi, guava etc.

Mode of feeding and symptoms of infestation

Adult female girdle tender branches by chewing off strips of bark. They lay eggs on the wound caused due to girdling. The eggs are then gradually covered with exudate. After completion of incubation period the larvae starts coming out during late August. The neonates live under the bark for some time and then bore into the xylem. Thus, the larvae may create tunnels up to 60 cm long.Tunnelling by the larvae may kill branches and sometimes the whole trees. Ring-barking of tender twigs by the ovipositing female adults usually causes the shoot tips to die and snap off. The beetle has one generation per year.

Management

i. Regular inspections of the orchard during peak adult activity to be done. This may enable growers to remove the beetles. Simultaneously, the immature stages may also be collected and destroyed.

ii. Infested shoots should be clipped off and destroyed.

iii. Clean the hole and pour kerosene/petrol/crude oil or formalin into the stem borer hole and subsequently close entrance of the tunnel by plugging with cotton wool and paste the mud.

iv. Use light trap @1/acre may be installed for suppressing the pest population.

v. In severe infestation, the tunnels to be sealed with clay after injecting Dichlorvos solution (Zhang, 1997).

7. Leaf roller, *Dudua aprobola* (Meyr.) (Tortricidae: Lepidoptera)

Distribution

The pest is a cosmopolitan one and has been reported in different fruit crops in South Africa, Hawaii, China and Australia.

Hosts of commercial importance

The pest has also been reported to feed on kathjamun (*Eugenia jambolana*), Chhota amaltas (*Cassia tora*) and Rambutan (*Nephelium lappaceum*).

Identification and Biology

Wingspan of the adult moth is about 2 to 2.5 cm. The wings are of fawn or rusty-brown colour and possess a prominent light spot on the coastal margins near the middle of the forewings, as well as other irregularly placed spot. The female leaf roller moth lays eggs underside the newly emerged leaves and hatch within 2-8 days. Pupation takes place inside small leafy structures made up of clipped leaves. Life cycle duration varies. It takes about 21.6-31.0 days during July-August and 46.0-46.5 days during February-March (Kulkarni *et al.*, 1967; Maiti and Sahoo, 1991).

Mode of feeding and symptoms of infestation

The larvae roll the tender leaves feed from within the rolled leaf. The infestated twigs distort and wither away. When infestation takes place on foliage, very poor flowering is usually noticed. In case of younger trees, flowering is affected drastically. Hence, the production gets reduced considerably. As per Lal and Malick (1976) leaf injury by the pest varied between 16.70 to 71.60% while plant infestation varied from 12.88% to 53.54% during August to February.

Seasonal incidence

Leaf roller infestation on litchi is mostly observed during July to February.

Management

i. **Cultural and Mechanical:** Deep summer ploughing to remove pupae from the field. The rolled leaves along with the larvae may be removed during the period of light infestations to check further infestation.

ii. **Biological:** Numerous parasitoids *viz.*, *Trichogramma* sp., *Apanteles* sp., *Brachymeria obscurata* (Wlkr.), *Phaeogenes* sp. and *Nemorillafloralis maculosa* Meig., have been reported to act as good biocontrol agents.

ii. **Chemical:** If required, Carbaryl 2g/l or Chlorpyriphos 2ml/l may be used when about 20% of leaf flushes are infested. This is for minimization of plant damage at critical periods of leaf growth both in young as well as older trees (Kuldeep *et al.*, 2015).

8. Fruit piercing moth, *Eudocima fullonia* (Clerck), *Eudocima salaminia* (Cramer), *Eudocima jordani* (Holland) (Noctuidae: Lepidoptera)

Susainathan (1924a, 1924b) mentioned 17 species of fruit piercing moths. On the other hand, Ayyar (1944) listed 20 species from Eastern Andhra Pradesh. Again, Sundarababu and David (1973) recorded 16 species of fruit piercers from Coimbatore. As per Rakshpal (1945) and Atwal (1963), the dominant fruit piercing moths at Gwalior (Madhya Pradesh), Nagpur (Maharashtra) and Punjab, respectively were *Eudocima fullonia* and *E. materna*. Bhumannavar and Viraktamath (2001) also recorded 29 species of fruit piercing moths on guava and pomegranate in Karnataka. Among them, *E. fullonia, E. materna* and *E. homaena* were the major fruit piercing moths on pomegranate in Karnataka (Baliki *et al.*, 2011) and on guava in Coimbatore as noted by Saravanan *et al.* (2005).

Distribution

The pest is well distributed in different litchi growing tracts of India.

Hosts of commercial importance

In addition to litchi, the pest also reported to infest some other fruits including guava, pomegranate etc.

Identification and biology

Eudocima fullonia

Rakshpal (1945) studied biological aspects of the pest in Madhya Pradesh on *Tinospora cordifolia* Miers. The authors found duration of egg, larval and pupal stages of *E. fullonia* as 3 4, 15 and 21 days, respectively. However, Atwal (1963) reported the duration of egg, larval and pupal stages as 2, 4 and 2 weeks respectively on the same host. Patel and Talgeri (1956) had also mentioned general biology of *Eudocima* spp. Morphological description of *E. fullonia* larva was given by Sevestopulo (1940). Comstock (1963, 1966) illustrated and described the life stages of *E. fullonia*.

Eudocima salaminia

Sands and Schotz (1989) studied the biology of *E. salaminia* on *Stephania japonica* Miers in Australia and observed that, *E. salaminia* bred only in cool, sheltered areas. This was because the immature stages are susceptible to desiccation. In India, Bhumannavar (2000) collected the adults of *E. Salaminia* from Mudigeri (Karnataka), however the author failed to trace its larval host plant.

Mode of feeding and symptoms of infestation

The mouth part is long and stout that enables the species to penetrate through tough-skinned fruit. The moth puncture skin of the fruit and starts feeding the fruit juices. Fruit flesh or pulp gets spoiled, become soft and mushy. Secondary infections with also takes place due to fungal and bacterial pathogens at the site of wound. During adult abundance, immature green fruits also get infested. This cause premature ripening and dropping of fruit. Incidence of this moth is usually low in Indian condition. However, when outbreaks occur, the crop gets affected badly.

Seasonal incidence

The moths are usually remains active during rainy season only. However Ramachandrachari and Padmanabham (1960a; 1960b) also found abundant population of fruit piercing moths were during October.

Management

i. Mechanical

a. Capturing and destruction of moths by light trapping after sunset.

b. Bagging or screening the fruits with brown paper or transparent oil paper bags. However, the method is labour intensive but it provides good management.

c. Regular monitoring, collection and destruction of infested and spoiled fruits. These procedures dissipate the odour coming out from the spoiled fruit. Thus, the infested fruits cannot serve as attractant of the moths.

ii. Physical

a. Smoking in the orchard helps in reducing fruit infestation. This practice masks the odour of ripening or affected fruit that attracts the moth. The smoking process is started a half an hour before dusk and continued for 2 to 3 hours after evening. This period represents the time in which the moths are seeking their night time feeding grounds.

iii. Chemical control

The efficacy of chemical management is very much variable and it remains ineffective in most of the cases.

9. Litchi nut borer, *Blastobasis* sp., *Gatesclarkeana* sp. (Blastobasidae: Lepidoptera)

Hosts of commercial importance

The pest is a polyphagous one. It infests as many as 33 fruit crops.

Nature and symptoms of infestation

The larva feeds on the fruit by making tunnels towards the seed. As a result, the fruits drop. In this case, the larva is most likely to be able to grow within the fallen fruits. Larva may also drown in the fruit juice when the skin of fruit is penetrated near the equatorial position. In this position of fruit, the flesh is thickest. Sometimes the damage resembles fruit fly damage as the rind tissue around the entry hole appears to be scalded. When nut borer damages the mature fruit, fruits get stained in cluster. In the immature fruits, the young larvae bore directly into the seed and eat completely. Oozing out of juice from the maturing fruits is an indication of the pest infestation.

Management

i. Regular monitoring of the orchard is the pre-requisite for successful management of this pest. If more than 5% panicles were found infested by the pest, chemical inseciticide may be applied.

ii. The nut borer is parasitised by a number of egg, larval as well as pupal parasitoids. However, these do not always able to keep the borer population below economic threshold level.

iii. The panicles can be covered with paper bags.

iii. Spraying of Carbaryl @2 g/l has been found effective in keeping the pest population under control. Full cover spray at 40 days before harvest or two sprays at fortnightly interval are recommended.

10. Litchi semilooper, *Anisodes illepidaria* Guen. (Geometridae: Lepidoptera)

Hosts of commercial importance

Besides litchi so many host plants viz. longan, rambutan, mango, and castor have been reported for this pest (Kumar *et al*., 2013).

Identification and biology

The larvae of different instars are blackt to brown in colour with banded appearance. Full grown larvae are approximately 1.7-2.2 cm long. Size of the pupae is about 0.8-0.9 cm and wing span of the adults are about 2.1-2.3 cm. Larval period is about 8-9 days. Life cycle is completed within 15-19 days depending on the environment. Newly formed pupa are green in colour that gradually turns brown before adult emergence. Numeraous pupa are seen on uper surface of leaves in scattered manner. About 5-6 overlapping generations are completed in a year.

Mode of feeding and symptoms of infestation

Larvae make silken threads that hang vertically between tree branches. A few may hang from the threads. The silken threads may serve as the purpose to move from the fully eaten twigs to the fresh one.

Seasonal incidence

In Bihar, India the pest is found to infest litchi from September to November. Peak period of infestation of the pest has been recorded during October.

Management

i. Larvae can be hand picked to reduce the incidence. Loopers pupate at the leaf surface and can be seen easily therefore, may also be removed manually.
ii. Spraying of neem based pesticides at recommended dosages.
iii. Spraying of aquous solution of *Bt* based biopesticide @ 2g/l.
iv. Need based application of may be done of Novaluron10EC @1.5ml/l or Spinosad 45 SC @ 0.4ml/l or Fipronil 5SC @ 2ml/l.

11. Litchi mite, *Aceria litchi* (Eriophyidae: Acarina)

The litchi erineum mite was reported by Mishra (1912) and Fletcher and Mujtaba (1917) under the genus *Eriophyes*. Keifer (1943) described and named the litchi erineum mite as *Eriophyes litchi* Keifer. Further, on the basis of morphological characters genus *Eriophyes* is divided into two genera namely, *Eriophyes* and *Aceria*. The mite is known as erinose mite or litchi mite (Chennabasavanna, 1966; Lal and Rahman, 1975).

Distribution

Litchi mite has been reported from all the litchi growing countries of the world including India, where litchi mite has been found in almost all litchi is growing areas (Kuldeep *et al*., 2015).

Hosts of commercial importance

Litchi

Identification and biology

The adult mites are small, white coloured insect that live in the velvety erinose produced on leaves as a reaction to their feeding. As trees produce new flushes, the mites migrate to these where they establish new pockets of erinose for food and shelter. The female mite lays eggs singly at the base of the hair on lower surface of leaves. The eggs are very small, about 0.04 mm in diameter, round and whitish in colour. Incubation period is about 2-3 days depending on prevailing environmental condition. Newly emerged nymphs feed on soft and tender leaves of the plant. The nymphs and adults are similar in appearance. They are whitish and four legged creatures. However, nymphs being smaller and have less number of lateral setae. Life cycle is completed within 8-20 days with 10-12 generations per annum (Prasad and Singh, 1981).

Mode of feeding and symptoms of infestation

Both nymphs and adults suck sap from the leaves, inflorescence and developing fruits. Both mature as well as immature stages of the mite puncture and lacerate the tender leaf tissues and suck the cell sap. As a result of infestation, the affected leaves show abnormal growth of epidermal cells in the form of chocolate brown colour velvety growth at its underside. The mites also cause formation of galls or wart-like swellings or depressions on the upper surface of leaves. Presence of chocolate brown velvety growth on ventral surface of the leaves is the indication of mite presence. The infested leaves show curling, thickening, withering and ultimately fall off. Additionally, the mite also feed on the newly formed bud, inflorescence as well as fruit epicarp. The affected bud fails to bear flower or fruit.

Young plants in the nurseries are highly susceptible to mite infestation. In extreme case, the young plant may even be killed. Usually, dispersal or spread of erineum mite takes place through old infested leaves, by winds or mechanically during crop management. Infestations in fresh plants mostly occur through direct contact between plants or carried around by orchard workers, tools and implements, wind and even by the bees.

Management

i. Pruning of the infested foliage from the tree as much as possible and destruction thereof. Use of clean, mite free planting material.

ii. Planting material to be collected from mite free orchards and dipping of the planting material in miticide before planting.

iii. Predatory mites, especially the Phytoseids have been recorded with *A. litchii* (Wu *et al*., 1991; Waite, 1992*). Agistemus exsertus* Gonzalez (Stigmaeidae), has also been used for control of this dreaded mite in China. In India, Lall and Rahman (1975) reported that *Phytoseius intermedius* Evans and *Macfarlane, Phytoseius* sp., *Typhlodromus fleschneri* Chant and *Cunaxa setirostris* were found in association with *A. litchii*. Somchoudhury *et al*. (1987) added a further six species, namely *Amblyseius largoensis Muma, Amblyseius syzygii, Amblyseius herbicolus* (Chant), *Typhlodromus sonprayagenis, Typhlodromus homalii*, and *Agistemus* sp.

iv. Satisfactory management of litchi mite can be achieved with a strict program of three successive sprays of a suitable miticide targeting the growth flushes. The first spray should be applied to infested trees and their neighbours as a new flush begins to emerge.

The second spray is applied when the flush has fully emerged and just before the new leaves start to expand. The third spray should be applied after the new leaves have fully expanded but have not hardened off.

Large numbers of insect pests occur on litchi tree at various growth stages, but a small number are a real threat to the cultivation of this crop. The major limitations in the production of litchi is the damages caused by various insect pests resulted low production but also drastically spoil the quality, rendering them unfit for human consumption, reducing marketable yield and thereby posing a serious threat to fruit industry. Integrated Pest Management strategy is the only approach to manage the pest efficiently. This should be implemented on the basis of a systems approach by considering the whole orchard ecosystem. When executing an IPM approach, growers should decide the ways to lowering down overall pest population in their orchard and ensure that the pest management tactics are compatible with other crop management strategies.

References

Atwal, A. S. 1963. Insect pests of citrus in the Punjab IV. Biology and control of fruit-piercing moth, *Ophideres fullonica* L. (Lepidoptera). Punjab Horticultural Journal, 3(1):43-45.

Atwal, A. S. and Dhaliwal, G. S. 2010. Agricultural pests of south asia and their management. Kalyani Publishers, Ludhiana. 616 pp.

Ayyar, T. V. R. 1944. Notes on some fruit sucking moths of the Deccan. Indian Journal of Entomology, 5(I and II):29-33.

Baliki, R. A., Kotikal, Y. K. and Prasanna, P. M. 2011. Status of pomegranate pests and their management strategies in India. Acta Horticulturae, 890:569-583.

Bhumannavar, B. S. 2000. Studies on fruit piercing moths (*Lepidoptera: Noctuidae*) – species composition, biology and natural enemies. Ph. D. Thesis submitted to University of Agricultural Sciences, GKVK, Bangalore, pp-181.

Bhumannavar, B. S. and Viraktamath, C. A. 2001. Proboscis morphology and nature of fruit damage in different fruit piercing moths (*Lepidoptera: Noctuidae*). Pest Management in Horticultural Ecosystems, 7(1):28-40.

Chennabasavanna, G. P. 1966. A contribution to the knowledge of Indian Eriophyid mite. Eriophyoidae: Trambideterm. Acarina, Bangalore, University of Agricultural Sciences, Hebbal.

Choudhary, J. S., Prabhakar S. C., Moanaro, Das, B., and Kumar, S. 2013. Litchi stink bug (*Tessaratoma javanica*) outbreak in Jharkhand, India, on litchi. Phytoparasitica, 41: 73-77.

Comstock, J. A. 1963. A fruit-piercing moth of Samoa and the South Pacific Islands. Canadian Entomologist, 95:218-222.

Comstock, J. A. 1966. Lepidoptera of American Samoa with particular reference to biology and ecology. Pacific Insects Monograph, 11. pp-74.

DAC, 2013. State of Indian Agriculture, 2012-13, Govt of India, Ministry of Agriculture, Department of Agriculture & Cooperation, New Delhi, India Offs et Press, New Delhi.

Fletcher, T. B. and Mujtaba 1917. Report of the Proceeding of the Second Entomological Meeting held at Pusa, pp-229-230.

Hameed, S. F. Sharma, D. O. and Agarwal, M. L.1999. Studies on the management of litchi pests in Bihar, RAU Journal of' Research, 9(1):41-44.

Hameed, S. F., Singh, P. P. and Singh, S. P. 2001. Pests. In: Chauhan, K. S. (ed). Litchi: Botany, Production, Utilization. Kalyani Publishers, Ludhiana. pp-228.

Hameed, S.F., Sharma, D.D. and Agarwal, M.L. 1992. Integrated Pest management in litchi. *Proceedings of National Seminar on Recent Developments in Litchi Production* held at R.A.U., Pusa. pp-38.

Han, S., Liu, W., and Chen, Q. 1999. Mass releasing *Anastatus japonicus* to control *Tessaratoma papillosa* in Hong Kong. Chinese Journal of Biological Control, 15(2):54-56.

Jaipal, C. S., Chandra, S. P., Bikash, D. and Shivendra, K. 2013. Litchi stink bug (*Tessaratoma javanica*) outbreak in Jharkhand, India, on litchi. Phytoparasitica, 41: 73-77.

Keifer, H. H. 1943. *Eriophyied studies*. XIII. California Department of Agriculture Bulletin, 32(3):212-222.

Khurana, A. D., and Gupta, O. P. 1972. Bark eating caterpillars pose a serious threat to fruit trees. Indian Farmers Digest, 5:51-52.

Kulkarni, S. M., Malhotra, C. P. and Ghos, M. 1967. *Platypepla aprobola* Meyer, a leaf roller on *Moghnia macrophylla* (Wild) O. Ketze. Indian Forest, 93: 111-119.

Kumar, K. K. and Kumar, A. 2007. Managing fruit borer complex in Litchi. Indian Horticulture, 56(6):22-23.

Kumar, V, Venkataramireddy, P., Anal, A. K. D. and Nath, V. 2013. Outbreak of the looper, *Perixera illepidaria* (Lepidoptera: Geometridae) on litchi, *Litchi chinensis* (Sapindales: Sapindaceae) - a new pest record from India. Florida Entomologist, 97(1):22-29.

Kumar, V., Kumar, A. and Nath, V. 2011. Emerging pests and diseases of litchi (*Litchi chinensis* Sonn.). Pest Management in Horticultural Ecosystems. 17:11-13

Lal, B.S. and Sharma, D.O. 1978. Studies on the bionomics and control of cocoa (*Acrocercops cramerella*), *Pesticides*, 12(12):40-42.

Lall, B. S. and Rahman, M. F. 1975. Studies on the bionomics and control of the erinose mite *Eriophyes litchii* Keifer (*Acarina: Eriophyidae*). *Pesticides*, 9:49–54.

Leksawasdi, P., and Kumchu, C. 1991. Mass rearing and releasing of the parasitoid *Anastatus* sp. Kasetsart Journal (Natural Science), 25:47-53.

Li, P., Chen, B., Dong, Y., Yao, Q., Xu, S., Chen, K. and Chen, G. 2014. Effects of temperature on emergence dynamics of *Conopomorpha sinensis* (*Lepidoptera: Gracillariidae*). Florida Entomologist, 97(3).

Lu, F. P., Zhao, D. X., Liu, Y. P., Wang, A. P., and Chen, Q. 2006. Toxicity of neem seed extract to *Tessaratoma papillosa* (Drury) relative to its allozyme genotypes. *Acta Entomologica Sinica*, 49, 241–246.

Maiti, B. and Sahoo, A.K. 1991. Biological observations of *Platypepla aprobola* Meyer. on *Polyalthia longifolia* Var. Pendula in West Bengal. *Environment and Ecology*, 9(3): 647-649.

Mazumder, N., Dutta, S. K., Bora, P., Gogoi, S. and Purnima, D. 2014. Record of litchi weevil, *Myllocerus discolor* (*Coleoptera: Curculionidae*) on litchi (*Litchi sinensis* Sonn. (Sapindaceae) from Assam. *Insect Environment*, 20(1):29-31.

Menzel, C. 2002. The lychee crop in Asia and the Pacific. Food and Agriculture Organization of the United Nations, Regional Office for Asia and the Pacific, Bangkok, Thailand. www.tistr.or.th/rap/publication/2002/2002_16_high.pdf. Accessed 23 March 2012.

Mishra, C. B. 1912. Litchi leaf curl. *Indian Agricultural Journal*, 7:93.

Patel, G. A. and Talgeri, G. M. 1956. Crop pests and how to fight them. Fruit crops. Directorate of Publicity, Govt. of Maharashtra, Bombay, pp-131-157.

Paunikar, S. 2015. *Myllocerus* spp., Serious Pest of Tree Seedlings In Forest Nurseries of North-Western and Central India. Biolife, 3(1):353-355.

Prasad, V. G. and Singh, R. K. 1981. Prevalence and control of litchi mite, *Aceria litchi* Keifer in Bihar. *Indian Journal of Entomology*, 43(1):67-75.

Purbey, S. K. and Nath, V. 2013. NRCL-technologies. NRCL-EB-11 pp-1-24.

Rakshpal, R. 1945. Citrus fruit-sucking moths and their control. *Indian Farming*, 6:441-443.

Ramachandrachari, C. and Padmanabham, V. 1960a. An assessment of incidence of different fruit moths on sweet oranges in Cuddapah district. *Andhra Agricultural Journal*, 7(2): 121-128.

Ramachandrachari, C. and Padmanabham, V. 1960b. A method for attracting and capturing fruit sucking moth, *Achaea janata* L. *Andhra Agricultural Journal*, 7 (5):197-198.

Rami Reddy, P. V., Srivastava, K. and Vishalnath, 2016. Litchi fruit borer. *Current Science*, 110 (5):758-759.

Rammurthy, V. V. and Ghai, S. 1988. A study on the genus *Myllocerus* (Coleoptera: Curculionidae). *Oriental Insects*, 22:377-500.

Sands, D. P. A. and Schotz, M. 1989. Advances in research on fruit piercing moths of Subtropical Australia. In: Proc. 4th Australian Conference on tree and nut crops, Lismore, 15-19 August, 1988 (eds.) Batten, D., pp-378-382.

Saravanan, P. A., Duraimurugan, P., Muthuraman, M. and Shanmugam, P. S. 2005. Species composition and biology of fruit piercing moths in Coimbatore District. Journal of Applied Zoological Researches, 16(1):15-16.

Sevastopulo, D. G. 1940. The early stages of Indian Lepidoptera: part VI. *Ophideres fullonia* L. and *Argadesa materna* L., Journal of Bombay Natural History Society, 421: 290-292.

Sharma, D. D. and Kumar, H. 1986. How to control bark-eating caterpillars. Indian Horticulture, 31(1):25.

Sharma, D. O. 1985. Major pests of litchi in Bihar. Indian Farming, 35(2):25-26.

Singh, J. P., Jaiswal, A. K., Monobrullah, M. D., and Patamajhi, P. 2009. Bioefficacy of insecticides against pentatomid bugs, *Tessaratoma javanica* Thunb.: A sporadic pest of Kusum, *Schleichera oleosa*. Indian Journal of Entomology, 71:259–273.

Somchoudhury, A. K., Sarkar, P. K., Singh, P. and Mukherjee, A. B. 1987. Bio efficacy of some pesticides against *Aceria litchii* (Keifer) (Acari: Eriophyidae). First National Seminar on Agricultural Acarology, Kalyani, West Bengal. pp-48.

Sundarababu, P. and David, B. V. 1973. A note on unprecedental occurrence of fruit-piercing moths on grapevine. South Indian Horticulture, 21:134-136.

Susainathan, P. 1924a. Fruit-sucking moths of South India. Proceedings of 5th Entomological Meeting, Pusa, pp-23-27.

Susainathan, P. 1924b. The fruit moth problem in the Northern Circars. Agricultural Journal of India, 19:402-404.

Thomas, M. C. 2000. Pest Alert: *Myllocerus* sp. (near) *undecimputulatus* Faust, a weevil new to the Western Hemisphere. http://doacs.state.fl.us/~pi/enpp/ento/weevil-pest-alert.htm.

Verma, A.N. and Khurana, A.D. 1974. Further new host records of *Indarbela* species (Lepidoptera: Metarbelidae). Haryana Agricultural University, Journal of Research, 4(3):253-254.

Wadhi, S. R. and Batra, H. N. 1964. Pests of tropical and sub- tropical fruit trees. Entomology in India. Silver Jubilee Number. Entomological Society of India, pp-227-260.

Waite, G. K. 1992. Pest Management in Lychees. Final Report, DAQ 81A, Australian Rural Industries Research and Development Corporation, Canberra, Australia, 42 pp.

Wu, W. N., Lan, W. M. and Liu, Y. H. 1991. Phytoseiid mites on litchis in China and their application. *Natural Enemies of Insects*, 13:82–91.

Zhang, Z. W., Yuan, P. Y., Wang, B. Q. and Qui, Y. P. 1997. *Litchi Pictorial Narration of Cultivation*. Pomology Research Institute, Guangdong Academy of Agricultural Science.

Web reference

http://tmnehs.gov.in/writereaddata/chap-14.pdf

http://www.daf.qld.gov.au/plant/fruit-and-vegetables/a-z-list-of-horticultural-insect-pests/lychee-erinose-mite.

http://nhb.gov.in/PDFViwer.aspx?enc=3ZOO8K5CzcdC/Yq6HcdIx C0U1kZZen FuNVXac DLxz28=

6

Pests of Guava and Their Management

Atanu Seni, Jaydeep Halder and Samrat Saha

Guava, *Psidium guajava* Linn. was originated in tropical America and is presently cultivated in the tropical and subtropical world (Al-Fwaeer, 2013). It is considered one of the healthiest fruit because it has many medicinal properties. Right from strengthening our immune system to regulating a healthy and ideal blood pressure, Guava provides multiple health benefits. It is one of the richest sources of Vitamin C, minerals, dietary fiber and fulfills about 20% of our daily folate (Vitamin-B9) requirement which is not produced in our body and is essential for DNA repair and synthesis (Kevat, 2013). But its steady production is hampered due to a number of biotic and abiotic stresses. Amongst the biotic stresses, insect pests contribute to the maximum strength for declining its production.

A large number of insects (almost more than 80 species) and mite pests have been reported to infest on guava at various growth stages. However, a few of the pests cause severe damage in guava fruit production both qualitatively and quantitatively (Firake *et al*., 2013).Among them, tephritid fruit flies, *Bactrocera dorsalis, B. cucurbutae* and *B. zonata*, bark eating caterpillars; *Indarbela quadrinotata* and *I. tetraonis,* trunk borer, *Aristobia testudo*; fruit borers; *Virachola isocrates, Conogethes punctiferalis,* scales; *Aonidiella aurantii, Chloropulvinaria psidii*, mealy bugs; *Ferrisia virgata, Plannococcus citri, P. lilacinus*, Tea mosquito bug, *Helopeltis antonii* etc. are some of the important pests which cause serious damage the guava plant. The occurrence and intensity of damage caused by them varies from different crop growth stages, regions and seasons.

Table 1: Pests of guava fruit trees.

Sl. No.	Common name	Scientific name	Family	Order
1.	Fruit flies	*Bactrocera dorsalis* (Hendel), *Bactrocera diversus* (Coq.), *Bactrocera cucurbitae* (Coq.), *B. zonata* (Sounder)	Tephritidae	Diptera
2.	Fruit borers	*Deudorix* (*Virachola*) *Isocrates* (Fab), *Conogethes punctiferalis* Guenee	Lycaenidae Pyralidae	Lepidoptera
3.	Bark-Eating Caterpillars	*Indarbela tetraonis* Moore, *I. quadrinotata* (Walker)	Metarbelidae	Lepidoptera
4.	Trunk borer	*Aristobia testudo* Voet.	Cerambycidae	Coleoptera
5.	Guava shoot borer	*Microcolona technographa* Meyrick	Cosmopterygidae	Lepidoptera
6.	Tea Mosquito bug	*Helopeltis antonii* Signoret	Miridae	Hemiptera
7.	Spiralling whitefly	*Aleurodicus disperses* Russell	Aleyrodidae	Hemiptera
8.	Aphids	*Aphis gossypii* Glover	Aphididae	Hemiptera
9.	Mealy bug	*Ferrisia virgata* Cockerell	Pseudococcidae	Hemiptera
10.	Scales	*Aonidiella aurantii* Newstead, *Chloropulvinaria psidii* Maskell, *Drepanococcus chiton* Green	Diaspididae	Hemiptera
11.	Thrips	*Selenothrips rubrocinctus* (Giard)	Thripidae	Thysanoptera
12.	Scarlet Mite	*Brevipalpus phoenicus* (Geijskes)	Tenuipalpidae	Acari

A number of chewing pest insects such as fruit flies, fruit borer, trunk borer, bark eating caterpillar, guava shoot borer causes losses in guava orchard by eating foliage, bark, shoot and fruits.

1. Fruit fly, *Bactrocera* spp. (Tephritidae: Diptera)

The insect is also known as "peacock" fly due to their habit of strutting about, vibrating their wings and displaying their elaborate wing and body markings (Nath, 2007). It is a severely destructive insect pest for guava, especially in the rainy season. The fruits are infested by various fruit fly species including *Bactrocera cucurbitae*, *B. zonata, B. dorsalis*, *B. diversus* and *B. correcta.* Incidence of flies in fruits of guava ranged from 10 to 50 per cent in different guava growing areas of India (Rai *et al*., 2016).

Distribution

These pests are distributed throughout the country. But under North Indian condition, *B. correctus* and *B. zonata* are dominant (Anon., 2012).

Hosts of commercial importance

They have a wide host range on over 150 fruit and vegetable crops, the most common being: citrus, guava, mango, papaya, avocado, banana, loquat, tomato, cherry, rose-apple, passion fruit, persimmon, pineapple, peach, pear, apricot, fig, apple, melon and coffee (Panwar, 1995).

Identification and biology

Adults are light brown with transparent wings and are fast growing (life cycle completes in about 16 days in summer), long lived and have high reproductive potential (females typically lay 1500 eggs in their lives, but can lay up to 3000). Females lay eggs on the ripening fruits in clusters of 2 - 15 eggs and these hatches in 2 - 3 days during March and 1 - 2 days during April. Maggot duration is 6 days in summer and extends up to 19 days with the fall in temperature. Pupation usually takes place 80 - 160 mm below the soil surface and pupal period ranges from 6 days (summer) - 44 days (winter). In North Indian condition, the pest shifts to guava, after harvesting of mango, attaining the peak population during July-August. Thereafter, the population dwindles and the insects hibernate during winter in pupal stage (Panwar, 1995, Atwal and Dhaliwal, 2002; Anon., 2014).

Mode of feeding and symptoms of infestation

Ovipositional injury takes place as minute depressions over fruit surface that can be seen from outside. As a result, the fruits become soft at the site of infestation. On hatching, the maggots feed on fruit pulp and infested fruits start rotting due to the secondary infection caused by several pathogens and drop down prematurely.

Management

a. With a view to prevent the pest carryover on next season, the fallen fruits should be collected and destroyed. This reduces the number of developing fruit fly maggots.

b. Encourage to cultivate Lucknow-46 cultivar, as it exhibits resistance against fruit fly (Rai *et al.*, 2016).

c. Frequent raking of the orchard soil also helps in reducing the pest population by killing the hibernating larvae and pupae through natural enemies and scorching sun.

d. Wrapping of fruits in paper is helpful to avoid fruit fly attack. It is found that the yield of bitter gourd and angled *Luffa* was increased by about 45% when the fruits were wrapped with two layers of paper bags, replaced every 2-3 days interval (Fang, 1989).

e. Conserve the parasitoids such as *Opius compensates, Spalangia philippinensis*, *Dirhinus giffardii*, *Pachycrepoideus vindemmiae*, *Diachasmimorpha kraussi* etc. (Nath, 2007).

f. Installation of Methyl Eugenol bottle traps with 0.1% methyl eugenol and 0.1% Malathion or DDVP is effective in suppressing pest population. This also helps in early monitoring and detection of the pests. The traps can be hanged at a height of 5-6 ft from the soil surface (@ 10/ha), well before fruit ripening. The practice of trapping should be started during second week of April, if mango orchard remains nearby. The trap may be recharged with Methyl Eugenol at weekly intervals (Anon., 2014).

g. Bait spray of *viz* Malathion 50 EC, 50 ml +0.5 kg molasses in 50 litres of water/ha is helpful in reducing the infestation (Nath, 2007).

2. Pomegranate butterfly, [*Deudorix (Virachola) isocrates* (Fab)] (Lycaenidae: Lepidoptera)

Distribution

It is widely distributed in India and common in other adjoining Asian countries.

Hosts of commercial importance

The pomegranate butterfly is primarily severe in pomegranate, but is causing havoc in the guava growing regions of Uttar Pradesh and other North Indian states. It also attacks on tamarind, sapota, amla and citrus (Chhetry, 2015).

Identification and biology

The caterpillar is hairy, brownish in color with a conspicuous pale yellow or orange spot on first and second thoracic segments, and third and fourth abdominal segments. The adult is moderate size butterfly showing sexual dimorphism. Females pale violet-brown with an orange patch on forewings. Males are deep violet blue. The adult female lays single shiny-white eggs on flower calyx and fruits. The incubation period is 7-10 days. On emergence, young larvae bore into the developing fruit and feed on the flesh and seeds. They take almost 15-23 days to complete their larval stage. Later, larvae emerge and pupate near the exit hole or inside the fruit. Pupal period ranges from 7-34 days. Lifecycle completes within 44 to 65 days, generally with 4 overlapping generations per year. The pest attack is mostly severe in fruiting season both in the rainy and winter seasons (Panwar, 1995; Chhetry, 2015).

Mode of feeding and symptoms of infestation

a. Lose cultivation of guava and pomegranate should be avoided.

b. The caterpillars bore inside the fruit and usually feeding on pulp and seeds.

c. The excreta can be seen in the form of exudation peeping out from the infected fruit.

d. The bore holes made by the larvae also invite secondary infection by a number of saprophytic pathogens which later causes fruit rotting. The affected fruits ultimately fall prematurely and emit offensive smell.

Management

a. Destruction of infested plant parts.

b. Removal of flowering weeds especially of Compositae family.

c. Bagging of fruits with polythene or paper bags or cloth bags soon after the fruit set prevents the pest attack.

d. Conserve the larval parasitoid such as *Trichogramma chilonis*, *Brachymeria euploeae etc.*

3. Castor capsule borer, [*Conogethes punctiferalis* Guenee] (Pyralidae: Lepidoptera)

Distribution

It is present in India and Pakistan through south-east Asia to Australia. It has been reported from various parts of the world, mainly Hawaii, Great Britain and the Netherlands.

Hosts of commercial importance

This is a polyphagous insect pest that damages the guava in larval stage. Although the insect primarily attacks castor, it also attacks more than 120 wild and cultivated plants including guava, sorgum, maize, ginger, mango inflorescence, peaches, pear, pomegranate, sunflower, tamarind, and other fruits (Panwar, 1995; Ganesha, 2013).

Identification and biology

The mature larva is pinkish in colour having minute black spots and pale stripes speckled on the lateral side. The moths are medium sized having orange yellow coloured wings with numerous black spots. Female moth lays 80-110 pinkish oval, flat eggs singly or in groups on tender parts of plant. Incubation period is 6 to

7 days. Apart from boring fruits larvae also attack buds and tender shoots. Larval period is 12- 21 days. The pupa develops in a chamber within the basal end of the fruit. Pupal period is 7-10 days. Total life history takes 25-33 days with three generations in a year. Maximum infestation of by the pest takes place in rainy season guava (Panwar, 1995; Ganesha, 2013).

Mode of feeding and symptoms of infestation

The larvae feed on pulp and developing seeds and the infested fruits drop prematurely. The infested fruits show deformity at the point of larval entry. Larval faeces and frasses may be seen exuding out from the bore hole. Such fruits become deformed, rotten due to secondary microorganism infection and drop down later.

Management

a. Collection and destruction of infested fruits may be done regularly to check further spread of the pest population.

b. Conserve parasitoids such as *Trichogramma chilonis* (egg), *Tetrastichus* spp. (egg), *Telenomus* spp. (egg), *Chelonus blackburni* (egg-larval), *Carcelia* spp. (larval-pupal), *Campoletis chlorideae* (larval), *Bracon* spp. (larval) etc.

c. Predators such as *Chrysoperla zastrowii sellimi*, coccinellids, King crow, common mynah, wasp, dragonfly, spider, robber fly, reduviid bug, preying mantid, fire ants, big eyed bugs (*Geocoris* sp), pentatomid bug (*Eocanthecona furcellata*), earwigs, ground beetles etc. (Anon., 2014).

4. Bark eating caterpillar, *Indarbela tetraonis* Moore and *I. quadrinotata* (Walker) (Metarbelidae: Lepidoptera)

Distribution

The insect is distributed all over the Indian subcontinent. Although, this pest is found in several parts of India like Bihar, Odisha, Haryana, Rajasthan, Madhya Pradesh, Maharashtra, Andhra Pradesh and Tamil Nadu, it is more prevalent and destructive on guava trees especially in Punjab, Uttar Pradesh and South India. The incidence of this pest varied from 56 to 82 per cent in orchards of Uttar Pradesh (Rai *et al.*, 2016).

Hosts of commercial importance

It is polyphagous in nature with wide range of host plants including mango, guava, ber, litchi, orange, pomegranate, bauhinia, loquat, mulberry and rose (Abrol, 2015).

Identification and biology

Adult moth is a stout, yellowish brown in colour with wavy brown stripes on the forewings. Hind wings are white in colour. Larvae are stout and dirty brown in colour. Adults emerge out in summer season and females lay about 15-25 eggs in clusters under the loose bark. Incubation period is about 8-10 days. The larvae remain hidden in the bore hole during daytime. At night, it comes out and feed on the bark of the tree. They also make zigzag galleries filled with excreta and frass by boring inside the wood. Larval period is about 9 -11 months. The mature larva pupates inside the stem. Pupal period is about 21- 41 days and life cycle completes in 4-5 months under south Indian condition, but may take more than a year in northern India (Anon., 2014).

Mode of feeding and symptoms of infestation

a. Infestation is identified by the presence of irregular tunnels and patches on the bark covered with silken web consisting of excreta and chewed up wood particles.

b. Faecal materials are often seen hanging from the bore holes present on trunk, together with wood dust and silken web.

c. Shelter holes may also be observed particularly at the junctions of main trunk and branches.

d. The young shoots wither and die giving the plant a sickly appearance. Under severe infestation, normal flow of plant sap is hindered. As a result, plant growth arrested and fruit formation is badly affected (Abrol, 2015).

e. Severe damage by the pest may cause death of the infested stem, where larvae can be found underneath the fresh webbings.

Management

a. Keeping the orchard clean and healthy to prevent infestation by this pest.

b. Detection of early infestation by periodically monitoring for drying young shoots.

c. Mechanically the caterpillars may be killed by inserting an iron spike in the stem holes at early stages of infestation.

d. Cleaning the tree trunks by removing all the webs and placing cotton or wool soaked in chloroform, kerosene, petrol, carbon di-sulphide and sealing the same with mud.

e. Pasting of trunk with Bordeaux paste (1:1:10) + 2g/litre of Carbaryl, during December- January (Firake *et al.*, 2013).

5. Trunk borer, *Aristobia testudo* Voet (Cerambycidae: Coleoptera)

It is a very destructive pest of litchi in China, introduced in India and recorded infesting litchi, guava, pigeon pea in North-Eastern (NE) region of the country (Firake *et al.*, 2013). This pest attained the status of major pest in NE India on guava, causing 80-100 per cent damage in different guava varieties (Rai *et al.*, 2016).

Distribution

It is present in North-east India (Sikkim, Assam), Bangladesh, China, Myanmar, Thailand, Laos, Vietnam.

Hosts of commercial importance

It is feed on many important crop plants such as peach, litchi, Custard apple, Japanese Plum etc. *Tetrastichus rhipiphorothripsidis*

Identification and biology

The adults are large long horned beetles; black in colour with yellow spots. Females lay eggs in the bark by making an inverted U-shaped puncture on the branches and main trunks. Grubs initially feed on the bark for about a month and later enter in to the wood, making zig-zag single long tunnels. The grubs push away straw coloured faecal matter as pellets from the holes making at the base of the plant. There is only one generation per year.

Mode of feeding and symptoms of infestation

a. The infestation by this pest can be identified by the presence of exit holes on the stem containing straw coloured faecal matter in the form of pellets.

b. The young shoots dry and die away later.

c. Under severe infestation, plant growth is badly affected with reduced fruit setting.

Management

a. Keep the orchard clean and hygienic for preventing insect infestation.

b. Early infestation can be periodically detected by looking at emergence holes on trunk.

c. Cleaning the tree trunks by removing infested plant parts or placing cotton or wool soaked in chloroform, kerosene, petrol, carbon di-sulphide and sealing the bore hole with mud.

d. Pasting of trunk with Bordeaux paste (1:1:10) + 2g/litre of Carbaryl, during December- January is also helpful for reducing trunk borers infestation (Firake *et al.*, 2013).

6. Guava shoot borer, *Microcolona technographa* Meyrick (Cosmopterygidae: Lepidoptera)

In Punjab, the infestation of guava shoot borer, varied from 46–66 per cent on full-grown trees whereas, it was 63–97 per cent on nursery plants. For infestation, they prefer apical buds of full-grown trees whereas in nursery plants they prefer most on auxiliary buds, leaf petiole and leaf mid-rib. The active period of this pest is mid April to November with its peak period of infestation from mid July to mid August (Sharma *et al.*, 2004).

Distribution

It is present in India and its neighboring countries.

Identification and biology

The wingspan is about 10 mm. The forewings are brownish-fuscous, with a slender blackish-grey streak along the basal third of the costa, interrupted in the middle by a whitish dot. The hind wings are dark grey with a pale or whitish base (Meyrick, 1928). Larvae bore in the tender shoots of their host plant and pupate within the feeding tunnel.

Mode of feeding and symptoms of infestation

a. The infested shoots dry up which can be identified from a distance by the presence of fine black growth on the leaves.

Management

a. Removal and destruction of the infested plant parts.

b. Monitor the orchard regularly.

c. Grow moderately resistant varieties *i.e.*, Red Flesh, Sardar Guava, Dharwar and Annu Ishkwala in severe infested region (Sharma *et al.*, 2004).

d. In severe cases spraying of Carbaryl 50 WP @ 2.5-3 g/l or Chlorpyriphos 20 EC @ 2ml/l of water is helpful to reduce the pest incidence.

7. Tea Mosquito bug, *Helopeltis antonii* S. (Miridae: Hemiptera)

Distribution

It is found in different parts of India mainly Andhra Pradesh, Telengana, Kerala, Karnataka, Goa, Maharastra and Tamil Nadu. It is serious on tea in Kerala.

Hosts of commercial importance

It feeds on tea, coffee, guava, avocado, cashew, cacao, pepper, tamarind, cinnamon, apple and neem.

Identification and biology

The insect is slender, about 6-8 mm in length with a black head, whitish to black abdomen, a dark red thorax, a knobbed process arises mid dorsally on the thorax. Eggs are elongated sausage shaped with two filamentous long processes. Eggs hatch out in 5-27 days depending on environmental condition. The newly hatched nymphs undergo five moults with duration of almost 14-16 days. The insect complete one generation in two to three weeks during June and about eight weeks or more in colder months. Peak population of the insect pest found during summer months and usually disappears at the onset of monsoon (David, 2001; Atwal and Dhaliwal, 2002).

Mode of feeding and symptoms of infestation

a. Both nymphs and adults suck plant sap from fruits of all sizes, tender twigs and leaves. Injury on the fruit results in the development of brownish spot which develops into a raised pustule known as scab. It is also reported that, a bug may probe in about 50 places on a single fruit. On tender fruits, the adjacent spots may coalesce with each other and finally crack. Such tender fruits fall down on ground (David, 2001).

b. When the bug pierces into the twig, black, linear scars are formed and if tender leaves are pierced small reddish streaks symptoms appear and gradually curl up (Patil and Naik, 2004).

Management

a. Wherever possible, destroy the alternate host plants.

b. Maintain adequate aeration and sunlight by proper pruning.

c. Collect and destroy the damaged plant parts.

d. Cultivate tolerant varieties.

e. Do not interplant guava with crops that are host for *Helopeltis* bugs, such as cotton, tea, sweet potato, cashew and mango.

f. Conserve the predators *i.e.* Reduviid bug, *Oxyopes* sp. and Green lacewing

g. Weekly spraying of fungal pathogen *Beauveria bassiana* @ 1x10^9 spores/ml with suitable adjuvants reduce the population

h. Spray NSKE 5% or azadirachtin 0.03% (300 ppm) neem oil based WSP @ 1000-2000 ml in 200-400 l of water/acre or azadirachtin 5% W/W neem extract concentrate @ 80 ml in 160 l of water/acre

8. Spiralling whitefly, *Aleurodicus disperses* Russell (Aleyrodidae: Hemiptera)

Distribution

The insect is the native of America and distributed all over the world. It was introduced in India from Sri Lanka in 1996 (Nath, 2007). In India, it is mostly distributed in Kerala, Karnataka, Tamil Nadu, Andhra Pradesh and Telengana states. It caused 80% guava fruit yield loss in Taiwan (Wen *et al.*,1995).

Hosts of commercial importance

It attacks more than 481 plant species belonging to 295 genera and 90 families including different important fruit and vegetable crops *i.e.*, guava, papaya, banana, citrus, pomegranate, chilli, brinjal, sapota, soybean, fig etc. (Sathes, 1999).

Identification and biology

Adult spiralling whitefly is bigger in size than the other aleyrodids. Female lay almost 200 eggs in groups of 15-25 on the undersurface of the leaf in circular fashion, hence the name is spiralling whitefly. Incubation period is 9 to 11 days (Waterhouse and Norris, 1989). They took almost 15-20 days to complete the nymphal phase and 10-11 days for pupal stage. Body of nymphs is covered with white cottony mass. Unmated females produce only male progeny while mated females produce both sexes. Populations built up occur during warm and dry weather. Heavy rains and cool temperatures may result in a temporary reduction of populations. Only the adult stage disperses beyond the leaf on which the egg is laid. They are most active during the morning hour (Waterhouse and Norris, 1989).

Mode of feeding and symptoms of infestation

a. The milky white minute adults and nymphs suck cell sap from the tender leaves. It causes leaves crinkle and turn to red in colour.

b. They excrete the honeydew which serves as a substrate for the growth of black sooty mold. The mold reduces photosynthesis which reduces the vitality of the plant and lesser fruit production.

Management

a. Installation of yellow sticky traps@ 5/ha is helpful to monitor this pest at early stage.

b. Many weed plants harbor whiteflies, removal of weed hosts found to reduce both the incidence of whiteflies and associated viral diseases.

c. Conservation of natural enemies like *Encarsia guadeloupe*, *Eretmocerus corni*, *Eretmocerus mundus* are some of the parasitoids, *Chrysoperla zastrowi sillemi, Mallada boninensis*, *Coccinella septempunctata* are the predators can suppress the pest population.

d. Release of *Chrysoperla zastrowi sillemi* @ 8,000 larvae/acre

e. Spraying of *Beauveria bassiana* or *Lecanicillium lecanii* or *Isaria fumosorosea* 2 x 10^9conidial/ml two times with a week interval reduce the spiraling whitefly nymphs and adult (Boopathi *et al*., 2015).

f. Spray NSKE 5% or azadirachtin 0.03% (300 ppm) neem oil based WSP @ 1000-2000 ml in 200-400 l of water/acre or azadirachtin 5% W/W neem extract concentrate @ 80 ml in 160 l of water/acre.

9. Aphid, *Aphis gossypii* Glover (Aphididae: Hemiptera)

Distribution

It is found in tropical and temperate regions throughout the world except extreme northern areas. It is common in North and South America, Central Asia, South East Asia, Africa, Australia, Brazil, Mexico and Hawaii and in most of Europe (Berim, 2003).

Hosts of commercial importance

It has a very wide host range with at least 700 host plants world-wide including many important fruit and vegetable crops *i.e.* guava, watermelon, cucumbers, pumpkin, pepper, okra, brinjal, asparagus.

Identification and biology

The nymphs are soft-bodied and greenish brown in colour. They congregate on tender plant parts mostly underneath the young leaves. Adults are generally wingless but high population density induces the production of winged individuals which can migrate to new food sources. They reproduce parthenogenetically and viviparously. Single female may give birth to 8-20 nymphs in a day. The nymphs moult four times to become adults in 7-10 days. Dry weather conditions are favorable to aphids whereas heavy rainfall decreases their numbers.

Mode of feeding and symptoms of infestation

a. Both nymphs and adults suck sap from new flushes and cause damage.

b. Severe infestation by the pest results in stunted growth, curling of leaves and gradual withering and death of tender region of the plants.

c. They produce copious amounts of honeydew, a sweet and watery excrement which serves as a medium on which sooty mold grows. Sooty mold blackens the leaf and decreases photosynthetic activity which hampers the quality fruit production (Halder *et al.*, 2013).

d. It also transmits more than 50 plant viruses. Some important are Watermelon Mosaic Virus 1 (WMV-1), Watermelon Mosaic Virus 2 (WMV-2), Cucumber Mosaic Virus (CMV) (Martin and Mau, 2007).

Management

a. Destroy the infested plant parts.

b. Follow clean cultivation.

c. Maintain adequate aeration by proper training and pruning.

d. Conserve the Parasitoid, *i.e. Aphidius colemani* and Predators *i.e.* Anthocorid bugs/pirate bugs (*Orius* spp.), mirid bugs, syrphid/hover flies, green lacewings (*Mallada basalis* and *Chrysoperla zastrowi sillemi*), predatory coccinellids (*Stethorus punctillum*), staphylinid beetle (*Oligota* spp.), cecidomyiid fly (*Aphidoletes aphidimyza*), gall midge (*Feltiella minuta*), earwigs, ground beetles, rove beetles, spiders, wasps etc.

e. Release first instar larva of *Chrysoperla zastrowi sillemi* @ 15 / flowering branch (four times) at 10 days interval from the time of flower initiation.

f. Spray azadirhachtin 5% (w/w) neem extract concentrate @ 80 ml in 160 l of water/acre

10. Mealy bug, *Ferrisia virgata* Cockerell (Pseudococcidae: Hemiptera)

Distribution

The pest is also known as white tailed mealy bug or striped mealy bug. It is a widely distributed species and found in almost all tropical and subtropical countries.

Hosts of commercial importance

They are polyphagous in nature and attacks plant species belonging to some 150 genera in 68 families. Many of the host species belong to the Leguminosae and Euphorbiaceae. Among the hosts of economic importance are avocado, banana, betel vine, black pepper, cassava, cashew, cauliflower, citrus, cocoa, coffee, cotton, custard apple, egg-plant, grape-vine, guava, jute, lantana, *Leucaena*, litchi, mango, oil palm, pigeon pea, pineapple, soyabean and tomato (Anon., 2016).

Identification and biology

Female bug is wingless with two long prominent waxy filaments at the posterior end and a number of waxy hairs over the body covered with waxy powder. In the posterior end of the body, the dorsum has a prominent blackish patch. It has the habit of encircling itself by secreting thin glassy threads of wax. Males are having one pair of wings. They reproduce in both sexually and parthenogenitically, the latter being more common.

A female lays about 109-185 eggs usually in an ovisac beneath her body. The eggs hatch in about 3-4 hours. The newly hatched nymphs (crawlers) crawl out of the ovisac and start feeding on tender parts of the plant. The females pass through three nymphal instars in 31-57 days. On the other hand, the male mealy bug passes through four nymphal instars in 26-47 days. The longevity of adult female is longer than the male. In dry weather they may move down below ground and inhabit the roots. This mealy bug is favoured by dry weather and heavy infestation occurred after periods of prolonged drought (Nath, 2007).

Mode of feeding and symptoms of infestation

a. Both nymphs and adults suck plant sap from tender parts of the plant including twigs, fruits, buds, flower, petioles and even from the stem of the plants and causing curling and distortion of plant tissue.

b. Infested buds may not open and fruits can be deformed and fall down on the ground.

c. In severe cases, they produce copious amount of honeydew which helps to develop sooty mould affecting photosynthetic ability of the plants.

d. It acts as a vector of Cacao swollen shoot virus (CSSV) in West Africa and Cocoa Trinidad virus in Trinidad (Thorold, 1975).

   Other mealybug species which are also causes damage to guava plants are *Paracoccus marginatus*, *Planococcus citri, Rastrococcus iceryoides, Nipaecoccus viridis* and *Drosicha mangifera.*

Management

a. Destruction of the infested plant parts with mealybug colony.

b. Removal of weeds adjacent to orchard, roadways sides and waste lands reduces the mealy bug development in off season.

c. Conserve the parasitoids such as *Aenasius advena, Blepyrus suturalis, Spalgis epius etc.* and release the predators; *Chrysoperla zastrowi sillemi, Cryptolaemus montrouzieri* (@ 10/tree), *Pullus* sp. suppresses the natural population (Anon., 2014).

11. Red scale, *Aonidiella aurantii* Newstead (Diaspididae: Hemiptera)

Distribution

It is a native of South China (Longo, 1995) but now has been widely distributed in California, Australia, New Zealand, Mexico, India, Chile, Argentina, Brazil, Israel, the eastern Mediterranean islands, and South Africa (Ebeling, 1959 and Bedford, 1998).

Hosts of commercial importance

It infests on different important fruit crops *i.e.*, citrus, guava (Hayes, 1970), papaya, olive etc.

Identification and biology

The female scale insect has a circular, brownish-red cover about 1.8 mm in diameter. The insect itself is visible through the cover and has an oval body which becomes kidney-shaped at the last instar stage. The female moults twice and during her development, it secrets the cover material which form a concentric ring in the center each time. The female is viviparous with the eggs hatching internally (Parry-Jones, 1936). She produces 100 to 150 young ones altogether and live nymphs or crawlers emerge from under their mother's cover at the rate of two to three per day.

When they first hatch the nymphs are a yellowish colour and search for a suitable place to settle in depressions on twigs, leaves or fruits. The male insect develops similarly but after the second moult it becomes oval and darker in colour than the female. After four moults, adult male emerge out from under its elongated cover. They are small, yellowish two-winged insect and lives for about 6 hours and its sole purpose is to mate. It locates unmated females by detecting the pheromones they release (Roelofs, 1978).They multiply more in dry season.

Mode of feeding and symptoms of infestation

The nymphs and adults suck the sap from the leaves, twigs, and fruits which causes discolouration, shoot distortion and leaf drop. The fruit may become pitted and unmarketable. In severe cases the affected twigs and branches may die back and production hampers drastically.

Management

a. Cutting and burning of infested plant parts.

b. Clean cultivation.

c. Regular monitoring is needed as early detection is helpful to suppress the population in time.

d. Rubbing off scales with cotton soaked in kerosene, diesel or methylated spirit.

e. Release of *Cryptolaemus montrouzieri* @ 20 adult beetles/tree is effective.

f. Conserve the parasitoid *i.e.*, *Aphelinus* sp., *Ahytis* sp., *Encarsia* sp. and predators *Cryptolaemus montrouzieri, Chilocorus nigrita.*

Beside this, there are many other scale insects have been reported to infest leaves, shoots and fruits of guava. The common species are green shield scale, *Chloropulvinaria psidii* Maskell, *Drepanococcus chiton* Green, *Aspidiotus destructor* Signoret.

Chloropulvinaria psidii also known as green shield scale is widely found throughout India and causes enormous damage to guava in South India. They are small, oval-shaped, greenish yellow sessile insects found in large numbers on lower surface of tender leaves, twigs and shoots. Nymphs and adults suck sap from infested parts affecting growth, flowering and production of the crop (Anon., 2014 and Rai *et al.*, 2016).

12. Thrips, *Selenothrips rubrocinctus* (Giard) (Thripidae: Thysanoptera)

It is also known as red banded thrips or cocoa thrips. It was first described from Guadeloupe, West Indies, where it was causing considerable damage to cacao. As a result, it was referred to as the cacao or cocoa thrips.

Distribution

It is present in tropical-subtropical region of Asia, Africa, Australia, North America, South America, West Indies and thought to have originated in northern South America (Chin and Brown, 2008).

Hosts of commercial importance

It attacks many important fruit crops such as mango, guava, avocado, cacao and other ornamentals and shade trees.

Identification and biology

The female is about 1.20 mm in length and has a dark brown to black body underlain by red pigment chiefly in the first 3 abdominal segments. The male is similar, but smaller in size. The nymph and pupa are light yellow to orange in colour with the first three and last segments of the abdomen bright red. Females lay up to 50 eggs, inserted into the lower leaf surface and covered with a drop of fluid, which dries to form a black, disc-like cover (Astridge and Fay, 2005). The eggs hatch within four days (Chin and Brown, 2008).

After hatching, there are two nymphal stages lasting nine to ten days. Fully-grown second stage nymphs are about 1 mm long. The two nymphal stages are followed by two resting stages (pre-pupal and pupal stages). The resting stages last 3 to 5 days before adults emerge (Chin and Brown, 2008). It takes almost 21days to complete its life cycle and there are several overlapping generations in a year.

Mode of feeding and symptoms of infestation

a. Both nymphs and adults lacerate the tissues and suck oozing sap. On the occasion of severe infestations, the insects feed on the entire leaf surface turning it into characteristic silver colour.

b. Faecal deposits may be seen on upper surface of infested leaves, eventually they turn reddish brown to black.

Management

a. Sufficient irrigation as thrips thrive well in dry conditions and watering will increase moisture and inhibit their development.

b. Using black silver mulch – light reflected from the silvery surface illuminating the undersides of leaves can repel thrips.

c. Yellow sticky traps or yellow pan water traps are helpful for monitoring and control of this pest.

d. Conservation of natural enemies, *Ciranisus maculates*, *Tetrastichus rhipiphorothripsidis* etc. (Rai *et al*., 2016)

13. Scarlet mite, *Brevipalpus phoenicus* (Geijskes) (Tenuipalpidae: Acari)

Distribution

It is cosmopolitan in distribution and present in Netherlands, Spain, Portugal, Sicily, Italy, Kenya, Tanganyika, Ethiopia, Mauritius, India, Malaysia, Taiwan, Syria, USA (Hawaii, California, Texas, District of Columbia, Florida), Cuba,

Trinidad, Argentina, Brazil, Venezuela, Japan (Okinawa), Philippines and Australia (Jeppson *et al.*, 1975).

Hosts of commercial importance

It is infested on different important crop plants such as citrus, tea, coffee, peach, papaya, loquat, coconut, apple, pear, guava, olive, fig, walnut and grape (Jeppson *et al.*, 1975).

Identification and biology

They are flat, oval, and have a dark green to red-orange colour. The adult males are more wedge-shaped than females. This species has two pairs of legs that extend forward and two extending back. Larvae have six legs, and are bright orange-red when newly emerged, later becoming opaque-orange in colour. The protonymphs and deutonymphs are somewhat transparent, and like the adults, they are eight-legged. Mite lays 50-60 eggs on stalks of fruits, calyx or leaves. Their developmental period varies with temperature and humidity. Incubation period is 9-22 days; larva and protochrysalis 3-10 days; protonymph and deutochrysalis 2-8 days; deutonymph and teleochrysalis 2-8 days; total development time is 16-48 days (Lal, 1979).

In tropical regions, at least 10 generations can occur each year. Ideal conditions for this species are a temperature range of 25-30 °C (77–86 °F) with high relative humidity. Their incidence appears to be higher in the drier months or after a prolonged drought (Laycock and Templer, 1973; Danthanarayana and Ranaweera, 1972).

Mode of feeding and symptoms of infestation

Both nymphs and adults suck cell sap from the fruits that results in browning of nodal regions and appearance of brown patches on calyx and surface of fruits. In case of severe infestation, these symptoms cover the entire surface of fruits leading to splitting of fruits (Nageshchandra and Channabasavanna, 1974). It is also a vector of both citrus leprosis and the coffee ring spot virus.

Management

a. Destruction of infested plant parts can reduces the mite population.

b. Grow healthy crops; avoid water and nutrient stress.

c. Conservation and release of predatory mite such as *Amblyseius* sp. can efficiently regulates the pest under field conditions.

d. Botanicals like *Kochea* and *Calotropis* leaf extract showed strong acaricidal action under field conditions (Rai *et al.*, 2014).

e. In case of severe infestation, spraying of Propargite @ 3 ml/lit (Singh and Singh, 2004). Dusting with sulphur dust or spray wettable sulphur 80WP @ 2g/lit also suppress the pest population. Spraying preferably should be done during off season or non-fruiting stage and proper waiting period must be maintained.

References

Abrol, D. P. 2015. Pollination Biology, Vol.1: Pests and pollinators of fruit crops. Springer Publication, Switzerland. Pp 448.

Al-Fwaeer, M., abo-abied, I., Abo-allosh, A., Halybih, M., Obeidat, K., Eiz-Eldeen Atawee, Alhawamleh, H. 2013. Study of pests attacking guava in Jordan. Angewandten Biologie Forschung, 1(3):43-48.

Anonymous, 2012. "IPM schedule for guava pest" published by National horticulture mission, Ministry of Agriculture, Department of Agriculture and Cooperation, Krishi Bhavan, New Delhi. pp-19.

Anonymous, 2014. "AESA based IPM package – Guava" published by Department of Agriculture and Cooperation, Ministry of Agriculture , Government of India.

Anonymous, 2016. *Ferrisia virgata* (*Striped mealybug*). www.cabi.org/isc/datasheet/23981.

Astridge, D. and Fay, H. 2005. Red-banded thrips in rare fruit. Department of primary industries and Fisheries, Queensland. http://www2.dpi.qld.gov.au/horticulture/5064.html.

Atwal, A. S. and Dhaliwal, G. S. 2002.Agricultural pests of South Asia and their management. Fourth edition. Kalyani Publishers, Ludhiana. Pp-498.

Bedford, E. C. G. 1998. "Red scale *Aonidiella aurantii* (Maskell)". In: E. C. G. Bedford, M.A. Van den Berg and E. A. De Villiers (eds.), *Citrus pests in the Republic of South Africa*. Dynamic Ad., Nelspruit, South Africa: 132–134.

Berim, M. N. 2003. *Aphis gossypii* Glov. - Cotton Aphid. http://www.agroatlas.ru/en/content/pests/Aphis_gossypii/

Boopathi, T., Karuppuchamy, P., Kalyanasundaram, M. P., Mohankumar, S., Ravi, M. and Singh, S. B. 2015. Microbial control of the exotic spiralling whitefly, *Aleurodicus dispersus* (*Hemiptera: Aleyrodidae*) on eggplant using entomopathogenic fungi. African Journal of Microbiology Research, 9(1):39–46.

Chhetry, M., Gupta, R. and Tara, J. S. 2015. Bionomics of *Deudorix isocrates* Fabricius (*Lepidoptera: lycaenidae*), a new potential host of sweet orange, *Citrus sinensis* L. Osbeck in J & K, India. *International Journal of Science and Nature*, 6 (2): 238-241.

Chin, D. and Brown, H. 2008. Red-banded thrips on fruit Trees. *Agnote*. http://www.nt.gov.au/dpifm/Content/File/p/Plant_Pest/719.pdf (19 August 2008).

Danthanaraya, W. and Ranaweera, D. J. W. 1972. The effect of rainfall and shade on the occurrence of three mite Pests of tea in Ceylon. *Annals of Applied Biology*, 70:1-12.

David, B. V. 2001. Elements of economic entomology. Popular book depot, Chennai. pp-589.

Ebeling, W. 1959. *Subtropical Fruit Pests*. University of California, Los Angeles. pp-436.

Fang, M. N. 1989. A non-pesticide method for the control of melon fly, *D. cucurbitae* Coq. *Special Publication of the Taichung District Agricultural Improvement Station*, **16**:193-205.

Firake, D. M., Behere, G. T., Deshmukh, N. A., Firake, P. D. and Azad thakur, N. S. 2013. Recent scenario of insect-pests of guava in North East India and their eco-friendly management. Indian Journal of Hill Farming, 26 (1):55-57.

Ganesha, Chakravarthy, A. K., Naik, M. I., Basavaraj, K. and Naik, C. M. 2013. Biology of castor shoot and capsule borer, *Conogethes punctiferalis* Guenee (*Lepidoptera: Pyralidae*). Current Biotica, 7(3): 188-195.

Halder, J., Rai, A.B. and Kodandaram, M. H. 2013.Compatibility of neem oil and different entomopathogens for the management of major vegetable sucking pests. National Academy Science Letters, 36(1):19-25.

Hayes, W. B., 1970. Fruit growing in India. Thirdedition. Kitabistan, Allahabad. 512 pp

Jeppson, L. R., Keifer, H. H., Baker, E. W. 1975. Mites Injurious to Economic Plants. Berkeley, USA: University of California Press. Pp-614.

Kevat, D. 2013. Guava Fruit 8 Health Benefits and Nutrition Facts You Will Love It. Health. http://wiki-fitness.com/guava-health-benefits-and-nutrition-facts/

Lal, L. 1979. Biology of *Brevipalpus phoenicis* (Geijskes) (*Tenuipalpidae: Acarina*). Acarologia, 20(1):97-101.

Laycock, D. H. and Templer, J. C. 1973.Pesticides for East African tea. Pamphlet, Tea Research Institute of East Africa, No. 23/73.15 pp.

Longo, S., Marotta, S., Pellizzari, G., Russo, A. and Tranfaglia, A. 1995. "An annotated list of the scale insects (*Homoptera: Coccoidea*) of Italy". Israel Journal of Entomology, 29: 113-130.

Martin, J. L. and Mao, R. F. L. 2007. *Aphis gossypii* (Glover). Crop knowledge master.

Meyrick, E. 1928. Exotic microlepidoptera. Taylor and Francis, 3 (13):391-392.

Nageshchandra, N. and Channabasavanna, G. P. 1974. Host plants of *Brevipalpus phoenicis* (Geijskes) (*Acarina: Tenuipalpidae*) in India. Acarology News Letter, 2:3.

Nath, P. 2007. Emerging pest problems in India and critical issues in their management. In: *Entomology: Novel Approaches*. Eds. P. C. Jain and M. C. Bhargava. New India Publishing Agency., New Delhi, pp 43-96.

Panwar, V. P. S. 1995. Agricultural insect pests of crops and their control. Kalyani publishers, Ludhiana. pp 286.

Parry-Jones, E. 1936. "Bionomics and ecology of red scale in Southern Rhodesia". *Publication Mazoe Citrus Experimental Station,* **5:**11–52.

Patil, G. R. and Naik, L. K. 2004. Studies on the seasonal incidence of tea mosquito bug on guava cultivars. Karnataka Journal of Agricultural Science,17(2):339-340.

Rai, A. B. Halder, J. and Kodandaram, M. H. 2014. Emerging insect pest problems in vegetable crops and their management in India: An appraisal. Pest Management in Horticultural Ecosystems, 20(2):113-122.

Rai, D., Rajkumar, M. B. and Punithavalli, M. 2016. Insect pest of guava (*Psidium guajava* L.) and their management. In: *Insect pests management of fruit crops* (eds: A. K. Pandey and P. Mall). Biotech books, New Delhi. Pp-71-82.

Roelofs, W. L., Gieselmann, M. J., Carde, A. M., Tashiro, H., Moreno, D. S., Henrick, C. A. and R. J. Anderson, 1978."Identification of the California red scale sex pheromone". Journal of Chemical Ecology, 4:211-224.

Sathe, TV. 1999. Whitefly, *Aleurodicus disperses*- a new pest of guava, *Psidium guajava* in Kolhapur, Maharastra. Indian Journal of Entomology, 16(2):195-196.

Sharma, D. R., Batra, R. C., Dhaliwal, G. S. 2004. Seasonal incidence, varietal susceptibility and chemical control of shoot borer, *Microcolona technographa* Meyrick (*Lepidoptera: Cosmopterygidae*) on guava. Journal of Research, 41(1):39-46.

Singh, J. and Singh, R.N. 2004. Major harmful mites of vegetables and other crops and their control. Bulletin.Courtesy: Dhanuka Group, New Delhi., pp.5.

Thorold, C. A. 1975. Diseases of cocoa. Oxford Press, UK, pp-423.

Wen, H. C., Tung, C. H. and Chen, C. N. 1995. "Yield loss and control of spiraling whitefly (*Aleurodicus dispersus* Russell)". Journal of Agricultural Research of China, 44:147–156.

Waterhouse, D. F. and Norris, K.R. 1989. *Aleurodicus dispersus* spiraling whitefly. pp. 12-23. In: Biological Control Pacific Prospects - Supplement 1. Australian Center for International Agriculture Research, Canberra.

7

Pests of Pineapple and Their Management

Jagdish Jaba, Kalmesh Manganavi and Siva Kumar Golla

Pineapple *Ananas comosus* (Linn.) Merr. (Family – Bromeliaceae) is a tropical plant with edible fruit comprising of coalesced and the most economically important plant. The pineapple is native of southern Brazil and its wild relatives present at Paraguay. It was introduced in India by the Portuguese. Before arrival of Columbus the Indians deported it to West Indies. The Spanish introduced pineapple in Philippines. It was also introduced in Hawaii during early 16th Century. The fruit reached in England during 1660 and began to grown in greenhouses in around 1720.

Pineapple is an herbaceous perennial plant which grows to about 1.0-1.5 meters tall. The plant has a short and stocky stem with tough and waxy leaves. Adequate aeration is required to pineapple roots for sustained growth and yield (Joy and Sindhu, 2012). The plant starts flowering within 20–24 months and fruiting take place in following 6 months. During reproductive growth phase, it usually produces about 200 flowers. In some large-fruited cultivars the number may exceed. Flower colors vary depending on variety from lavender to light purple to red. As soon as the fruit is produced, the 'suckers' (side shoots) come out from leaf axils of the main stem. In addition to the suckers, the crown cutting of fruits can also be used as planting material for cultivation. During first year of growth, the axis grows, subsequently thickens and bears numerous numbers of leaves.

World Scenario

Major pineapple growing countries in the world are Brazil, Thailand, Philippines, China and India. In the world 909.84 thousand ha area is under the cultivation of pineapple and approximate production is 19412.91 thousand tons (Anonymous, 2011). In India, total area under pineapple production is 88.7 thousand ha and the production is 1415.4 thousand tons. Assam has maximum area under pineapple cultivation (14 thousand ha) with moderate productivity whereas West Bengal has higher production and productivity. Productivity is much lower in Kerala, Karnataka and Meghalaya states.

Like any other plant, pineapple is also infested by a number of pest insects in its different stages of growth. Among them, the notable pests are mealybugs, scale insects, fruit borer, bud moths, termites, midges, fruit flies etc. Mites also infest the plant. Mealybug is one of the most important pest insects of pineapple in various pineapple growing countries. Other pests occasionally reach threshold levels if favorable environmental conditions arise causing serious crop loss. Edaphic and environmental factors, crop phenology and crop management practices also affect intensity of infestation by the pest. To protect the crop from insect and mite pests, integrated approach comprising mechanical, cultural, biological including need-based application of chemical need to be done. Before going to describe elaborately the list of pests along with their systemic position is follows:

Table 1: Insect and mite pests of pineapple.

Sl. No.	Common Name	Scientific Name	Family	Order
1.	Mealy bug	*Dysmicoccus brevipes* Cock.	Pseudococcidae	Hemiptera
2.	Scales	*Diaspis bromeliae* Kerner	Diaspididae	Hemiptera
3.	Thrips	*Holopothrips ananasi* Costa Lima	Phlaeothripidae	Thysanoptera
4.	Fruit borer	*Strymon megarus* Godart	Lycaenidae	Lepidoptera
5.	Fruit fly	*Melanoloma canopilosum* Hendel, *M. viatrix* Hendel	Richardiidae	Diptera
6.	Sugarcane midget	*Elaphira nucicolora* Guenee	Noctuidae	Lepidoptera
7.	Bud moth	*Opogona sacchari* Bojer	Tineidae	Lepidoptera
8.	White grub	*Phyllophaga* spp.	Scarabaeidae	Coleoptera
9.	Fig beetle	*Cotinis mutabilis* Gory and Percheron	Scarabaeidae	Coleoptera
10.	Pineapple weevil	*Diastethus bromeliarum* Champion	Curculionidae	Coleoptera
11.	Termite	*Mastotermes darwiniensis* Froggatt	Mastotermitidae	Blattodea
12.	Red mite	*Dolichotetranychus floridanus* Banks	Tetranychidae	Acarina

1. Mealybug, *Dysmicoccus brevipes* Cockerell (Pseudococcidae: Hemiptera)

Distribution

The pest has been reported worldwide and infestation visible wherever pineapple is cultivated, including many African countries and Australia (Anonymous, 2014 and Beardsley, 1993).

Hosts of commercial importance

The pineapple mealy bug is primarily a pest of pineapple, although it does not feed on this single host plant to complete its life cycle. Due to its polyphagous nature, the mealy bug has been reported on more than 100 plant genera in 53 families. Wide range of other plants like *Annona*, avocado, banana, carrot, celery, citrus, cocoa, coconut, coffee, cotton, *Euphorbia*, ginger, *Gliricidia*, *Hibiscus*, mulberry, orchids, taro, pumpkin and many perennial grasses (Anonymous, 2014).

Identification

They are cottony, small and oval-shaped, soft-bodied and fluid feeding insects. The pest is found in a variety of forms. Among the different forms, pink ones are generally referred to as 'pineapple mealybug'. The mealybug usually found on propagating materials like suckers (Bartholomew *et al.*, 2003). They have mobility only during early stage of their life (Carillo, 2011). Body is covered with waxy mealy secretions that give them brown chalky appearance. The adult females are soft, convex in appearance and pinkish in colour. Their body is covered by seventeen (17) pairs of mealy wax filaments (Tanwar *et al.*, 2007). Lateral wax filaments are usually less than one fourth of breadth of the body. The wax filaments of those towards the back are half as long as the body. During adult stage the females also produce a fluffy waxy mass which helps in holding the eggs. Mealybugs first appear on roots. For that reason, it is very much difficult to manage the pest during early stages of infestation. The mealy bug also found to infest on the aerial region of the plant, preferably on leaf axils as well as on the developing fruit.

Biology

The species is ovo-viviparous one. Adult females lay ready-to-hatch eggs. Various biological aspects of *D. brevipes* was studied extensively by several workers (Ito, 1938). Life cycle of the species passes through 3 nymphal stages before becoming adult. Total life cycle duration is about 95 days. Eggs are minute, about 0.3 to 0.4mm in length. The early nymphal stage contributes to major dispersal of the bugs and hence they are called as crawlers. The nymphal period may extend up to about 40-50 days depending upon environmental condition. The 1st, 2nd and 3rd nymphal stages last for about 10-26, 6-22 and 7-24 days respectively.

Longevity of the adult female ranges from 31-80 days with an average of 56 days as reported by Balachowsky (1957). On an average, about fifteen overlapping generations are completed per year. Number of antennal segments in adult males is eight. The species undergo hibernation in colder months usually in egg stage, rarely in other stages. Hibernation takes place on both host plant as well as in soil. During warmer conditions, the pest remains active, reproduce and multiply throughout the year (Tanwar *et al.*, 2007).

Mode of feeding and symptoms of infestation

In favorable environmental condition the pest population increases tremendously. Both nymphs and adults start feeding by sucking the plant sap from tender leaves leads the plant to wilt. The mealy bugs are covered by many wax filaments and are attended by numerous numbers of ants. Period between 6th and 12th weeks after planting is the most critical stage of mealy bug occurrence on the crop as in this period the pineapple "eyes" are open to pests. They are usually found in colonies infesting the plant. The most active stage of their life cycle is the crawler stage. During this stage they move around the host plant in search of a suitable feeding place and at times they get static after finding it. Secretion of honey dew by the pest in large quantities leads to development of sooty mould fungus, *Capnodium* sp. that cover the affected area and thus photosynthetic activity of the host influenced badly and it also causes rotting and leaking of fruits. The sugary honeydew secretions of mealybug attract ants and they also carry them and thus help in their spread. When the fruits are extremely infested by the pest the fruits are entirely covered with white, waxy coating and thus making it unfit for marketing as well as consumption. Under heavy infestation the plant wilts quickly.

Management

Mealybug infestation on pineapple is an international problem. It badly affects commercial pineapple production in a number of ways (Beardsley, 1993). The greatest impact of this mealy bug infestation is associated with wilt disease of pineapple (Rohrbach *et al.*, 1988). The pest may also infest developing plants and fruits and thereby it is becoming serious pest of the crop. It also reduces the fruit quality and quantity (Carter, 1933). Following integrated approach may be adopted to protect the plant from this dreaded pest.

i. Biological management

There are a number of bio-agents available in the environment for suppressing mealybug population. The bio-agents may also be introduced in the crop ecosystem with a view to suppress the pest population. A few of them are being discussed as follows.

a. ***Rhinoleucophenga***: The species is a potential bio-agent (predator) of the mealybug infesting pineapple (Culik, 2009). It has also been found to act as a larval predator of scale insects as noted by Vilela (1990) and Grimaldi (1993). These may be introduced to manage pineapple mealybug.

b. ***Cryptolaemus montrouzieri***: This bioagent is commonly known as ladybird beetle. It destroys mealybug. It has already been introduced in Karnataka, India to suppress pink mealybug population. The adult ladybird beetle lay eggs on mealybug egg masses. After hatching, the grubs feed on mealybug eggs and crawlers. It has feeding potential of about 5,000 mealy bugs during their life stages.

c. ***Anagyrus kamali*:** In Karnataka, India, the species has been introduced from China for the control of pink mealybugs. It has good potentiality to parasitize the mealybug and effectively lower down their population.

d. ***Verticillium lecanii* or *Beauveria bassiana*:** Spraying of entomopathogenic fungal formulation, *V. lecanii* or *B. Bassiana* ($2x10^8$cfu/ml) @ 5 g/ml per liter of water have been found efficacious during high humid and hot months in suppressing the mealybugs population (Tanwar *et al.*, 2007).

ii. Cultural

a. Avoid using 5-6 years old suckers for planting purpose.

b. Mealybug spread can be checked by destroying ant colonies. If required, ant infestations to be suppressed by using suitable technique.

c. The mealybug infested fields should be prepared after removal and destruction of the crop debris.

d. Remove weed hosts from the field as they act as alternate food sources.

e. Destruction of alternate hosts such as *Hibiscus*, guava etc. in the vicinity of field.

f. Use of sterilized equipment when taking up planting and intercultural operations in an un-infested field.

iii. Chemical

Chemical management practices are the final attempt for the management of mealy bug. A few of them are as follows:

a. Application of Chlorpyriphos @ 2.5 ml/l or Imidacloprid 17.8 SL @ 0.3 ml/ lor Quinalphos 25EC @ 2 ml/l. Requirement of spray fluid @ 500-600 lit/ ha.

b. Indirect management of mealybugs can be achieved by soil drenching either with Chlorpyriphos 20 EC (@ 2.5 ml/l) or by applying 5% Malathion dust (@ 25 kg/ha) as suggested by Tanwar *et al.* (2007). This treatment will manage the ant colonies.

c. As a prophylactic measure, basal portion of the suckers to be dipped in 0.02-0.04% solution of Methyl parathion.

2. Scale insects, *Diaspis bromeliae* (Kerner) (Diaspididae: Hemiptera)

Distribution

The pest is distributed in different countries of the world.

Hosts of commercial importance

D. bromeliae is a polyphagous and recorded from several host plants belonging to 10 plant families (Borchsenius, 1966). However, Miller and Davidson (1990) attribute it with hosts in 10 genera belonging to two families. Hosts include species of *Aechmea*, *Ananas*, *Anthericum*, *Billbergia*, *Brassia*, *Bromelia*, *Bromeliaceae*, Canna, *Chamaerops*, Hedera, Hibiscus etc.

Identification and biology

The pest is generally found to infest on upper leaf and fruit surfaces of pineapple (Waite, 1993). It shows variation in appearance. Some individuals are very small (1-2 mm) and shiny pearl like objects (5 mm) and body is covered with mealy wax. Adult females, in most of the time are immobile and permanently attached with the plant. They also secrete a waxy coating from the body which acts as defense mechanism. This coating over the body resembles scales found over the body of reptiles or fishes. Yellow spots appear on infested leaf surface when scale insect population is low (Waite, 1993). The other scales that are reported to infest pineapple normally are not hazardous to the plant. The appearance of brown/ red pineapple scale, *Melanaspis bromeliae* is similar with *D. bromeliae*. However, it is chocolate brown in colour and the centre is elevated (Carter, 1967). Zimmerman (1948) also detected Nigra scale, *Parasaissetia nigra* on pineapple.

The scale insects show high degree of sexual dimorphism. Female individuals of the scale insects retain external morphology of the immatures even after attaining sexual maturity. This morphological condition is popularly known as neoteny. The male adults usually have wings but they never feed. They usually die within a day or two after mating.

The mealy bug passes through three distinct life stages *i.e.*, egg, immature and adult. Several overlapping generations of the species are completed in a year. The neonates of the species emerge out from the egg with functional legs are known as crawlers. Immediately after emergence, the crawlers crawl around in search of a suitable site to settle down and starts feeding by inserting their stylet. There remain variations on such themes. The scale insects that are associated with ants and those act as herders and carry the young individuals to a safer site of feeding. In either case, many such species of crawlers lose their legs when they change their skins. However, the males retain their legs which they use during mating. Most of the scale insects are hermaphrodites and some are parthenogenetic.

Sexually mature female individuals produce eggs beneath the covering of scale or in a cottony material. Very small sized crawlers emerge out of the eggs. The crawlers move to tender growth of the plant, insert their mouthparts and begin to feed plant sap by sucking. Adult scale insects can easily be identified by their colour and shape of covering. The covering protects them from natural enemies. The shape of male scale is slightly different from the female. Females are apterous and as they insert their mouthparts into plant tissues, get settled in that particular place.

Mode of feeding and symptoms of infestation

Both nymphs as well as adults feed on tender parts of the plants by inserting their sucking mouthparts. Tiny six legged crawlers (nymphs) emerge from the eggs move to tender parts of the plant and starts feeding by inserting their piercing and sucking mouthparts. Rust coloured spots appear on the site of infestation. Protective cover of the scale insect is made up of a waxy secretion of their body and molted skins. Heavily infested plants become covered with abundant population of the insect and the affected part of the plant become fully encrusted with insects.

Management

i. Biological

Biologically the pest remains under control by a number of natural enemies (Waite, 1993).Tiny wasps like, *Aphytis chrysomphali* (Mercet), *Aspidiotiphagus citrinus* (Craw) and *Aphytis diaspidis* (Howard) (Hymenoptera: Aphelinidae) etc. parasitize the scales that results in considerable mortality of the scale (Zimmerman, 1948). Coccinellid beetles like *Rhyzobius lophanthae* Blasid and *Telsimis nitida* Chapin also found to prey upon the species (Carter, 1967, Waite, 1993).

ii. Chemical

A light, oil and insecticide mixture if applied just as plants begins to grow, the pest population become satisfactorily suppressed during crawling stage. The oil-insecticide mixture should be applied before opening of leaves. The perfect time of application of any insecticide is when the insects are in crawling stage. At this stage the insect do not have any protective covering and therefore vulnerable to almost all kinds of chemical insecticides. Several conventional insecticides and insecticidal soaps have been recommended for the management of crawlers.

The plants should thoroughly be covered by the spray fluid to kill the crawlers with one application of insecticide. Recommended insecticides having systemic action may be applied as foliar spray to manage the pest.

3. Thrips, *Holopothrips ananasi* (Phlaeothripidae: Thysanoptera)

The pest causes serious damage to pineapple plant as noted by Cavalleri and Kaminski (2007). In addition to this pest, *Frankliniella schultzei* and *Thrips tabaci* are also considered as key pests of the crop. The thrips feeds on tender parts of the plant like, flowers as well as buds. Feeding by the thrips results in "dead-eye" symptom on fruit.

Distribution

Distributed worldwide in pineapple growing areas.

Identifying features

The thrips is small (about 1.5 mm long) in size, slender and brown in colour. Hind wings are pale yellow in colour. Adults have characteristic wings that are provided with numerous fringes on all the margins. Eggs are elliptical in shape and white in color, roughly 0.02 cm in length. These immature stages do not feed and remain in soil. Larval stages last for about 4-7 days. Adults are about 0.2 cm long and are pale yellow to dark brown in colour.

Biology

The life history of thrips is unique. It reproduces very fast. The females have a saw-like ovipositor which is used in making incision on the plant tissue for laying eggs. Eggs are laid singly just beneath the epidermis of tender leaves and stems. Eggs are initially whitish in colour and in course of pre-embryonic development change to orange colour. Incubation period is about 4-10 days. Immediately after hatching, the neonates begin to suck plant sap containing nutrients. Immatures and the adults have similar mode of feeding. After completion of larval stage, they pupate in soil. The adults are about 1 mm in length and a yellow brown in colour with dark cross stripes over the body. Complete development of larva takes about 9 days. Pre-pupal and pupal development is completed in about 4-7 days. Entire life cycle of the pest lasts for about 3 weeks and usually 5-10 generations are completed in a year.

Mode of feeding and symptoms of infestation

Thrips are fluid feeder and they feed on the tender parts of plant by lacerating and then sucking plant juice. The typical symptom of infestation is the development of silvery-flecked leaf surface. In severe cases, the injured plant parts turn brown and brittle. Small sized black spots are also appeared on the affected leaves. Presence of excreta on affected plant parts is the sign of thrips injury. The pest usually found to rest near the leaf veins or in crevices. Primarily they become active during day time.

In addition to cause direct damage by sucking plant sap, they also transmit a number of fungal and viral diseases in plants. The affected leaves lose their normal colour and appearance. When many adjacent cells are injured, whitish or silvery spots or streaks developed on the affected plant parts. In case of advanced injury, the leaves take blasted appearance.

Management

i. **Prevention:** Thrips are most aggressive during dry period. Irrigation sometimes helps reducing its population and they also drifted by heavy wind. ETL (Economic threshold level) of the pest is 20% of infested plants in case of small-scale cultivation.

ii. **Botanicals and Biological:** Pyrethrum, citronella, garlic etc. are natural repellants to thrips. A spray fluid made of pepper and garlic help in suppressing thrips population. For this purpose, two bulbs of garlic and a few numbers of hot chilli peppers to be blended in water. After blending, the solid particles to be filtered. About 5 liters of water to be added in that solution to make the spray fluid.

    About 2 kg of fresh plant material of *Andrographis paniculata* need to be taken in 250 ml of water. After grinding it thoroughly approximately 21 liters of cow urine and 10 gm of crushed dried chilli fruits to be added to the mixture. The solution thus made, to be kept for sometime after addition of 10 liters of water. Then the solution to be filtered and will be ready for spraying on the vegetation.

iii. **Cultural:** Removal of weeds, crop rotation, mulching etc. are effective in reducing thrips population. Deep ploughing after harvest may also be practiced to destroy the pupa.

iv. **Chemical:** Spraying of soap water will suppress the thrips that should to be repeated at least twice a week.

4. Pineapple fruit borer, *Strymon megarus* Godart (Lycaenidae: Lepidoptera)

Distribution

It is also one of the major pests of pineapple. It is present in almost all pineapple producing regions causing heavy yield losses to the crop.

Identifying features and biology

Eggs are white in color, slightly flat, circular and measures about 0.8 mm in diameter. During flowering period, the eggs are laid on flowers. Larval stages are completed within the fruit. Adults are reddish in colour. Caterpillar enters into the inflorescence and feed from within by tunneling and destroying it. Larval stages completed within it and it takes about 15 days. After completion of larval stage, the full-grown larvae pupate at the base of peduncle. Pupa is about 12 mm long and 5 mm in width. Pupa is brown in colour with some dark spots. Pupal period is about 7 to 10 days. The caterpillar damages the tissues of the inflorescence by feeding. As a result, the injured inflorescence secrets a resin-colored liquid gum between the fruitlets, that ultimately turns into dark brown colouration.

Upper wing surface of the moth is greyish and lower surface is cream coloured. Wing span of the moth is about 28-35 mm. The adult moths fly in a rapid and haphazard manner during the day or night.

Mode of feeding and symptoms of infestation

Almost all pineapple varieties are affected by this pest. Recent evidence shows that, the insect is able to feed on fruits even after dry petal stage. The larvae bore into the fruit making holes and results in uneven development fruit. Damage by this pest varies depending on climatic variation greatly and extend of infestation can go up to 90% or more than that. Dry environmental condition seems to favour borer infestation. It generally prefers to infest the plant during flowering and fruit formation stage.

Galleries are found in the pulp because of boring by the caterpillar that causes oozing resulting in "gummosis" on outer surface of the fruit. Damage by the pest makes the fruit unfit for human consumption as well as marketing.

Management

i. **Cultural:** Fruit borer management begins prior to establishing the plantation. Land with forest influence is more prone to this pest. After flowering the fruit should be covered with red or grayish bags impregnated with an adjuvant along the plantation perimeter to trap adults.

ii. **Biological:** Biological control has given excellent results in managing borer larvae. *Bacillus thuringiensis* var. *kurstaki* have been found to show good efficacy in managing the pest. This organism only acts through oral ingestion. Hence, it should be used in prophylactic way before emergence of the larvae.

iii. **Chemical:** Application of Chlorpyriphos 20EC (2.5 ml/l) at 500 lit/ha spray fluid may be applied to manage the borer. Other conventional pesticide molecules viz., Lambda cyhalothrin also work well against this pest.

5. Banana bud moth, *Opogona sacchari* (Bojer) (Tineidae: Lepidoptera)

Distribution

It is the native of the humid tropical and subtropical regions. The pest was reported from the Mascarene Islands of the Indian ocean by Bojer in 1856 (Davis and Pena, 1990).

Hosts of commercial importance

Major hosts plants of the pest are *Ananas comosus* (pineapple), *Bambusa* (bamboo), *Dracaena*, *Musa* (banana), *Saccharum officinarum* (sugarcane), *Strelitzia*, Yucca, *Zea mays* (maize). Some minor hosts are, Alpinia, Begonia, *Bougainvillea spectabilis* (Bougainvilla), Bromeliaceae, Cactaceae (cacti), *Capsicum* (peppers), *Chamaedorea*, *Cordyline*, *Dieffenbachia* (dumbcanes), *Dioscorea* (yam), *Euphorbia pulcherrima* (poinsettia), *Ficus*, *Heliconia*, *Hippeastrum*, *Maranta* (arrowroot), *Solanum melongena* (aubergine) etc.

Identification and biology

The pest can easily be identified with its typical characteristic features of developmental stages. The larvae are dirty white in colour and somewhat transparent in appearance. They have a bright reddish-brown head and possess one lateral ocellus at each side. Thoracic and abdominal plates are brown in colour and are clearly visible. The larvae measure about 21-26 mm in length and about 3 mm in diameter.

The presence of mature larvae can be detected by the presence of characteristic masses of bore meal and excreta on the bore holes. The brown pupae are about 10 mm in size formed in a cocoon of 15 mm. The adult is bright yellowish and nocturnal in habit. Wingspan of the adult is 18-25 mm. Forewings have longitudinal dark brown banding and in males, a dark-brown spot towards the apex is also present. The hind wings are pale but brighter in appearance (Suss, 1974; Aguilar and Martinez, 1982). When the insect is in rest, their antennae remain pointed forwards.

This is oviparous, incubation period is about 12 days depending on environmental condition. Fecundity is about 200 eggs. The eggs develop to full fed larvae within 50-60 days. The insect passes through seven larval instars in its life cycle. Pupal stage lasts for about 20 days. Adult life is of about 6 days (Veenenbos, 1981). Adult longevity totally depends on prevailing environmental conditions and it gets shortened at warmer conditions. During hot and humid months, the insect completes only eight generations per year (Giannotti *et al.*, 1977; Heppner *et al.*, 1987).

Mode of feeding and symptoms of infestation

Pest infestation depends exclusively on optimal soil moisture, type of planting material and its age, prevailing temperature, relative humidity (RH), insecticide usage etc. The caterpillars usually feed on decaying parts of the plant. However, they gradually start infesting surrounding healthy tissues. The pest also infests leaves and thus destroys the xylem tissues. As a result of which the leaves become wilted. Overall growth and vigour of the plant gets retarded and eventually the whole plant died. Secondary infection by a number of fungal pathogens makes the condition more critical. The larvae bore into the peel of the mature fruit and cause exudation of secondary metabolites.

Management

i. Management of the pest is mainly depends upon the application of suitable insecticide. The chemicals having contact and stomach poisonous action with short residual toxicity may be recommended for managing the pests in extreme infestation.

ii. Population of adult moths can be suppressed by placing Dichlorvos strips. One strip per 30 m^3 may be kept for three months. By using this strip, the adult moth population can be minimized before they start laying eggs.

6. Sugar cane midget, *Elaphria nucicolora* (Noctuidae: Lepidoptera)

Distribution

The pest is basically found to occur in South America, South-Eastern United States, Oahu, and Hawaii regions (Py *et al.*, 1987 and Sanches, 1999). Another species, *T. basilides* is the major pest of pineapple found in tropical regions of the Western Hemisphere, including Mexico, Central and South America.

Hosts of commercial importance

Its main hosts are sugarcane and pineapple. The larvae are polyphagous in nature and recorded from several host plants of different families (Zimmerman, 1958). Robinson (2010) reports Cucurbitaceae and Gramineae as host plants of the pest.

Identification and biology

Females lay eggs singly on the inflorescence. Egg laying takes place before anthesis. Life cycle involves egg, larva, pupa and adult stages. Incubation period is about 5 days. The caterpillar is more or less uniform in colour with dark fuscous variegations. Larval period is about23 days. The pupa is dark brown in colour and measures up to 7.6 cm in length. Pupal period is about 5 days and adult longevity is about 10 days. Adults have about 12.70 cm wingspan.

Mode of feeding and symptoms of infestation

The caterpillar scrapes the surface of fruit, producing a translucent colour of the pulp and frequently leads to external "gummosis". The affected fruit become hard after coming in contact with air. They dig holes of varying depth in the developing fruits. This causes deformation of fruit (Py *et al.*, 1987).They also form numerous galleries in the fruit. This physical injury on the fruit exposes it for secondary microbial infection.

Management

i. Forests or mountainous environment favours multiplication of the pest. Strict monitoring is required for chalking out effective management strategy.
ii. Biological control has found to provide excellent results in managing this pest viz., *Bacillus thuringiensis*.
iii. Chlorpyriphos or Lambda cyhalothrin may be recommended for management of the larval stages of the pest (Carrillo, 2011).

7. Pineapple fruit flies, *Melanoloma canopilosum, M. viatrix* (Richardiidae: Diptera)

Distribution

The pests are found various pineapple growing regions of the World mainly in Peru and Paraguay. They are mostly neo-tropical in nature.

Hosts of commercial importance

Their main host is pineapple, *Ananas comosus*.

Identification and biology

Eggs are white in colour and tapering at the ends. The larvae are apodous, yellowish white in colour, vermiform. The larvae measures about 9.5 mm in length. Head is cone shaped, small and retractable. Mouthparts are represented by a pair of mandibular hooks. The first thoracic segment has a pair of spiracles with short extensions.

The larvae usually inhabit inside the flesh of fruit. Pupae are reddish brown in colour and cylindrical in shape. Their spiracles are distinctive in nature and have well developed cephalic area. Adult flies grow up to 5.0- 6.5 mm long, wingspan is about 1.0 cm. Black coloured adults having abundant small pubescence. They have wide and short scutellum and with thorny hind femora. Wings are membranous and clear, devoid of any protrusions. A dark spot is present from base to apex of the wing. The radial vein appears dark and cubito-anal cell is almost round in shape (Arevalo and Osorio, 1995).

Mode of feeding and symptoms of infestation

The pest is a fruit infester. The female flies puncture the fruit with its sharp and long ovipositor and lay eggs therein. Through the puncture injury, secondary infection takes place by a number of bacterial pathogen and decaying of fruits starts. The larvae after hatching feed on the fruit tissues. Simultaneously, pathogens start multiplying within the affected fruit. Also, the larvae groove into the pineapple fruit by creating burrows. This phenomenon causes discoloration of the fruits, uneven ripening, fermentation and further decaying of the affected fruit (Martínez *et al.*, 2000).

Management

i. Collection and destruction of infested fruits is an important tactic of fruit fly management. Affected fruits should be cleared up or to be exposed to the sun to kill the larvae.

ii. Left over fruits of fruit stalls become the breeding places of the pest. Hence, sanitation need to be maintained for its effective suppression.

iii. Traps containing fermented fruits covered with an inverted funnel may be placed near the breeding sites of fly. Some insecticidal chemical need to be mixed with it.

iv. Bagging and netting of fruits are also advisable to protect the fruit from this dreaded pest. Fruit fly adults require sufficient quantities of protein for their reproduction. Hence, protein bait or beer wastes impregnated with insecticides can also be recommended to manage the fly.

8. White Grubs, *Phyllophaga* spp. (Scarabaeidae: Coleoptera)

Distribution

Scarab species have been reported feeding on pineapple roots round the world including Australia, Asia, South Africa, Hawaii, China etc.

Identification and biology

The adults are heavy bodied, robust, oblong and hard-shelled beetle. Most of them are with long and spindly legs. Colour ranged from light reddish brown to shiny black. Size ranges from 12-25 mm in length. Grubs are white in colour with brown head and legs. Tip of the abdomen is darker in colour. When the grubs get disturbed, they curl up in a typical 'C' shaped structure. Full grown grubs are about 2.5 cm or more in length.

The adult females lay eggs which are dull coloured, spherical in shape. Colour of eggs become darker before hatching. Grubs have 3 pairs of thoracic legs. They are tan to dark brown in colour having well developed head capsule (Waite, 1993). Pupa is creamy white, pale yellow or dark brown in colour. Adult beetles vary considerably in their shape, size and colour.

May beetle adults are yellow to dark reddish brown to black shiny in colour. The June beetles are dull velvety green on top, brownish yellow on the sides and shiny green and orange yellow ventrally. Life duration (adult to adult) of the beetles is very long. It is about 2 to 4 years depending upon various agro-climatic factors. Grubs are usually numerous in numbers and feed on the second season crop following a large beetle flight.

Eggs are found encased in soil aggregates (Waite, 1993). Grubs of older scarab develop within the soil adjacent to the roots of their host plants. The grubs also feed on the organic matter in the soil. They don't disperse far from site of egg laying.

Mode of feeding and symptoms of infestation

White grubs feed and destroy roots of their host plants. Infested parts of the root become spongy. Infestation by the pest leads the sod to roll back like a piece of carpet. Symptoms of damage become visible during April-May and September-October.

Management

i. **Biological management:** Biocontrol agents of these pests usually found in the field. However, satisfactory suppression of pests does not take place as they remain hidden within the soil.

ii. **Cultural management:** Crop rotation need to be practiced with non-host crop.

iii. **Chemical management:** Preventive control requires application of insecticides having long residual toxicity. Chlorpyriphos, Dichlorvos etc. may be applied as drenching so as to lower down population development of the pest. These insecticides give satisfactory control of newly emerged grubs. The best time of application is during the month or so before egg hatching starts. It needs to be continued till the availability of very young grubs.

9. Fig beetle, *Cotinis mutabilis* (Scarabaeidae: Coleoptera)

Distribution

The fig beetle is native to Southwestern U.S. and Mexico. But recently increased their range and noticed in coastal Southern California.

Hosts of commercial importance

Pineapple (Camino-Lavin *et al.*, 1996), peach, grapes etc. are the preferred hosts of the pest. The adults usually feed on fruit whereas, the larvae infest roots (Moron and Deloya, 1991 and Camino-Lavin *et al.*, 1996).

Identification and biology

Beetles are large sized, white-coloured and measure approximately 12-50 mm (Stone, 1982). They feed on organic matter present in soil (Coviello and Bentley, 2000). The full fed larvae are cream coloured, stiff body with brown hairs on the thorax. They pupate in soil by making hollow earthen cocoon or cells (Coviello and Bentley, 2000). Initially the pupae are whitish in colour and gradually changes into cream coloured and ultimately become green just before emergence. Adults are velvety green in colour having brownish bands around the apical margin of wings. Ventrally the wings are bright metallic green in colour. Adult females are larger in size than the males. Head is equipped with hornlike projections that helps in penetrating the fruit (Stone, 1982) and they are strong fliers (Chappell, 1984).

Females lay eggs during August and fecundity is about 60. Incubation period is about 12 days. Eggs are whitish in colour and are found over the soil (Stone, 1982). Grubs are attracted to dead organic matter particularly compost and manure pits. There remain 3 larval instars in their life cycle. Pupal size varies from12-50 mm and pupal duration ranges from 25-27 days.

Mode of feeding and symptoms of infestation

The plant parts affected by the pest is flower including pollen, nectar and petals. They also found to feed on roots as well as fruits.

Management

In addition to some cultural practices like management of white grubs, conventional insecticides having contact and stomach toxicity may be applied for its effective management.

10. Pineapple weevil, *Diastethus bromeliarum* Cham. (Curculionidae: Coleoptera)

Hosts of commercial importance

It is one of the most common insect pests infesting a number of plants under the family Bromeliaceae.

Identification and biology

The weevil has the characteristic head provided with snout. Oval, dull coloured and semi-transparent eggs are laid by the female. The larvae are white with brown head and approximately 16.5-22 mm in length. The weevils are poor flier and body is without scales.

The female oviposits eggs singly in shallow excavations made by the snout usually in the fruit stalk or near the junction of fruit and stalk. Rarely, they also lay eggs at the base of crown or on the basal shoots as noted by O'Brien (1994). Incubation period is about 8-10 days. The newly emerged grubs tunnel upward within the root or fruit stalk. Larval stage lasts for about 8 to 10 weeks depending on environmental condition. They move upward and/or downward destroying the inner tissues of stalk. Pupation takes place at the extremity of the tunnel. Pupal period is about 18-24 days. The pest prefers humid environment and dense vegetation for comfortable growth and multiplication. The life cycle completed within 3-4 months (Gowdey, 1922).

Mode of feeding and symptoms of infestation

The whole life cycle of the weevil completed in a single plant. The larvae move within the stem by making tunnels. Gelatinous exudation comes out from the affected parts that also protect weevil slits. The symptoms of infestation include feeding marks on the leaves. The injured leaves become brown. Decomposition also takes place at the base of affected central leaves (Barbra and Frank, 2010). Sometimes the infested fruits get rotten (Salas and O'Brien, 1997 and O'Brien, 1994).

Management

i. Planting good quality sucker stock.

ii. Monitoring of the pest may be done using pheromone traps @ 5/ha. Additionally, this technique may become a mass adult trapping method by placing 4-6 traps/ha.

iii. Sanitation should be maintained in the field.

iv. The planting to be done in shade free areas.

v. Replanting should be done after every two crop cycles.

vi. Dipping planting materials Malathion or Imidachloprid is helpful in eliminating hidden infestation.

vii. Crop rotation with non-host plants to be practiced.

viii. Biocontrol agents like *Bacillus thuringiensis* and *Lixadmontia franki* can manage the weevil with satisfaction. Application of *Beauveria bassiana,* has also given excellent results applied in large volumes of water (4000 l/ ha).

ix. In severe case of infestation, Clorpyriphos in high volumes of water (4000 l/ha) may be used to lower down the pest population.

11. Termites, *Mastotermes darwiniensis* (Mastotemitidae: Isoptera)

It is one of the important pests of pineapple. The species is the only living member of the genus *Mastotermes*.

Identification and biology

The termite is a social insect and they undergo incomplete metamorphosis. The life stages include egg, nymph and adult. The functional female is known as Queen. The queen lay eggs en masse and approximately 20 in a mass. During nymphal stage they undergo through a number of moults (4-7) before becoming a mature one (worker, soldier or winged reproductive). As they are hemimetabolous, the nymph resembles the adults. Nymphal stage lasts for about 2-3 months depending on the availability of food and existing climatic conditions. Soldiers live for about 1-2 years. Wing buds are observed during fourth nymphal stage and eye pigment appears during sixth nymphal stage. There are 4 different castes in the society of termite as follows:

Queen

The queen is the winged reproductive of the colony (alates). When anew colony is formed it feed and cares for the young and continued till sufficient number of soldiers and workers are developed.

Nymphs (Pseudergate)

These are 'worker like' individuals and found abundantly in the colony. They perform almost all the activities of a colony except reproduction and defense. The nymphs cause damage to the crop.

Soldiers

Soldiers are the most distinctive caste of the colony. Role of the soldiers is to protect the colony from their enemies. Their reproductive organs are not well developed. They are also susceptible to desiccation and they seldom leave the colony. They contain a special structure, known as fontanelle which is used to emit a special secretion that helps in defending the colony. Basically, it acts as repellent to ants and other enemies.

Reproductives

They have darker compound eyes and denser cuticle than workers and soldiers. These are larger in size and also have well developed wings. The individuals fly when humidity and temperature of the environmental conditions approximate the conditions inside the colony. This normally happens during summer months.

Mode of feeding and symptoms of infestation

It is one of the most destructive pests of pineapple. They feed by cutting and chewing of the plant parts. In case of extreme infestation, the plant gets died. Nymphal stages are mainly responsible for causing crop loss.

Management

i. Selection of low-risk sites and use of suitable varieties resistant to damage by the termites.

ii. Biocontrol of termites is largely focused on the utilization of entomopathogenic fungi (e.g., *Metarhizium*) and nematodes. However, success is not easily achieved in the field condition.

iii. Maintaining sanitation in the field is helpful. In areas of high termite activity, the nest soil needs to be treated with Chlorpyriphos before or during planting.

iv. Drenching with Chlorpyriphos 20EC @ 2.5 ml/l is advocated in serious case of termite infestation.

12. Pineapple mite, *Dolichotetranychus floridanus* (Tenuipalpidae: Acarina)

In addition to this, pineapple fruit mite and blister mite also found to infest pineapple in Hawaii (Carter, 1967).

Distribution

The pest is distributed in different pineapple growing region of the world including Mexico, Florida, Cuba, Panama, Honduras, Hawaii, Central America, the Philippine Islands, Japan, Java, Australia, Brazil, Philippines and USA (Jeppson *et al.*, 1975).

Identification and biology

This is the largest mite found to infest on pineapple. The mites are bright orange to red in colour. Adults are very small and about 0.3-0.4 mm in length and 0.1mm mm in width. They prefer to feed on the white basal portion of leaves particularly on the crown region (Petty, 1978). The nymphs are pale and almost translucent.

Mode of feeding and symptoms of infestation

The mite cause damage to tender leaves and fruits by sucking plant sap. Severe infestation by the mite produces large, dark brown lesions on the affected parts of the plant. Ultimately the lesions cover basal tissues and that lead to necrosis of the leaves. In pineapple growing areas of the world, the mite frequently causes severe damage to the young plants. The plants that are get infested in early stages of growth remain stunted and fruit production hampered badly. Heavily affected plants may even die before fruiting (Jeppson *et al.*, 1975). Epidermal tissues of the affected plant become dried and cracked that allows secondary infection by a number of fungal and bacterial pathogen.

Management

i. Management strategies of mite are based on close monitoring of the basal white tissues of leaves and "rusty" feeding sites.

ii. The entomopathogenic fungus *Hirsutella thompsonii* has been reported to use successfully in managing the pest.

iii. The best management strategy is to plant mite free planting material as suggested by Jeppson *et al.*(1975).

iv. Population densities of *D. floridanus* has been found to be reduced by routine acaricide application and or minimal fertilizer treatments.

iv. Need based use of acaricides is also advocated for its effective management at fortnightly intervals.

References

Aguilar, J., Martinez, M. 1982. *Opogona sacchari* présent dans les cultures sous serres en France. Bulletin de la Société Entomologique de France, 87(1-2):28-30.

Anonymous, 2014. *Dysmicoccus brevipes* (pineapple mealybug). Invasive Species Compendium, CABI.

Anonymous, 2011. Indian Horticulture data base, 2011.

Arevalo, E. A., and Osorio, M. 1995. General Considerations *Melanoloma viatrix* Hondel a new pest of pineapple. Revista Colombiana De Entomología, 21(1):1-8.

Barbra, L. and Frank, J. H. 2010. EENY-161 (IN318), One of a series of Featured Creatures from the Entomology and Nematology Department, Florida Cooperative Extension Service, Institute of Food and Agricultural Sciences, University of Florida. (http://entomology.ifas.ufl.edu/creatures).

Bartholomew, D. P., Paull, R. E. and Rhrbach, K. G. 2003. The pineapple botany, production and uses. CABI Publishing. New York, United States. pp-301.

Balachowsky, A. 1957. Observation biologique et systématique sur la cochenille de lananas (*Dysmicoccus brevipes* Ckll.) la Martinique. *Revue de Pathologie Végétale et d'Ent. Agric. France,* 36(4):188-197.

González-Hernández H., Reimer N. J., Johnson M. W. 1999. Survey of the natural enemies of *Dysmicoccus* mealybugs on pineapple in Hawaii. Biocontrol, 44(1):47-58.

Beardsley, J. W. 1993. The pineapple mealy bug complex; taxonomy, distribution, and host relationships. *In:* First International Pineapple Symposium, D. P. Bartholomew and K. G. Rohrbach, (eds), Acta Horticulture, 334:383–386.

Borchsenius, N. S. 1966. A catalogue of the armoured scale insects (Diaspidoidea) of the World.'Nauka', Moscow -Leningrad, pp-449.

Camino-Lavín, M., Jiminez-Perez, A., Castrejon-Gomez, V., Castrejon-Ayala, F. and Figueroa-Brito, R. 1996. Performance of a new trap for melolonthid scarabs, root pests. Southwestern Entomologist, 21(3):325-330.

Carter, W. 1933. The pineapple mealy bug, *Pseudococcus brevipes* (Ckl.) and wilt of pineapple. Phytopathology., 23:207-242.

Carter, W. 1948. The effect of mealy bug feeding on pineapple plants grown in finely atomized nutrient solution. Phytopathology, 38(8):645-657.

Carter, W. 1963. Mealy bug wilt of pineapple; a reappraisal. Ann. N.Y. Acad. Sci. 105:741-764.

Carter, W. 1967. Insects and Related Pests of Pineapple in Hawaii. Pineapple Research Institute of Hawaii, Honolulu, pp:105.

Carillo, E. V. 2011. Guide for integrated pest identification and management in pineapple, pp:6-8.

Cavalleri, A. and Kaminski, A. L. 2007. A new *Holopothrips* species (Thysanoptera: Phlaeothripidae) damaging *Mollinedia* (*Monimiaceae*) leaves in Southern Brazil Zootaxa.1625:61-68.

Cave, R. D. 1997. *Admontia* sp., a potential biological control agent of Metamasius callizona. JBrom Soc. 47:244-249.

Chappell, M. A. 1984. Thermoregulation and energetics of the green fig beetle (*Cotinus* [*Cotinis*] *texana*) during flight and foraging behaviour. Physiological Zoology, 57:581-589.

Coviello, R. and Bentley, W. 2000. UCIPM Pest Management Guidelines: Fig beetle (*Cotinis texana*). (http://www.ipm.ucdavis.edu/PMG/r261300511.html).

Culik, M. P., and Ventura, J. A. 2009. New species of *Rhinoleucophenga,* a potential predator of pineapple mealybugs. *Pesquisa Agropecuária Brasileira*, 44(4):417-420.

Davis, D.R., and Pena, J.E. 1990. Biology and morphology of the banana moth, *Opogona sacchari* (Bojer) and its introduction into Florida (Lepidoptera: Tineidae). Proc. Entomol. Soc. Wash., 92:593-618.

Frank, J. H. and Cave, R. D. 2005. *Metamasius callizona* is destroying Florida's native bromeliads. *In*: Hoddle MS. USDA Forest Service Publication FHTET-2005-08. Vol 1.Second International Symposium on Biological Control of Arthropods; 2005 Sep 12-16; Davos, Switzerland. Washington D.C.: USDA Forest Service, Pp-91-101.

Giannotti, O., Oliveira, B. S., Ioneda, T., Fell, D. 1977. Observations on the development and sexual behaviour of *Opogona sacchari*in the laboratory. Arquivos do Instituto Biologico Sao Paulo, 44:209-212.

Grimaldi, D. A. 1993. Amber fossil Drosophilidae (Diptera). Part II: review of the genus *Hyalistata*, new status (Steganinae). American Museum Novitates, no. 3084, pp-1-15.

Gowdey, C. C. 1922. The principal Agricultural Pests in Jamaica. Dept. of Agriculture Jamaica, *Entomological Bulletin*, (2):68.

Heppner, J. B., Pena, J. E., Glenn, H. 1987. The banana moth, *Opogona sacchari* in Florida. Entomology Circular No. 293. Division of Plant Industry, Florida Department of Agriculture and Consumer Services, Gainesville, USA.ipm.ait.asia, retrieved on 5 Sep 2013.

Ito, K. 1938. Studies on the Life Histories of the Pineapple Mealy bug, *Pseudococcus brevipes* (Ckll.). J. Econ. Ent., 31(2):291-198.

Inclana, Diego, Felipe J. Bermudeza, E, Alvaradoa, Mike Ellis b, Roger N, Williamsb, Nuris A. 2008. Comparison of biological and conventional insecticide treatments for the management of the pineapple fruit borer, *Strymon megarus* (Lepidoptera: Lycaenidae) in Costa Rica. Ecological engineering, 34(4):328-331

Jeppson, L. R., Keifer, H. H., and Baker, E. W. 1975. Mites Injurious to Economic Plants. University of California Press, Berkeley, California.

Joy, P. P. and Sindhu, G. 2012. Diseases of Pineapple (*Ananas comosus*): Pathogen, symptoms, infection, spread and management. Pineapple Research Station, Vazhakulam, Muvattupuzha, India.

Le Roux, W. 1992. Pineapple Production in the Eastern Cape: a Technical Manual. Gem InstantPrint, Cape Town.

Martínez, N. B. de, Rosales, C.J., and Godoy, F. 2000. La mosca del fruto la piña *Melanoloma viatrix* (Diptera: Richardiidae) nuevo insectoplagaen Venezucla. *Agronomía Tropical* 50(1):135-140. (Spanish). megapib.nic.in/pppineapple.htm., retrieved on 11 Oct 2013. ncof.dacnet.nic.in/Books_on_Organic_Package_of_Practices/POP_NERegion.pdf, retrieved on4 Sep 2013.

Miller, D. R. and Davidson, J. A. 1990. A list of the armored scale insect pests, pp. 299-306 In D. Rosen, D. [ed.], Armored Scale Insects, Their Biology, Natural Enemies and Control [Series title: World Crop Pests, Vol. 4B]. *Elsevier*, Amsterdam, the Netherlands, Pp-688.

O'Brien, C. W. 1994. Two new species in the *Cholus spinipes* group (Cholini, Curculioninae, Curculionidae). Transactions of the American Entomological Society, 120:412-421.

Petty, G. 1978. Pineapple Pests: White Grubs in Pineapple. Pineapple Series H: Diseases and Pests, Government Printer, Pretoria, Republic of South Africa, pp-4.

Rohrbach, K. G., Beardsley, J. W., German, T. L., Reimer, N. J. and Sanford, W. G. 1988. Mealybug wilt, mealybugs and ants of pineapple. *Plant Disease*, 72(7):558-565.

Py, C., Lacoeuilhe, J. J. and Teisson, C. 1987. The Pineapple: Cultivation and Uses. G.P.

Robinson, G. S., Ackery, P. R., Kitching, I. J., Beccaloni, G. W. and Hernández, L. M. 2010. HOSTS - A Database of the World's Lepidopteran Hostplants. NHM, London. http://www.nhm.ac.uk/hosts.

Sanches, N. F. 1999. Pragas e seu Controle. In: Cunha, G. A. P., Cabral, J. R. S. and Souza, L. F. S. (eds) *O'Abacaxizeiro: Cultivo, agroindústria e economia*. EMBRAPA ACT, Brasilia, Brazil, pp-307-341.

Salas, J. and O'Brien, C. W. 1997. *Cholas vaurien* O'Brien (Coleoptera: Curculionidae), nueva plaga de la piña en el estado Lara, Venezuela. Boletin de Entomologia Venezolana 12:157–158.

Sipes, S. B. 2000. Crop profile for pineapples in Hawaii, NSF Centre for integrated pest management, North Carolina State University.

Stone, M. W. 1982. The peach beetle, *Cotinis mutabilis* (Gory and Percheron) in California (Coleoptera: Scarabaeidae). Pan-Pacific Entomologist, 58:159–161.

Suss L, 1974. *Opogona sacchari* (Bojer) (Lepidoptera, Lyonetiidae), a new pest of ornamental plants in greenhouses. Bollettino di ZoologiaAgraria e di Bachicoltura, 12:1-28

Tanwar, R. K., Jeyakumar, P. and Monga, D. 2007. Mealy bugs and their management, National Centre for Integrated Pest Management, New Delhi, (19):4-13.

Veenenbos, J. A. 1981. *Opogona sacchari*, a pest risk from imports of ornamental plants of tropical origin. Bulletin OEPP/EPPO Bulletin 11:235-238.

Vilela, C. R. 1990. On the identity of Drosophila gigantea Thomson, 1869 (Diptera, Drosophilidae). Rev. Bras. Entomol, 34:499-504.

Vorsino, A. E., Taniguchi, G. Y., and Wright, M. G. 2005. Opogona sacchari (Lepidoptera: Tineidae), a New Pest of Pineapple in Hawaii, Proc. Hawaiian Entoml. Soc., 37:97-98

Waite, G. R. 1993. Pests. In: Broadly, R. H., Wassman, R. C. and Sinclair, E. R. (eds) Pineapple Pests and Disorders. Department of Primary Industries, Brisbane, Queensland, Australia..

Wood, D. M. and Cave, R. D. 2006. Description of a new genus and species of weevil parasitoid from Honduras (Diptera: Tachinidae). Florida Entomol. 89(2):239-244.

Zimmerman, E. C. 1948. Insects of Hawaii. A Manual of the Insects of the Hawaiian Islands Including Enumeration of the Species and Notes on the Origin, Distribution, Hosts, Parasites, etc. University Press of Hawaii, Honolulu, Hawaii.

Zimmerman, E. C. 1958. Insects of Hawaii. Volume 7, Macrolepidoptera. Hawaii: University of Hawaii Press, xi + 542 p. http://hdl.handle.net/10125/7336.

8

Insect Pests of Pomegranate and Their Management

P. Venkata Rao, Riju Nath and Adrish Dey

Pomegranate (*Punica granatum* Linn.) is an important fruit crop cultivated in arid and semi-arid regions, serving nutritional, medicinal and aesthetic need of human beings. It is grown in Iran, Spain, India, Moracco, Egypt, Afghanistan, Arabia, West Pakistan, Bangladesh, Burma, China, USA and Japan.World pomegranate production is currently estimated at 1.5-2.0 million tonnes, 90% of which is in Iran and India. The rest is distributed amongst various Mediterranean countries, China, the United States, Armenia, Azerbaijan, Pakistan, and Argentina (CBI, 2014). In India, it is found from Kashmir to Kanyakumari cultivated in Maharashtra,Gujarat, Uttar Pradesh, Tamil Nadu,Karnataka andAndhra Pradesh.Among the different states growing pomegranate, Maharashtra is the largest producer occupying 2/3rd of the cultivated area in India followed by Karnataka, Andhra Pradesh, Gujarat and Rajasthan.

Pomegranate is traditionally known for its human health benefits that have also been confirmed recently. It is a good source of vitamins, antioxidants, calcium, iron, potassium, magnesium and zinc (Jurenka, 2008 and Haidari *et al.*, 2009). The consumption of pomegranate as fresh fruit or juice has recently increased due to its many recently discovered medicinal properties. The cultivation of pomegranate has also recently been revaluated for the crop's positive agronomic characteristics, such as the high adaptability to various climatic or soil conditions and the limited water requirements that suit its cultivation on marginal land (Chandra and Jadhav, 2008).

Pomegranate has many pests that, if not well managed can seriously affect commercial fruit production.The present chapter details the pests of pomegranate describe in their harmfulness, their main morphological and biological characteristics and the techniques used in their integrated management. Also, discussion will be there on the most damaging species occurring in the main pomegranate production areas of the world and the risks of their possible introduction to the Indian sub continent.

Pomegranate trees are reported to be infested by about 45 species of insect pests (Butani, 1979 and Mir *et al.,* 2012) whereas Ksentini *et al.* (2011) reported 91 insects have been found to infest the crop in India. Balikai *et al.* (2011) also confirmed that a total of 91 insect species, with six mites and one snail pest feed on pomegranate in India causing economic losses. Although, all the pests do not cause considerable harm to the crop.

Table 1: Pests of pomegranate

Sl.No.	Common name	Scientific name	Family	Order
1.	Anar butterfly	*Deudorix* (*Virachola*) *isocrates* Fab., *D. epijarbas* Moore and *D. livia* Klug.	Lycanidae	Lepidoptera
2.	Fruit sucking moths	*Eudocima maternal* Clerk, *E. fullonia* and *E. homaena* Hub.	Noctuidae	Lepidoptera
3.	Fruit borer	*Conogethes punctiferalis* (Guen.)	Pyralidae	Lepidoptera
4.	Bark eating caterpillar	*Indarbela* spp. (Moore)	Cerambycidae	Coleoptera
5.	Shot hole borer	*Xyleborus perforans* Wollastan	Scolytidae	Coleoptera
6.	Stem borer	*Coelosterna spinator* Fab.	Cerambycidae	Coleoptera
7.	Ash whitefly	*Siphorinus phillyreae* Haliday	Aleyrodidae	Hemiptera
8.	Thrips	*Rhipiphorothrips cruentatus* Hood	Thysanoptera	Thripidae
9.	Aphids	*Aphis punicae* (Passerini)	Aphididae	Hemoptera
10.	Mealy bug	*Pseudococcus lilacinus* (Cockerell)	Pseudococcidae.	Hemiptera
11.	Scale insects	*Parasiassetia nigra* (Nietner)	Coccidae	Hemiptera
12.	Fruit fly	*Ceratitis capitata* (Wied.)	Tephritidae	Diptera
13.	Leaf eating caterpillar	*Achea janata* (Linn.)	Noctuidae	Lepidoptera
14.	Mites	*Aceria granati* Can. & Massal	Eriophydidae	Acarina

1. Anar butterfly, *Deudorix* (*Virachola*) *isocrates* Fab., *D. epijarbas* Moore and *D. livia* Klug. (Lycanidae: Lepidoptera)

This is a serious pest found all over India and most of the world. Out of three species of fruit borers, mostly two species viz., *Deudorix* (*Virachola*) *isocrates* Fab. and *D. epijarbas* Moore are observed in India. *D.livia* Klug first observed in Tunisia in the year 2006 (Ksentini, 2011).

a. *Deudorix* (*Virachola*) *isocrates* Fab.:

This is considered to be a harmful species to pomegranate in India.

Distribution

Pomegranate fruit borer is distributed all over Asia including India. The pest also reported from Iran, Spain, Afghanistan, Pakistan, Iraq and Mediterranean countries, the Arabian Peninsula and Ukraine.

Hosts of commercial importance

Pomegranate (*Punica granatum* Linn.), *Tamarindus indica, Psidium guajava, Emblica officinalis and Citrus* spp. etc. It may also infest amla, annona, apple, ber, litchi, mulberry, peach, plum, tamarind, wood apple, soap nut and *Citrus* spp.etc (Chhetry *et al*., 2015).

Identifying features

The adult males are glossy bluish and brown-violet in colour; and females have conspicuous orange coloured patches on forewings. The adults are mostly brown with reflex bluish in males and orange in females.

Biology

The larvae (15-20 mm long at maturity) are dark brown in colour with short hairs and white patches on body. The eggs are laid singly on various parts of the plant, including flower buds. The pre-imaginal period is completed in 18-47 days, whereas the total life cycle lasts 1-2 months, depending on the seasonal climatic trend. Four generations completed in a year.

Mode of feeding and symptoms of infestation

The pest has been found boring into young fruits of pomegranate. They feed on internal contents (seeds and pulp). The holes are often plugged by the anal segment of the caterpillar or its excreta are seen on infested fruits. Infestation by the pest causes fruit rotting and dropping. Infestation on pomegranate may cause losses up to 50 % of the fruit. Severe damage by the pest is noticed in the 30-50 days old fruits. Fourth and fifth instar larvae cause the maximum damage to fruits and total rejection of fruits is not uncommon on infestation by these insects.

The extent of damage in fruit was 40-70 % and 50-90 % in India as reported by Bose (1985) and Ramkrishna Ayyar (1984) respectively.

Seasonal incidence

The severity of damage is usually noticed between April to August months in India.

b. *Deudorix epijarbas* (Moore):

Distribution

The pest has been reported to cause damage to the crop in East Asia from India to the Fiji including Philippines. In India it is one of the most important pests of pomegranate in Jammu and Kashmir and Himachal Pradesh.

Hosts of commercial importance

In addition to pomegranate, it has also reported to feed on apple, litchi, citus etc.

Identification and biology

The female butterflies are larger in size with longer wing expanse than their male counterparts. The females are dull brown in colour measuring about 22.97 mm. Average width of female butterfly including wing span is observed to be 42.26 mm with longevity of about 13.20 days. The males are brick red in colour having an average body length of 18.54 mm and longevity of 9.20 days (Insha and Abdul, 2020).

The adult females, after mating, lay pale green coloured eggs on the flower, flower buds, and stems. The newly hatched larvae startfeeding on the flower buds and flowers, goes inside soil for pupation. Pupal hibernation normally starts from October and ends during May-June (Mohi-Ud-Din *et al.*, 2018).

Mode of feeding and symptoms of infestation

The female butterfly lays eggs on flowers, buds and the calyx of developed fruits. It is a regularly occurring pest of pomegranate and the caterpillars bore into developing fruit feeding on the internal content including pulp and seeds. The bore hole made by the larva for entering the fruit invites secondary infection by fungi and bacteria causing fruit rot. Larval excreta can be seen coming out of the entry holes and ultimately leading to fruit rot and damage of full mature fruit. The extent of crop loss in pomegranate varies between 50-90 per cent affecting quality of fruits and market price.

Seasonal incidence

Maximum occurrence and incidence of the pest is seen during July to August.

c. *Deudorix livia* Klug:

Distribution

This is an important and destructive pest of pomegranate distributed all over the world including the Middle East, North Africa and the Arabian Peninsula.

Hosts of commercial importance

Pomegranate (*Punica granatum* Linn.), *Acacia* spp., *Prosopis farcata* etc.

Identification and biology

The wings are normally brownish orange in males, and bluish brown in the females; but varies between species. The larvae are reddish-brown in colour. It takes around 44 days for complete larval development when reared on pomegranate fruits under laboratory conditions. The fertilised females live for about 12 days. The female lay about 42-50 eggs on the fruits of pomegranate. The neonate larva enter inside the fruit, feed on the ripening seeds till maturity of the fruit, and exits making a hole on the fruit. They pupate in the soil. Larvae 17-20 mm length and is dark brown with whitish patches all over with short hairs. Full-grown larvae are dark brown in colour having short hair and white patches on the body and measures about 16-20mm long. Due to larval feeding inside fruit, rotting occurs which ultimately leads to fruit drop.

Mode of feeding and symptoms of infestation

The adult female lays eggs on flower buds, calyx and young fruits. Upon hatching, the caterpillars bore inside the developing fruits and are usually found feeding on pulp and seeds just below the rind (Butani, 1979). Subsequently fungi and bacteria causing fruit rot make them unsuitable for human consumption. The pest has been found to attack pomegranate all over the growth period. A single fruit may be infested by as many as 6-8 larvae.

Seasonal incidence

It is mostly prevalent during the 'mrig' bahar (June to September). Fruit injury revealed at the age of 30- 50 days.

Integrated management of pomegranate fruit borers

An integrated pest management (IPM) technology has been developed against the fruit borer with the following components.

i. Planting alternate host crops like sapota, tamarind, guava andaonla in the pomegranate orchard should be avoided.

ii. Clippping off of calyx cup soon after pollination following two applications of neem oil @ 3% at the time of butterfly activity may be done.

iii. Spraying NSKE 5% or neem formulations 2 ml/1 during flowering may be practised.

iv. Before maturity, bagging of the young fruits with coarse cloth or muslin cloth or butter paper or polythene of 300 gauge thickness for hindering egg laying by butterflies and boring by larva is effective in small scale cultivation. Physical barriers like fruit cover at 30–50 days after fruiting, clipping off the calyx cup immediately after pollination help to reduce the egg laying and fruit damage.

v. The affected fruits (fruits with exit holes) should be collected and destroyed continuously till final harvest.

vi. Release of *Trichogramma chilonis* @ 2.5 lakhs/ ha four times at 10 days interval.

vii. Kakar *et al.* (1987) suggested the use of Cypermethrin (@ 150 g a.i./ha) and Deltamethrin (@ 7.5 g a i/ha) for effective and profitable control of fruit borer.

viii. Spraying Indoxacarb 14.5%SC @ 1.0 ml or Spinosad 45 %SC @0.5 ml or Cypermethrin 25%EC @ 1.0 ml per liter of water may be done to reduce pest population.

2. Fruit sucking moth, *Eudocima* (=*Othreis*) *maternal* (Linn.), *Eudocima fullonia* (Clerck), and *Eudocima homaena* Hub.

Distribution

Fruit sucking or fruit piercing moths are serious pests of pomegranate in the tropical and subtropical belts spanning from Africa to the Pacific Islands.

Hosts of commercial importance

The fruit sucking moths, *Eudocima* spp. are polyphagous pest. It feeds on a wide varietyof important fruits including pomegranate, mango, citrus, carambola, papaya, litchi, grapes, guava etc. (Sundarababu and David, 1973).

Identifying features

Othreis fullonia adults are covered with orange colored scales and an inverted “C” shaped marking is seen at the centre of hind wings in both the sexes. *O. maternal* is covered with orange colored scales and a black dot marking is seen at the centre of hind wings in both the sexes. *O. homaena* is covered with orange colored scales and fore wings are with parrot green color band, and an inverted “C” shape marking is seen at the centre of hind wings in both the sexes.

Newly emerged larvae are about 5 mm in length and pale green in color. Full grown caterpillars are 50-60 mm long and are bright colored with orange, yellow and blue spots on speckled body and are semiloopers. Pupation takes place in a transparent pale whitish silken cover and is shiny dark brown.

Biology

The adult moth can live up to 25-30 days and lay 200-400 spherical eggs on climbers like *Cocculus pendulus, C. hirsutus* and *Tinospora cardifolia*. Incubation period is 8-10 days. The larval stage passes through five instars which take 3-5, 3-5, 2-3, 2-5 and 4-10 days respectively. Pupation last for about 13 days and then adults emerge out.

Mode of feeding and symptoms of infestation

Moths normally attack at night time penetrating the ripened skin of fruits with their proboscis and suck the juice. These holes form the avenues for entry of rot causing microorganisms. The damaged fruits soon start rotting from the punctured regions and ultimately the fruits drop occurs.

In mature fruit a large, round, brown spot develops on the skin around the piercingsite. Internal injury is generally recognized by bruised dry area underneath the fruit skin. Secondary rots sizeable area beneath the skin is bruised and often honeycombed. Fruit rot organisms rapidly invade the wound causing fermentation and breakdown.

Secondary feeders take advantage of the bore hole originally drilled by this fruit sucking moth. Single moth normally attack one fruits in a night. Damaged fruits become unfit for marketing and must be removed at packing to avoid contamination.Serious infestation on pomegranate was recorded in different parts of India including Maharashtra (Mote *et al.*, 1991), Andhra Pradesh (Ayyar, 1944) and Tamil Nadu (Cherian and Sundaram, 1936). Mote *et al.* (1991) observed that the damage wasmaximum on ripe pomegranate fruits (21.06 to 47.62%) than on unripe fruits (2.86 to 13.86%) at Rahuri (Mahashtra).

Seasonal incidence

The moth remains active throughout the year. However, maximum infestation on the crop has been recorded during rainy season, *i.e.*, during hot and humid months.

Management

i. Cultural

a. Removal of eggs from calyx and collection and destruction of affected fruits.

b. Destroy host plants (wild creepers) of larvae viz., *Tinospora cordifolia, T. sinensis, T. smilacina, Cocculus hirsutus, Anamirta cocculus, Tiliacora acuminate, Diploclisia glaucescens, Cissampelos pareira, Convolvulus arvensis, Trichisia pattens* and *Pericampylus glaucus* present in and around the field.

c. Adjust fruit season to escape the peak activity of moths. Avoid taking *mrigbahar* in pomegranate, as the fruit sucking moth remains active from August to October.

d. Moth attack occurs mostly along the outer rows of the orchard. So, fruit crops should be planted in square blocks and not in a few long rows.

ii. Mechanical

a. Periodical observation to detect early infestation in the drying branches.

b. Pest monitoring using light trap @ 1/ acre above 15 cm above the crop canopy. It also helps in mass trapping of the adult insects.

c. Collection and destruction of adult moths using torch or other light source at night is a usual practice adopted by the growers.

d. Covering the fruits with newspaper, butter paper or polythene bags to protect from juice sucking.

e. Create smoke in the fields by burning the weeds and crop waste during evening hours in the orchard at different spots. Smoke masks the odor of the mature fruit, so that, the moths fail to detect fruits and do not enter the orchard.

iii. Biological

a. Egg parasitoids, *Telenomus* sp., *Ooencyrtus* sp. and *Trichogramma* sp. are among the most successful biological control agents of FSM. Sontakay (1944) recorded 30 per cent egg parasitisation by a chalcid parasitoid at Nagpur (Maharashtra). Per cent egg parasitisation by *Telenomus* sp., *Ooencyrtus* sp., and *Trichogramma* sp. was 47.6, 15.5 and 0.6 %, respectively, on solitary eggs (n=2170) and 54.6, 23.9 and less than 0.1 % on egg masses (235 egg masses; 7508 total eggs), respectively (Denton *et al.*, 1989).

b. *Euplectrus indicus* Ferr. (Eulophidae) parasitised the last instar larva of *E. maternal* whereas *E. Maternus* Bhatnagar parasitised first three larval instars of *E. materna*, *E. fullonia* and *E. Homaena* (Bhumannavar, 2000). Jayanthi and Verghese (2010) recorded up to 40% parasitisation of *E. maternal* larvae by *Euplectrus* sp. and *Winthemia* sp.

c. *Polistes olivaceus* (De Geer) (Vespidae) was found to prey on larvae of *E. fullonia.* Several birds including *Megalurulus mariei*, *Zosterops lateralis griseonota* and the Indian mynah, *Acridotheres tristis* (L.) at times destroyed large numbers of *Eudocima* larvae.

iv. Chemical

a. Use of repellents like citronella oil should be tried.

b. For managing semiloopers of fruit sucking moth, spraying with Dichlorovas is helpful.

c. Set up poison baits (5% Malathion +95% jiggery/molasses) in flat earthen pots and install compact fluorescent lamps (CFL) bulb to attract the adults.

d. Use of poison baits containing Malathion 20 ml along with jaggary (100g) and vinegar (6 g) litre of water.

e. Spraying Chlorpyriphos 20 EC @ 2.0 ml or Indoxacarb 14.5%SC @ 1.0 ml or Spinosad 45 %SC @0.5 ml or Cypermethrin 25%EC @ 1.0 ml per liter of water.

3. Fruit borers, *Conogethes punctiferalis* (Guen.) (Pyralidae: Lepidoptera)

Distribution

Conogethes punctiferalis Guen. is an important pest not only in South and South East Asia and Australia but also reported to be a new and introduced pest in Britain and Europe.

Hosts of commercial importance

These are highly polyphagous pests, found to infest about 36 crop plants belonging to 23 families in India. The host plants include guava, mango, peaches, pomegranate, jack, avocado, mulberry, loquat, peach, plum, pear, sorghum, sunflower, cotton, cocoa, castor, turmeric, ginger, tamarind, amaranthus, soapnut, hollyhocks, Annona and other Zingiberaceous plants.

Identification and biology

Larvae are whitish green with pinkish tinge having dark head and fine hairs all over the body. Adults are yellowish moth with numerous black spots on the wing and body.

Eggs are pinkish laid singly on young capsules or flower buds. Incubation period lasts for about 2-6 days. Larvae are pale reddish brown with black blotches and tubercles on body. Larval period is up to about 60-80 days. Pupation takes place on plants in stem or fruit. Pupal period lasts for about 7-10 days. Adults are yellow colored moth with black dots and lives for about 3-4 days.

Mode of feeding and symptoms of infestation

Larvae bore into the stem causing drying and damages capsules. Caterpillar bores into the tender fruits and feeds on the internal contents (seeds and pulp). Infected fruits dry up and drop without ripening. The affected fruits are normally deformed at the larval entry point. Larval faeces are often seen exuding out of the bore hole.

Seasonal incidence

Conogethes spp. activity is high during rainy season, *i.e.*, June–July and September-November.

Management

i. Cultural

a. Periodic monitoring for detection of early infestation.

b. Maintaining adequate aeration in the orchard by seasonal pruning of plants.

c. Avoid cultivation of pomegranate close to guava, an preferred alternate host.

d. Collection and destruction of the infested fruits regularly.

e. Set up light trap@ 1/acre and operate between 6 pm and 10 pm.

ii. Biological

a. Hymenopteran parasitoids, *Chelonus blackburni* is very effective in managing *Conogethes* spp.

b. Conservation of parasitoids such as *Trichogramma chilonis, T. pretiosum* (egg), *Tetrastichus* spp. (egg), *Telenomus* spp. (egg), *Chelonus blackburni* (egg-larval), *Carcelia* spp. (larval-pupal), *Campoletis chlorideae* (larval), *Goniophthalmus halli* (larval), *Bracon* spp. (larval) etc.

c. Injection of *Bacillus thuringiensis* (*Bt*) preparation into the borehole of stem or fruit effectively checked the pest resurgence. The native isolates of entomopathogenic fungi (EPF) *Metarhizium* sp. and *Beauveria bassiana* and entomopathogenic nematode (EPN) *Heterorhabditis indica* effective against the larvae of *C. punctiferalis* in the laboratory.

d. Conservation of predators such as *Chrysoperla zastrowii* Sellimi, coccinellids, King crow, common mynah, wasp, dragonfly, spider, robber fly, reduviid bug, preying mantid, fire ants, big eyed bugs (*Geocoris* sp), pentatomid bug (*Eocanthe conafurcellata*), earwigs, ground beetles, rove beetles etc.

e. Natural enemies can be released against fruit borer like, *Trathalaflavo orbitalis, Brachymeria atteviae, Chelonus blackburni, Epitranus erythrogaster* etc. and predators Pentatomid bugs, Reduviid bugs (Assassin bugs) etc.

iii. Chemical

a. Flubendamide + thiocloprid was more effective on cardamom panicle borer, *C. pinicolalis* at 30 days after anthesis.

b. Flubendiamide and Spinosad @ 0.003% were effective against *Conogethes* spp.

4. Bark eating caterpillar, *Indarbela* sp. (Metarbelidae: Lepidoptera)

There are 13 species of the genus *Indarbela* of which, *I. quadrinotata* (Walker), *I. tetraonis* (Moore) and *I. dea* (Swinhoe) are very much economically important.

Distribution

The Bark eating caterpillar is a serious pest of a large number of plant types throughout Asia including India, Spain, China, Mediterranean countries, Iran, Iraq, the Arabian Peninsula and Ukraine.

Hosts of commercial importance

It is a polyphagous pest infesting a large number tree species including fruit trees such as Mango, pomegranate, citrus, guava, jujube lychee or litchi, and cashews and forest trees including *Casuarina* spp., *Acacia* spp., and *Albizia* spp. The pest has been reported to infest 70 plant species across fruits, forests and avenue plantations recorded as 70 plant species are the host plants of *Indarbela* sp. from Haryana.

Identification and biology

Eggs are pale brownish and spherical in shape. Length of the eggs varied from 0.67 to 0.73 mm (average 0.69 ± 0.05 mm) and width 0.42 to 0.46 mm (average 0.44 ± 0.04 mm) (Satyanarayana and Arunakumara, 2016). Caterpillars are about 40mm long and pinkish white in colour with brown spots. Pupae are chestnut brown in colour. The adult moths are white with numerous small black spots and streaks on fore wings.

Eggs are laid by the moths under loose bark or in cracks and crevices in clusters of 15-25 from April to June. Eggs hatch in about 8-11 days. Larval duration is of 8-10 months. Pupal period is 21- 41 days. Total life cycle lasts for about 4-5 months in south India and more than a year in north India. The insect completes one generation per year.

Mode of feeding and symptoms of infestation

The early instar larva is usually observed in groups and later moved out for searching the concealed places on the shoots and bored a short tunnel downward in to the wood. Caterpillars bore into the stem making zig zag tunnels and feeds inside the bark. The infestation makes the plants weak also affecting vitality and vigour of the infested trees. Surrounding the affected portion of the stem, wood dust and excreta pellets are found hanging in the form of a web. Sometimes, brownish larvae can be seen beneath fresh webbing.

The pest infestation can be identified by presence of patches covered with silken web full of larval excreta and irregular tunnels on the stem and branches.Severe damage can result in the death of attacked stem. The young shoots wither and die making plants sickly. Fruiting capacity is severely arrested.

Management

i. Cultural

a. Keeping the orchard clean, healthy and avoiding overcrowding of trees will help to minimize attacks of the pest.

b. Removal and destruction of alternate hosts and severely affected branches of the tree.

ii. Mechanical

a. Scraping the loose bark to prevent oviposition by adult beetles.

b. Hooking out the caterpillar from the bore hole and killing them mechanically.

c. Insertion of cotton plug soaked in kerosene or petrol into the holes and closing them with mud or cement.

iii. Biological

a. The larvae are parasitised by entomogenous fungus *Beauveria bassiana*in nature. Hence, it can be used as a potential bio-control agent.

b. The entomopathogenic nematode *Steinernema carpocapsae* (Weiser) have been found most effective in the control of bark-feeding moth *Indarbela dea* (Swinhoe) of litchi.

iv. Chemical

a. Different methods have been suggested from time to time for the management of bark caterpillar which include granule placement in the galleries (Verma, 1985) and spot treatment through insecticide sprays (Mathew and Rugmini, 1998).

b. For taller trees under forest situations, insecticide sprays using power sprayers have been recommended (Sangha and Makkar, 2005).

c. The mixing of petroleum spray oils at the recommended rates on the containers and painting the tree branches, limbs and trunk will give a sustained long-term solution.

d. Spraying Chlorpyriphos 20 EC @ 2.0 ml or Dichlorvos @ 2ml or Indoxacarb 14.5%SC @ 1.0 ml or Spinosad 45 %SC @0.5 ml or Cypermethrin 25%EC @ 1.0 ml per liter of water are also effective.

5. Shot hole borer, *Xyleborus perforans* Wollastan (Scolytidae: Coleoptera)

Distribution

Shot hole borer is the major insect pests of forest trees in India and abroad. This is an original pest of tea and coffe plantations in Sri Lanka, South India and Assam. Since 1981 onwards it has become major pest of pomegranate in many parts of Maharashtra and Karnataka.

Hosts of commercial importance

It is actually a pest of tea and coffee. Pomegranate (*Punica granatum* Linn.) has been recorded as a new host of shot hole borer in Maharashtra (Mote *et al.*, 1991).

Identification

Eggs are shiny oval or round, white in colour. Larvaeare white and apodous and can be up to 4 mm in length. Adultis about 2-3 mm long, reddish-brown to black and cigar shaped. The adultsare short, incapable of flight, with stubbed head capsule and chewing mouthparts.

Biology

The egg emergence period is nearly 6.7 ± 0.3 days, the larval stage (three instars) completed with an average of 10.6 ± 0.5 days and pupal emergence took an average of 7.3 ± 0.33 days. Adults lived longer with an average of 18.10 ± 0.53 days. Distinct sexual dimorphism is observed among adult beetles. Usually, females were more in number than males with a sex ratio of 13: 1 (female: male) (Kamala Jayanthi *et al.*, 2020). Two generations are completed per year.

Mode of feeding and symptoms of infestation

Females generally bore through the basal part of the stem and roots. Initial infestation in an orchard begins with a mild yellowing of lateral branch on one or more trees, generally in a contiguous patch. Within a week the whole tree becomes yellow followed by drying of branches. Some infested trees have shown heavy bearing but reduced size and immature ripening. On careful examination, the main trunk just about a foot above the soil shows small pinholes, which may or may not be seen with frass coming out of it. However, if the infestation is of shot hole borer, subterranean (below soil) holes in the root region are a symptom. If the pest is endemic in that area, care should be taken during new infestations and as well as during replanting as per guidelines.

Management

Early diagnosis with symptoms is a must. Therefore, regular visits to the orchards by growers are suggested.

i. Cultural

a. Keeping proper distance between plants so as to avoid overcrowding of trees.

b. Avoiding water logging and keeping the soil raked and aerated.

c. Uprooting the infested trees and burning especially to get rid of the infested root.

ii. Mechanical

a. Affected buds and fruits of the plant should be collected and destroyed.

b. Detection of early infestation by periodically looking for drying branches.

c. Pruning of the affected plant parts and destruction thereof.

iii. Biological

a. *Trichogramma, Tetrastichus* spp., *Telenomus* spp., *Chelonus blackburni, Carcelia* spp. *Campoletis chlorideae, Bracon* spp. parasitise the pest.

b. *Chrysoperla* sp. rove beetles, spiders, parasitic wasp, coccinellids, robber fly, dragonfly, Beduviid bug (*Geocoris*sp.), praying mantis, fire ants, pentatomid bug (*Eocanthecona fucellata*), earwigs, ground beetles, common mynah and king crow etc play important role by predation on the pest.

iv. Chemical

Signs of lateral branch yellowing to quick drying of full tree, should be immediately treated as under.

a. Drenching the soil with Chlorpyriphos 20 EC @5 ml per litre of water around all uninfested trees prophylactically once in 6 months, followed by Azadirachtin 6 ml per litre. Where the pest is a serious threat to pomegranate cultivation, repeat the above drenching after a month.

b. If infestation is low, drench with Azadirachtin (0.15%) @ 3 ml per litre around the main trunk @2-3 litres of mixture per tree.

c. Jagginavar and Naik (2005) reported that, Chlorpyriphos + Carbendazim during *Ambebahar*, and drenching with Chlorpyriphos + Carbendazirn, Imidachlorprid+ Carbendazim during *mrigbahar* season emerged as best treatments by recording higher reduction in percent shotholes.

d. Spray Chlorpyriphos 20 EC @ 2.0 ml or Indoxacarb 14.5%SC @ 1.0 ml or Spinosad 45 %SC @0.5 ml or Cypermethrin 25%EC @ 1.0 ml per liter of water.

6. Pomegranate stem borer, *Coelosterna spinator* Fab. (Cerambycidae: Coleoptera)

The pest once considered as aminor pest of pomegranate. But gradually it is assuming major status and is becoming major problem in combination with shot-hole borer.

Distribution

India, Spain, China, Mediterranean countries, Iran, Iraq, the Arabian Peninsula and Ukraine.

Hosts of commercial importance

It is polyphagous pest and in addition to pomegranate it has also been reported to cause damage to *Mangifera indica* and *Emblica officinalisi* etc.

Identication and biology

Adult beetles are pale yellowish-brownin colour with light grey elytra and are about 30-35 mm long.

Eggs are laid on stems under the bark, normally in between 20-40 in number. Egg period lasts for 12 to 15 days. Upon hatching the larva feeds on the soft tissues around the oviposition cavity and bores into the stem and roots. The length of the larval period is about nine or ten months. Grub stage last for 9 to 10 months. Pupal period is about 16 to 18 days. The beetle emerges by making a circular hole in the bark. Only one generation is completed per year and adults live for 45 to 60 days.

Mode of feeding and symptoms of infestation

The grubs bore into the cambium then girdle the stem or branch causing death of the tree. The caterpillar bore inside the bark and feeds on the tissue materials. Several holes can be seen on the infested trunk. Fecal mater hanging with wood dust in web form from trunk is an indication of the borer activity. It prefers breeding in dead wood but also attacks the living branches.

Adult insects also scrape the growing shoots leading to drying. Adult beetles remain active by the day time and feed by gnawing the green barks. The stem borer infestation varies from 5.00 to 37.00%.

Management

i. Cultural and mechanical

a. Hand picking and killing of adults in kerosene water during July-September.

b. The grubs can be killed by inserting a hooked wire down the hole.

c. Pruning of dead branches during pre-monsoon and before taking *bahar.*

d. Avoid heaping dead or uprooted wood within or near orchards.

e. Attract and kill by installing the light traps.

ii. Biological

Tachinid flies, spiders, braconid wasp, big eyed bugs (*Geocoris*sp) etc. have been found as natural enemies of the pest.

iii. Chemical

a. Application of paste of mixture containing 10 liters of water + red soil (4kg) + Chlorpyriphos 20 EC (20ml) + Copper oxychloride (25 g) with the help of brush up to 1-2 feet above from second year crop onwards is effective.

b. Jagginavar *et al.* (2008) reported that grape stem borer is effectively controlled by Dichlorvos 76EC (80 ml/l) stem injection which recorded hundred percent reduction of live tunnels.

c. This pest can be managed by injecting 8% Dichlorvos 76 SC (80 ml/L) with a squeeze bottle till the tunnel is filled and closed with wet mud (Krishna Naik *et al.*, 2011).

7. Ash whitefly, *Siphorinus phillyreae* Haliday (Aleyrodidae: Hemiptera)

Distribution

India, Spain, China, Mediterranean countries, Iran, Iraq, the Arabian Peninsula and Ukraine.

Hosts of commercial importance

It is a cosmopolitan and polyphagous insect species reported on more than 50 botanical species in various families (Nguyen and Hamon, 2011). Important hosts are pomegranate (*Punica granatum* Linn.) and *Citrus* spp.

Identification

The ash whitefly is smaller in size. A white waxy coverings found on the body and wings of adults. Nymphs are distinguished by the presence of two longitudinal tufts of white wax and by the production of typical droplets of a glassy wax that cover the body.

Biology

Eggs are laid underside of apical leaves often in circles or small groups. Incubation period is about 3-30 days. Nymphs have short wax mealy along the sides of the body. The nymphal stage lasts for 9 to 19 days. Pupae are convex in shape and possess deep yellow patches on the abdomen. The crawlers seetle on leaf tissue for sucking the sap and forms "scales" throughout the remaining part of their life. In about 2-8 days, the pupae change into adult whiteflies. Adults become active during early morning hours.

Mode of feeding and symptoms of infestation

The nymphs and adults of both species suck sap from the tender leaves. Infestation can be detected by foliar yellowing and abundant production of honeydew on which sooty mould develops, covering the plant, including fruits. Heavy infestation on pomegranate can cause leaf wilt, early leafdrop and smaller fruit size. A sudden infestation by these insects is frequently due to improper agricultural practices or to the use of insecticides that eliminate natural enemies or alter the equilibrium of the agro-ecosystem, favouring the uncontrolled development of pest populations.

The damage can be manifested by stunted growth and yellowing of leaves, in severe cases leading to shedding of leaves. Severe incidence of ash whitefly was also recorded first time on pomegranate by Nair (1978) in northern Karnataka. Whiteflies have been observed in serious form on the leaves in pomegranate orchard where there is crowding of trees due to close planting.

Seasonal incidence

Kotikal *et al*. (2011) showed that population density of the whitefly was maximum during the second fortnight of February and the first fortnight of March. Adults and nymphs made early appearance during the month of February resulting in maximum damage during March and April in pomegranate.

Management

i. Cultural

a. Maintenance of field sanitation and removal of alternate host plants from the orchard.

b. Maintaintenance of adequate aeration by proper training and pruning of the plant.

c. Installation of yellow sticky traps @ 4-10 traps/acre. White flies can be trapped by hanging bright yellow sticky traps coated with polybutene adhesive at the height of the crop canopy. Locally available empty tins can be painted yellow and coated with grease/Vaseline/castor oil on outer surface may also be used.

d. Spraying water with high volume sprayer by focussing the nozzle towards the under surface of leaves helps in washing out the honeydew, eggs, larvae, pupae and adult whitefly.

ii. Biological

a. The bioagents like, *Encarsia haitierris*, *E. gaudeloupae, Cryptolaemus montrouzieri* and green lace wing, *Mallada astur* etc. are effective in suppressing population of the pest.

iii. Chemical

Exclusive use of insecticides is never fully effective for the control of whiteflies (Gyeltshen *et al*., 2014). However, some pesticidal recommendations are as follows.

a. Spraying Dimethoate 30 %EC @1.0 ml or Imidacloprid 17.8% SL@ 0.3ml or Thiamethoxam 25% WG @0.25 g per litre of water.

b. Cyantraniliprole 10.26% OD @ 360 ml in 400 l of water / acre may also be applied.

c. Thimmaiah (2002) noticed Dimethoate + Fish Oil Rosin Soap (FORS) to be highly effective against all sages of whiteflies recording 90% nymphal, 80% pupal, 95 per cent adult mortality on guava.

8. Pomegranate thrips, *Rhipiphorothrips cruentatus* Hood. (Thripidae: Thysanoptera)

Other species of thrips recorded on pomegranate are *Retithrips syriacus* (Mayet), *Anaphothrips oligochaetus* Karny and *Scirtothrips dorsalis* Hood. Among these the first one feeds on leaves and the other three infest the flowers.

Distribution

India, Mediterranean countries, Iran, Iraq, the Arabian Peninsula and Ukraine.

Hosts of commercial importance

In addition to pomegranate the pest also found to infest on grapes, rose, *Lagestoemia indica, Punica granatum* etc.

Identification and biology

Eggs are dirty white in colour and bean shaped in appearance. Newly emerged nymphs are reddish in colour and turn yellowish brown as they mature. Adults are soft bodied, slender, minute insects with heavily fringed wings, blackish brown with yellowish wings and measuring about 1.4 mm long.

Female lays an average of 50 white bean-shaped eggs on the under surface of leaves.The incubation period is 3-8 days. Pupal period lasts for about 2-5 days.

Mode of feeding and symptoms of infestation

Both nymphs and adults feed on underside of the leaves by rasping the surface and sucking the oozing cellsap. Leaf tip turn brown and get curled, drying and shedding of flowers and scab on fruits which reduces market value of fruits. Leaf tips turn brown and get curled due to the feeding of thrips on the underside of the leaves by rasping the surface and sucking the oozing cellsap leading to drying and shedding of flowers.

Thrips infestation is often seen on leaves and also on young fruits causing characteristic scab on fruits.

Seasonal incidence

The incidence of this pest is mainly seen from July to October with the peak period in September. If the morning relative humidity increases then the population of sucking pests could be higher during the next two weeks.

Management

i. Cultural

a. Maintenance of adequate aeration in the orchard by proper training and pruning.

b. Intercultivation of alternate host crops like chilly, onion, garlic, brinjal and tomato in pomegranate orchard to be avoided.

c. Removal and destruction of affected plant parts.

ii. Mechanical

a. Setting up of blue sticky traps at 15 cm above the canopy for monitoring thrips (@ 1trap / 10 plants or 4-10 traps/acre). Locally available empty tins can be painted blue and coated with grease/Vaseline/castor oil on outer surface may also be used.

iii. Biological

a. Predators like Syrphid fly, minute pirate bug, praying mantis, predatory thrips, damsel bug, lacewing, coccinellid, spider etc. effective in suppressing the pest population.

iv. Chemical

a. Spraying Imidacloprid 17.8% SL@ 0.3ml/l or Thiamethoxam 25% WG@0.25 g/l of water prior to flowering is important. In case of species other than *R. cruentatus*, Acephate 0.075% may be sprayed. Spraying should be carried out preferably during evening hours.

b. Spraying with Acetamiprid 20 sp @ 0.005% to 0.01% or Spinosad 45 SC @ 0.25 ml/l or NSKE 5% or *Verticillium lecani* (2x108 cfu/g) @ 2g/l starting from prior to flowering at the interval of 10 days is effective in managing the pest.

9. Pomegranate aphid, *Aphis punicae Passerini*. (Aphididae:Homoptera)

The pomegranate aphid is one of the key pests in pomegranate orchards.

Distribution

It is common in the Mediterranean area, Asia, Africa, and the Indian subcontinent (Bhagat, 2012 and Lee *et al*., 2015).

Hosts of commercial importance

This species is common on pomegranate but has also been recorded from *Lawsonia inermis* (Lythraceae), *Duranta plumier*, *Lantana camara* (Verbenaceae), *Bignonia* sp., *Campsis radicans* (Bignoniaceae), and *Plumbago capensis* (Plumbaginaceae) (Blackman and Eastop, 2006).

Identification

Both winged and wingless forms breed parthenogenetically. Body is oval or slightly elongated, with six segmented antennae. The morphology and colour of *A. punicae* may vary with environmental conditions as in many aphid species. In spring, the apterous viviparous females are green, 1.38- 1.73 mm long, with blackish siphunculi and a tongue shaped pale brown to pale green cauda. In summer, the subsequent generations have pale green colour and generally significantly smaller bodies (0.80-1.17 mm), pale brown to pale green siphunculi with greyish brown apexes and pale green or yellowish cauda (Sugimoto, 2011).

Biology

Both winged and wingless forms breed parthenogenetically. The nymphal period lasts for about 7-9 days and the entire life cycle takes about 22-25 days. On an average 12-14 generations are usually completed per year as reported by Blackman and Eastop (2006). High humidity favours the multiplication of aphid.

Mode of feeding and symptoms of infestation

Both adults and nymphs feed on tender leaves, inflorescences and fruits. They suck the cell sap from the lower surface of the leaves and devitalize the plant. Infestation by the aphid results in pale and curled leaves, retarded growth and development of plants. The infestation causes yellowing of leaves and sticky to touch. Besides that it also transmit viral diseases and secret honey dew on which fungi grow.

Seasonal incidence

This pest has a high reproductive potentiality and it is observed from April to June. Kotikal *et al.* (2011) reported that the peak activity of pomegranate aphids was observed during the second fortnight of December.

Management

i. Cultural

a. Collection and destruction of the damaged plant parts along with the pests.

b. Maintenance of adequate aeration by proper training and pruning.

c. Use of yellow sticky traps @ 4-10 traps /acre.

d. Setting up of yellow pan water/sticky traps 15 cm above the canopy for monitoring aphid.

ii. Biological

a. The aphid has many efficient natural enemies belonging to Coleoptera (Coccinellidae), Diptera (Syrphidae and Cecidomyiidae), Neuroptera (Chrysopidae) and Hymenoptera that often can greatly reduce the aphid population. It has been found to preyed by three coccinellids viz., *Scymnus castaneus* Sic, *S. latemaculatus* Motsch, *C. sexmaculata* F. and the syrphid, *Paragus serratus* F. Four parasitoids *Aphidius* sp., *Aphelinus* sp., *Trioxys* sp. and sp. were reared from the aphids. The predators play a significant role in suppressing the pomegranate aphids (Mani and Krishnamoorthy, 1995).

b. Release first instar larva of *Chrysoperla zastrowisillemi* @ 15 / flowering branch (four times) at 10 days interval from flower initiation during April is beneficial.

iii. Chemical

a. Thiamethoxam and Imidacloprid proved significantly superior in controlling the aphids on pomegranate and were found to reduce the incidence of aphid by 85.9 and 83.5%, respectively compared to the control (Misra, 2002 and Ananda *et al.*, 2009).

b. Spraying with Dimethoate 30 %EC@1.0 ml/l or Malathion (0.1%) at 15 days interval also effectively manage the aphid population.

10. Tailed mealy bug, *Ferrisia virgata* Cockerell. (Pseudococcidae: Hemiptera)

About 20 species of mealy bugs and scales have been recorded damaging pomegranate in India (Mani and Krishnamoorthy, 1990). Among the various species of mealy bugs *Ferrisia virgata* Cockrell is predominant in pomegranate orchards in Maharashrta, India. Incidence of mealy bug is now-a-days causing serious concern in many pomegranate orchards in Maharashtra.

Distribution

It is distributed throughout Asia, Central America, Pacific Islands, West Indies and South America.

Hosts of commercial importance

It is polyphagous and has a very wide range of host plants. The most notable hosts of mealy bugs are Pomegranate (*Punica granatum* L.), *Vitis vinifera*, *Mangifera indica* and *Citrus* spp.

Identification

First and second instar nymphs are light yellow with six numbers of antennal segments (Highland, 1956). Third instar nymphs have seven antennal segments. Males begin to differentiate from females in the third instar when the body color turns dark and the body shape begins to resemble the adult form.

Wing buds are present in third instar males. Faint dorsal stripes and small caudal tassels gradually become apparent in females with each molt. Adult males are heavily sclerotized, dark gray in color, winged, and have ten antennal segments. Adult females are greenish-yellow in color with two dorsal, dark gray stripes, and antennae with eight segments (Kaydan and Gullan 2012). Additional identifying features of the mealybug are the two posterior waxy tails or tassels with a length half that of the body.

Biology

The egg period is about 28-32 days. Adult females and males live for about 23 to 28 days, respectively. Oviposition period is 8-9 days. The total life span of female and male 46-49 and 23-29 days respectively. The mode of reproduction is parthenogenetic and amphimictic, and females lay an average of 155 eggs during its life time. Eggs are yellow to pale white in colour. Total nymphal period lasts for 21-29 days. Females are apterous, long, slender covered with white waxy secretions.

Mode of feeding and symptoms of infestation

The nymphs and adults suck cell sap from leaves, twigs, flowers and fruits. The damage is more pronounced on flowers and fruits. The nymphs prefer to settle at the attachment of fruit, inside the calyx (styler or distal end) of fruit and within the cluster of the fruits. In case of severe incidence, the flowers and fruits drop down from the tree. Similarly, the insects form thick encrustation on developed fruits, which hamper the quality of fruits. The mealy bugs secrete honeydew on which the sooty mould (*Capnodium* sp.) develops and affects the photosynthetic activity of plant adversely. Black coating appears on fruits and hence quality gets reduced.

Fruit infestation with the mealy bugs ranged from 25 to 100% with a mean of 56.5% in South India (Karuppuchamy, 1994).

Seasonal incidence

Moist and warm conditions are favourable for growth and multiplication of these pests.

Management

i. Cultural

a. Pasting a grease band of 5 cm width on the main stem prevents the crawlers from reaching the branch.

b. Collection and destruction of the infested plant parts and alternate hosts.

d. Providing light irrigation in the orchard as heavy irrigation enhances their build up.

e. Overlapping and overcrowding branches may also be pruned to check spread of these pests in the orchards.

ii. Biological

a. *Menochilus sexmaculatus, Rodolia fumida, Cryptolaemus montrozieri* etc. are good predators of the mealybug. Activity of predators increased from April to June and then declined from July onwards (Karuppuchamy, 1994). Apart from above listed natural enemies three species of praying mantids were also observed in the pomegranate ecosystem, the specific feeding habits were not ascertained.

b. Parasitic wasps like, *Leptomastidea abnormis* (Girault) is also effective in managing the pest.

iii. Chemical

a. Spraying Chlorpyriphos (0.02%) or Malathion (0. 2%) with fish oil rosin soap has been found to manage this insect population.

b. In case of heavy infestation of *F. virgata*, spraying of 0.045 per cent Dimethoate in early fruiting season or non-fruiting season is suggested (Ghule and Dhumal, 1992).

c. Spraying Chlorpyriphos 20 EC @2.0 ml or Dimethoate 30 %EC@1.0 ml or Imidacloprid 17.8% SL@ 0.3ml or Thiamethoxam 25% WG@0.25 g per litre of water for effective control of mealybug on pomegranate.

11. Pomegranate scale, *Parasaissetia nigra* (Nietner) (Coccidae: Homoptera), *Icerya purchasi* Maskell (Margorodidae: Hemiptera)

Ananda *et al.* (2009) reported two species of scale insects, *Parasaissetia nigra* (Neitn.) (Coccidae: Homoptera) and *Icerya purchasi* Maskell (Margorodidae: Homoptera) on pomegranate in Karnataka, India.

Distribution

India, Spain, China, Mediterranean countries, Iran, Iraq, the Arabian Peninsula and Ukraine.

Hosts of commercial importance

The pestis polyphagous, feeding on hosts from 77 plant families, especially on ornamental plants of tropical origin e.g., *Ficus* and *Hibiscus*. Several agricultural crops are also infested that include avocados, *Annona cherimola*, citrus, coffee, cotton, *Croton tiglium*, guavas, mangoes etc.

Identification and biology

Early stages of *P. nigra* can be difficult to separate from those of several other soft scale species. Immature and newly formed adults of *P. nigra* are translucent yellow and occasionally mottled. Crawlers are about 0.35 mm long with two black eyes placed antero-laterally. The adults are up to 5.5 mm long and 4 mm wide, yellow initially, often becoming shiny, dark brown to purple-black with age. The adult female is elongate-oval, slightly narrowed anteriorly and 3-4 mm long.

Reproduction is entirely by parthenogenesis and oviposition may occur over a long period in cooler climates. Eight hundred or more eggs are laid under the body of the female and are protected there for 1-3 weeks until hatching. On hatching, the first nymphal instars (crawlers) move away from the female to another part of the plant, where they fix themselves and start feeding.

Mode of feeding and symptoms of infestation

Adults and nymphs suck cell sap from the fruit and tender shoots causing drying of branches. In case of severe infestation, the whole tree dries up. The insects secret honey dew like substance which helps in developing black sooty mould. As a result, all the leaves and the branches turn blackish affecting the growth of the plant. This restricts photosynthesis, weakening the plant and sometimes stunting new growth and causing defoliation. The honeydew produced may attract ants that deter natural enemies from attacking the scales.

A heavy infestation of *P. nigra* may devitalize the host directly by sap depletion and by injecting toxins. In India, *P. nigra* damages pomegranates and together with the coccid *Saissetia coffeae* is a pest of *Santalum album* trees, causing severe leaf and fruit fall in successive years until the tree dies.

Seasonal incidence

Peak activity is observed during March-April (Ananda *et al*., 2009) in pomegranate growing areas of India.

Management

i. Cultural

a. Removal and destruction of alternate hosts, which harbour the insects.

ii. Biological

a. Application of *Verticellum lecanii* (2x 10 ^{5} CFU/g) @ 3 g per litre of water is effective.

b. *Sculellisa cyanea* is an effective egg parasitoid of this mealybug

iii. Chemical

a. Use of systemic insecticides for spraying.

b. Malathion at 0.1% has been used to control *P. nigra* chemically on pomegranates in India.

12. Fruit Fly, *Ceratits capitata* (Wied.) (Tephritidae: Diptera)

In addition to *C. capitata*, the fruit fly spp., *B.dorsalis*, *B.correcta* and *B. zonata* are highly polyphagous in nature feeding variety of horticultural crops in India including pomegranate.

Distribution

India, Pakistan, South-East Asia, Malaysia, Indonesia, Formosa, Philippines, Australia, China, Hawaii Islands, China and Taiwan.

Hosts of commercial importance

Mango, guava, peach, apricot, cherry, pear, ber, citrus, banana, papaya, avocado, passion fruit, coffee, melons, jack fruit, strawberry etc.

Identification and biology

The adult fly is brown or drak brown with hyaline wings and yellow legs. Females lay eggs in small crevices in the epicarps of fully mature fruit. The larvae develop while feeding on the pulp and this activity, combined with the subsequent development of rot, render the fruit unmarketable. The egg period is about 22-23 days. The maggot feeds on pulp and become full grown in about 7 days. It pupates 3-7 inches below the soil.

Mode of feeding and symptoms of infestation

The female lays eggs under the rind of the fruits by puncturing. After hatching the maggots feed on the pulp. The affected fruits cease to develop and drop. The maggots destroy and convert the pulp into bad smelling, discoloured semi liquid mass unfit for human consumption. Infestation results in fruit drop and liquid oozes out from the fruit upon pressing.

The intensity of infestation depends on various factors, including the susceptibility of the pomegranate variety, the seasonal climatic trend and the presence of other suitable fruit crops in the cultivated area.

Seasonal incidence

Infestation by this pest become prominent during the rainy season.

Management

i. Cultural

a. Tillage operation of the orchard soil helps in checking the pest population as the pupae and hibernating larvae are destroyed by natural enemies and the scorching sun.

b. The infested and fallen fruits should be carefully disposed off.

ii. Behavioral

a. Since the pest remains inside the fruit chemical control measures are ineffective. Using fly traps containing Methyl Eugenol and an insecticide (attracticide) is effective in monitoring as well as suppressing the pest population.

iii. Biological

a. Conservation of parasitoids such as *Opius compensates, Spalangia philippinensis*, *Diachasmimorpha kraussi* etc. is helpful in checking the pest population.

b. Entomopathogenic nematode species like *Heterorhabditis bacteriophora, Steinernema riobrave, S. carpocapsae* can also be used to manage the pest.

iv. Chemical

a. Application of bait sprays of Malathion 50 EC @ 2 ml/l with molasses or jaggery (10 g/l) before ripening.

b. Low residue Organophosphate insecticides can also be used against the fly on pomegranate.

13. Leaf eating caterpillar or Castor semilooper, *Achea janata* (Linn.) (Noctuidae: Lepidoptera)

Distribution

Itis widespread throughout the tropical and subtropical region including India, Spain, China, Mediterranean countries, Iran, Iraq, the Arabian Peninsula and Ukraine.

Hosts of commercial importance

Many workers have reported that the pest is a polyphagous one and can cause damage on various crops like pomegranate, grapevine, guava, rose, ber, citrus, cocoa, coconut etc. It is also reported to cause considerable damage to ber crop by feeding on the foliage and young plants of grafted ber as well as pomegranate nursery in Rajasthan, India.

Identification and biology

Moth are stout, pale reddish brown with wavy lines on fore wings and black hind wings having a medial white bank and large white spots on the outer margin, anal and apical margins of the wings. The wings are also fringed with hair. A monkey face like appears on the dorsum of thorax is present. The caterpillars are semiloopers with various colours. Some are grey with red or brown lateral stripes while others are bluish grey specked with blue-black and having yellowish lateral stripes. Pupae are glistening dark green, later becoming brown. Eggs are hemispherical bluish green and ridged with 40 to 45 striae.

A female lay on an average 400 eggs singly on tender leaves. Incubation period is about 3 to 5 days. Larval period lasts for 9 to 23 days depending on environmenatal condition. Pupation takes place in the soil, fallen leaves or in folds of leaves. Pupal period lasts for about 7 to 26 days. Pre-oviposition period is 6 to 21 days. Total life cycle is completed within 48 to 50 days. On an average 5 to 6 overlapping generations are completed in a year.

Mode of feeding and symptoms of infestation

Caterpillar feeds voraciously on leaves, tender petioles and young fruits. In severe foliage infestation, complete defoliation of plants takes place leaving only veins of the leaves.

Management

i. Collection and destruction of the caterpillars mechanically in small scale cultivation.

ii. Spraying the crop with Chlorpyriphos 20EC@ 2.0 ml/l as prophylactic spray or on initiation of insect infestation.

iii. Spraying Indoxacarb 14.5%SC @ 1.0 ml or Spinosad 45 %SC @0.5 ml or Cypermethrin 25%EC @ 1.0 ml per liter of water have also found effective in managing the pest.

14. Pomegranate mite, *Aceria granati* Can. & Mass. (Tenuipalpidae: Acarina)

Other mites found to infest the crop are,*Oligonychus punicae* (Hirst.) *Tenuipalpus granati* Sayed. etc.

Distribution

These species are largely distributed in the Mediterranean area and Middle East including India, Spain, China, Mediterranean countries, Iran, Iraq, the Arabian Peninsula and Ukraine.

Hosts of commercial importance

Vitis vinifera, *Pistacea vera*, *Prunus armeniaca*, *Ficus carica* and *Olea europea.*

Mode of feeding and symptoms of infestation

The mites can occasionally reach pest status on pomegranate. The species infest preferably pomegranate leaves that became first yellowish and then dry up. These species cause rarely serious damage to pomegranate, because they are well controlled by numerous natural enemies, mainly predatory mites (Phytoseiidae). The rare infestation of *A. granati* can be easily noticed for the evident leaf roll on shoots. The blemish due to mite infestation occurs anywhere on the fruit.

Seasonal incidence

Ananda *et al.* (2009) reported peak activity of the mite during March- April in South Indian condition.

Management

i. Healthy, well-maintained trees will tolerate higher mite populations than weak or stressed trees.

ii. Several species of predatory insects and mites attack all stages of plant feeding mites to keep populations below damaging levels in most tree fruits.

iii. Chemicals applied for controlling other pests and diseases may upset the ratio of plantfeeding to predatory mites, reducing the effect of biological control.

iv. Application of Sulphur dust as prophylactic measure against mite may be advised. Wettable Sulphur also works.

v. Spraying of Propargite 57%EC @0.5 ml, Abamectin 1.9%EC @1.0 ml or Azadirachtin 1% @2.0 ml per litre of water are also effective.

References

Ananda, N., Kotikal, Y. K. and Balikai, R. A. 2009. Management practices for major sucking pest of pomegranate. I*n: Karnataka J.Agric. Sci.*, 22(4):790-795.

Ayyar, T. V. R. 1944. Notes on some fruit sucking moths of the Deccan. Indian Journal of Entomology, 5 (I & II):29-33.

Balikai, R. A., Kotikal, Y. K. and Prasanna, P. M. 2011. Status of pomegranate pests and their management strategies in India, Acta Horticulturae. 890:489-492.

Bhagat, R. C. 2012. Aphid (Insecta) of agricultural importance. In J.and K.state, India: a check-list and biodiversity. International Journal of Food. Agriculture and Veterinary Sciences 2:116-125.

Bhumannavar, B. S. 2000. Studies on fruit piercing moths (Lepidoptera: Noctuidae) – species composition, biology and natural enemies. Ph. D. Thesis submitted to University of Agricultural Sciences, GKVK, Bangalore, pp-181.

Blackman, R. L. and Eastop, V. F. 2006. Aphids on the world's herbaceous plants and shrubs, volume 1, Hosts lists and key. Museum (UK): John Wiley and Sons, The Natural History, pp-1415.

Bose, T. K. 1985. Fruits of India, Tropical and sub-tropical. Naya Prokash, Calcutta, India. Pp-637.

Butani, D. K. 1979. Insects and fruits. Periodical Expert Book Agency, New Delhi. pp-415.

CBI (Centre for the Promotion of Imports from Developing Countries) 2014. Product Fact Sheet: fresh pomegranates: in the Europe market. http:// www.cbi.cu/market-.information/fresh-fruit-vegetables/pomegranates/Europe.

Chandra, R. and Jadhav, V. T. 2008. Pomegranate (*Punica granatum* L.) biodiversity and conservation. In :Biodiversity and agriculture. Uttar Pradesh State Biodiversity Board, 63-64 pp.

Cherian, M. C. and Sundaram, C. V. 1936. Fruit-sucking moths on tomatoes and their control. Madras Agricultural Journal, 24:360-363.

Chhetry, M., Gupta, R. and Tara, J. S. 2015. Bionomics of *Deudorix isocrates* Fabricius (Lepidoptera: Lycaenidae), a new potential host of sweet orange, *Citrus sinensis* L. Osbeck in Jandk, India. International Journal of Science and Nature, 6:238-241.

Denton, G. R. W., Muniappan, R., Marutani, M., McConnell, J. and Lali, T. S. 1989. Biology and natural enemies of the fruit-piercing moth, *Othreisfullonia* (Lepidoptera: Noctuidae) from Guam. Pp 150-154 in Johnson M. J., Ullman, D. E. and Vargo, A. (eds) ADAP Crop Prot. Conf. Proc., Honolulu, Hawaii, U. S. A.

Ghule, B. D. and Dhumal, V. S. 1992. Chemical control of scale insects on pomegranate.Journal of Maharastra Agric. Univ.,17:322-323.

Gyeltshen, J., Hodges, A. and Hodges, G. S. 2014. Orange spiny whitefly, *Aleurocanthu sspiniferus* Quaintance (Insecta: Hemiptera: Aleyrodidae), University of Florid. Institute of Food and Agricultural Sciences.

Haidari, M., Ali, M., Casscells, S. W., Madjid, M. 2009. Pomegranate (*Punica granatum*) purified polyphenol extract inhibits influenza virus and has a synthetic effect with oseltamivir. Phytomedicine 16(12):1127-1136

Highland, H. A. 1956. The biology of *Ferrisia virgata*, a pest of azaleas. Journal of Economic Entomology 49:276-277.

Insha, Y. and Abdul A. B. 2020. An updated data on the bionomics of pomegranate fruit borer, *Deudorix epijarbas* (Moore, 1858) (Lepidoptera: Lycaenidae), infesting pomegranates in Kashmir. Acta agriculturae Slovenica, 116(1):1477-1487.

Jagginavar, S. B., Sunita, N. D. and Patil, D. R. 2008. Management strategies for grape stem borer *Celosterna scrabrator* Fabr. (Coleoptera: Cerambycidae). Indian Journal of Agricultural Research 42(4):307-309.

Jagginavar, S. B., Naik, L., Krishna 2005. Management of shothole borer, *Xyleborus perforans* (Wollastan) (Coleoptera: Scolytidae) in pomegranate. Indian Journal of Agricultural Research 39(2):133-137.

Jayanthi, P. D. K. and Verghese, A. 2010. Natural parasitization of larvae of fruit piercing moth, *Eudocima (=Othreis) materna*. Insect Environment, 16(2):67.

Jurenka, J. M. T. 2008. Therapeutic Applcations of Pomegranate (*Punica granatum* L).: A Review. Altern. Med. Rev. 13(2):128-144.

Kakar, K. L., Gora, G. S. and Nath, A. 1987. Incidence and control of pomegranate fruit borer, *Virachola Isocrates* (Fab.). Indian J. Agril. Sci. 57(10): 749-752.

Kamala Jayanthi, P. D., Raghava, T., Nagaraja, T. and Sreedevi, K. 2020. In vitro rearing and gallery tunnelling pattern of Island pinhole borer, *Xyleborus perforans* (Wollaston), a scolytid associated with pomegranate wilt complex., Current Science, 118(2,25):194-198.

Karuppuchamy, P. 1994. Studies on the management of pests of pomegranate with special reference to fruit borer, *Virachola Isocrates* Fab., Ph.D thesis , TNAU, Coimbatore, pp-181.

Kaydan, M., Gullan, P. 2012. A taxonomic revision of the mealybug genus *Ferrisia* Fullaway (Hemiptera: Pseudococcidae), with descriptions of eight new species and a new genus. Zootaxa, 3543:1-65.

Kotikal, Y., Ananda, N. and Balikai, R. 2011. Seasonal incidence of major sucking pests of pomegranate and their relation with weather parameters in India. Acta Horticulturae, 890(890):589-596.

Krishna Naik, L., Jagginavar, S. B. and Biradar, A. P. 2011. Beetle enemies of pomegranate and their management. Acta Hort. (ISHS) 890:565-568.

Ksentini, I., Jardak, T. and Zehgal, N. 2011. First record on *Virachola livia* Klug.(Lepidoptera: Lycaenidae) and its effects on different pomegranate varieties in Tunisia, *E*PPO Bulletin 41: 178-182.

Lee, Y., Lee, W., Kim, H. and Lee, S. 2015. A new record of *Aphis punicae* Passerini, 1863 (Hemiptera: Aphididae) from Korea. Journal of Asia-Pacific Entomology, 18: 157–163.

Mani, M. and Krishnamoorthy, A. 1990. Evaluation of the exotic predator, *Cryptolaemus montrouzeri* Muls, (Coccinellidae: Coleoptera) in the suppression of green shield scale, *Chloropulvinaria psidii* (Maskell) (Coccidae:Hemiptera), Entomon, 15:45-48.

Mani, M. and Krishnamoorthy, A. 1995. Natural enemies of *Siphoninusphillyreae*and*Aphis punicae*on pomegranate. Entomon, 20:31-34.

Mathew, G. and Rugmini, P. 1998. Control of bark-eating caterpillar, *Indarbela quadrinotata* in forest plantation of *Paraserianthes falcuturia.* Indian Journal of Environmental Toxicology, 8(1):37-40.

Mir, M. M., Umar, I., Mir, S. A., Rehman, M. U., Rather, G. H., Banday, S. A. 2012. Quality Evaluation of Pomegranate Crop – A review. Int. J. Agric. Biol.14:658-667.

Misra, H. P. 2002. Field evaluation of some newer insecticides against aphids and jassids on okra. Indian J. Ent., 64:80-84.

Mohi-Ud-Din, S., Zaki, F. A., Mir, S. A. and Wani, R. A. 2018. Biology of anar butterfly, *Deudorix epijarbas* infesting pomegranate under field conditions. Indian Journal of Entomology, 80(3):808-811.

Mote, U. N., Tambe, A. B. and Patil, C. S. 1991. Observation on incidence and extent of damage of fruit sucking moths on pomegranate fruits. J. Maharashtra Agric. Univ., 16(3), 438–440.

Nair, M. R. G. K. 1978. Insects and mites of crops in India. Indian Council Agril.Res., New Delhi, pp-40.

Nguyen, R. and Hamon, A. B. 2011. Ash Whitefly, *Siphorinus phillyreae* (Haliday) (insect: Hemiptera: Aleyrodidae: Aleyrodinae). University of Florida Cooperative Extension Service. Gainesville. FL., USA:IFAS. Available from: http://edis.ifas.ufl.edu/in304.

RamkrishnaAyyar, T. V. 1984. Handbook of Economic Entomology for South India. International Books and Periodical Supply Service, Karol Bagh. New Delhi. pp-528.

Sangha, K. S. and Makkar, G. S. 2005. Field evaluation of different insecticides against bark eating caterpillar, *Indarbela quadrinotata (*Walker) on Populus deltoids. Indian Forester, 131(5):94-700.

Satyanarayana, C. and Arunakumara, K. T. 2016. Biology and management of guava bark eating caterpillar (*Indarbela tetraonis* Moore). Agric. Sci. Digest., 36(3):197-201.

Sugimoto, S. 2011. The taxonomic identity of Aphis punicae Hemiptera: Aphididae Shinji. Entomological Sceience, 14:68-74.

Sundarababu, P. and David, B. V. 1973. A note on unprecedental occurrence of fruit-piercing moths on grapevine. South Indian Horticulture, 21: 134-136.

Thimmaiah, A. G. 2002. Use of insecticides in combination with oils for the management of spiralling whitefly *Aleurodicus disperses* Russell on Guava. *M.Sc.(Agri.)* Thesis, Univ. Agric. Sci., Dharwad (India).

Verma, T. D. 1985. Incidence and chemical control of bark eating caterpillar, Indarbela quadrinotata Walker on plum trees, Indian Journal of Agricultural Sciences, 55(2):131-132.

than 1/4th the length of the body present around the margin. Adult males are pink in colour, particularly during the pre-pupal and pupal stages. However, they appear yellow during first and second instars. Adult males are about 1.00 mm in length, elongate oval body which is widest at thoracic region (0.30 mm). Antenna of adult males is ten segmented. A distinct aedeagus, a heavily sclerotized thorax, head and well-developed wings are present.

There remain two important characteristic features that distinguish *P. marginatus* adult females from all other species of *Paracoccus.* The features are, presence of oral-rim tubular ducts dorsally restricted to marginal areas of the body and absence of pores on the hind tibiae.

The female mealybug can easily be distinguished by the presence of eight antennal segments. Ovisac is 3 to 4 times of the body length and develops ventrally beneath female body. Yellow coloured body fluid of comes out when it is pressed.

Papaya mealybugs are most active during warm and dry weather. Females are wingless and move by crawling short distances or by being blown in air. Fecundity is about 100 to 600 on an average. Eggs are greenish yellow in colour. Egg laying is usually continuous over a period of 1-2 weeks. Incubation period is about 10 days. Immediately after hatching the nymphs or crawlers begin to search for suitable feeding sites.

In adult males, stout fleshy setae on the antennae are present. These setae are absent on the legs. Female passes through 3 instars, whereas males have four instars. Males have longer developmental time (25-30 days) than the females (24-26 days). Adult females attract the males with sex pheromones. Under greenhouse conditions reproduction occurs rapidly round the year.

Mode of feeding and symptoms of infestation

Infestation of the mealybug appears in clusters of cotton like masses on the above ground portion of the plants. The insect sucks plant sap by inserting their stylets into the epidermis of tender leaf, fruit as well as stem. During feeding, they also inject a toxic chemical substance within the plant that results chlorosis, stunting, leaf deformation or crinkling, early leaf and fruit drop and even death of plants. When pest population becomes high, the affected plants covered with honeydew secreted by the insects. As a result of which, formation of black sooty mould takes place which interferes with the photosynthetic activities and causes further damage to the crops. Heavy infestation by the pest renders the fruit inedible due to the development of thick white waxy coating.

The pest assumed major status in India during 2009. During that period of time the pest caused severe damage to several commercially important crops and huge losses to farmers in Tamil Nadu, India. It has also recorded that, in the same year over 1,500 hectares standing mulberry crop in Tirupur was destroyed by this pest leading to enormous financial losses to silk worm producers across the district.

Responsible factors for high population buildup

With rapid development, high survival rate and high reproductive potentiality, *P. marginatus* population may reach a high level. Wax layer and waxy fibres over the ovisac and body of mealybug nymphs and adult females protect them from adverse environmental conditions and routine chemical pesticides. Availability of alternate hosts / weeds around fields not cared by the growers favours population development of the pest. Movement of crawlers through air, irrigation water or farm equipment helps in fast spread of the mealybug from infested field to healthy fields. Free movement of infested fruits, vegetables and other material between States.

Association with ants providing the mealybugs protection from parasitoids and predators and aiding in dispersal of the pest. Piecemeal pruning of mulberry crop provides sufficient time for migration and settlement of crawlers from the old infested crop to the pruned crop. In certain crops like tapioca or cotton, stems which often carry mealybug infestation are stocked in the farm for propagation or other purposes. These stocks, near the newly planted crop act as reservoirs of papaya mealybug.

Mode of dispersal

Healthy plants can be infested from mealybug infested plants. The crawlers can crawl from an infested plant to another. Small crawlers also get readily dispersed by wind, rain, irrigation water, birds, ants as well as vehicle etc. The wax that sticks to each ovisac and nymphs also facilitates passive dispersal of the pest. The female mealybug is not active and unable to fly. In fact, human beings greatly facilitate spread or dispersal of the pest. Long-distance movement is aided through transport of infested planting materials, fruits and vegetables from one end of a farm to the other or even across the country. Ants, attracted by the honeydew, have also seen to carry mealybugs from plant to plant.

Association with ants

Ants are get attracted to the sugary secretion *i.e.*, honeydew of the mealybugs. In return, ants help mealybugs in spreading. They also protect mealybugs from predator ladybird beetles, parasites and other natural enemies. Ants also keep the papaya mealybug colony clean from detritus that accumulate in the secreted honeydew. Otherwise it may be harmful to the colony. Species of ant, *Oecophylla smaragdina* has been found attending papaya mealybug, feeding on honeydew on papaya and other plants.

Management

Mealybug management often involves the control of attendant ants that help in proper development of mealybugs colony. In absence of ants, mealybug populations usually remain small and invade slowly in new areas and the field would be free of a serious mealybug infestation.

It is important to detect the species for chalking out successful management programs. Due to the presence of waxy covering over their body conventional plant protection chemicals show limited efficacy. Considering these aspects, the management of mealybug may be chalked out in the following ways:

a. Cultural and mechanical

i. Monitoring and scouting to detect early presence of the mealybug.

ii. Pruning of infested branches and burning them.

iii. Removal and burning of crop residues and weeds/alternate host plants like *Hibiscus*, *Parthenium* etc. in and nearby crop.

iv. Avoiding the movement of planting material from infested areas to other areas and avoiding flood irrigation.

v. Prevention of movement of ants and destruction of already existing ant colonies.

vi. Sanitization of farm equipment before moving to the uninfested crop.

vii. Application of sticky bands or alkathene sheet or a band of insecticide on arms or on main stem to prevent movement of crawlers.

b. Biological management

i. Natural enemies of the papaya mealybug include commercially available mealybug destroyer, Cryptolaemus *montrouzieri*, ladybird beetles, lacewings, hover flies, *Scymnus* sp. and certain hymenopteran and dipteran parasitoids. Conservation of these natural enemies in nature plays important role in reducing the mealybug population.

ii. In nature, lepidopteran predator, *Spalgis epius* (Lycaenidae) is a well known representative of carnivorous butterfly feeding on various species of pseudococcids and coccids. *S. epius*, being the dominant predator, feeds efficiently on the ovisacs, nymphs and adult of papaya mealybug. When high activity of S. epius and other natural enemies is observed, care should be taken to delay spraying operations and measures should be taken to conserve them.

iii. Exotic parasitoids/predators such as *Anagyrus loecki* Noyes and Menazes, *Acerophagous papayae* Noyes and Schauff are very important in Sri Lanka. There is a need to introduce such exotic parasitoids in India to contain the pest without harming the environment.

e. Chemical management

i. Destruction of nearby ant colonies them with drenching of Chlorpyriphos 20 EC @ 2.0 ml/litre of water.

ii. Spot application of insecticide immediately after noticing mealybug on some plants in the crop field.

iii. If the activities of natural enemies are not observed, use of botanical insecticides such as Neem oil (1 to 2%), NSKE (5%), or Fish Oil Rosin Soap (25g/litre of water) should be the first choice.

iv. Application of recommended chemical insecticides as the last resort such as Profenophos 50 EC (2 ml/l), Chlorpyriphos 20 EC (2ml/l), Buprofezin 25 EC (2 ml/l), Dimethoate 30 EC (2 ml/l), Thiomethoxam 25 WG (0.6 g/ l), Imidacloprid 17.8 SL (0.6 ml/l).

v. Drenching soil with Chlorpyriphos around the collar region of the plant help prevent movement of crawlers and activity of ants.

2. Grasshopper or painted grasshopper, Poecilocerus pictus (Pyrgomorphidae: Orthoptera)

Distribution

Akk grasshopper, *Poecilocerus pictus* is one of the common grass hopper species occurring in Pakistan, India, and Afghanistan (Cejchan, 1969 and Sheri, 1976).

Identification and biology

The adults are blue-black or greenish in colour with bluish green strips on head and thorax-Abdomen is yellowish with transverse blue-black bands. Forewings are red or bluish-green with yellow veins, while hind wings are hyaline. Antennae are bluish black with yellow rings.

Male and female mates after emergence, Mating lasts for 5 – 7 hours. Female starts laying egg three to four weeks after mating, for egg lying female thrust its abdomen deep into the soil and deposit 145 to 170 eggs, about 18-20 cm beneath the surface. The eggs are orange coloured, elongated and are laid in spiral fashion. Female covers the egg mass with frothy secretion. During autumn the nymph hatches out from the eggs in about a month and become adult in 55 to 60 days. The eggs laid in September-November hatches into nymph after about four months and become adult after a nymphal period of about 2 1/2 months.

Mode of feeding and sysmptoms of infestation

It is primarily a defoliator of *Calotropis gigantea*, but also causes damage to papaya, ber, guava, melon, peach, citrus, some other plants and vegetables. Both adult and nymphs is voracious feeder. They defoliate the plant. In severe infestation they even eat the bark of the papaya tree.

Seasonal incidence

The infestation starts in April and continues till October. Maximum damage is caused during July-August.

Management

Dusting of plants with 0.1% Malathion is effective if infestation reaches economic threshold level. Some other insecticides having contact and stomach toxicity may also be advised to reduce its population.

3. Whitefly, *Bemisia tabaci* Genn. (Aleyrodidae: Hemiptera)

Distribution

Bemisia tabaci was first recorded over 100 years ago in tobacco crop in Greece (Anonymous, 1989). It is one of the world's top 100 invasive organisms found all over the World. It is currently recognized as a complex of cryptic species with worldwide distribution.

Hosts of commercial importance

The pest is a highly polyphagous one and found to feed on over 900 host plants all around the world.

Identification and biology

Adult

Adult is soft-bodied, very small and moth-like insect. Wings are covered with powdery wax and the body is light yellow in color. The wings are held over the body like a tent when the insect is in rest. Males are slightly smaller than the females. Adult longevity is about 1 to 3 weeks.

Egg

Female lays eggs mostly near the veins on the underside of tender leaves. Fecundity is about 300. Eggs are very small in size (about 0.25 mm), pear-shaped and attached vertically with the leaf surface by pedicel. Freshly laid eggs are white in colour that gradually turns brown during the course of pre-embryonic developmental stage.

Nymph

After hatching, the newly emerged first instar larva (nymph) moves on the leaf surface in search of suitable feeding site. After getting a suitable feeding site, they insert their piercing and sucking mouthparts and begin to suck plant sap from the phloem. Adults emerge from puparium is through a T-shaped slit, leaving behind empty pupal cases or exuviae.

Mode of feeding and symptoms of infestation

The pest is liquid feeder. Both the adults and nymphs suck the plant sap with the help of their piercing and sucking mouthparts. Due to de-sapping, the vigor and vitality of the affected plant get reduced drastically. In severe infestations, the leaves turn yellow and drop off. When the population is high, they secrete large quantities of honeydew over the plant surface. This favours growth of sooty mould fungus and reduces the photosynthetic efficiency of the plants.

The host plants are indirectly damaged by the transmission of more than 50 Gemini viruses (Markham, *et al.*, 1996). Papaya leaf curl disease is transmitted by this whitefly.

Management

i. Water spraying may be useful in dislodging the adults.

ii. A small, hand-held, battery-operated vacuum cleaner has also been recommended for vacuuming adults off leaves. Vacuum in the early morning or other times when it is cool and whiteflies are sluggish. Kill insects by placing the vacuum bag in a plastic bag and freezing it overnight. Contents may be disposed of the next day.

iii. Fumigating with a small petrol socked cotton ball. For biological control follow common practices.

iv. Foliar applications of systemic insecticides in the neonicotinoid class such as Clothianidin, Dinotefuran, Imidacloprid, Thiamethoxam, Chlorantraniliprole and Spinosad can provide sufficient population suppression of whitefly (Shinde *et al.*, 2018 and Avery *et al.*, 2019).

4. Oriental scale, *Aspidiotus destructor* Sign. (Diaspidae: Hemiptera)

Distribution

It appears to be the native of South Asia but has spread round the world. The pest has been recorded worldwide in tropical and subtropical areas including China, Southeast Asia, India, Pakistan, Russia, Brazil, Central America and Caribbean, Africa and North America (Tao, 1999 and Claps *et al.*, 2001).

Hosts of commercial importance

It is a higly polyphagous pest infesting over 60 plant families (Davidson and Miller 1990, Ben-Dov, 2014). Common perennial hosts include tropical fruits and ornamental plants, including banana, coconut, camellia, guava, mango, palm, papaya, ginger, bird of paradise, sugarcane, ficus, apple, avocado, citrus, grape and palms.

Identification and biology

This scale resembles other armored scales in that the body is protected by a waxy cover. Infestations may be noted by the formation of closely packed colonies composed of what resemble miniature fried eggs. Females develop through two nymphal stages, while males have an additional non-feeding pre-pupal stage (altogether four immature stages). Positive identification of the scale requires that adult females be slide mounted and the diagnostic characters observed at high magnification.

Eggs

Freshly laid eggs are smooth, elongate and whitish, becoming pale yellow over time. Eggs are laid under the scale cover around the body of the female. Mean length and width of eggs are 0.22 mm and 0.09 mm, respectively (Salahuddin *et al.*, 2015).

First instar nymphs

Newly hatched nymphs (also called crawlers) are free-moving and recognized by the presence of legs, antennae, and a pair of bristles at the tip of abdomen. Crawlers are light green to yellowish brown, translucent and somewhat oblong, with an average length and width of 0.23 mm and 0.11 mm, respectively (Salahuddin *et al.*, 2015).

Second instar female

Females remain pale yellow, circular and somewhat transparent in the second instar, which lasts for 8-10 days. Female nymphs are 0.6-1.1 mm long (Williams and Watson, 1988).

Male nymphs

Development of male characteristics starts in the middle of the second instar. Male covers become reddish brown and more elliptical in shape, and then transform in stages into the pre-pupal, pupal and adult stages. The second instar lasts for 5-8 days for males (Williams and Watson, 1988). The male pre-pupal and pupal stages are spent under the scale produced by the second instar.

Adult females

Adult female coconut scales have a circular or broadly oval cover that is 1.5-2.0 mm in diameter. The cover is flat and translucent with a subcentral pale exuviae. Bodies of slide-mounted specimens are 0.7-1.2 mm in length with a pyriform and membranous body. Diagnostic characteristics include: three pairs of lobes with no indication of a fourth pair; plates long, flat and fringed; and with sclerotization on the dorsum of the pygidium (Williams and Watson, 1988).

Adult males

Adult male coconut scales are small, two-winged, reddish, gnat-like insects with eyes, antennae, three pairs of legs and long appendages. Adult males do not feed and are short lived.

Adult females lay 28-65 eggs in concentric circles under the scale cover over a period of 11-13 days, and may produce 3 or 4 consecutive batches in their lifetime (Taylor, 1935). Crawlers may emerge while the female has already started laying the next batch of eggs (personal observation). The crawlers move over the leaf surface for 2-48 hours to find a feeding site (Waterhouse and Norris, 1987).

After settling, females remain sessile throughout their development; adult males undergo a pseudo-pupation, develop a pair of wings and can disperse by flying to find mates (Ghauri, 1962). Females release pheromones through their anus to attract males (Moreno, 1972). Sexual reproduction is influenced by food availability, but typically the sex ratio is 1:1 (male: female); although both male and female colonies have been observed (Ahmad and Ghani, 1972).

Crawlers are the primary dispersal stage of coconut scale within and between host trees. Coconut scale development depends upon climatic conditions: at a mean temperature of 30°C and 65% relative humidity, egg to adult development was 35.5 days for females and 28.5 days for males, with 3 or 4 generations from July to December on mango plants in Pakistan (Salahuddin *et al.*, 2015).

Mode of feeding and symptoms of infestation

Both nymphs and adults of the pest cause damage to the crops by sucking plant sap. The insect feeds on plant sap from tender leaves, stems and fruits, causing yellowing, tissue distortion and die back. The scale is a pest of concern on coconut and other perennial crops due to its relatively short life cycle of around 35 days and multiple overlapping generations per year. Coconut scale is known to be dispersed by birds, bats and insects as well as wind.

Management

a. Cultural management

i. Pruning and training of fruit trees and proper disposal of infested leaves, branches and twigs may effectively help control scale insects on nursery plants and trees. Excessive use of plant fertilizers contributes to scale outbreaks.

b. Biological management

i. Over 40 species of parasitoids and predators and several fungal pathogens, are known to infect the scale (Ben-Dov *et al.*, 2014). In the absence of these natural enemies, population explosions of this pest may occur. *Aphytis melinus* DeBach and *Aphytis lingnanensis* Comp. (Hymenoptera: Aphelinidae) are the most common parasitoid species that suppress the scale populations (DeBach 1974, Watson *et al.*, 2015).

ii. *Aphytis melinus* and/or another parasitoid, *Comperiella bifasciata* Howard (Hymenoptera: Encyrtidae) were introduced into Hawaii, California, Argentina and other regions for the control of the pest. In Pakistan, *Pakencyyrtus pakistanensis* Ahmad (Encyrtidae: Hymenoptera) was reported as an important parasitoid in mango (Ahamad and Ghani, 1972).

iii. Coccinellid ladybird beetle predators of the scale include *Chilocorus* spp., *Telsimia nitida* Chapin, *Pseudoscymnus anomalus* Chapin and *Cryptognatha* spp.

c. Chemical management

i. Chemical applications should be done only when parasites are not economically effective. The application of pesticides may kill natural enemies of the scale and result in a resurgence of the pest.

ii. Various insecticides are registered for control of scale insects in ornamental and fruit crops. Crawler stages are generally the most susceptible to insecticides. Contact insecticides, including horticultural oils, become progressively less effective once the scale insects develop their waxy cover.

iii. Insect growth regulators may be effective, provided they are applied when the immature stages are present. Several spray applications at 15-20 day intervals may be necessary for complete management of a heavy infestation.

iv. The toxicity of insecticides to parasitoids and other beneficial insects should be considered before starting a spray program for scale insects. Since the scale is a pest of quarantine importance, phytosanitary treatment with gamma irradiation has been developed as a potential control measure for this scale (Follet, 2006).

5. Green peach aphid, *Myzus persicae* Sulzer (Aphididae: Hemiptera)

Distribution

This aphid has a worldwide distribution. It was first reported on Oahu in 1910 and is now present in various countries of the world. In addition to green peach aphid, *Aphis gossypii* Glover also found to feed on papaya.

Hosts of commercial importance

The aphid has several host plants. The crops include papaya, broccoli, cabbage, carrot, cauliflower, eggplant, green beans, lettuce, sweet potato, tomato etc. It also infests many ornamental crops such as carnation, chrysanthemum, poinsettia, rose etc.

Identification and biology

Nymphs are pale yellowish green in colour with 3 dark lines on the back of the abdomen. However, dark lines are absent in adults. Usually there are 4 nymphal stages in their life cycle. Nymphal period is completed in about 6 to 11 days depending on environmental condition.

Adults are wingless, small to medium sized and vary in colour from green to pale yellow. Winged adults are green with black or dark brown markings on their abdomen. Antennae are $2/3^{rd}$ as long as their body. Adult females give birth to about 50 nymphs during their life.

Parthenocarpic reproduction have been observed in their life cycle. As a consequence of this type of reproduction, populations are composed solely of females and there are no males. Average life span of the adults is about 18 days. Longevity may be affected by prevailing environmental conditions and availability of host plants. Studies in cooler temperatures report the life cycle lasting up to 50 days as revealed by Toba (1964). Overwintering observed during extreme cool temperature.

Mode of feeding and symptoms of infestation

Aphids feed by sucking sap from tender parts of their hosts. Seedlings or transplants of crops may be stunted as a result of infestation by large populations of this pest. Green peach aphids seldom cause economic damage on older crops.

In fruit crops, extensive feeding causes distortion of young leaves and shoots and results in premature dropping of fruits.

Aphids vector a number of plant viruses and thus also indirectly cause damage to the crops. The green peach aphid vectors viral diseases in more than 30 plant families, including beans, sugar beet, sugarcane, citrus and tobacco (Hill, 1983). The aphid transmits both P (PRSV-P) and W (PRSV-W) strains of Papaya Ringspot Virus. PRSV-P infects papaya. PRSV-W does not infect papaya.

Both wingless and winged aphids are able to transmit viral particles. However, more transmissions are takes place with the wingless forms (Toba, 1962).

Seasonal incidence

Higher population is observed during rainy months and lowest during dry weather.

Management

a. Cultural

Removal and destruction of previous crop residues and weed hosts before lanting new crops is helpful in managing aphid population.

b. Biological

In nature, there exist a number of natural enemies of this aphid. Some of the natural enemies are specific to this aphid and others are general to all aphids. In Hawaii the most effective natural enemies are the predatory syrphid maggots, *Allograpta sp.,* lady bird beetles and some parasitic wasps.

c. Chemical

The green peach aphid has developed resistance to certain insecticides. Several systemic insecticides like Imidachloprid 17.8SL, Thiomethoxam 0.3SG are effective in suppressing aphid population.

6. Fruit fly, *Bactrocera dorsalis* (Tephritidae : Diptera)

In addition to *B. dorsalis* the crop is also infested by several other tephritid pests occasionally. They are, *B. cucurbitae, B. tryonis* etc.

Distribution

The tephritid fruit fly found to infest various crops in different countries of the World. It is distributed all throughout India which is a major tropical and sub-tropical fruit producer in the World.

Hosts of commercial importance

The pest was earlier reported to infest more than 150 host plants. After resolution of the dorsalis complex, its host range has now been reduced to 117 species belonging to 76 genera and 37 families. The pest has also been recorded on the weed Solanum indicum Linn. in Eastern India.

Identification and biology

Adult flies are noticeably larger than a house fly. Face marked with a dark spot in each antennal furrow. Body of the fly is about 8.0 mm in length, wings are mostly hyaline and about 7.3 mm length. Prominant yellow and dark brown to black markings on the thorax are found in adult fly. Usually, the abdomen has two horizontal black stripes and a longitudinal median stripe extending from the base of the third segment to the apex of the abdomen. The markings form a T-shaped pattern. However, the pattern varies considerably. Ovipositor of the adult female is slender and sharply pointed.

All abdominal tergites are separate (view from side to see overlapping sclerites); tergite five with a pair of slightly depressed areas (ceromata). Males are with a row of setae (the pecten) on each side of tergite three (White and Elson-Harris, 1992).

Development from egg to adult hot and humid conditions requires about 15-16 days. Pupation takes place in soil. The maggots, after getting maturity emerges from the fruit, drop down on the soil and form a tan to dark brown puparium. About 9-10 days are required for attainment of sexual maturity after adult emergence. The developmental periods usually vary considerably depending upon prevailing environmental condition. Under optimum conditions, a female can lay more than 3,000 eggs during her lifetime. Apparently, ripe fruits are preferred for oviposition by the female. But immature ones may also be infested.

Mode of feeding and symptoms of infestation

The female adult flies puncture the fruits with the help of ovipositor and lay eggs in batches just beneath the fruit surface. A watery fluid oozes out from the puncture, which becomes slightly concave with seepage of fluid and transforms into a brown resinous deposit after coming in contact with the air. Sometimes pseudo-punctures (punctures without eggs) have also been observed on the fruit skin. These punctures also serve as an entrance for various microorganisms.

Within a day or so, the egg hatches and tiny first instar maggot come out. The adult females prefer to lay the eggs in soft tender fruit tissues. After emergence from the egg, the maggots bore into the fruit pulp, feed and make the feeding galleries. The fruit subsequently starts rotting and becomes distorted. Gradually the larvae leave necrotic region and move forward to healthy tissues. Thus, they often introduce various pathogens and hasten decomposition of fruits. The affected fruits fall off prematurely from the plant. These fruits are not fit for human consumption.

There are three larval instars completed within the fruits. The full-grown larvae come out of the fruit by making exit holes for pupation in the soil. After completion of pupal period in soil the adult flies are come out and again starts infestation on the crops.

Management

Non-chemical

a. Cultural

i. The general recommendations include collection and destruction of affected fruits, ploughung and raking the area under and between the trees during summer and early harvest of mature fruits. Infested fruit should be buried 3 feet under soil surface.

ii. Bagging of fruits that prevents oviposition by the adult females. The use of protective covering is effective but costly. In spite of its cost, protective coverings are still used to a certain extent largely by home gardeners.

iii. In an experiment conducted in guava orchard, soil raking once a day, once in three days and weekly intervals resulted in 80, 70 and 43% pupal mortality respectively.

b. Mechanical and behavioural

The common mechanical method of control is use of baited traps. Baited traps are used to destroy adults. Shrubs within 100 yards of larval hosts may be used advantageously in placing traps. Methyl eugenol traps are also very effective in suppressing the fruit fly population.

c. Biological

Biological control of some fruit fly species has been tried but it has been noted that the introduced parasitoids have had little impact in managing the pest. Though many species of parasitoids have been reported, their utilisation in management of fruit flies has not yet been attempted extensively in India. Impact of some entomopathogenic nematodes (EPN) have also been tried in different countries.

d. Chemical

i. **Cover spray:** It is nothing but the conventional insecticide application on crop vegetation. Several conventional insecticidal chemicals are usually recommended for this purpose.

ii. **Bait spray:** It works on the principle that, both male and female tephritid flies are strongly attracted to protein source from which ammonia comes out. Bait sprays have the advantage over conventional cover sprays that, they can be applied as a spot treatment. The flies are attracted to the insecticide-bait mixture and get killed. Hence, there remain minimal impact on natural enemies and residual toxicity in the produce.

    Molasses, protein hydrolysate etc. may be utilised to make the spray fluid as bait. Baits encourage the adults (especially females) to feed on the spray fluid and can provide good kill of the pest. However, both cover and bait sprays are used for their management.

iii. **Soil application of insecticides:** The mature maggots come in contact with soil for pupation and hence soil application of toxic chemicals have also found effective in minimizing fruit fly population. Dichrotophos (@ 600 gm a.i. /ha) and Trichlorphon (@ 1920 gm a.i. /ha) may be recommended for this purpose.

7. Grey weevil, *Myllocerus viridians* Faust. (Curculionidae: Coleoptera)

Distribution

The pest is native to Sri Lanka and India.

Hosts of commercial importance

The pest is a polyphagous one and found to feed on Acacia, papaya, drumstick, citrus, jute, mango, hibiscus etc.

Identification and biology

This species is a medium sized insect. Body is black and sometimes uniform light green in colour. Colour varying to pale greenish white with chalky-white efflorescence. Head with yellow and with metallic green at the apex of the rostrum. Head narrowed from back to front. Rostrum evidently longer than the head, mandibles reddish brown. Antennae black or piceous. Prothorax sub-conical. Elytral striae are very narrow and covered with fine longitudinal punctures.

Mode of feeding and symptoms of infestation

Adult weevils have been observed in numerous plants as they are known to defoliate the tender leaves and shoots extensively. Grubs feed on roots resulting in wilting of plants. Hence, it is considered as serious polyphagous pest of economic importance.

Management

In case of heavy infestation by the pest any insecticide having contact and stomach poisonous toxicity may be applied in reducing pest population.

8. Yellow peach moth, *Conogethes punctiferalis* Guen. (Crambidae: Lepidoptera)

It's listed as a minor pest of papaya in India.

Distribution

This species is found in Southern and Eastern Asia, Australia, Indonesia, and New Guinea. The species is also present in Hawaii (Nishida, 2002).

Hosts of commercial importance

The species is highly polyphagous and has been recorded on about 65 different host plants from 30 different families. Many of the hosts are economically important (Devasahayam and Abdulla Koya, 2005).

Identification and biology

Adults are medium sized, pale straw yellow with numerous small black spots on the wings. Wing span is about 18 to 24 mm (Chong *et al.*, 1991). Eggs are round and light yellow in color. Incubation period is about 6–7 days. The larvae have a black head and a pale greenish body with a pinkish suffusion dorsally. Coloration of the larvae may vary depending on type of the food. Full grown larvae are 16 to 26 mm in length and are stout, pale or reddish-brown with numerous flattened horny warts that have short bristly hairs. Pupation usually occurs within the larval tunnels in a silken cocoon, surrounded by shelters of webbing and frasses.

Mode of feeding and symptoms of infestation

Larva of the pest is the damaging stage. It bores into stems, shoots, buds, fruits, and seeds of many plants. Boring by this species can predispose the fruits to secondary pathogens (Chong *et al.*, 1991). There are a limited number of reports describing the amount and type of damage caused by this species and most reports are limited to a specific host crop (Korycinska, 2012).

Management

Spraying Malathion 0.1% at 30 day intervals during July to October is effective in controlling the pest infestation. The spraying has to be initiated when the first symptom of pest infestation is seen on the plant.

Other minor pests of papaya are fruit spotting bug, fruit piercing moth, false spider mite, two spotted mite etc. In different agro-climatic regions those pests sometimes cause minor damage to papaya.

References

Agarwal, M. L., Sharma, D. D. and Rahman, O. 1987. Melon Fruit-Fly and Its Control. Indian Horticulture, 32(3):10-11.

Ahmad, R., Ghani, M. A. 1972. Studies on *Aspidiotus destructor* Sign. (Hemiptera: Diaspididae) and its parasites, *Aphytus melinus* Debach (Hymenoptera: Aphelinidae) and *Pakencyyrtus pakistanensis* Ahmad (Hymenoptera: Encyrtidae) in Pakistan. Commonwealth Institute of Biological Control Technical Bulletin, 15:51-57.

Anonymous, 1989. "Management of whitefly, *Bemisia tabaci* Genn. on cotton". Andhra Pradesh Agriculture University, Rajendranagar, Hyderabad.

Avery, P. B., Kumar, V.; Skvarch, E. A.; Mannion, C. M.; Powell, C. A.; Mckenzie, C. L.; Osborne, L. S. 2019. An ecological assessment of *Isaria fumosorosea* applications compared to a neonicotinoid treatment for regulating invasive ficus whitefly. J. Fungi, 5:36.

Ben-Dov Y., Miller, D. R., Gibson, G. A. P. 2014. Scale Net *Aspidiotus destructor* Signoret. (12 February 2014)

Carvalho, F.A. and Renner, S.S., 2014. The phylogeny of the Caricaceae. In *Genetics and genomics of papaya* . Springer, New York, New York. pp-81-92.

Cejchan, A. 1990. Beitraze Zuo Kenntinis der Fauna Afghanistan Acridoidaem orth. Cas. Morav. Mus. Brane. Sci. Net. 54:229-276.

Chong, K. K., Ooi, P. A. C. and Tuck, H. C. 1991. Crop Pests and Their Management in Malaysia. Tropical Press Sdn. Bhd, Malaysia, pp-242.

Claps, L. E., Wolff, V. R. S., González, R. H. 2001. Catálogo de las Diaspididae (Hemiptera: Coccoidea) exóticas de la Argentina, Brasil y Chile. Revista de la Socieded Entomológica Argentina 60:9-34.

Davidson, J. A., Miller, D. R. 1990. Ornamental plants. In Rosen, D (ed.) Armored scale insects, their biology, natural enemies and control. Elsevier, Amsterdam, the Netherlands, 4 (B): pp-603-632.

DeBach, P. 1974. Coconut scale in Fiji. In Biological Control by Natural Enemies. DeBach P (ed.). Cambridge Univ. Press, Cambridge UK, pp-133-135.

Devasahayam, S. and Abdulla Koya, K. M. 2005. Insect Pests of Ginger. In P. N. Ravindran & K. N. Babu (Eds.), Ginger: The Genus. Zingiber, pp-367-390.

Follet, P. A. 2006. Irradiation as a phytosanitary treatment for *Aspidiotus destructor* (Homoptera: Diaspididae). Journal of Economic Entomology 99:1138-1142.

Ghauri, M. 1962. The morphology and taxonomy of male scale insects (Hemiptera: Coccoidea). British Museum (Natural History). Adlard and Son, Dorking, UK. Pp-221.

Heppner, J. B. 1989. Larvae of Fruit Flies. V. *Dacus cucurbitae* (Melon Fly) (Diptera: Tephritidae). Fla. Dept. Agric. & Consumer Services, Division of Plant Industry. Entomology Circular No. 315. 2 pages.

Heu, R. A. 2002. Distribution and host records of agricultural pests and other organisms in Hawaii. USA: State of Hawaii Department of Agriculture.

Hill, D. S. 1983. *Dacus cucurbitae* Coq. pp. 391. In Agricultural Insect Pests of the Tropics and Their Control, 2nd Edition. Cambridge University Press. 746 pages.

Hill, D. S. 1983. *Myzus persicae* (Sulz.). pp. 202. In Agricultural Insect Pests of the Tropics and Their Control, 2nd Edition. Cambridge University Press. 746 pages.

Korycinska, A. 2012. Rapid assessment of the need for a detailed Pest Risk Analysis for *Conogethes punctiferalis* (Guenée). The Food and Environment Research Agency, pp-7.

Kumari, D. A., Anitha, V., Lakshmi, B. K. M. 2014. Evaluation of insecticides for the management of scale insect in mango (*Mangifera indica*). International Journal of Plant Protection 7:64-66.

Lall, B. S. 1975. Studies on the Biology and Control of Fruit Fly, *Dacus cucurbitae* COQ. Pesticides. 9(10):31-36.

Liquido, N. J., Cunningham, R. T. and Couey, H. M. 1989. Infestation Rate of Papaya by Fruit Flies (Diptera: Tephritidae) in Relation to the Degree of Fruit Ripeness. J. Econ. Ent. 82(10):213-219.

Lockwood, S. 1957. Melon Fly, *Dacus cucurbitae*. Loose-Leaf Manual of Insect Control. California Department of Agriculture.

Markham, P. G., *et al*. 1996. "The transmission of geminiviruses by biotypes of *Bemisia tabaci* (Genn.)". In: Gerling D, Mayrer RT (Eds.), *Bemisia*: Taxonomy, Biology, Damage, Control and Management, Intercept, Andover (1996):69-75.

Marsden, D. A. 1979. Insect Pest Series, No. 9. Melon Fly, Oriental Fruit Fly, Mediterranean Fruit Fly. University of Hawaii, Cooperative Extension Service, College of Tropical Agriculture & Human Resources.

Metcalf, C. L., and Flint, W. P. 1962. Destructive and Useful Insects Their Habits and Control, Fourth Edition. Revised by: R. L. Metcalf. McGraw-Hill Book Company, Inc. New York, San Francisco, Toronto, London. pp-754.

Ming, R., Hou, S., Feng, Y., Yu, Q., Dionne-Laporte, A., Saw, J. H., Senin, P., Wang, W., Ly, B. V., Lewis, K. L. and Salzberg, S. L., 2008. The draft genome of the transgenic tropical fruit tree papaya (Carica papaya Linnaeus). *Nature*, 452(7190):991-996.

Moreno, D. S. 1972. Location of the site of production of the sex pheromone in the yellow scale and the California red scale. Annals of the Entomological Society of America. 65:1283-1286.

Nakahara, S. 1982. Checklist of the Armored Scales (Homoptera: Diapididae) of the Conterminous United States. Washington, USA: USDA, Animal and Plant Health Inspection Service, Plant Protection and Quarantine, pp-110.

Namba, R. and Higa, S. Y. 1981. Papaya Mosaic Transmission as Affected by the Duration of the Acquisition Probe of the Green Peach Aphid - *Myzus persicae* (Sulzer). Proc. Hawaiian Entomol. Soc. 23(3):431-433.

Narayanan, K., Jayaraj, S. and Subramaniam, T. R. 1975. Control of the Diamond Back-Moth, *Plutella xylostella* Linn. and the Green Peach Aphid, *Myzus persicae* Sulzer with Insecticides and *Bacillus thuringiensis* var. *thuringiensis* Berliner. Madras Agric. J. 62(8):498-503.

Nishida, G. M. 2002. Hawaiian Terrestrial Arthropod Checklist. Fourth Edition, pp-313.

Nishida, T and Bess, H. A. 1957. Studies on the Ecology and Control of the Melon Fly *Dacus* (Strumeta) *cucurbitae* Coquillett (Diptera: Tephritidae). Hawaii Agric. Exp. Station Tech. Bull. No. 34. pp- 2-44.

Nishida, T. and Haramoto, F. 1953. Immunity of *Dacus cucurbitae* to Attack by Certain Parasites of *Dacus dorsalis*. J. Econ. Ent. 46(1):61-64.

Purcifull, D. E., Hiebert and J. Edwardson. 1984. Watermelon Mosaic Virus 2. CMI/AAB Descriptions of Plant Viruses No. 293 (No. 63 revised).

Purcifull, D., Edwardson, J., Hiebert, E., Gonsalves, D. 1984. Papaya Ringspot Virus. CMI/AAB Descriptions of Plant Viruses, No. 292 (No. 84 revised).

Salahuddin, Rahman, H. U., Khan, I., Daud, M. K. 2015. Biology of coconut scale, *Aspidiotus destructor* Signoret (Hemiptera: Diaspididae), on mango plants (*Mangifera* sp.) under laboratory and greenhouse conditions. Pakistan Journal of Zoology.

Sheri, A. M. 1976 Reproduction of a common grass hopper *Poecilocerus pictus* Fabricious. Pak. J. Agric. Sci., 13:37-40.

Shinde, B. D., Mokal, A. J.; Narangalkar, A.; Naik, K. V. 2018. Chemical management of whiteflies infesting chili. *I*nt. J. Chem. Stud., 6:2813-2816.

Tanwar, R. K., Jeyakumar, P. and Monga, D. 2007. Mealybugs and their management, Technical Bulletin 19, August, 2007, National Centre for Integrated Pest Management, New Delhi.

Tao, C. 1999. List of Coccoidea (Homoptera) of China. Taichung, Taiwan: Taiwan Agricultural Research Institute, Wufeng, 1-176.

Thangamalar, A., Subramanian, S. and Mahalingam, C. A. 2010. Bionomics of papaya mealybug, Paracoccus marginatus and its predator Spalgius epius in mulberry ecosystem, Karnataka J. Agric. Sci., 23(1):39-41.

Timberlake, P. H. 1918. Notes on Some of the Immigrant Parasitic Hymenoptera of the Hawaiian Islands. Proc. Hawaiian Entomol. Soc. 3(5):399-407.

Toba, H. H. 1962. Studies on the Host Range of Watermelon Mosaic Virus in Hawaii. Plant Dis. 46:409-410.

Toba, H. H. 1963. Vector-Virus Relationships of Watermelon Mosaic Virus and the Green Peach Aphid, *Myzus persicae.* J. Econ. Ent., 56:200-205.

Toba, H. H. 1964. Life-History Studies of *Myzus persicae* in Hawaii. J. Econ. Entomol. 57(2):290-291.

UK CAB International, 1966. *Aspidiotus destructor.* Distribution maps of plant pests, June. Wallingford, UK: CAB International, Map 218. (10 February 2015)

Van Emden, H. F, Eastop, V. F., Hughes, R. D. and Way, M. J. 1969. The Ecology of *Myzus persicae*. Ann. Rev. Ento. 14:197-270.

Vargas, R. I. and Carey, J. R. 1990. Comparative Survival and Demographic Statistics for Wild Oriental fruit Fly, Mediterranean Fruit Fly, and Melon Fly (Diptera: Tephritidae) on Papaya. J. Econ. Ent., 83(4):1344-1349.

Waterhouse, D. F., Norris, K. R. 1987. *Aspidiotus destructor* Signoret in Biological Control Pacific prospects. Inkata Press, Melbourne. Chapter 8:454.

Watson, G. W., Adalla, C. B., Shepard, B. M., Carner, G. R. 2015. *Aspidiotus rigidus* Reyne (Hemiptera: Diaspididae): a devastating pest of coconut in the Philippines. Agricultural and Forest Entomology 17:1-8.

White, I. M. and Elson-Harris, M. M. 1992. Fruit flies of economic significance: their identification and bionomics. C.A.B. International, United Kindom.

Williams, D. J., Watson, G. W. 1988. The scale insects of the tropical South Pacific region. Part 1, the armored scales (Diaspididae). CAB International, Wallingford, UK. pp-290.

Zimmerman, E. C. 1948. *Myzus persicae* (Sulzer). pp. 116-118. In: Insects of Hawaii, A Manual of the Insects of the Hawaiian Islands, including Enumeration of the Species and Notes on their Origin, Distribution, Hosts, Parasites, etc. Volume 5, Homoptera: Aphididae. University of Hawaii Press, Honolulu.

10

Pests of Sapota and Their Sustainable Management

Jaydeep Halder and Atanu Seni

Sapota, *Manilkara zapota* (Linn.) (Family: Sapotaceae), commonly known as sapodilla or chiku, is a perennial and evergreen fruit tree. It is native to Southern Mexico, Central America as well as Caribbean. This fruit was introduced during 1888 in Thane district of Maharashtra in India (Cheema *et al.*, 1954). Now, it is an important tropical fruit grown almost throughout the country except in hilly areas. The fruit is a rich source of sugar, minerals, vitamins and dietary fibers. Presently the India is considered to be the largest producer of Sapota in the world (Nandre and Shukla, 2014).

Sapota trees are ravaged by a number of insect and mite pests throughout the year (Sandhu *et al.*, 1974 and Bhutani, 1979). To manage these pests Indian farmers are commonly using many synthetic insecticides which may lead to problems like resistance to insecticides, resurgence of target insects and secondary pest outbreak in addition to these residues to food and beverages, contamination of ground water, adverse effect on human health and wide spread killing of non-target organisms (Halder *et al.*, 2013). It is therefore need of the hour to adopt integrated approaches to suppress the pest population through proper and judicious use of these chemicals. This chapter therefore critically reviews the work done on different components of IPM for the ecofriendly management of major insect pests of sapota in India and abroad.

The major pests infesting sapota crop at various growth stages in India includes sapota leaf webber (*Nephopteryx eugraphella* Ragonot), Parijatha hairy caterpillar (*Metanastria hyrtaca* (Cramer)), Green scale (*Coccus viridis* (Green.)), Fruit fly (*Bactrocera* (*Dacus*) *dorsalis* Hendel.), Bud borer (*Anarsia achrasella* Bradley.), Sapota Seed Borer (*Trymalitis margarias* Meyrick.), Mealy bug (*Planococcus citri* Risso., *P. lilacinus* (Cockerell.), *Ferrisia virgata* Cockerell), sapota aphid (*Toxoptera aurantii* (B.)) and blossom thrips (*Franklinieella dampfi* Priesner.) the details of which have been enumerated below.

Knowledge on biology of any insect pest is imperative as it will help us to identify the weakest link of its life-cycle and accordingly suitable management practices can be adopted. Information on host-range of the pest helps to develop ideal cropping sequence as cultural measures whereas identifies the damage symptoms and proper diagnostics of insect prerequisite for any IPM programme. Therefore, timely and suitable measures should be initiated to avoid economic damage. So, the details distribution, host-range, life-cycle, identification of the pest and damage symptoms are briefly enumerated here under.

Table 1: List of important pests of sapota.

Sl. No.	Common name	Scientific name	Family	Order
1.	Sapota leaf webber or chiku moth	*Nephopteryx eugraphella* Rag.	Pyralidae	Lepidoptera
2.	Hairy caterpillar	*Metanastria hyrtaca* Cram.	Lasiocampidae	Lepidoptera
3.	Sapota Seed Borer	*Trymalitis margaritas* Meyrick.	Tortricidae	Lepidoptera
4.	Bud borer	*Anarsia achrasella* Lat.	Gelechidae	Lepidoptera
5.	Leaf miner	*Acrocercops syngramma* (Meyrick)	Gracillaridae	Lepidoptera
6.	Aphid	*Toxoptera aurantii* (Boy. De. F.)	Aphididae	Hemiptera
7.	Mealy bug	*Planococcus citri, P. lilacinus, Ferrisia virgata*	Pseudococcidae	Hemiptera
8.	Green scale	*Coccus viridis* Green.	Coccidae	Hemiptera
9.	Fruit fly	*Bactrocera dorsalis* (Hendel) *B. zonata* (Saunders)	Tephritidae	Diptera
10.	Sapota fruit mite	*Tuckerella kumaoensis* Gupta	Tuckerellidae	Acarina

1. Sapota leaf webber or chiku moth, *Nephopteryx eugraphella* Ragonot (Pyralidae: Lepidoptera)

Distribution

It is a regular and key pest of sapota throughout India (Sandhu *et al.*, 1974). The pest was first recorded on sapota in the year 1919 at Pusa (Fletcher, 1920).

Hosts of commercial importance

Sapota is considered as a major host plant of this pest. It has also reported to infest on cured tobacco.

Identification and biology

Freshly laid eggs are soft, pale yellow in colour but semi-transparent in nature. The fertile eggs turn pink in colour within 24 hrs of laying. Neonate larvae are initially light pink in colour, about 1.35 mm in length and later turn into reddish brown. Three dorso-lateral stripes are visible on either side. Pupae are obtect and pupation takes place in plant debris or deep inside the soil. Moths are grayish brown with setaceous antennae. Forewings are brownish with four black transverse wavy lines whereas hind wings are semi-hyaline and membranous white. The wings are provided with fringes at the outer margins.

Gravid females lay naked eggs singly or in small batches of 3-7, mostly along the mid-rib of the underside of young leaf, leaf-petioles or sometimes on tender branches. Up to 374 eggs were laid under controlled conditions (Nair, 1995). The egg, larval, pupal periods and total life-cycle completed in 2-5, 13-26, 8-13 and 26-38 days, respectively, during summer in North India (Gupta and Gangrade, 1964). There are 7-9 generations per year. Recently Shukla and Patel (2011) studied biology of this pest at Fruit Research Station, Navsari Agricultural University, Gujarat, India. They reported that pre-mating, mating, pre-oviposition, oviposition and post-oviposition period on an average, were1.26 ± 0.45, 0.44 ± 0.10, 3.1 ± 0.73, 3.0 ± 0.47 and 0.95 ± 0.37 days, respectively. The authors also recorded longevity of male and female was 5.43 ± 0.63 and 7.83 ± 0.44 days respectively. Fecundity was observed as 72 ± 24.15. The larval stage lasts for 15.76 ± 0.84 days. Pre-pupal and pupal stages and total life cycle of the insect was recorded as 1.13 ± 0.44, 6.93 ± 1.16 and 44.46 ± 2.82 days respectively.

Mode of feeding and symptoms of infestation

Damaging stage of this pest is caterpillar. The caterpillar webs the leaves and feed by scrapping chlorophyllous content from within. They also feed on growing shoots, flower buds, flowers and fruits thorough out the year. The larvae skeletonize the tender terminal leaves that are usually joined together by silken threads produced by them. Due to their feeding photosynthetic activity of the plants drastically reduced. They remain hiding in between the leaves and under lose web of excreta. Leaves of the affected plants dries up in clusters and show dry, webby appearance visible from a far distance. Later on, the larvae devour buds, flowers and also bore into the fruits. Thus, reducing the yield considerably. Damage to flowers and/or buds varied from 1.0 to 6.6%.

Seasonal incidence

The pest has been found active throughout the year with the peak in May-June on flower/bud and in the month of February on leaves (Shukla and Patel, 2011).

2. Parijatha hairy caterpillar, *Metanastria hyrtaca* Cram. (Lasiocampidae: Lepidoptera)

Distribution

It is commonly available throughout India.

Hosts of commercial importance

The pest is a polyphagous one and found to infest a number of host plants *viz.*, cashew, drumstick, sapota, guava, jamun, saal (*Shorea robusta*), seuli or Parijat (*Nyctanthes arbor-tristis*), *Madhuca longifolia*, etc. Apart from these, it also reported to feed members of several Citrus species.

Identification and biology

Moths are stout with greyish-brown wings. Females are bigger in size with light and dark wavy transverse bands. Caterpillars are cylindrical with greyish dark brown in colour with black spots. Hind wings are white in colour. They also have long radiating hairs sparsely distributed thorough out the body.

The life-cycle is completed in about 70-80 days depending upon the weather conditions.

Mode of feeding and symptoms of infestation

Damaging stage of the insect is larva. Larvae feed gregariously by scrapping the chlorophyll of the leaves during night and during the day time hide in the bark. Large number of black excreta deposited in the base plants indicating occurrence of this pest.

3. Sapota seed borer, *Trymalitis margarias* Meyrick (Tortricidae: Lepidoptera)

Distribution

The pest is available in Sri Lanka and later introduced in India. It causes havoc damage to Sapota. In India, it was first noticed in "Kankradi" area of Thane district of Maharashtra.

Hosts of commercial importance

Sapota is recorded as its major host.

Identification and biology

The first instar larvae are tiny, white in colour having pinkish tinges whereas, fully developed larvae are pinkish red in colour. Before pupation the full grown larvae used to come out from the seed and made a silken cocoon with within the leaf fold. The pupa is small, dark brown in colour and obtect. The adult moths are small in size, whitish forewings with grayish spots and fringed along the margins. Hind wings are cream coloured with thicker fringes at the outer margins.

The mating of male and female moths was happened generally during evening hours between 7.00 to 11.00 p.m. and the period of coitus lasted for 8 to 13 minutes (Shukla, 2009). Gravid female moth lays about 85 - 240 eggs during her entire life. The average pre-oviposition, oviposition and post-oviposition ranged for 1.83±0.185, 1.44±0.373 and 1.61±0.443 days, respectively. The duration of egg, larval and pupal stages lasted for an average of 11.16±0.715, 12.33±0.832 and 13.15±1.663 days, respectively (Shukla, 2009). The adults survived for 3.5 to 5 days. The total life cycle from egg laying to adult emergence complete in 35-41 days.

Mode of feeding and symptoms of infestation

Damaging stage of the insect is caterpillar. The caterpillars feed exclusively on endosperm of the seed. Full grown larvae prepare a tunnel to come out from the affected fruit for pupation. Through the injury of exit hole, the fungus as well as ants enters inside the fruits and thus secondary infection takes place. The affected fruit becomes unfit for human consumption. As high as 40-80% damage to the mature fruits have been noticed by Wagh *et al.* (2011). Due to the infestation by this pest, quality of the fruits deteriorates badly and hence the product fetch low market price.

The Sapota crop suffered badly during 1999-2000 in Dahanu, India to the extent of 40-90% during September to March (Patel, 2001). In southern India, 25% incidence in Bangalore during 2006-07 (Anonymous, 2008) and 60% incidence was recorded in Periyakulam in Tamil Nadu during 2008-09 by Jayanthi and Verghese (2010). Presently its infestation has been reported from Gujarat also.

4. Aphids, *Toxoptera aurantii* (Boy. De. F.) (Aphididae: Hemiptera)

Distribution

Cosmopolitan in distribution.

Hosts of commercial importance

Polyphagous and reported to infest crops like citrus, jackfruit, litchi, tamarind, peach, pear, tea, coffee etc.

Identification and biology

Adult apterous female has oval body, about 1.2-1.5 mm long. Body colour varied to yellowish green to greenish brown where as adults are darker in appearance. Presence of prominent paired cylindrical, long and blackish "Siphuculi" or "Cornicles" at the dorsum of the 5 or 6th abdominal segments is the identifying character of this aphid. Winged female is about 1.2-1.4 mm long with a fusiform body. Antennae are longer than the apterous female.

Both the alate as well as apterous females can multiply by means of parthenogenesis and viviparously. In a single day a female may give birth to 8-22 nymphs which moult four times to become adults completing the life cycle in 7-11 days.

Mode of feeding and symptoms of infestation

Damaging stage of the insect are both nymphs and adults. Small soft-bodied greenish brown aphids found on lower surface of the leaves, buds and growing shoots in groups. The adults and nymphs of the aphid feed on the underside of leaves or on the growing tips of shoots and flower buds sucking juices from the plant. The affected portion may become chlorotic, droop down and dry prematurely. Honeydew is secreted by the aphids and this favours sooty moulds to grow, resulting in a decrease in the economic value of the produce and also hindering the efficient photosynthesis of the plants.

5. Mealybug, *Planococcus citri*, *P. lilacinus*, *Ferrisia virgata* (Pseudococcidae: Hemiptera)

Several species of mealybugs infest Sapota. These are polyphagous and feed on a number of crops.

Distribution

Cosmopolitan in distribution.

Hosts of commercial importance

P. citri is reported to infest crops like citrus, coffee and cacao whereas *P. lilacinus* feeds on cashew, fig, cacao, coffee etc.

Identification and biology

Adults are elongated to broadly oval shaped insect, body covered with characteristic whitish waxy powdery coating. Nymphs are smaller in size.

Reproduction is mostly by means of parthenogenesis. Adult female lay around 150-400 eggs in ovisac. Eggs hatch within 3 to 9 days; first instar nymphs are highly mobile and called as crawler. The nymphal stage lasts for about 19-26 days. On an average 10-15 generations are completed in a year.

Mode of feeding and symptoms of infestation

Damaging stage of the insect are both nymphs and adults. They suck plant sap from the tender twigs, buds, petals, growing fruits resulting plants lost its vitality. Affected twigs and buds become wrinkled. Beside these, mealy bugs also secret sugar rich honeydew deposited on the leaves that creates black sooty-mould and there by hindering the photosynthesis activity of the plants. Honey dew also attracts the ants and through this symbiotic association by they get protection from their natural enemies. Ants also help to mealy bugs to transport from one plant to another.

6. Fruit fly, *Bactrocera dorsalis* (Hendel) (Tephritidae: Diptera)

Distribution

Cosmopolitan in distribution.

Hosts of commercial importance

Polyphagous in nature and have been recorded from more than 150 kinds of fruits and vegetables including crops like mango, guava, sapota, apricot, avocado, banana, citrus, coffee, fig, loquat, roseapple, papaya, peach, pear etc.

Identification and biology

Eggs of *B. dorsalis* are white to yellow-white in colour. Fully grown maggots are about 10 mm in length and creamy white in colour. Before pupation, the mature larva drops to the ground and forms a dark brown to tan coloured spindle shaped hard puparium of about 4.9 mm in long. The adult fruit fly can easily be recognized by the presence of two horizontal black stripes and a longitudinal median stripe extending from the base of the third segment to the tip of abdomen.

The female fly makes a cavity with the help of her sharp ovipositor and thrusts the white cigar shaped elongate eggs mostly singly or in groups. Ripe fruits are generally preferred for laying egg. However, immature ones also sometimes infested. After laying the eggs, the female secretes a gummy secretion on the oviposition puncture and makes the entrance waterproof. The incubation period is 3-9 days depending up on weather conditions. The freshly hatched maggots bore into the pulp, forming galleries. Depending on weather, the larval period extends from 4 days in summer to 21 days in winter. The full-grown maggots come out of the fruits and drop down on soil and pupate therein at a depth of 1.5 to 15 cm. The pupal period is 5 to 9 days in summer and may extend up to 30 days in cold weather. The total life cycle completes in 12- 34 days depending on weather conditions. Several overlapping generations are completed in a year.

Mode of feeding and symptoms of infestation

Damaging stage of the insect is the maggot. Due to egg laying by gravid females, there is oozing of fluid from the fruits and several brown rotten patches appeared on the fruits which reduced the market prices. After hatching the maggots make tunnel in the fruit pulp, contaminating them with their frass and providing entry points for fungi and bacteria, which cause the fruit to rot. Due to feeding on pulp, there is a pre-mature dropping of fruits and make them unfit for consumption. The young infested fruits can be destroyed in a few days, but older fruits show less obvious symptoms but on splitting open, cluster(s) of whitish maggots in pulp is found. The infestation varies from 30-100 per cent in different fruit crops.

7. Sapota fruit mite, *Tuckerella kumaoensis* Gupta (Tuckerellidae: Acarina)

Both the red-coloured nymphs and adults of this mite feeds on fruit surface from marble sized fruit stage, resulted the fruit surface became rough and thereafter suck the cell sap. The affected fruit surface becomes rough and corky in appearance or black or dark coloured which results in qualitative loss of harvested products (Shukla *et al.*, 2013).

Apart from these insect pests, other pests like bud borer (*Anarsia achrasella*) webs together flower buds as well as flowers. Extend of infestation has been reported to range from 2-15% (Wagh *et al.*, 2011). Tender leaves of sapota are twisted by the leaf twister weevil, *Apoderus* sp. and the incidence was observed from September to December with a peak activity during second fortnight of October as reported by Satish *et al.* (2014).

Seasonal incidence of major sucking pests of sapota

Seasonal incidence of an insect gives an idea about the peak period of activity of the insect and accordingly suitable control measures may be adopted. Recently Nandre and Shukla (2014) studied the population dynamics of *B. dorsalis*. Their study revealed that, fruit fly population prevailed throughout the year with maximum activity (172.1 flies/ trap) during March to August and lower during the month of December and January (11.1 to 21.3). The pest population coincided with the maturity and harvesting period of the fruits. Correlation studies revealed that, the *B. dorsalis* population showed significantly positive correlation with temperature and morning R. H. whereas negative correlation with the maximum R. H. Again, Satish *et al.* (2014) from Mudigere, Karnataka, India revealed that, the fly (*B. dorsalis*, *B. zonata*) damage was observed during October to January with maximum infestation in second fortnight of December. Shukla (2009) documented that sapota seed borer infestation fluctuated throughout the year with its peak in October, 2007 (10.5% fruit infestation), while in year 2008-09 the highest fruit infestation (7%) was recorded in the month of November. Thereafter it declined and reached only 1.5% cent in January 2007-08 and 0.5% in February 2008-09, respectively. The data also revealed that the pest has one peak in a year corresponding to the crop harvesting periods (October or November).

The data on the seasonal activities of bud borer, *A. achrasella* revealed a varying degree of infestation throughout the year with the peak during March (11.29 %). The incidence of scale insects on sapota was maximum during first and second fortnight of April with an average of 34.60 and 27.10 scales per 5 infested leaves per plant, respectively as noted by Satish *et al.* (2014). Mani and Krishnamoorthy (2008) from IIHR, Bangalore, Karnataka reported that, higher population of mealy bugs occurred during April. In another findings Satish *et al.* (2014) documented that population of mealybugs (*P. lilacinus* and *P. citri*) was increased from first fortnight of March and reached the peak on first fortnight of April (15.40 mealy bugs/ 30 cm shoot).

Integrated management of Sapota pests

To save the sapota crop from ravages of various insect pests, it is very essential to adopt suitable, need-based and quick result-oriented control measures, keeping in mind the resurgence of pests as a result of development of pest resistance to insecticides, residual problem, minimum environmental pollution as well as less harm to natural enemies such as parasitoids, predators and pollinators. For this, an integrated approach should be adopted to suppress the pest population below economic threshold level (ETL) through proper and judicious use of suitable control measures at the right time. Prior to understanding sapota IPM comes there is a need to know the crop plant and its growth throughout the season and adequate knowledge of the threshold of economic damage. Crop growth stage is the most important criterion because the relationship between insect injury and crop damage is dependent on the stage when the maximum injury occurs.

Research has shown that injury during the vegetative stages is usually not as detrimental to the plant as that during reproductive stages (Dey and Halder, 2011). Periodical survey and monitoring of insect pests is an important task to obtain information on population dynamics of all pests and natural enemies and to estimate damage to the plants. Integrated pest management there by involves various methods of control viz., mechanical, cultural and use of pesticides both bio-pesticides and chemicals.

Mechanical control

Deep summer ploughing of fields to expose the soil-inhabiting stage of the insects particularly, pupae of fruit fly, mealy bugs hiding in the cracks and crevices in the soil to the natural enemies. Collection and destruction of developing stages (eggs/nymphs or caterpillars/adults) of serious pests and infested plant parts from the fields will help in reducing the pest incidence. Installation of yellow sticky traps attracts aphids, thrips etc. By using such traps in the fields, first appearance of the pest and pest densities can be monitored for taking timely measures to control the pests (Dey and Halder, 2011). Erecting of bird perches @ 20/acre for encouraging predatory birds such as King crow, common mynah etc. is also advisable. Burning the mass of hairy caterpillar larvae found on tree trunks is recommended (Anonymous, 2014).

Cultural management

Clean cultivation, crop rotation, judicious use of nitrogenous fertilisers are important for the management of insect pests of sapota. Periodical removal of weeds should be done. Ants act as transporters for mealy bugs so destruction of ants' colony in and around the field restricted its migration. Uprooting and burning of severely infested (mealy bugs, mites) twigs minimizes the future pest infestation. Nitrogen plays an important role on the intensity of insect pests. So, judicious use of nitrogenous fertilizers during the crop growth stage must be followed. Neem cake should be incorporated @ 100 kg/ha to protect the crop from pest infestation. Similarly, growing of forage crops as a mixed crop is advisable. The forage crops help in maintaining ecological balance and thereby suppressing pest population on the crop. Collection and destruction of the off-season stray mature sapota fruits after main harvest till November help in bringing down the Sapota seed borer incidence (Anonymous, 2014).

Use of resistant varieties

Some lines or varieties of a crop possess some inherent genetic characters that prevent them to attack of certain pests. These resistant varieties can be cultivated without much change in normal practice of cultivation and these can also be incorporated into insect pest management practices. Resistant varieties are being continuously developed by the various ICAR Institutes, State Agricultural Universities and State Agriculture Departments of that particular region. Recently, Shukla and Radadia (2015) screened eighteen sapota varieties for their reaction against fruit mite, *T. kumaonensis* during 2009-10 to 2012-13. They observed that fruit mite remains active round the year on all sapota varieties. However, the varieties *viz.*, Cricket Ball, Kalipatti, Murabba, Challa Collection-3 and Paria Collection were most susceptible to mite infestation. On the other hand, the variety Zumakhiya has found tolerant as it harboured very less fruit mite throughout the experimental period.

Biological management

Biological control can be defined as "The action of parasites, predators or pathogens in maintaining another organism's population density at a lower average than would occur in their absence" (Paul DeBach, 1964). All biological control involves the use, in some manner, populations of live natural enemies to suppress pest populations to lower densities, either permanently or temporarily with an active human role (Van Driesche *et al.*, 2008). Biological control of pests is a method of controlling pests (including insects, mites, weeds and plant diseases) that relies on predation, parasitism, herbivory or other natural mechanisms. Biological control of insect pests is gaining importance due to their target specificity, self-perpetuity and obvious safety to the environment (Halder *et al.*, 2013). Some parasitoids and predators have been reported controlling the major insect pests of sapota (table-1).

Table 2: Different predators and parasitoids of major insect pests of sapota.

Name of the bioagents	Taxonomic position (Family: Order)	Target pests	Preferred stage(s)
Parasitoids			
Coccophagus cowperi Girault	Aphelinidae: Hymenoptera	Green scale, *Coccus viridis*	Nymphs
Fopius arisanus (Sonam)	Braconidae: Hymenoptera	Fruit fly, *Bactrocera dorsalis*	Eggs
Diachasmimorpha kraussi (Fullaway)	Braconidae: Hymenoptera	Fruit fly, *Bactrocera dorsalis*	Maggots
Diaeretiella rapae (M'Intosh)	Braconidae: Hymenoptera	Black bean aphid, *A. craccivora*	Mostly nymphs
Predators			
Cryptolaemus montrouzeiri (Mulsant)*	Coccinellidae: Coleoptera	Green scale, *Coccus viridis* and mealy bugs (*Planococcus citri*, *P. lilacinus*, *Ferrisia virgata*)	Nymphs and adults
Chrysoperla zastrowi sillemi Henry	Chrysopidae: Neuroptera		Nymphs and adults
Cheilomenes sexmaculata (Fabricius)	Coccinellidae: Coleoptera		Nymphs and adults
Scymnus coccivora Ayyar	Coccinellidae: Coleoptera		Nymphs and adults
Brumoides suturalis Fab.	Coccinellidae: Coleoptera		Nymphs and adults
Nephus regularis (Sicard)	Coccinellidae: Coleoptera	Black bean aphid, *A. craccivora;* Solenopsis mealy bug, *Phenacoccus solenopsis*	Nymphs and adults
Mallada boninensis (Okamoto)	Chrysopidae: Neuroptera	Soft-bodied insects	Eggs, nymphs and adults

*For the management of mealy bugs and scale insects after two weeks of release 20 predator beetles *viz., Cryptolaemus. montrouzieri* beetle per tree is beneficial (Anonymous, 2014).

Botanicals as a tool for IPM

Pesticides derived from plants have the potential to play a major role in pest management in sustainable agricultural production. These are renewable, non-persistent in the environment and relatively safe to natural enemies, non-target organisms as well as human beings (Halder *et al.,* 2010). So, plant derived insecticides plays a vital role in the insect pest management of the sapota. However, very little work has been done on evaluating different botanicals for the management of insect pests of Sapota.

Neem cake may be incorporated @ 40 kg/acre, to protect from pest infestation (Anonymous, 2014). Similarly, to control sapota leaf webber application of neem seed kernel extract (NSKE) 5% is advisable.

Chemical management

In outbreak and/or resurgence of certain insect pests, the mechanical, cultural and biological control measures are may not be of much value to quickly suppress the insect pest population. Under such conditions, farmers will always prefer chemical control measures for quick results. For the management of fruit fly, monitoring and mass trapping the flies with methyl eugenol traps is recommended. Similarly, use of bait spray is also found effective.

Future strategies

The detailed analysis of past research works on major insect of sapota management revealed that the focus on cultural and biological control which lack in depth basic information. Besides, in host plant resistance multidisciplinary and consortia based collaborative approaches to be initiated. The interspecific resistance needs to focus. The role of agroecosystem analysis and ecological engineering and biological control with special reference to entomopathogens and plant derived toxins need to be explored.

- Surveillance is the one of the important tools in pest management. Hence, pest survey and monitoring should be done at weekly interval.
- Development of suitable forecasting method for these borer pests under different agroclimatic zones has to be formulated.
- Use of pheromone and light traps for monitoring and mass trapping of the pest population to be emphasized.
- Search for the potential biocontrol agents, newer plant origin insecticides and novel biorational molecules with green chemistry to reduce the pesticide contamination in the environment.

- There is a need for development genetically improved strains of the bioagents with longer shelf-life and higher field persistence.
- State Governments should ensure the quality control of the bioagents available in the markets.
- Development of suitable low-cost mass production technology of the potential bioagents and standardization of field release methods need to be addressed.
- All the pesticide dealers should be given training programme in the plant protection so that they must understand the importance of IPM and hazards associated with chemical pesticides.
- Periodical validation and refinement IPM modules for the management of these pests are the need of the hour.
- Steps to be taken to promote the IPM among the farmers thorough method demonstrations, farmers' field school, radio, television, print media etc.

References

Anonymous, 2008. Annual Report of Indian Institute of Horticultural Research-2008, Bangalore, Karnataka, India.

Anonymous, 2014. "AESA based IPM – Sapota" published by National Institute of Plant Health Management, Rajendranagar, Hyderabad, Telangana, pp-23-25.

Bhutani, D. K. 1979. Insects and fruits. Periodical Export Book Agency, Delhi, pp-87-94.

Cheema, G. S., Bhatt, S. S. and Naik, K. C. 1954. Commercial fruits of India. Macmillan and Co., pp-422.

De Bach, P. 1964. Biological control of insect pests and weeds (1st edition), Chapman and Hall, London.

Dey, D and Halder, J. 2011. Integrated management of Soybean Insect Pests for Sustainable Agriculture.*In*. Potential Plant Protection Strategies (Ed. D. Prasad and R. Sharma, ISBN: 9789380578293), I. K. International Publishing House, New Delhi, pp-133-152.

Fletcher, T. B. 1920. Life histories of Indian microlepidoptera. Mem. Department of Agriculture. India, 6(6):151.

Gupta, R. L. and Gangrade, G. A. 1964. Life-cycle and seasonal history of chikoo moth, *Nephopteryx eugraphell*a Rag. Indian Journal of Entomology, 17:326-336.

Halder, J., Rai, A. B. and Kodandaram, M. H. 2013. Compatibility of neem oil and different entomopathogens for the management of major vegetable sucking pests. *National Academy Science Letters*, 36(1):19-25.

Halder, J., Srivastava, C. and Dureja, P. 2010. Effect of methanolic extracts of *Vinca rosea* and *Callistemon lanceolatus* alone and their mixtures against neonate larvae of gram pod borer, *Helicoverpa armigera* (Hubner). Indian Journal of Agricultural Sciences, 80(9):820-823.

Jayanthi, P. D. K. and Verghese, A. 2010. Establishment of sapota seed borer, *Trymalitis margarias* Meyrick, an invasive species in India: Exigencies involved in limiting the spread. Karnataka Journal of Agricultural Science, 23(1):165.

Mani, M. and Krishnamoorthy, A. 2008. Field efficacy of Australian ladybird beetle *Cryptolaemus montrouzieri* Mulsant in the suppression of *Maconellicoccus hirsutus* (Green) on sapota. Journal of Biological Control, 22(2):471-473.

Nair, M. R. G. K. 1995. Insect and mites of crops in India. Published by the Indian Council of Agricultural Research, New Delhi, pp-252.

Nandre, A. S. and Shukla, A. 2014. Population dynamics of fruit fly [*Bactrocera dorsalis* (Hendle)] on sapota. Agriculture Science Digest, 34(1):70-72.

Patel, Z. P. 2001. Record of seed borer in Sapota, *Manilkara achras* (Mill.) Forsberg. Insect Environment, 6(4):149.

Sandhu, G. S., Singh, H., Singh, A. and Bhalla, J. S. 1974. Insect pest of sapota and their control. Punjab Horticultural Journal, 14 (3):134-136.

Sathish, R., Naik, D. J., Veerendra, A. C. and Murali, R. 2014. Pest complex of sapota [*Manilkaraachras* (Mill.)Forsberg] under hill zone of Karnataka.Pest Management in Horticultural Ecosystems, 20(1):86-88.

Shukla, A., Radadia, G. G., Patel, K. A. and Patel, K. G. 2013. Population dynamics of sapota fruit mite, *Tuckerella kumaoensis* Gupta (Acari: Tuckerellidae) in Gujarat, India. *Pest* Management in Horticultural Ecosystems, 19(1):95-98.

Shukla, A. and Patel, P. R. 2011. Bionomics of chiku moth, *Nephopteryx eugraphella* Ragonot (Lepidoptera: Pyralidae) on sapota. Pest Management in Horticultural Ecosystems, 17(1): 6-10.

Shukla, A. 2009. Seasonal incidence and biology of sapota seed borer, *Trymalitis margarias* Meyrick. *Pakistan Entomologist*, 31(2):107-110.

Shukla, A. and Radadia, G. G. 2015. Reaction of sapota varieties against fruit mite, *Tuckerella kumaonensis.* Indian Journal of Plant Protection, 43(3):338-340.

Van Driesche, R., Hoddle, M., Center, T. 2008. Control of Pests and Weeds by Natural Enemies: An Introduction to Biological Control. Blackwell, Malden, MA, USA.Pp-473.

Wagh, V. H., Ghule, T. M, Solanki, R. D., More, K. A., Kharat, A. P., Dhote, V. W. 2011. The insect pest status of sapota in India. *In*: Proceedings of the International Symposium on Minor Fruits and Medicinal Plants for Health and Ecological Security (ISMF & MP), West Bengal, India, 19-22nd December, 2011, pp-254-256.

11

Pests of Amla and Their Management

Roshna Gazmer, Nripendra Laskar and Atanu Maji

The Indian gooseberry or amla (*Phyllanthus emblica* Linn.; Family: Euphorbiaceae) is a deciduous fruit largely cultivated in the Indian subcontinent for its fruits. Amla is native to the tropical South Eastern Asia, particularly the Central and Southern India (Mortan, 1960), where the plants are often seen to grow in the wild up to an elevation of 1800 m (Firminger, 1947). In India, amla is mostly cultivated in the semi-arid regions of Maharashtra, Gujarat, Rajasthan, Andhra Pradesh, Karnataka, Tamil Nadu, and the Aravalli ranges in Haryana, Punjab and Himachal Pradesh (Pathak, 2003). The tree attains a height of 8 to 18 m, with a crooked trunk and spreading pattern of branching. The branchlets are glabrous or finely pubescent; leaves are simple, nearly stalkless, set along the branchlets resembling pinnate type. The mature trees can tolerate a large variation of temperature ranging from freezing temperature to as high as 46°C, and prefer well-drained loamy soil to heavy clay soils. The amla fruits are rich source of Vitamin C, even when they are in dried or powdered state. The fruits are spherical, having a smooth appearance, light greenish to yellowish in colour, with six vertical stripes or furrows on it. Further, amla is known to have medicinal values, and is used against cough, bronchitis, jaundice, diabetes, dyspepsia, diarrhea, fever, hair fall etc.

Like any other plants, this is also infested by a number of insect pests that reduces quality and quantity of production. The insect pests causing major economic damage in amla will be discussed in this chapter.

Table 1: Insect pests of amla

Sl. No.	Common name	Scientific name	Family	Order
1.	Bark eating caterpillar	*Indarbela tetraonis* Moore *I. quadrinotata* Walker	Metarbelidae	Lepidoptera
2.	Apical twig gall maker	*Betousa stylophora* Swinh.	Thyrididae	Lepidoptera
3.	Fruit sucking moth	*Achaea janata* Linn.	Noctuidae	Lepidoptera
4.	Anar butterfly	*Virachola isocrates* Fab.	Lycaenidae	Lepidoptera
5.	Amla aphid	*Setaphis bougainvilleae* Theob.	Aphididae	Hemiptera
6.	Mealy bug	*Nipaecoccus viridis* (Newstead)	Pseudococcidae	Homoptera
7.	Hairy caterpillar	*Euproctis* sp.	Lymantriidae	Lepidoptera
8.	White fly	*Aleurolobus diacritica* Regu & David	Aleyrodidae	Homoptera
9.	Leaf miner	*Phyllocnistis citrella* Stainton	Gracillariidae	Lepidoptera
10.	Leaf roller	*Gracillaria theivora* Wism.	Gracillariidae	Lepidoptera
11.	Stone borer	*Curculio* sp.	Curculionidae	Coleoptera
12.	Fruit midge	*Clinodiplosis* sp.	Ceratopogonidae	Diptera

Besides, there are some minor pests that can also inflict potential damage under certain conditions, and they are listed as follows:

Sl. No.	Common name	Scientific name	Family	Order
1.	Bagworms	Eumeta crameri Westwood	Psychidae	Lepidoptera
2.	Tussock moth	Euproctis subnotata Walker	Lymantridae	Lepidoptera
3.	Cowbug	Otinotus oneratus (Walker)	Membracidae	Hemiptera
4.	Bihar hairy caterpillar	Spilosoma (Spilarctia) obliqua (Walker)	Arctiidae	Lepidoptera

1. Bark eating caterpillar, *Indarbela tetranois* Moore (Metarbelidae: Lepidoptera)

Distribution

The bark eating caterpillars viz., *Indarbela tetranois* and *I. quadrinotata* are widely distributed in the tropical and sub-tropical regions of Indian subcontinent. To date, about 14 species of *Indarbela* have been reported from India and the most common species are *Inderbela tetraonis I. quadrinotata, I. campbelli, I. minima* and *I. watsoni* that are present throughout the country.

Hosts of commercial importance

They are highly polyphagous pests, attacking a number of fruit trees which includes jamun, pomegranate, loquat, falsa, pear, guava, ber, fig, mulberry, bael, karonda, custard apple, litchi, citrus, mango etc. In any given locality, usually more than one species remains predominant, and *Inderbela* spp. are more severe in neglected orchards where the trees have grown old and overcrowded.

Identification

The nocturnal females lay around 15-25 eggs in cracks and crevices under loose bark usually near the forks. The larva after hatching has a dirty brown appearance and nibbles the live bark superficially for 2 to 3 days before boring into the stem. The full grown larva appears smooth, with sparse hairs on body and measures about 3.5-4 cm in length. Sometimes light brown sclerotized patches are present dorso-laterally on the larval body. Abdominal legs are present on 6th to 9th segments. The pupa is light brown in colour measuring up to 1.5 cm and with two short pointed cephalic processes. The adult moths emerge in May-June. They are stout, pale brown in colour with rufous head and thorax. The forewings of the moths are pale brown with numerous brown spots and bands on them, and have sub-apical brown spots. The hind wings are light black in colour. Head is depressed and antennae uniformly pectinate (Mathew, 1997).

Biology

Single female lays up to 2000 eggs during summer months. The incubation period is 8 to 11 days and, the larval development completes in 9 to 11 months. Pupation occurs within the larval tunnel, where the cephalic end of pupa slightly protrudes out and lasts for 3 to 4 weeks. Adults have a short life span of less than a week (Sontakay,1945 and Patel *et al*., 1966).

Seasonal incidence

The adult moths become active in May to July, while larvae are mostly active during July to November (Jhala *et al*., 2005) completing single life-cycle in a year.

Mode of feeding and symptoms of infestation

The caterpillar feeds underneath the bark and as a result when the bark withers damage symptoms become visible. The larvae, while feeding, destroy the phloem vessels and interrupt the flow of nutrients in plants. Further, boring into the branches and stems also weakens the trees. These damages affect the vigour, vitality and fruit bearing capacity of the trees. Severely infested branches, or even the main trunk, often break off under the pressure of strong winds. The presence of caterpillars is usually indicated by winding ribbon-like webbings, made up of excreta, bark pieces and frassy material of varying colour. The caterpillars construct these silken galleries while feeding, and extend around the feeding sites. The bore holes, normally formed in a zigzag manner, can be easily located from a distance (David *et al*., 1963).

Management

The best time of control of bark eating caterpillar is during the egg hatching period when the caterpillars are small. The standard management strategies are as follows:

i. Keeping the orchard clean and avoiding overcrowding of trees.

ii. Killing the caterpillars mechanically by inserting a thin wire/iron spike into the tree holes, but the process is tedious and labour expensive.

iii. Removal of the webbings during February-March and injection of around 5 ml emulsion prepared by diluting 2 ml Dichlorvos or 5 ml methyl parathion or 40g Carbaryl in 10 litres of water, in the bore holes. Mixture of kerosene, soap and water can also be used after soaking in a cotton swab and plugging in the holes with mud.

iv. Two species of entomogenous fungi, *Aspergillus candidus* and *Beauveria bassiana* have been reported to cause mortality of this insect in the field.

v. Use of mixture of petroleum spray oils at recommended rate as paint on branches and trunk provides sustained long term control of the insect species.

vi. Injection of ethylene glycol and kerosene oil at a ratio of 1:3 into the tunnels and then sealing the opening with mud also give good result.

2. Apical twig gall maker, *Betousa stylophora* (*Hypolamprus stylophora*) Swinh. (Thyrididae: Lepidoptera)

The apical twig gall maker is a sporadic and specific pest of amla. The newly grown branches suffer maximum damage due to feeding by the sluggish caterpillars that form galls at the tip of growing twigs. The heavily affected branches have an arrested growth temporarily. However, normal growth of twigs is restored in subsequent years.

Distribution

This is a major pest reported from Gujarat, Rajasthan, Madhya Pradesh and Maharashtra of India (Meshram, 2003, Dadmal and Pawar, 2002).

Hosts of commercial importance

Though the insect is polyphagous in nature infesting amla, ber, falsa, jamun, fig, guava, grapevine, pomegranate, mulberry trees etc., they generally prefer to feed on amla and grapevine.

Identification

The females lay eggs on the surface of the leaf, and young larvae tunnels through the apical shoot resulting inbulged appearance or galls. Small sized caterpillars, black in colour become visible when the galls are split open (Bose, 1935; Beeson, 1941; Mathur and Singh, 1954; Kumar, 1990). The full grown caterpillar is cylindrical in shape and tapering gently at both ends. Larvae have grey head and jet-black thorax and abdomen. It has 10-11mm length and 2.25-2.50 mm width. The pupa is beautiful red and measures about 6 mm in length and 2.75mm in width (Haldhar *et al.*, 2019).

Biology

The eggs are laid singly on the twig surface and shoots of new flush, specifically near or at the growing tips of branches. Within a week tiny caterpillars hatch out from the eggs and bored into twig or shoot. The larva develops slowly. About 2 days are taken by the larvae to prepare the chamber that is well closed on all sides and later within the chamber it prepares a cocoon. In the absence of chamber, larva fails to pupate. The pupal stage extends upto 8-10 days in May-June. It completes only a single generation per year. The adult moths emerge out through the anterior end of pupal chamber and live for a short period (Haldhar *et al.*, 2019).

Seasonal incidence

The insect attacks amla nurseries and orchards in rainy season between June and August, and completes one generation per year (Meshram, 2003). The activity of pest population is majorly guided by weather conditions (Kumar *et al.*, 2007).

Mode of feeding and symptoms of infestation

The insect mostly infests in the nursery stage, but old bearing trees are also infested. The larvae attack apical region of the shoot and produce gall on the branch terminal and make tunnel. Damaged region bulges out abruptly into a gall. Thus, apical growth is prevented, side shoots develop below the gall level and hamper the subsequent growth in the following season. In the beginning of the infestation terminal shoots swell, which increases in size with the passage of time. The young insects penetrate from apical portion of the shoot after the monsoon and feed inside the tissue, changing into galls.

The larvae, while feeding, secrete some metabolites inducing gall formation that resembles "snake charmer's flute" at the tip of growing twigs (Bharpoda *et al.*, 2009) The infestation adversely affected the growth of twigs. Growth of shoot is hampered and few lateral shoots develop below galls. In the month of October-November full size galls can be observed. Favourable condition is rainy season. The affected shoot did not grow in the following season.

Management

i. Pruning of the twigs along with galls and destroying the larvae inside by crushing the galls mechanically or by burning the twigs. Overcrowding of branches in a plant should be discouraged and timely pruning should be done.

ii. If the pest occurs regularly, Dimethoate 0.05% can be sprayed at the beginning of the season, and may be repeated at fortnightly intervals, if needed.

iii. Prophylactic spray can be done with any systemic insecticide to reduce the pest damage.

3. Fruit sucking moth, *Achaea janata* Linn. (Noctuidae: Lepidoptera)

Distribution

This insect is native to the Indo-Malaysian region, and was first reported in Hawaii. It is widespread throughout Australia, China, India, Japan, Korea, Philippines, Papua New Guinea and Thailand.

Hosts of commercial importance

The fruit sucking moth is a polyphagous pest and infests on different fruit crops, including apples, apricots, bananas, breadfruit, coffee, figs, grapefruit, guava, kiwifruit, litchi, longan, mandarins, mangoes, nectarines, oranges, papaya, passion fruit, peaches, persimmons, pineapple, plums, and star fruit (Waterhouse and Norris, 1987).

Identification

The eggs are hemispherical with about 1 mm in diameter, yellowish green in colour and laid on the lower side of leaves, but can be found on the bark also. First stage larvae are cylindrical, and when fully grown reach up to 2 inches in length (Tryon, 1898). During the 2nd instar the last abdominal segment become humped and dark green to black in colour (Comstock, 1963, Hargreaves, 1936). Generally, when larval density is high a dark coloration occurs, and the light colored larvae are seen when they are isolated. The 2nd and 3rd abdominal segments bear paired lateral markings that resemble eye spots. Pupation takes place in a silken cocoon which is spun between the leaves. The leaves that contain the cocoon may remain attached with the host plant or fall on the ground (Waterhouse and Norris, 1987).

Adult moths are large and robust with a wingspan of 4 inches (Waterhouse and Norris, 1987). The base of the abdomen is pale brown and the tip is yellow-orange in colour. Forewings of females are look as a leaf and olive to purple-brown in colour with greenish flecks. The outer edges of the female's forewings are scalloped or toothed, whereas for males they are evenly curved.

Biology

A female moth lays up to 100 eggs in batches, whereas it may reach to several thousand when moth populations are high. Eggs hatch within 3 to 4 days (Kumar and Lal, 1983). The insect contains 5 larval instars, and after each molt the newly emerged caterpillars eat away the discarded skins. Pupal period continues for 14 to 21 days, but dry condition may adversely affect the adult emergence (Hargreaves, 1936). After emergence, female adult passes through a preoviposition period of 4 to 8 days and its longevity is 27 to 30 days. In contrast, the longevity of males is 26 to 28 days (Kumar and Lal, 1983). The life cycle generally completes within 35 to 49 days, but may be take 30 to 33 days under warmer condition. Continuous generations are completed throughout the year.

The adult moths are strong fliers and in search of food they can travel a long distances away from their breeding grounds (Waterhouse and Norris, 1987). Adult flies mainly at the period between dusk and 11:00 PM (Tryon, 1898). Usually they are not attracted by the light (Kumar and Lal, 1983). When moths feel any danger or get disturbed, it flares its forewings and exposes its conspicuous hind wings (Waterhouse and Norris, 1987).

Seasonal incidence

The fruit sucking moths are prevalently found during the rainy season only, *i.e.* from the first week of July to the last week of September in the central parts of India (Rakshpal, 1945). Similar trend was observed in southern Andhra Pradesh, Maharashtra and Punjab. But the larvae mostly become inactive after October.

Mode of feeding and symptoms of infestation

Both caterpillar and adult moth cause damage to the host plant. Caterpillars mostly feed between evening to morning hours, and are usually found at the inner side or on the leaf edges (Hargreaves, 1936). Young larvae when feel any danger they drop to the ground, while the older larvae remain hanging from the plant using their hind legs and swaying their rest of the body from side to side (Mosse-Robinson, 1968). The adult moth punctures the skin of the fruit, and feeds upon the fruit juices. Fruit flesh damaged by this moth becomes soft and mushy (Heu *et al.*, 1985).Simultaneous damage by the fungal and bacterial pathogens also develops at and around the wounded site. The moth also transmits a fungus *Oospora citri*, that causes fruit rots. Other microorganisms that gain entry into the fruit and cause rotting may include *Fusarium* sp., *Colletotrichum* sp. and several types of bacteria (Hargreaves, 1936). Premature ripening and dropping of fruits are a major symptom of the insect damage.

Management

i. Biological management

a. *Trichogramma chilonis* and *Euplectrus plathypenae* efficiently manage the population of fruit sucking moths under natural condition (Heu *et al*., 1985).

ii. Cultural and mechanical management

a. When fruits are easily accessible and insect population is low, netting and killing of moths can be accomplished after sunset.

b. The fruit sucking moth avoid light, so illumination of orchards has been tested as a possible mean of deterring the moth attack.

c. Covering of the fruits using transparent oil paper or brown paper bags can protect the fruits against moth infestation. Brown paper bags last for about a month in field whereas the transparent oil paper bags may be used for two consecutive seasons if the weather is not very wet. This method is practical when each individual fruit is of significant economic value or when the fruits are easily accessible. Although this method is labor intensive.

d. Smoking of the orchard may be done.This method involves obscuring the odour of mature and ripened fruit that attract the moth with smoke. Containers filled with inflammable material like tar, oil and green plant trimmings can be used to generate smoke and placed within the orchard @ of 2 to 4 per acre. Smoking is done from dusk to dawn.

e. Regular collection and disposal of spoiled and fallen fruits. This approach can reduce fruit sucking moth damage to a large extent.

ii. Chemical management

a. Poison bait that include molasses 200 g + Malathion 50 EC @ 2 ml + 2 litres of water or molasses 1kg + vinegar 60g + Malathion 50ml + water 10 litres can be used to attract and kill insects.

b. Spraying trees with Carbaryl during maturity of fruits can also reduce the damage. Citronella oil can be used as an insect repellant.

4. Anar butterfly, *Deudorix* (*Virachola*) *Isocrates* Fab. (Lycaenidae: Lepidoptera)

Distribution

Anar butterfly is distributed throughout the India Particularly it is found in areas where pomegranate is grown.

Hosts of commercial importance

It is a polyphagous pest infecting a number of host plants like amla, ber, apple, citrus, guava, loquat, litchi, mulberry, plum, peach, pear, sapota, pomegranate, tamarind etc.

Identification and biology

The adult males are glossy bluish and brownish violet and in case of females a conspicuous orange patch on the forewings is seen. The adults are mostly brown with reflex bluish in males and orange in females.

Preferred site for oviposition is the lower middle region of the fruit, which may protect the eggs from direct exposure to predators. Total larval period may vary according to the type of fruits, and ranges from 15-25 days. Larvae pupate inside fruit, but occasionally may remain outside attaching to stalk of fruits, pupal period 7-34 days. Males are glossy, bluish violet in colour, and females are brownish violet with an orange patch on the forewings. Total life cycle completes in 44 to 65 days and has 4 generations in a year.

Mode of feeding and symptoms of infestation

The insect remains active throughout the year. Affected fruits are generally deformed at the point of larval entry. Frass materials may be seen exuding out from the borer hole in tree trunk. Such fruits weaken, rot and fall down before maturation. Attack of this insect mostly occurs during September-October that coincide with the fruiting season of the plant. The female butterfly lays eggs, singly on young fruits. The larva bores the fruit and feeds on seeds and pulp, making them hollow from inside. Damaged fruits subsequently get infected by bacteria resulting in rotting, and emit a bad smell. Larval excreta come out from entry holes.

Management

i. Cultural management

a. Clean cultivation and maintaining health and vigour of plants should be followed.

b. Cultivation of pomegranate and guava close to amla plantation should be discouraged as these are major alternate host plants of the pest.

ii. Mechanical management

a. Removal of weeds belonging to the Compositae family.

b. Collection and destruction of infested fruits can prevent further spread of infestation.

c. The fruits which are screened with paper or polythene bags can avoid infestation.

iii. Biological management

a. Release of *Trichogramma chilonis* @ 1.0 lakhs/ acre four times at 10 days interval.

iv. Chemical management

a. Spraying of Dimethoate (0.04%) or Carbaryl (0.2%) at pea size stage of amla fruits. The spray may be repeated after two weeks, depending upon the intensity of attack.

b. Spraying of Carbaryl 0.2 % or Fenvalerate 0.01 % five times at an intervals of three weeks commencing at initiation of setting of fruits.

c. Immediately after pollination, clipping off calyx cup followed by the applications of neemoil @ 3 % two times.

d. Spraying of NSKE 5% or neem formulations @ 2 ml/l at flowering stage.

e. Spraying of Deltamethrin 2.8 EC @ 1.5 ml/l of water at fortnight interval between flowering to fruit development stage.

f. Spraying of Azadirachtin 1500 ppm @ 3.0 ml/l or Malathion 50 EC 0.1% at an interval of 15 days commencing from initiation of the flowering upto the harvesting.

5. Amla Aphid, *Setaphis bougainvillea* Theob. (Aphididae: Hemiptera)

Distribution

The pest has been reported from countries like India, Pakistan, Sri Lanka. Four species of aphids, *Cerciaphis emblica* Patel and Kulkarni, *S. bougainvilliae* Theob., *Setaphis viridis* Van der Goot and *Schoutedonia emblica* Patel and Kulkarni have been reported feeding on amla trees (Patel and Kulkarni, 1954).

Hosts of commercial importance

This pest is reported to infect a wide range of fruit crops like amla, apple, guava, apricot, ber, jack-fruit, cherry, fig, citrus, litchi, grapevine, jamun, papaya, peach, pomegranate, plum etc.

Identification

Freshly laid eggs are light green in colour, cylindrical and 0.6 mm long. Later it turns into shiny black. Nymphs are oval or slightly elongated, light greenish or reddish brown with six segmented antennae. Adults which feed on leaves are green in colour while those feeding on the bark are chocolate coloured.

Biology

Eggs hatch after one or two days. They reproduce sexually as well as parthenogenetically. A female has a life span of 2 weeks between which it gives birth to about 50 young ones. During March-April young individuals take only 10 days to mature and start reproducing. All these young don't have wing. After completing 3-4 generations the aphids migrate to its alternate hosts to pass the summer season. Life-cycle consists of three nymphal instars with a total nymphal duration of 8-9 days. The adult aphid lives up to 12.8±1.58 days with a fecundity rate of 28.4±2.63 (Devi and Rajashekaran, 2011).

Seasonal incidence

It is found throughout the year except during the leaf fall stage or mid-winter and peak seasonal activity during the spring and summer seasons. The incidence of this pest reaches its peak during September but it is observed from July to October (Patel and Kulkarni, 1952).

Mode of feeding and symptoms of infestation

The nymphs and adults remain on the ventral surface of leaves and suck the sap (Patel and Kulkarni, 1954). Yellowing of leaves with honey dew is frequent and it encourages the presence of ants. The infested leaves become yellow and dry up. Infested shoots at the growing points become bended and twisted. The new shoots also infested at its growing points.

Management

a. Clipping off and destruction of affected leaves and shoots.

b. Collection and destruction of alternate weed hosts.

c. Erection of yellow pan traps @ 4-5 trap/acre.

d. Predators like *Scymnus* spp., *Chrysoperla zastrowi sillemi, Cheilomenes sexmaculatus* and parasitoids like *Dicyphus hesperus, Aphidius colemani* can also be used.

6. Mealy bug, *Nipaecoccus viridis* (Newstead) (Pseudococcidae: Homoptera)

Distribution

This pest is distributed throughout India, China, Pakistan and Bangladesh. In India, this pest is mainly found in the Indo-Gangetic plain.

Hosts of commercial importance

It is reported to infest a number of fruit crops like apple, amla, apricot, ber etc.

Identification

Adult female is spherical, 2.3-4 mm long, slightly flattened in shape and covered with creamy white wax. The body beneath the wax is purplish in colour. The adult female remains almost hidden by her large white domed egg sac.

Biology

Incubation period is very short, *i.e.*, about3 to 4 hours. The developmental period of nymph of male and female varies from 31-57 and 26 - 47 days respectively. Longevity of male is 1-3 days and female is 36-53 days. After fertilization, the adult gravid females crawl down to the ground along the tree-trunk and start to lay eggs in clusters of 300-400 eggs at 5 - 15 cm depth. They generally oviposit in a confined area around the tree base. Male mealy bugs die soon after they mate and the females die soon after they oviposit. Hatching of the eggs is influenced by temperature and moisture status of the soil and hatching takes quite a few months.

After hatching the young nymphs crawl in to search for a suitable food-plant. Thereafter, the nymphs start to move upward along the tree-trunks and this migration continues for several weeks. As the nymphs reach the fresh twigs, they congregate there and start to suck the sap. The nymphs moult three times and depending on the environmental condition the nymphal period lasts for three months or more. Incase of males they undergo some sort of pupation which transform them into winged adults. There is no pupa like stage incase of females, so they don't undergo any appreciable change except their size. Thus, they have only one generation per year.

Mode of feeding and symptoms of infestation

Nymphs and adults infest new sprouts and suck plant sap by inserting their long proboscis into the plant tissue. When they occur in a large number, they devitalize the plant. They also secret honeydew which result in development of sooty mould that give an unhealthy appearance to the plant as a whole and consequently the infested branches give blackish appearance. With times, they are found clustering in masses on young shoots, like fungal outgrowths. Insect covers tender growing points with white mass and suck the sap. Nymphs and adult female remain clustered on ventral surface of leaves, terminal shoots and sometimes on fruit and suck sap and reduce the plant vitality with yellowing of leaves and premature shedding of fruits. In case of severe infestation, twigs become leafless and dry leading to complete death of the tree.

Seasonal incidence

The activity of *N. viridis* is mostly seen during spring and following the dry season.

Management

i. Cultural management

a. Crop residue has to be removed and burnt, as they may harbor populations which may infest the new crop.

b. Field borders should be free from weeds and debris that may support mealybug between plantings. Remove alternate host plants like *Hibiscus*, okra, custard apple, guava etc. in and nearby crop field.

c. Equipment should be sanitized before moving to uninfested portions in a crop field to avoid contamination.

d. Apply sticky bands like 'Track-trap' or 'Bird Tangle Foot' on arms or on main stem to prevent crawlers of mealybugs reaching the bunch.

e. Eggs of mealybugs can be destroyed by raking the soil around the tree base which expose them to the sun and heat.

ii. Mechanical management

a. Physical barriers such as ant fences can be applied parallel to the field periphery to keep ants away from field and subsequently help in managing mealybug populations.

b. Manual picking can be employed in plants that are not severely infested in small conservatory or kitchen gardens or strong jet of water can be applied to remove the insects.

iii. Biological management

a. The coccinellid beetles such as *Rodolia fumida*, *Scymnus coccivora, Cheilomenes sexmaculata* and *Nephus regularis* are important predators of mealybug nymphs. Among the biological control agents, introduction of *Anagyrus pseudococci*, *Cryptolaemus montrouzieri* (Australian Ladybird), *Leptomastix dactylopii, Hypoaspis* sp., *Beauveria bassiana* and *Verticillium lecanii* are effective in managing the infestation. *Hypoaspis* is a small mite that feeds on crawlers.

iv. Chemical management

a. Spraying Dichlorvos 76 EC @2 ml/l, Chlorpyriphos 20 EC @2 ml/l, Imidacloprid 200 SL @1ml/l or Malathion @2.5ml/l of water at 15 days intervals.

b. Removal of loose bark to expose the hiding population of mealybugs and swab stem and twigs with Dichlorvos 76 EC @ 2 ml + 2 g of fish oil resin soap in a litre of water is effective.

c. Spraying of Chlorpyriphos 0.05% in areas where hatching begins or is expected to begin can kill the just-hatched nymphs.

7. Hairy caterpillars, *Euproctis sp.* (Lymantriidae: Lepidoptera)

Distribution

Over 50 species of *Euproctis* occur in India and Pakistan of which *Euproctis fraterna, E. lunata, E. subnotata* and *E. flava* are common, but *Euproctis fraternal* is the predominant one. The caterpillars are distributed in the Palaearctic, African, oriental, Australian region and the Indian subcontinent.

Hosts of commercial importance

Host of hairy caterpillar include plum, jujube, jamun, phalsa, pomegranate, amla, fig, custard apple, loquat, grapevine, guava, peach, pear, citrus, mango, plum etc. (Sah *et al.*, 1972).

Identification

Eggs are spherical and covered with yellow hair from anal tuft of the female moth. Full grown caterpillar measures about 18-25 mm in length, reddish brown colour with plenty of hair all over the body which leads to skin irritation and itching to workers. Adult males are stout, dull coloured. Adult females are robust, yellowish with brownish forewing.

Biology

Female moths lay about 92 to 241 eggs in clusters on the ventral surface of tender leaves. Egg, larval and pupal stage last for about 9-14, 25-56 and 7-15 days respectively. Pupation takes place in silken cocoon in the leaf fold. The adult longevity is 5-7 days. Total life cycle gets completed in 44-84 days. The pest completes up to six generations per year on certain arid fruits and the duration of development varies considerably on different hosts (Kuswaha and Bhardwaj, 1968).

Seasonal incidence

Larva develops faster during spring and summer, slows down during winter and again becomes active in February.

Mode of feeding and symptoms of infestation

Newly hatched caterpillars feed gregariously on ventral surface of tender leaves by scrapping the chlorophyll, consuming the leaf margins and eventually skeletonizing them completely. The 3rd instar larvae spread out and feed voraciously, resulting in defoliation of twigs and branches and sometimes the whole tree. The caterpillars also nibble the epidermis of fruits resulting in a corky, deformed surface making the fruits unfit for consumption. Attack is more pronounced to tender and succulent leaves than the older ones.

Management

i. Manual collection and destruction of egg mass and congregating larvae or dipping them in kerosenised water.

ii. Hymenopteran parasites like *Apanteles taprobanae*, *Brachymeria* sp., *Charops obtusus* and *Goryphus* sp. has been found to be effective against the caterpillar (Mani *et al.*, 2001).

iii. Verma *et al.* (1972) reported that all the larval instars were controlled by treatment with 0.1% Carbaryl.

iv. Bhatnagar and Lakra (1992) obtained the most effective control of 3rd to 6th instar larvae by 0.005 % Cypermethrin, 0.0014% Deltamethrin.

8. Whitefly, *Aleurolobus diacritica* Regu and David (Aleyrodidae: Homoptera)

Distribution

The insects are primarily endemic to Central and South America.

Hosts of commercial importance

It is a polyphagous pest infesting different fruit trees like amla, ber, sapota etc.

Identification

Adults are small, fly-like and often dull white in colour. Powdery wax present on their membranous wings. Adults measure around 1.0-1.5 mm long with two pairs of pure white wings. After the first moult, legs and antennae are atrophied, making the remaining instars sessile. The fourth instar is often referred to as pupa which is identified by the appearance of oval, black colour covered with white fine wax filaments, scattered on lower side of the leaves. However, it is not a true pupa as feeding occurs during the first stage and transformation into an adult takes place in the last stage without any pupal moult. The eggs are fixed firmly on the leaf tissue by a stalk or pedicel at one end. The eggs are generally pyriform or ovoid with pedicel, a peg like extension of chorion, providing a mean of attachment and serves as a guide for spermatozoa during fertilization (Sandhu and Singh, 1963, Byrne and Bellows, 1991).

Biology

Adult female lays about 80 eggs in straight rows on the underside of leaves. The average number of eggs laid varies with the season. Incubation period is 3-5 days during summer but can be extended up to 33 days during winter period. Nymphal stage lasts for 9-14 during summer and 17-81 days during winter. Pupal period continues for 2-8 days and longer during winter than in summer. Adult emerges through a 'T' shaped slit in the dorsum during the morning hours. Adults copulate immediately after emergence.

The lifecycle is completed in 32-44 days (Sandhu and Singh 1963). The conditions favouring *A. diacritica* populations are drought, nitrogen deficiency, water logging, and high humidity, heavy rains, and soil alkalinity and temperatures that are conducive for multiplication of this pest (David *et al.* 1962).

Mode of feeding and symptoms of infestation

Nymphs and adults are seen in between the veins on under surface of leaves and suck the sap from the plant parts. During the dry season the pest is highly active but its activity decreases during the monsoon period. As they suck the sup the affected parts become yellow, the leaves get wrinkled and curl downwards and finally shed. Beside feeding, they also secret honeydew which favours the growth and development of sooty mould.

In case of severe infestation, black fungal layers cover the surface area of the leaves which finally hamper the photosynthetic activity of the plants that result in stunted growth. It also acts as a vector leaf curl virus.

Management

i. Removal of infested leaves and application of neem based pesticide or spraying of Dimethoate or Malathion 50 EC @ 2 ml/l or application of Imidacloprid 17.8%SL @ 0.3 ml /l along with 2% urea..

ii. Spraying with 0.1% Acetamiprid effectively reduces white fly population.

9. Leaf miner, *Phyllocnistis citrella* Stainton (Gracillariidae: Lepidoptera)

Distribution

This pest is distributed in different parts of the world including Africa, Australia, Japan, China, Malaya Formosa, Philippines, East Indies and India (Rajput and Hari Babu, 1985).

Hosts of commercial importance

It is listed as a common and important pest for amla, citrus, jasmine, cinnamon etc.

Identification

Adults ofare minute moths with about 4 mm wingspread. They have white and silvery iridescent scales on the forewings, with several black and tan markings, often with a black spot on each wingtip. The hind wings are white, with long fringe scales extending from it. The head is very smooth-scaled and white and the haustellum has no basal scales. Larvae are minute (up to 3 mm), translucent greenish-yellow and are located inside the leaf mine. The pupa characteristically is in a pupal cell at the leaf margin. Adults generally are too minute to be easily noticed, and are active diurnally and in the evenings.

Biology

The greenish yellow eggs are usually laid singly on the ventral side of tender leaves near the mid rib. Incubation period is 3 to 6 days. The young larvae enter the epidermis and start mining. Larval period 1 to 2 weeks. Depending upon the season, peak period may last for 3 to 4 weeks. Total life cycle completed in 20-60 days and there are 9-13 generations per year.

Seasonal incidence

The pest is active throughout the year except during extreme cold. The peak activity of insects coincides with emergence of new flushes in February-March, June-July and October- November (Koli *et al.*, 1981 and Singh, 1984).

Mode of feeding and symptoms of infestation

The larvae make serpentine mines in leaves. A central line of frass fill these mines. The leafminer larvae only infest the younger, flushing foliage. Adults lay their eggs on both the upper and lower surface of the leaves. Larvae may also form mines in succulent stems and sometimes fruit. Severely damaged leaves typically become curled and distorted and mining often results in deformation, partial chlorosis, necrosis and dropping of leaves.

Management

i. Cultural management

a. Avoiding pruning of live branches more than once in a year, so that the cycles of flushing are uniform and short.

b. Application of nitrogenous fertilizer should be restricted when leaf miner populations endemic.

c. Setting up of yellow pan water trap/sticky traps 15 cm above the canopy for monitoring and management of leaf miner @ 8-10 traps/acre.

ii. Biological management

a. Natural enemies such as *Bracon phyllocnistoids*, *Tetrastichus phyllocnistoides*, *Simpiesis purpurea*, *Cirrospilus quadristriatus*, *Kratosysma* sp., *Pediobus* sp. *Sympiesis* sp. are important in regulating the pest population (Singh, 1997). It is reported that *Ageniaspis* sp. parasitizes 80% larvae of the pest (Atwal, 1964).

iii. Chemical management

a. Cypermethrin (0.015%), Permethrin (0.01%) and Decamethrin (0.0025%) have been recorded highly effective (Radke and Kandalker, 1990: Singh and Sharma, 1990).

b. Spraying of (0.05%) Dimethoate has also proved effective in managing this pest (Tandon, 1993).

10. Leaf rollers, *Gracillaria theivora* Wism. (Gracillaridae: Lepidoptera)

Hosts of commercial importance

Ornamentals and fruit trees, vines, apple, pear etc are the host of this pest.

Identification

The leafroller usually lays its egg mass on twigs or smaller branches. At first a dark gray or brown "cement" coats the mass and this later bleaches to white. Pinholes perforate this covering in spring as larvae hatch and emerge through it. Larva are cylindrical, yellow with thin scattered hairs and feed inside the protective shelter of leaves that they roll or web together. When disturbed, the larvae wriggle vigorously and quickly drops to the ground on a silken thread.

As the larva matures, its head turns dark brown, and the plate becomes tan to olive green. Adult are small, brownish moth. At maturity, larvae attain around 1 inch length. Leafroller larvae are pale green or light brown with brown or black heads and are a little more than 1/2 inch long when full grown. When mature, leafrollers pupate within a rolled leaf. Silk webbing lines the area around each pupa. Adult moths have a wingspan of about 5/8 to 7/8 inch. Wings have a bell-shaped outline when viewed from above.

Biology

The pest spends the winter in the egg stage and completes only one generation a year. Eggs hatch into tiny larvae from March to as late as mid-May in cooler areas.

Larvae feed on leaves for about 30 days then pupate in a loose cocoon, which they form in a rolled leaf or similar shelter. Eight to 11 days later the adult emerges from the pupa. The moths live for a week and during this time they mate and lay eggs.

Mode of feeding and symptoms of infestation

Infestation by the pest may be identified by webbing, withering and dropping of leaves. Larvae bind the leaves together and feed by remaining within the leaf. In case of sever incidence leaves dry up and drop leading to drying of twig also. Leaflets turn into pale brown or dark brown in colour. Upper epidermis detaches from the lower epidermis. Black excreta fill the intervening space. Folded portion become skeletonized and finally dries up. In heavy attack, the leaflets get twisted to form cocoons. If the larvae face any kind of disturbance, all larvae leave their mines.

Management

i. Clean cultivation, removal of weeds and collection and destruction of infested plant parts along with the pest. Rolled leaves are clipped off and destroyed with the larvae in the early stage of infestation.

ii. Avoiding overcrowding of branches and maintenance of sanitation in the orchard.

iii. The natural enemies often help to keep the leaf rollers at low, non-damaging levels. Certain tachinid flies, braconid wasps and ichneumonid wasps successfully parasitize the larvae. Lacewing larvae, assassin bugs, and certain beetles are also regarded as common predators for these insects. Birds sometimes feed on the larvae and pupae.

iv. In case of sever infestation, spraying of Carbaryl (0.2%) or Quinalphos (0.05%) is effective.

11. Stone borer, *Curculio* sp. (Coleoptera: Curculionidae)

Distribution

The pest occurs in different amla growing countries of Africa, Asia, Europe etc.

Hosts of commercial importance

Amla, blueberry, cowpea, pea are damaged by the insects. Other crops infested include cotton, soybean, strawberry etc.

Identification

Egg is oval in shape and white in color measuring about 0.9 mm long and 0.6 mm wide. The larva is pale yellow in color, although the head and prothoracic plate are yellowish brown. The larva is apodous, but bears deep furrows around its body. The body is thickest about one-third the distance between the head and anus, tapering gradually to a fairly pointed posterior end. The body bears stiff bristles. The pupa is yellowish white in color. The adult is oval and robust in appearance. It is black in color, with a faint bronze tint. The mouthparts are elongate, slightly longer than the thorax, and slightly curved. The thorax and elytra are marked with coarse punctures. The beetle measures around 4.8 to 5.5 mm in length.

Biology

The female deposits her eggs in feeding sites, with only a single egg deposited in each feeding puncture. Each female deposits about 30 to 280 eggs during an oviposition period of about 45 days. Duration of the egg stage is about four (range three to six) days. The larva attains a length of about 7 mm at maturity. There are four instars. At completion of the larval stage the insect drops to the soil and burrows to a depth of 2.5 to 7.5 cm. During a prepupal period of about 3-14 days the larva creates a pupal cell, and molts to the pupal stage. Duration of the pupa is about 5-19 days.

After transformation to the adult, the beetle remains in the pupal cell for two to three days while it hardens and then digs out to the surface to emerge. Adults are most active during the morning and early evening hours, seeking shade during the sunny day time. They pretend to be dead and drop to the soil when disturbed.

Mode of feeding and symptoms of infestation

Both adult and larval stages feed on seeds of fruits. Small cavities and shallow furrows are typical forms of injury. Adults also feed on the lower epidermis of foliage, but this damage is not so significant. Externally small brown patches have been seen on the fruit surface because of oviposition. Presence of exit hole is also a characteristic of infested fruits. During June with the onset of rains tiny weevil emerges out and it continues in July-August that coincides with the fruiting season of amla. After hatching the larva travels through the mesocarp and enters inside the stone, start feeding on it and destroy it completely.

Management

i. Cultural management

Deep ploughing of the orchards after harvesting exposes the diapausing larvae and is effective in bringing down the pest population.

ii. Chemical management

a. Insecticides are commonly applied by conventional spray technology or ultra-low volume techniques to protect against injury by *Curculio* (Dupree 1970, Chalfant and Young 1988). Adults are the usual target. Usually, one or more applications are made prior to pod development, followed by insecticide treatments at three to five day intervals during pod growth.

b. Insecticide directed to the soil beneath plant canopy is also beneficial as the adults often aggregate there during the day heat. Treatment of mature pods is not necessarily beneficial (Chalfant *et al.*, 1982). Spraying of 0.2% Carbaryl or 0.05% Quinalphos at pea size of fruit is usually effective. Second spray may be applied at a fortnightly interval with altering the insecticide, if needed.

12. Fruit midge, *Clinodiplosis* sp. (Ceratopogonidae: Diptera)

Identification

Adults are tiny, fragile flies approximately 2 to 3 mm long. Mature larvae are about 1 mm long and 0.3 mm wide, legless, and yellow to red in color. Immature larvae are creamish-white, while matured ones are pinkish-orange in colour.

Biology

The adult stage (during which mating and egg-laying take place) probably lasts only for one to few days. Eggs are oviposited between the scales of flower buds after the buds begin to expand, but totally dormant buds are not infested. Eggs hatch in two to three days. Larvae apparently feed on bud tissues and on the pedicels that hold the individual flower buds to the peduncle within the developing flower cluster. Larvae probably drop to the ground after feeding, then pupate and transform to the adult stage. Oviposition continues until the end of May. Full life cycle is completed in 20 to 28 days, with the last larval stage diapausing within the soil.

Mode of feeding and symptoms of infestation

Flower buds dry up and disintegrate within two weeks after infestation. High levels of flower bud abortion may occur during winter and early spring. The severity of damage varies from year to year and tends to be worse after mild winters. Vegetative meristems may also be infested and killed or damaged, leaving only very short shoots with a few highly distorted leaves. In the beginning, small grey black spots appear at the site of infestation, it darkens to brown at later stage. The emergence hole is minute but visible easily. The affected parts of fruit rot because of the damage done by the midge larvae and also fruits become susceptible to secondary infection by different pathogens.

Incidence of this pest occurs in the fruiting season in amla from September to January. Adults are seen on wings in the months of September-October in South Indian condition. The eggs are laid inside the fruits. Larvae after hatching feed on the content of the fruits.

Management

i. After harvesting of fruits, deep ploughing of soil expose the diapausing larvae and can effectively lower down the pest population.

ii. Spraying of Carbaryl (0.2%) or Quinalphos (0.05%) at the beginning of fruiting can effectively control the pest infestation.

References

Atwal, A.S. and Josan, G.S. 1962. Citrus pests round the year. *Punjab Hort J.*2(1):8-16.

Atwal, A.S. 1964. Insect pests of citrus in the Punjab. VI. Biology and control of Citrus leaf miner, *Phyllocnistis citrella Station* (*Lepidoptera: Phyllocnistidae*). Punjab Horticultural Journal, 4(2):100-103.

Beeson, C.F.C. 1941. The Ecology and Control of Forest Insects of India and neighbouring countries. Vasant Press, Dehra Dun.

Bharpoda, T.M., Koshiya, D.J and Korat, T.M. 2009.Seasonal occurrence of insect-pests on amla (*Emblica officinalis* Geartn) and their natural enemies. Karnataka J. Agric. Sci., 22(2) :314-318.

Bhatnagar, P. and Lakra, R. K. 1992. Biology and control of hairy caterpillar, *Euproctis fraterna* Moore (*Lepidoptera: Lymontriidae*) on jujube (*Ziziphus mauritiana* Lamk.). In Bio-ecology and Control of Insects Pests: Proceedings of the National Symposium on Growth, Development and Control Technology of Insect Pests, pp-150-155

Bose, B.B. 1935. Life histories of some Indian Thyrididae (Lepidoptera). Indian Journal of Agricultural Sciences, **5**(6):737-742.

Byrne, D.N., Bellows, T.S. 1991. Whitefly Biology.*Annual Review of Entomology*, 36:431 - 457.

Chalfant, R. B., Mullinix B, Nilakhe, S. S. 1982. Southern peas: interrelationships among growth stage, insecticide applications, and yield in Georgia. Journal of Economic Entomology 75:405-409.

Chalfant, R. B., Young, J. R. 1988. Cowpea curculio, *Chalcodermes aeneus* Boheman (Coleoptera: Curculionidae); insecticidal control on the southern pea in Georgia, 1980-1986. Applied Agricultural Research, 3:8-11.

Comstock, J.A. 1963. A Fruit-Piercing Moth of Samoa and the South Pacific Islands.Canadian Entomologist.95(2):218-222.

Dadmal, S.M., Pawar, N.P. 2002. Record of Gall caterpillar, *Betousa stylophora* Swinhoe on amla (*Emblica officinalis* Gaetrn.) in Akola, Vidarbha region.InsectEnvironment, 8(2):95.

David, A.L., Kalyanaraman, V.M., Narayanaswamy, P.S. 1962.The ecology and distribution of the sugarcane whitefly, *Aleurolobus barodensis* M. in Madras state and its control.Indian Journal of Sugarcane Research and Development.6:212 - 216.

David, B.V., Vijayaraghavan, S. and Narayanaswamy, P.S. 1963.*Indarbelatetraonis* Moore- A new bark borer pest on curry leaf tree and its control. Madras Agricultural Journal, 50:195-199.

Devi, S.J. and Rajasekharan, B. 2011.Biology and morphology of amla aphid, *Schoutedonia emblica*. Madras Agricultural Journal, 98(7/9):277-278.

Dupree M. 1970. Ultra-low-volume insecticide sprays for control of the cowpea curculio. Journal of the Georgia Entomological Society 5:39-41.

Firminger, T.A. 1947. Firmingers Manuals of Gardening for India (8thed), Thacker, Spink and Co. Ltd., Calcutta.

Haldhar, S., Ram, C., & Singh, D. 2019. Biotic stress (insect) of amla (Emblica officinalis)in arid region of India: a review. Journal of Agriculture and Ecology, *07*(01), 16–26. https://doi.org/10.53911/jae.2019.7102

Hargreaves, E. 1936.Fruit-Piercing Lepidoptera in Sierra Leone. Bull. Entomol. Res. 27: 589-605.

Heu, R., K. Teramoto, E. Shiroma, B. Kumashiro, and R. Dinker. 1985. Hawaii Pest Report. **5**(1-4):4-5.

Jhala, R. C., Bharpoda, T. M., Patel, M.G., Patel, V.B. and Patel, J.R. 2005. Survey on insect pests and their natural enemies occurring in amla orchards under middle Gujarat conditions. In: Amla in India.Amla Grower Association of India, Salem, Tamil Nadu, India, pp-139-143.

Koli, S.G.; Maker, P.V. and Choudhari, K.G. 1981.Seasonal abundance of citrus pests and their control.Indian J. Ent. 43(2):183-187.

Kumar, R., Ali, S. and Umesh, C. 2007.Seasonal incidence of insect pest of *Vigna munga* and its correlation with abiotic factors.Annals of Plant Protection Sciences, 15:366-369.

Kumar, K. and Lal,S.N. 1983. Studies on the Biology, Seasonal Abundance and Host-Parasite Relationship of Fruit Sucking Moth *Othreis fullonia* (Clerk) in Fiji. Fiji Agric. J.45(2):71-77.

Kumar, S. 1990. Insect pests of Socio-Economic tree species of tribal areas in Madhya Pradesh and their control. Recent Research in Ecology, Environment and Pollution, 5:53-67. Eds. Trivedi, R.N., Sen Sharma, P.K. and Singh, M.P. Today and Tomorrow's Publishers, New Delhi.

Kushwaha, K. S. and Bhardwaj, S.C. 1968. Biology and external morphology of forage pests.1. tussock caterpillars, *Euproctis* spp. (*Lymantriidae: Lepidoptera)*. Indian Journal of Agricultural Science, 37(2):93-107.

Mani, M., Gopalakrishnan, C. and Krishnamoorthy, A. 2001.Natural enemies of ber hairy caterpillar *Thiacidas postica* Walker (*Lepidoptera: Noctuidae*) in Karnataka. *Entomon*.26(3/4): 313-315.

Mathew, G. 1997. Management of the bark caterpillar *Indarbela quadrinotata* in forest plantations of *Paraserianthes falcataria*.KFRI Research Report 122.

Mathur, R.N. and Singh, B. 1954-61. *A list of pests of forest plants in India and the adjacentcountries.* Forest Bulletin, 171:1-10.

Meshram, P.B., Patra, A. K. and Garg, V.K. 2003. Seasonal history and chemical control of gall forming insect *Betousa stylophora* Swinh.(*Lepidoptera: Thyridida*) on *Emblica officinalis* Gae.Indian Forester, 129(10): 1249-1256.

Mortan, J.F. 1960. The emblic (*Phyllanthus emblic* L.).Economic Botany. 14:119-127.

Mosse-Robinson, I. 1968. Fruit-Sucking Moths (*Lepidoptera: Noctuidae*). Australian Zoologist. 14(3):290-293.

Patel, G. A. and Kulkarni, H.L. 1954.*Cerciaphis emblica* sp. nov. (Fam: Aphididae)- A new aphid pest on *Emblica officinalis*. Journal of the Bombay Natural History Society, 51: 597-607.

Patel, G.P. and Kulkarni, M.L. 1952.*Cerciaphis emblica* sp. *nov.,* a new aphid pest of Emblica officinalis.Current Science.21:350.

Patel, J.J.; Patel, J.R.; Valand, V.M.; Patel, B.H. and Patel, M.J. 1998.Bioefficacy of some of the new insecticides against leaf miner *Phyllocnistis citrella* and Psylla, Diaphorina citri infesting citrus. Indian J. Ent. 60(10: 50-56.

Patel, R.M., Sahay, G. and Patel, V.M. 1966. Biology of bark-eating caterpillar (*Indarbela quadrinotata* Wlk.) on guava in Gujarat. Indian J. Hort. 23:120-124.

Pathak, R.K. 2003. Status report on genetic resources of Indian Gooseberry- Amla (*Emblica officinalis* Gaertn.) in South and South East Asia.IPGRI. Pp-97.

Radke, S.G. and Kandalkar, H.G. 1990. Chemical control of citrus leaf miner.Indian J. Ent.52(3):397-400.

Rajput, C.B.S. and Hari Babu, R. 1985. Citriculture, Kalyani Publishers, New Delhi- Ludhiana, pp. 364.

Rakshpal, R. 1945. Citrus fruit-sucking moths and their control.*Indian Farming*, 6: 441-443.

Sah, B.N., Mehra, B.P. and Naqvi, A.H. 1972. Bionomics of *Euproctis fraterna* Moore (Lepidoptera: Lymantriidae) on *Ziziphus xylopyra* Willd. Indian Journal of Agricultural Science, 42(7):630-636.

Sandhu, J.S., Singh, S.1963. Studies in the biology of sugarcane whitefly *Aleurolobus barodensis* Maskell. Indian Journal of Sugarcane Research and Development.7(2):83 - 88.

Singh, S.P. 1984. Biological control of insect pests of national importance. In: Integrated Pest and Disease Management. S. Jayaraj (ed). TNAU, Coimbatore. pp. 309-318.

Singh, K. and Sharma, V.K. 1990. Chemical control of citrus leaf miner (*Phyllocnistis citrella* Stainton). Advances in Horticulture and Forestry, 1: 92-94.

Singh, S.P. 1997. Biological suppression of fruit pests. In; Fifteen Years of AICRP on Biological Control. Project Directortae of Biological Control, Bangalore, pp. 139-171.

Sontakay, K.R. 1945- The bark eating borer of orange. Indian Farming,6(2):74-75.

Tandon, P.L. 1993. Insect and mite pests of tropical fruits.In Advances in Horticulture – Fruit crops. Malhotra Publishing House, New Delhi. 3(3); 1530-1550.

Tryon, H. 1898. Orange-Piercing Moths - Fam. Ophiderinae. Queensland Agric. J. 2: 308-315.

Verma, A. N., Singh, R. and Khurana, A. D. 1972. Chemical control of hairy caterpillar.Indian Journal of Agricultural Science, 42(10): 928-931.

Waterhouse, D. F. and Norris, K. R. 1987. Chapter 29 *Othreis fullonia* (Clerk). pp. 240-249. In: Biological Control Pacific Prospects. Inkata Press; Melbourne., pp-454s.

12

Pests of Ber and Their Management

Suraj Sarkar and Puran Pokhrel

Ber (*Ziziphus mauritiana* Lam.) is one of the most ancient and important fruit crops. It is considered as the "King of arid zone fruits", due to its adaptations to tolerate the biotic and abiotic stresses prevailing under rainfed conditions (Anbu *et al.*, 2009). It is also called as desert apple, jujube, Badari (Sanskrit), Chinee apple, Ber (Hindi), Kul or Boroi (Bengali), Indian plum, Boroi, Beri, Permseret (Anguilla) etc. It is a tropical fruit tree that belongs to the family *Rhamnaceae*. Ber is rich in amino acids, protein, sugar, calcium, phosphorus, carbohydrates, iron, vitamin A, ascorbic acid. The fruit contains from 70 and 165 mg ascorbic acid per 100 gm of pulp, which is two to four times higher than the vitamin C content of citrus fruits. The mineral content of phosphorus, calcium and iron in *Z. mauritiana* fruit is also reported as being higher than in apples and even oranges (Morton *et al.*, 1987, Jawanda and Bal,1978).

There is a traditional Chinese proverb that "eating three jujubes a day keeps the doctor away" (Bao, 2008). Generally, they are eaten as fresh, however, may be pickled, dried and made into confectionery or juice can be extracted for drinks (DeKock, 2006).

In India, the crop is grown in about 22,000 ha area. In Indian arid and semi-arid region ber is a popular fruit crop (Jamandar *et al.*, 2009) and it is mostly cultivated in Rajasthan, Gujarat, Punjab, Haryana, Uttar Pradesh and Maharashtra. In Tamil Nadu, Karnataka, Andhra Pradesh, Bihar, Chhattisgarh, Assam, Madhya Pradesh and West Bengal ber is cultivated to some extent. Ber is a quite nutritious fruit containing higher amount of vitamin C (much higher than apple and citrus) after only aonla and guava (Khera and Singh, 1976).

The successful production of ber is hampered due to attack of various insect pests which reduce the yield, spread disease, injure or kill plants and adversely affect quality of the fruits. Though the crop is becoming popular among the growers but the avoidable loss is more due to insect pests and diseases. In India, more than 130 species of insect pests found to infest on ber (Lakra and Bhatti, 1985) but only few species have attained the pest status and cause substantial economic damage to ber. Moreover, the consumers are now becoming aware of quality of ber fruits and hence plant protection measures are needed to safeguard the infestation of prevailing insect pests.

Table 1: The insect pests causing major economic damage onber

Sl. no.	Common name	Scientific name	Family	Order
1.	Fruit fly	*Carpomyia vesuviana* (Costa)	Tephritidae	Diptera
2.	Jujube beetle	*Adoretus himalayanus* (Voss)	Curculionidae	Coleoptera
3.	Hairy caterpillar	*Euproctis* sp.	Lymantriidae	Lepidoptera
4.	Leaf webber	*Synclera univocalis* (Walker)	Pyraustidae	Lepidoptera
5.	Bark eating caterpillar	*Indarbela quadrinotata* (Walker)	Metarbelidae	Lepidoptera
6.	Mealy bug	*Perissopneumon tamarindus* (Green)	Margarodidae	Hemiptera
7.	Fruit borer	*Meridarchis scyrodes* (Meyr)	Carposinidae	Lepidoptera
8.	Leaf roller	*Psorosticha zizyphi* (Stainton)	Oecophoridae	Lepidoptera
9.	Hopper	*Qadria pakistanica* (Ahmed)	Cicadellidae	Hemiptera
10.	Lac insect	*Kerria lacca* (Kerr)	Lacciferidae	Hemiptera
11.	Termites	*Odontotermes* sp., *Microtermes* sp.	Termitidae	Isoptera
12.	Ber butterfly	*Tarucus theophrastus* (Fabricius)	Lycaenidae	Lepidoptera
13.	Green slug caterpillar	*Thosea* sp.	Limacodidae	Lepidoptera
14.	American bollworm	*Helicoverpa armigera* (Hubner)	Noctuidae	Lepidoptera
15.	Mite	*Larvacarus transitans* (Ewing)	Tenuipalpidae	Tetranychoidea
16.	Grey hairy caterpillar	*Thiacidas postica* (Walker)	Noctuidae	Lepidoptera
17.	Mite	*Eriophyes cernus* (Mssee)	Eriophyidae	Acari
18.	Leaf miner	*Cameraria* spp.	Gracillariidae	Lepidoptera

1. Ber fruit fly, *Carpomyia vesuviana* (Costa) (Tephritidae: Diptera)

Distribution

This pest is widely distributed in countries like India, Iran, Pakistan, China, Bangladesh, Georgia, Turkey, Turkumenia, Mauritius, Uzbekistan, Indian Ocean Island, Oman and South Europe.

Hosts of commercial importance

Carpomyia vesuviana was originally recorded from Italy on the slopes of Mount Vesuvius, hence the name given. Most destructive pest of ber is the fruit fly that causes a good amount of yield loss and reduces the quality of fruits. It is a monophagous pest of *Zizyphus* species and may cause loss upto 80% under severe infestation (Karuppaiah*et al*., 2014). All the wild and cultivated species of ber like *Z. zizyphus, Z. numularia, Z. mauritiana, Z. lotus, Z. spina-cheresti, Z. jujube, Z. rotuntifolia and Z. sativa* are attacked by this pest.

Identification

The eggs are dull-creamy white in colour, opaque and 0.9 to 1.0 mm long. Full grown maggots are creamy white and measuring about 6 mm long. Pupae are barrel shaped, 4 to 5 mm long and light brown when freshly formed and coriaceous in 3 to 4 days. The adult fly is brownish-yellow, smaller than the housefly; brownish longitudinal stripes are present on the thorax which are surrounded by black spots. Bristle hairs are present on the tip of abdomen. The wings having four yellowish or greyish-brown characteristic cross bands.

Biology

Female flies insert their eggs in clusters of 10 to 50 under the fruit skin about 1/25 to 1/8 inch below from the fruit surface. The eggs are white, elongate and elliptical measuring about 1/25 by 1/250 inch. Within 1-2 days they hatch. Larvae are white in colour, legless, and look like an elongated cone. The mouth is present at the pointed end of its body. Number of larval instars are 3. The third instar larva has a length of 2/5 inch. Entire larval stage continued for 11-15 days. Matured larvae drop to the ground and pupate in the soil. The puparium is yellowish-brown and seed-like. Adults emerge in about 10 days. Generally, two horizontal black stripes present on the abdomen and from the base of the third segment a longitudinal median stripe extending to the apex of the abdomen. These markings may form a 'T' shaped pattern, but the pattern considerably varies. Upon emergence from the puparium females starts laying eggs about 8 days after. Under optimum conditions, a female during her lifetime can lay more than 3,000 eggs, but under field conditions the egg laying capacity is approximately 1,200 to 1,500 eggs per female.

Seasonal incidence

The fly become active in Autumn and continues its activity throughout winter and spring, though maximum infestation is caused during February and March. External factors like rainfall, temperature and relative humidity influence the damage intensity of the pest (Lakra and Singh, 1985).

Mode of feeding and symptoms of infestation

Infestation initiates at the fruiting period; adult female fly lays eggs in the young developing fruits. After hatching, maggots start feeding on the pulp. Larvae make galleries which are filled with excreta and causes rotting of the infested fruits. The fruits become deformed, retardation of growth occurs and such fruits are easily drop off (Singh, 2008).

Management

i. Cultural management

a. Clean cultivation and sanitation of the field is an effective measure to prevent fruit fly infestation by reducing the reproductive cycle and minimizing the population build up (Singh, 2008).

b. Ploughing the soil during summer, rainy and spring period can destroy the hibernating pupal stage by exposing them to the sunlight and natural enemies like birds. Heating of the soil by burning grasses and summer irrigation can also kill the pupae (Singh, 2008).

c. Faroda (1996) performed cross between Seb x Tikidi variety of ber and the obtained F_1 generation showed resistance of 90% with poor quality fruit and backcrossed BC_1 line showed resistance of 87 to 90% with desirable fruit characters. Lakra and Singh (1983) reported that the adult flies show a tendency of avoiding egg laying in those fruits that have a size less than 9 x 4.5 mm and prefer to lay more than 50% of their eggs in fruits that have a size 20 x 9 mm. Arora *et al.* (1999) reported that infestation of fruit fly has a positive correlation with the weight of fruit, pulp stone ratio, TSS and total sugars whereas it also has a negative correlation with vitamin C, acidity and total phenol. According to Saxena and Rawat (1968) the early maturing cultivar that produce medium to big fruit size, juicy, sweet soft and thin skinned and attractive coloured fruits are more susceptible to fruit fly. Varieties of ber like Kekara, Mundia, Umran, Chhuara and Banarasi suffer less incidence of fruit fly whereas, incidence is higher in varieties like Gola, Ajmeri and Kaithali.

ii. Mechanical management

a. Male annihilation technique can be used by setting up fly trap using methyl eugenol. Prepare methyl eugenol 1 ml/L of water + 1 ml of Malathion solution. Between 6 and 8 AM place 10 ml of this mixture in every trap and keep them at 25 different places in one ha. Collect and destroy the adult flies.

iii. Biological management

a. Singh (1989) reported that Braconid parasitoid, *Biostresvanden boschi* Fullaway from India can control fruit fly but the proportion was very negligible. Other parasitoids like *Opius carpomyia, Bracon fletcheri* and *Omphalia* sp. were also noticed (Kavitha and Savithri, 2002). But reduction of the pest population was not significant (Saxena and Rawat, 1968).

According to Farrar *et al.* (2004) braconid parasitoid *Fopius carpomyia* infect the larval stage and its ovipositor was very suitable for parasitisation. The rate of parasitisation was 21 to 26.7%.

b. Rajaram and Siddeswaran (2006) reported that spraying of azadirachtin 1% extract and *Ocimum sanctum* 1% were effective for 10 days. Ahmed *et al.* (2005) reported the integrated management of fruit fly with dipterex 100g/ acre + 5% molasses baiting, followed by hoeing and collection of fruits fallen on the ground throughout the season proved to be better compared to single treatment applied trees. Mari *et al.* (2013) reported that application of tobacco extracts and neem powder can reduce the infestation of *C. vesuviana* significantly.

iv. Chemical management

a. Dashad *et al.* (1999) reported that insecticide schedule consisting of Fenthion 0.05% and Carbaryl 0.01% at an interval of 15 days were effective. Spraying with Dimethoate followed by Eco-neem can reduce the infestation (Singh *et al.*, 2008) and lowest incidence of fruit fly was observed when Lambda Cyhalothrin 0.0025% applied followed by Beta-cyfluthrin 0.0018% (Gyi *et al.*, 2003).

2. Jujube beetle, *Adoretus pallens* Arrow (Scarabidae: Coleoptera)

A number of beetles have been reported to attack ber tree, which include, *Adoretus himalayanus* Voss, *A. pallens* Arrow, *A. nitidus* Arrow, *A. versutus* Harold.

Distribution

This is recorded as an emerging pest of ber. In India it was first time reported from Andhra Pradesh. Later it was reported from Rahuri of Maharashtra and Jobner of Rajasthan in 1996 (Pareek and Nath, 1996) and recently in Bikaner district of Rajasthan, India during 2010. Apart from India, this pest has been reported to cause severe damage in Bangladesh and Pakistan also.

Hosts of commercial importance

Though the pest is polyphagous in nature infesting jujube, phalsa, jamun, fig, guava, grapevine, pomegranate, mulberry trees and many other ornamental and shrubs. They generally prefer to feed on ber and grapevine.

Identification

Eggs are smooth, white and elongated. Grubs are wrinkled, creamy-white and 'C' shaped. Full grown maggots measure about 15 mm in length. Adult beetles are stoutly built, hardy, dirty-yellowish coloured and convex shaped, measuring about 8 mm long. Tarsi on the tips of their legs are deep red and their size varies from 9 to 13 mm.

Biology

Eggs are laid singly in soil during May to August. These hatch in about 6 to 9 days. Grubs remain active throughout the summer months and make earthen cells in autumn to hibernate during winter. Pupation takes place in the soil during next spring. Pupal period lasts for 11-12 days. The pest is univoltine in nature and the total life cycle completes in 4-6 months.

Seasonal incidence

Large numbers of adult beetles emerge from the soil and appear early in spring, their activity reaches maximum during summer months and population declines thereafter with heavy monsoon showers.

Mode of feeding and symptoms of infestation

Fruits are directly damaged as the grubs make tunnels inside the seeds. Adult beetles make ovipositional punctures in the mature fruits. Adult beetle are active in the field during morning and evening time. Eggs are laid by the adult female beetle mostly on the stylar end; rarely on the distal end of the fruits and the oviposition punctures are covered with brown encrustation. After hatching, grub enters into seed by puncturing the endocarp and then starts to feed on soft seed coat. Later it move downward and enters into endosperm. Pupation takes place within the seed by making hollow galleries. The beetle completes its entire life cycle within a single fruit. The pest completely feeds on the developing seed. They can infect all stages of fruit; however, it is prevalent in fruits of pea to pebble size. Infested fruits are abnormal in shape and nearly half of the fruit towards stalk portion turns reddish brown with a rough surface and the remaining half towards stylar portion remains greenish colour. Often, infested fruits fail to mature.

Management

i. Cultural

a. Frequently ploughing up the soil around the orchard during summer months (july to September) and again during winter months (December-January) to expose the grubs, hibernating pupa and inactive adults to predatory birds and natural enemies.

b. If possible, some highly attractive host trees like drumstick (*Moringa oleifera*), karonda (*Carissa carandas* Auct.), neem (*Azadirachta indica* A. Juss.), khejri (*Prosopis cineraria*) should be planted at the border of orchard to attract the adult beetles and escape damage to ber plantation.

ii. Mechanical

a. Collection and destruction of infested fruits around the trees.

b. Manual collection of adult beetles by vigorously shaking the host tree at night started from the day first good pre-monsoon shower is received and should be continued daily at evening hours for 5-6 days.

iii. Biological management

a. Spraying of botanicals such as *Pongamia glabra* leaf extract and neem seed kernel extract (NSKE @ 5%) is reported to be effective against jujube beetle.

b. Use of fungi *Beauveria brongniartii* (Sacc.) petch and *Metarhizium anisopliae* (Metchnikoff) and bacteria, *Bacillus* spp. has shown some efficacy against the pest.

iv. Chemical

a. The beetles congregate at a large number in the host tree at night, this provides an ample opportunity to kill them by insecticidal spraying with Carbaryl 50 WDP at 0.1% just before setting of the fruit and repeat the sprays at three weeks interval.

b. Adults generally prefer manure pits, dung heaps, rotten organic matter for oviposition. Such manure pits can be treated with Methyl parathion dust (Metacid 2% dust) @25 kg per ha to kill the adults.

3. Bark eating caterpillar, *Indarbela quadrinotata* (Walker) (Metarbelidae: Lepidoptera)

Distribution

From India only about 14 species of these moths belonging to 3 genera have been reported. Out of these, genus *Indarbela* is the largest that contain 12 species of which, at least 5 species viz., *I. campbelli* Hampson, *I. minima* Hampson, *I. watsoni* Hampson, *I. tetraonis* Moore and *I. quadrinotata* Walker, occur in southern India. Of these, *I. quadrinotata* is the most common and widely distributed species.

Hosts of commercial importance

Indarbela quadrinotata is highly polyphagous in nature and the recorded hosts include *Acacia catechu, A. lenticularis, A. tortilis, Albizia chinensis, A. lebbeck, A. odoratissima, A. sini, A. procera, A. siris, Anogeissus* sp., *Artocarpus integra. Bassia* sp., *Bauhinia* sp., *Berrya* sp., *Bombax* sp., *Boswellia* sp., *Callicarpa* sp., *Cassia fistula, Casuarina equisetifolia, Citrus aurantifolia, Gmelina arborea, Lagerstroemia speciosa, Mangifera indica, Morus alba, Mytrogyna* sp., *Phyllanthus emblica, Psidium guajava, Populus nigra, P. deltoides, Prunus armeniaca, P. salicinae, Punica granatum, Samanea saman, Shorea robusta,*

Stephegyne sp., *Syzygium curnini, Strychnos* sp., *Tectona grandis, Terminalia arjuna, T. superba, Xylia xylocarpa* and *Zizyphus mauritiana* (Kumawat and Swaminathan, 1990; Patil *et al,* 1990; Singh and Bhandari 1987).

Identification

Eggs are laid in clusters on tree bark. The newly hatched larvae are light brownish, bare. The full grown larva is smooth having sparse hairs on body and measures about 3.5 to 4 cm in length. Light brown sclerotized patches are present dorso-laterally. They have simple thoracic legs with the last segment ending in a curved claw. Abdominal legs are present on segments 6-9. The pupa is light brownish in colour measuring 1.5 cm in length and with 2 short pointed cephalic processes. Abdominal segments bear transverse rows of spine-like processes, of which there are two rows each on segments 5 to 8. Only one row on 3, 4 and 8 segments and only two spines on segment 10. In the caudal end there are several small spine like processes. These spines and teeth like processes help the pupa to orient itself towards the tunnel mouth prior to eclosion. Pupal period lasts for 3 weeks. The moth is light brown in color and measures 15- 18 mm across the wings. Fore wing has a sub-apical brown spot and with several transverse rows of brown scales. Hind wing is light black in colour. Head is depressed and antennae strongly pectinate, with the pectination being uniform throughout (Mathew, 1997).

Biology

Complete life cycle of the insect has been worked out on various hosts by different workers. Life cycle is annual and has one generation per year. Eggs hatch in 15-25 days. The larval period lasts for 9- 10 months. Pupation takes place within the tunnel of larva, with the cephalic end of the pupa protruding outside slightly. The pupal period continues for about 15-25 days.

Seasonal incidence

With the onset of rains in late May, new generation start to appear. During this period, eggs and the newly hatched larvae could be seen on the bark of the trees. Usually, larva takes 2 to 3 months to make the tunnel in the wood. Until October fresh attack continued, after which the population remained more or less steady.

Mode of feeding and symptoms of infestation

Adults start emerging during May to July. In early June females oviposit under loose bark in clusters. The newly hatched larvae initially occur in groups browsing on the bark. Subsequently they migrate and get lodged in the axils of side shoots or lateral buds to initiate an infestation. As the larva grows, it slowly tunnels into the wood. This tunnel is used for shelter and is kept free of frass. Larva continuously maintains the dimensions of the tunnel in the wood to accommodate its growth.

For constructing the sleeve-like structure that extends from the tunnel mouth, larva use frass and excreta that are removed from the tunnel. The larva invariably feeds upon bark under concealment in the sleeve. It is a nocturnal pest. It feeds on the bark of the host plant and remains hidden inside the galleries. Presence of galleries formed by silk and frass that is loosely plastered on trunk and main branches, especially near the fork portion is the characteristic symptom of presence of bark eating caterpillar. Large patches of bark that are eaten away, expose the tree to various pathogens.

Compared to post monsoon months, larval feeding is brisk during the summer months. Mishra (1991) reported that larval feeding is correlated with temperature and relative humidity. He further noticed that food consumption increased with a rise in temperature, the maximum being 166 mg/day at 35°C and minimum 16 mg/day at 12.5°C. Similarly, the food consumption is maximum at 96% RH while it decreased with lowering of RH, the minimum being 3 mg/day at 56% RH.

Management

i. Mechanical management

a. Thrusting of flexible wire in the active holes and plugging the holes with mud has been quite effective, eco-friendly and economic way for management of this pest.

b. Removal of ribbon like frass from the younger tree can reduce the infestation up to 50 percent. If neglected, old trees are more prone to attack by this pest.

ii. Biological management

a. Two species of entomogenous fungi viz., *Aspergillus candidus* Link. and *Beauveria bassiana* (Bals.) Vuvill, have been reported to cause mortality of this insect in the field. Of these, *A. candidus* has been reported on *Indarbela* developing on pomegranate in Hissar, where about 5- 11% infection was recorded (Ramsingh *et al.*, 1982).

b. An insect parasite *Podagrionella indarbelae* (Chalcidoidea: Torymidae) was reared from the eggs of a related species viz., *I. tetraonis* collected on cashew from Kerala (Narendran and Sureshan, 1988).

iii. Chemical management

a. Injection of Chlorpyriphos into the borer tunnels and plugging the holes with soil soaked with insecticides (Sidhu and Poon, 1983).

b. Cotton swabs by soaking in 10 ml of Dichlorvos or petrol used to plug the bore hole. Dichlorvos gave 100 % pest control whereas, petrol provided 96.7% pest control after 7 days of plugging and no infestation will occur for 60 days after treatment (Patil *et al.*, 1990).

c. Surface application of chemicals is effective in trees like orange having coarse bark where 90-95% control has been recorded.

d. Application of the solution of 1 lit of kerosene and 100 g soap in 9 lit of water in the holes controls the bark eating caterpillar effectively.

4. Hairy caterpillars, *Euproctis fraterna* (Moore), *E. lunata* Walker, *E. subnotata* Walker and *E. flava* (Bremer) (Erebidae: Lepidoptera)

Distribution

Over 50 species of Euproctis occur in India and Pakistan of which *Euproctis fraterna, E. lunata, E. subnotata and E. flava* are common on ber, but *Euproctis fraternal* is the predominant one. The caterpillars are widely distributed in the Palaearctic, African, oriental, Australian region and the Indian subcontinent.

Hosts of commercial importance

Hosts include plum, jujube, jamun, phalsa, pomegranate, aonla, fig, mulberry, custard apple, loquat, grapevine, guava, peach, pear, citrus, mango etc. plum and castor are the most preferred hosts (Sah *et al.*, 1972).

Identification

Eggs are spherical shaped and covered with yellow hair from anal tuft of the female moth. Full grown caterpillar measures about 18-25 mm in length, reddish brown colour with plenty of hair all over the body. During harvest, body contact with urticating hair leads to skin irritation and itching to workers. Adult male moths are stout, dull coloured. Adult females are robust, yellowish moth having brownish forewing.

Biology

Female moth lay about 92 to 241 eggs on the ventral surface of tender leaves in clusters. Egg, larval and pupal stage last for 9-14, 25-56, 7-15 days respectively. The larva pupates in silken cocoon in leaf fold or on a bunch. The adult longevity is 5-7 days and complete the total life cycle within 44-84 days. The pest completes six generations in a year on certain arid fruits and the duration of development varies considerably on different hosts (Kuswaha and Bhardwaj, 1968).

Seasonal incidence

The larval development is faster during spring and summer, slows down during winter and again becomes active in February.

Mode of feeding and symptoms of infestation

Newly hatched caterpillars feed gregariously on lower surface of tender leaves by scrapping the chlorophyll, consuming the leaf margins and eventually skeletonizing them completely. After 3rd instar the larvae spread out and feed voraciously, resulting in defoliation of twigs and branches and sometimes the whole tree. The caterpillars also nibble the epidermis of fruits resulting in a corky, deformed surface rendering the fruits unfit for human consumption. The attack of this pest is more pronounced to tender and succulent leaves than the older ones.

Management

i. Mechanical management

a. Manual collection of egg mass and congregating larvae and destroy or dipping them in kerosenised water.

ii. Biological management

a. The hairy caterpillar is parasitised by *Apanteles taprobanae* (Hymenoptera: Braconidae), *Brachymeria* sp. (Hymenoptera: Chalcididae), *Charops obtusus* and *Goryphus* sp. (Hymenoptera: Ichneumonidae).

b. *Exorista* species has recently been found to be a parasitoid of the caterpillar in Karnataka (Mani *et al.*, 2001).

iii. Chemical management

a. Verma *et al.* (1972) reported that Carbaryl 0.1 % or Methyl parathion 0.05 % can effectively control all the larval instars.

b. Bhatnagar and Lakra (1992) reported that 3rd to 6th instar larvae are controlled most effectively by applying 0.005 % Cypermethrin, 0.0014 % Deltamethrin, 0.0075 % Fluvalinate and 0.0005 % Fenvalerate.

c. Spraying of 0.2 % Carbary 150 WP was found to be effective to control the pest (Pareek and Nath, 1996).

5. Ber leaf Webbers, *Synclera univocalis* (Walker) (Pyraustidae :Lepidoptera)

Distribution

The pest has been reported from a number of countries like India, Pakistan, Sri Lanka, Myanmar, Palestine, Syria, South Africa and South America. In India wide presence of this pest observed in the states of Punjab, Haryana, Uttar Pradesh, Rajasthan and Gujarat.

Hosts of commercial importance

About a dozen species of leaf webbers/rollers like *Syncler aunivocalis* (Walker), *Pagydatra ducalis* Zeller, *Ancylis* spp., *Acroclita corinthia* Meyrick, *Phycita* sp., *Limaecia* sp., *Pleurota* sp., *Psorosticha zizyphi* (Stainton) have been reported from different countries damaging ber tree (Verma*et al*., 2004). Cultivated ber (*Z. mauritiana*) is the most preferred host while wild ber (*Z. mauritiana* var. *rotundifolia*) is least preferred host.

Biology

Female moth lay on an average 75 eggs on tender leaves singly or in clusters. Incubation period is 2 to 4 days. Larval and pupal stage last for 13-17 and 8-14 days respectively. Adult longevity is 4-11 days. The pest is multivoltine and completes as many as 14 generations per year.

Seasonal incidence

Though the pest remains active throughout the year but cause greater damage to ber orchards during summer and rainy season (May to November). It has been observed that during peak vegetative flush, sometimes more than 50% tender leaves are attacked by one or the other species of leaf webbers/rollers.

Mode of feeding and symptoms of infestation

Newly hatched caterpillars attack the unopened or partially opened leaf and the latter instar larvae keep the leaf folded along the long axis with silken threads. The larvae feed by scrapping the chlorophyll leaving only the papery epidermis. Advanced stage larva may join two adjacent terminal leaves together longitudinally and feed within the leaf fold. Usually single larva is present in each web but sometimes 2 to 3 larvae may present. Severely attacked leaves curl up, turn brown and latter dry up giving an unhealthy appearance to the tree. Due to caterpillar attack small plants remain stunted, flowering and fruit set is adversely affected.

Management

With the appearance of the pest on new vegetative flush during May-November spraying of 25 ml Fenvalerate (Fenval 20 EC), 200 ml Quinalphos (Ekalaux 25 EC), 25 ml Cypermethrin (Cymbush 25 EC) or 200 gm Carbaryl (Sevin 50 WP) in 100 litres of water is recommended. Same schedule is to be repeated after 3 to 4 weeks if there is reappearance of the pest. (Mann and Brar,1980; Mann and Singh,1981; Singh and Mann,1982; Batra *et al*.,1988).

6. Ber mealy bug, *Perissopneumon tamarindus* (Green) (Margarodidae : Hemiptera)

Distribution

This pest is widely distributed in India and Pakistan. In India the pest is more common and destructive in the states of Punjab, Haryana and Rajasthan.

Hosts of commercial importance

The pest is polyphagous in nature and infests apple, banana, ber, citrus, fig, mulberry, tamarind etc.

Identification

Eggs are oval, shiny yellow and less than one mm long. Grubs are flattened-oval, dull, ash-brown colour with a slight mealy covering. Adult females are wrinkled, apterous, 10-14 mm long and completely covered with waxy secretions. Males are alate, having well developed single pair of wings.

Biology

Gravid females descend down from tree during September-October and lay eggs in the soil at 50-150 mm deep. A single female lays 100-300 eggs in 50-55 days. The eggs overwinter in soil and hatch in spring (April-may). Nymphal periods are 40-62 days and 42-65 days in case of male and females respectively. There are 3 nymphal instars in female and 2 for male followed by prepupal and pupal stages. Prepupal period lasts for 1-3 days and pupal stage varies between 10-14 days (june-july). Female longevity is 59-81 days and the male dies soon after fertilization. The pest is univoltine in nature.

Mode of feeding and symptoms of infestation

During summer months nymphs and adult females may be seen congregating on shoots, twigs and tender leaves sucking cell sap. As a result of continuous desapping, there is loss of vitality and in case of severe infestation growth of tree arrested and the fruiting capacity are badly affected. Sooty mould grows rapidly on the honeydew secreted by the bug covering the area with a superficial black coating.

Management

i. Frequent raking of the soil under the trees during winter months help to destroy the overwintering eggs and thereby reduces the pest population. Tying around the tree with 300 mm wide alkathene sheet (400 gauge) prevent the nymphs from ascending the tree.

ii. Grease painting may also be done. In case of severe infestation, spraying with 0.1% diazinon is recommended.

7. Fruit borer, *Meridarchis scyrodes* (Meyr) (Carposinidae: Lepidoptera)

Distribution

The pest inflicts serious damage to ber in Southern India and Gujarat. The pest is also reported from Pusa (Bihar) and Nagpur (Maharashtra) but in northern India this is not a serious problem.

Hosts of commercial importance

T. B. Flecher in the year 1916 first time observed that the larvae of this pest feed on the pulp of ber at Coimbatore. Besides ber, the pest is also recorded to damage jamun and olive trees.

Identification

Adult moth is small dark brown in color and having fringed wings. The early instar larvae is light yellowish and mature larvae is red in color.

Biology

Female moth lays an average of 23-35 eggs and has an incubation period of 5.30 ± 0.40 days. Pre-oviposition period is of 1 day. The larval period is of 16.30 ± 3.50 days covering five larval instars. Pre-pupal and pupal stage lasts for 1-2 and 7-8 days respectively. Adult male and female in the presence of food survive for an average of 2.65 ± 0.50 and 3.70 ± 0.27 days respectively. One generation is completed in about 30 days (Patil and Patil, 1997).

Seasonal incidence

The fruit borer is more active from July to August (Ayyar, 1923). Sonawane (1965) reported that the pest incidence prevailed only during fruiting period.

Mode of feeding and symptoms of infestation

Female moth lay eggs in the depression near the fruits talk. Larvae upon hatching bore into the fruits and start feeding on the pulp and eventually reach the seed surface and feed on pulp around the seed. The larval period is completed within the fruits. The infested fruit is filled with larval excreta and more than one larva may be found in a single fruit. Bored holes are the external symptom of borer damage and presence of excreta inside the fruit is the characteristic symptom of borer damage.

Management

i. Wild ber bushes present in the vicinity of ber orchard should be removed and clean cultivation has to be done.

ii. Collection and destruction of all infested and fallen fruits has to be done.

iii. Harvesting of fruits is to be made at early stage when they are still green but fully mature.

iv. Spraying of Malathion @ 2 ml or Carbaryl @ 4 g or Fenvalerate @ 0.5 ml /lit of water or Neem seed kernel extract (NSKE) @ 5% in sequence, in which first spray is done at marble stage, second spray after 15 days and third spray during fruit ripening stage along with alteration of insecticides would be effective against the fruit borer (Yelshetty *et al.*,1996, Karuppaiah, 2013).

8. Ber gall mite, *Larvacarus transitans* (Ewing) (Tenuipalpidae: Tetranychoidea)

Distribution

The pest is widespread in India exclusively on jujube (Pritchard and Baker, 1958) but till date its presence is found only in Pusa, Bihar, Indo-Gangetic plains, Delhi and Rajasthan. Latif and Wali (1961) reported that this species causes considerable damage to orchards, especially to grafted plants.

Hosts of commercial importance

Larvacarus transitans produces twig galls and considered as a major pest of ber in Rajasthan. Association of the galls and mites were first time reported by Ewing in 1922 but detailed biology of this pest is still unknown.

Identification and biology

Adults with three pair of legs, anterior margin of prodorsum is without notch or projection and smoothly rounded, 2 pairs of setae and 1 segmented palpi is present. Cuticle is strongly striated. Pritchard and Baker (1958) mentioned that unlike eriophyids, these mites approach the gall forming habit which shows only at early stages of adaptation to such life. The adoptions elongate the body of mites and reduce the rear pair of legs. The female can be differentiate from male having round opisthosoma, whereas the opisthosoma of male consists of a pair of terminal stylet like rods known as genital stylets. A male measure about 244 microns while female is of 270 microns.

The larva matures into females, which lay unfertilized eggs giving rise to males. This process takes about 25-30 days. One of the males fertilizes the mother female producing a normal male-female progeny.

Seasonal incidence

Sharma and Kushwaha (1989) reported that the incidence of mite started from second week of October and continues till second week of August. The maximum numbers of galls are found in May. Similarly, peak mite population occurs during May and gradually declines later in subsequent months due to prevailing high temperature and subsequent monsoon shower. The galls get ruptured during monsoon and the mites after emerging from galls hide themselves in cracks and crevices of tree trunk, and start fresh infestation during September when new plant growth takes place after rainy season.

Mode of feeding and symptoms of infestation

The mite produces lac-like, hemispherical, disc shaped, closed, soft and hollow galls on the cortical region of stems similar to black scales. At first these galls appear like minute specks less than 10 micron in length and width, giving the appearance of rust on the green twigs. They can achieve a maximum size of 6 to 7 mm in length and 4 to 4.5 mm in width. A single larva settles on the fresh twig and starts producing a covering of rusty secretions. The origin of this is not known and further study is needed. This secretion acts as growth regulator, starting the stimulation of plant tissue by altering the growth pattern of the twig to create a suitable micro-environment for the mite to develop in an enclosed gall. Generally, each gall is started by a single larva, but sometimes several larva settle close together producing several galls which coalesce into a single mass stretching along the twig, having a number of larvae within several partitions.

The mite larva escapes from gall through a hole bored by the larva of a species of *Anarsia zeller* (Glechiidae, Lepidoptera). Fresh infestations started by the larva which have escaped from gall. If a hole is not made by *Anarsia* sp. all stages of the mite will perish within the gall (Ghai and Baker, 1989).

Management

Pruning of the branches up to 60 cm length from apical bud provides effective control. The twig galls produced by the false spider mite, L. *transitans* may harbor *Phytoseius intermedius* Evans and Macfarlane (Phytoseiidae). This predatory mite and *Anarsia* larva feeds on different stages of *L. transitans* (Ghai and Baker, 1989). Spraying with Ethion (0.03%) and Dicofol (0.04%) in first week of October has found effective against the mite.

9. Some minor pests

Oriental fruit fly, *Bactrocera dorsalis* (Hendel), peach fruit fly, *B. zonatus* (Saunders) and guava fruit fly, *B. correcta* (Bezzi) are polyphagous in nature and reported to feed occasionally on ber tree (Agarwal and Kapoor, 1985).

Among the lepidopteran leaf feeders castor semilooper, *Achaea janata* (Linn.), *Thiacida spostica* (Walker), *Streblote siva* (Lefevre), wild silk moth *Cricula trifenestrata* (Helfer), caterpillars of tasar silk moth, *Antheraea paphia* (Linn.) and ber blue butterfly, *Tarucus theophrastus* (Fabr.), slug caterpillar, *Thosea* spp. are of minor importance.

A large number of grey and black weevils such as *Myllocerus discolor* (Boheman), *M. sabulosus* (Marshall), *M. unidecimpustuletus* (Faust), *Peltotrachelus pubes* (Faust.), *Phytoscaphus triangularis* (Olivier), *Sitones crinitus* (Olivier), *Xanthochelus faunus* (Olivier) and *X. supercilious* (Gyllenhal) feed on the tender leaves and are sporadically serious in nurseries and young plantations. Of these, only *Xanthochelus supercilious* (Gyllenhal) is of regular occurrence, sometimes causing severe damage. Both grub and adults do the damage. Grubs prefer to feed on roots while adults prefer tender leaves and sometimes flower buds are also attacked. Longicorn beetles, *Celosterna scabrator* (Fab.) and *C. spinator* (Fab.) reported boring the stems of ber tree and are minor pests. Other beetles found damaging ber tree include, spiny beetle, *Platypria andrewesi* (Weise) and tortoise beetles, *Oocassida cruentaa* (Fab.), *O. obscura* (Fab.) and *O. pundibunda* (Boheman).

Many species of homoptera/hemiptera like spittle bug, *Machaerota planitiae* (Distant), mealy bug, *Drosicha mangiferae* (Green), various armoured scales, *Aonidia zizyphi* (Rahman and Ansari), *A. aurantii* (Maskell), *A. orientalis* (Newstead), *Chrysomphalus ficus* (Ashmead), *Hemiber lesialataniae* (Signoret), *Phenacaspis megaloba* (Green), tree hopper *Otinotus oneratus* (Walker), green striped leaf hopper, *Eurybrachys tomentosa* (Fab.), leaf hopper, *Qadria pakistanica*(Ahmed), cotton jassid, *Amrasca lybica* (de Berg.) suck the cell sap and are of minor importance. Besides, lac insects namely *Kerria communis* (Mahdihassan), *K. lacca* (Kerr), *K. fici* (Green), *K. mysorensis* (Mahdihassan), *K. indicola* (Kapoor), *K. pusana* (Misra), *K. jhansiensis* (Misra) have been recorded from ber tree. Their presence was considered harmful unless and until these species are commercially reared for lac cultivation. The ber leaf gall midge, *Phyllodiplosis jujubae* (Grover) (Cecidomyiidae; Diptera) is a sporadic minor pest of nurseries and young plantations (Gangwar, 1983).

Besides *Larvacarus transitans*, another two mites namely gall mite, *Eriophyes cernus* (Massee) and *Eutetranychus* spp. are quite common and of minor importance. The termites, *Microtermes obesi* (Holmgren) and *Odontotermes obesus* (Rambur) are highly polyphagous in nature and their damage is more common in nurseries and new plantations.

References

Agarwal, M.L. and Kapoor, V.C. 1985. On a collection of Trypetinae (Diptera: Tephritidae) from northern India. Annals of Entomology, 3(2):59-64.

Ahmed, B., Anjum, R., Ahmed, A., Yousaf, M., Hussain, M. and Muhammad, W. 2005. Comparison of different methods to control fruit fly (*Carpomyia vesuviana*) on ber (*Zizyphus mauritiana*). Pakistan Entomol, 27(2):1-2.

Anbu, S., Balasubramanyan, S., Venkatesan, K., Selvarajan, M. and Duarisingh, R. 2009. Evaluation of Varieties and Standardization of Production Technologies in Ber (*Ziziphus mauritiana*) under Rainfed Vertisols. International Jujube Symposium, *Acta Horticulturae*; 840:55-60.

Arora, P.K., Kaur, N., Batra, R.C. and Mehrotra, N.K. 1999. Physico-chemical characteristics of some ber varieties in relation to fruit fly incidence. J. Appl. Hort (Lucknow), 1(2): 101-102.

Ayyar, T.V.R. 1923. Short note on some South Indian Insects. Reports of proceedings of 5thentomological meeting, pusa.pp 263-269.

Bao, W., 2008. Good juju for jujubes *In:* China Daily, 2008-12-15. Downloaded on: 13-02-2014 (http://www.chinadaily.com.cn/bw/2008-2/15/content_7303562.htm).

Batra, R.C., Sandhu, G.S. and Singh, S.N. 1988. Note on chemical control of ber leaf webber. *Entomon*, 50(1):53-54.

Bhatnagar, P. and Lakra, R. K. 1992. Biology and control of hairy caterpillar, *Euproctis fraterna* Moore (*Lepidoptera: Lymontriidae*) on jujube (*Ziziphus mauritiana*Lamk.). In Bio-ecology and Control of Insects Pests: Proceedings of the National Symposium on Growth, Development and Control Technology of Insect Pests, pp. 150-155.

Dashad, S.S., Chaudhury, O.P. and Rakesh 1999.Chemical control of ber fruit fly. *Crop Res.* (Hisar), **17**(3):333-335.

DeKock, C. 2006. Uses. *In:* Ber and other jujubes. Fruits for the Future 2 (Revised Edition) Edited by Azam, A. S., Bonkoungou, E., Bowe, C., DeKock, C., Godara, A. and Williams, J.T. International Centre for Underutilized Crops, Southampton, UK.

Ewing, H.E.1922. Proc. Entomol. Soc. Wash. 24:104-108.

Faroda, A.S. 1996. Developed resistance to fruit-fly in ber through hybridization. ICAR News-Sci. TechnolNewslett.**2**(4):23.

Farrar, N., Asadi, G. H. And Golestaneh, S.R. 2004. Damage and host ranges of ber fruit fly *Carpomyia vesuviana* Costa (*Diptera: Tephritidae*) and its rate of parasitism. J. Agric. Sci.1(5):120-130.

Gangwar, V.S. 1983. Biology of *Phyllodiplosis jujubae* Grover infesting leaves of *Ziziphus jujube*Lamk. (Cecidomyiidae: Diptera). Cecidologia Internationale, 4(1/2/3):57-73.

Ghai, S. and Baker, E.W. 1989.Notes on the gall forming mite, *Larvacarus transitans*(Tenuipalpidae) on *Ziziphus mauritiana*in India. In: *Progress* in *acarology,* Vol. 2(eds. G.P. Channa Basavanna and C.A. Viraktamath) Oxford and IBH, New Delhi, 81-89.

Gyi, M.M., Dikshit, A.K. and Lal, O.P. 2003. Residue studies and bioefficacy of lamdacyhalothrin and beta-cyfluthrin in ber (*Zizyphus mauritiana* Lamarck) fruits.Pestic Res J.15(1):26-27.

Jamadar, M.M., Balikai, R.A. and Sataraddi, A.R. 2009. Status of Diseases on Ber (Ziziphus mauritiana Lamarck) in India and their Management Options. Acta Horticulture, 840: 383-390.

Jawanda, J.S. and Bal, J.S. 1978. The ber-highly paying and rich in food value. Indian Horticulture, 23(3):19-21.

Karuppaiah, V. 2013. Managing fruit borers menace on ber. http://www.krishisewa.com/articles/disease-management/352-ber-fruit-borer.html.

Karuppaiah, V. 2014. Biology and management of ber fruit fly, *Carpomyia vesuviana* Costa (Diptera: Tephritidae): A review. African Journal of Agricultural Research, 9(16):1310-1317.

Kavitha, Z. and Savithri, P. 2002. Documentation of insect pests on ber. *South Indian Hort*, 50(1/3): 223-225.

Kavitha, Z. and Savithri, P. 2002. New record of some natural enemies on ber pests in Tirupati Region. *South Indian Hort*. 50(4/6):513-514.

Khera, A. P. and Singh, J.P. 1976. Chemical composition of some ber cultivars (*Zizyphusmauritiana* Lamk.). Haryana Journal of Horticultural Science, 5(112):21-24.

Kumawat, S.R. and Swaminathan, R. 1990. Relative preference of jujube (*Zizyphus mauritiana* Lamk.) cultivars by bark eating caterpillar, *Indarbela quadrinotata* (Walker). Indian Journal of Entomology, 52(2):336-338.

Kushwaha, K. S. and Bhardwaj, S.C. 1968. Biology and external morphology of forage pests.1. tussock caterpillars, *Euproctis* spp. (*Lymantriidae: lepidoptera*). Indian Journal of Agricultural Science, 37(2):93-107.

Lakra, R. K. and Bhatti, D. S. 1985. Insect pests of ber with special reference to fruitfly, *Carpomyia vesuviana* Costa Paper presented at 3rd National work shop on Arid Zone fruit Research, Rahuri.

Lakra, R. K. And Singh, Z. 1983. Oviposition behaviour of ber fruit fly, *Carpomyia vesuviana* Costa and relationship between its incidence and ruggedness in fruits in Haryana. Indian J.Entomol, 45:48-59.

Lakra, R. K. and Singh, Z. 1985. Seasonal fluctuations in incidence of ber fruit fly, *Carpomyia vesuviana* (Costa) (Tephritidae: Diptera) under agro-climatic conditions of Hisar. Haryana Agric. Univ. J. Res. 15:42-50.

Latif, A and Wali, M. 1961. Distribution, bionomics and description of *Larvacarus transitans*(Ewing) (*Acarina: Phytoptipalpidae*). *Pakistan J. Sci. Res:* 13: 77-87.

Mani, M., Gopalakrishnan, C. and Krishnamoorthy, A. 2001.Natural enemies of ber hairy caterpillar *Thiacidas postica* Walker (*Lepidoptera: Noctuidae*) in Karnataka. Entomon.26(3/4): 313-315.

Mann, G.S. and Brar, J.S. 1980. Record of *Synclera* sp. nr. *Traducalis* Led. And *Pleurota* sp. as a pest of jujube (*Zizyphus mauritiana* Lmk.) at Ludhiana (Punjab). Indian Journal of Horticulture.37(1&2):106-107.

Mann, G.S. and Singh, D. 1981. Population studies of jujube leaf-webbers on *Zizyphus mauritiana*Lmk. At Ludhiana (Punjab). Entomon, 6(4): 351-355.

Mari, J.M., Chachar, Q.L. and Chachar, S.D. 2013.Organic management of fruit fly in jujube ecosystem. International J. Agric Technol, 9(1):125-133.

Mathew, G. 1997. Management of the bark caterpillar *Indarbela quadrinotata* in forest plantations of *Paraserianthes falcataria*. KFRI Research Report 122.

Mishra, S.C. 1991. Effect of temperature and relative humidity on consumption and digestion of food (bark) by larvae of *Indarbela quadrinotata*Wlk. (*Lepidoptera: Indarbelidae*). Journal of Entomological Research, 15(4): 287-293.

Morton, J., Julia, F. and Miami, F.L. 1987. Indian Jujube. In: Fruits of warm climates. pp. 272–275.

Narendran, T.C. and Sureshan, P.M. 1988. A contribution to our knowledge of Torymidae of India (Hymenoptera: Chalcidoidea). Bollettino-del-Laboratorio-di-Entomologia-Agraria-Filippo- Silvestri, 45: 37-47.

Pareek, O. P. and Nath, V. 1996. Ber. In *Coordinated Fruit Research in Indian Arid Zone - A two decades profile* (1976-1995). *N*ational Research Centre for Arid Horticulture, Bikaner, India: 9-30.

Patil, P and Patil, B.V.1997. Studies on the biology of ber fruit borer *Meridarchis scyrodes* Meyrick; (*Carposinidae, lepidoptera*). Karnataka J. Agric. Sci. 10(1):56-59.

Patil, P.V., Patil, V.K. and Deshpande, A.D. 1990. Incidence and control of bark eating caterpillar, *Indarbela* sp. on guava trees. Journal of Maharashtra Agricultural University, 15(2): 230.

Pritchard, A.E. and baker, E.W. 1958. The false spider mites (*Acarina:Tenuipalpidae*). Univ. Calif. Publ. Ent. 14:175-274.

Rajaram, V. And Siddeswaran, K. 2006.Efficacy of insecticides and plant products against the fruit borer and fruit fly in ber under rainfed condition. Internat. J. Agric. Sci. 2(2):538-540.

Ramsingh, Singh, J., Singh, R. and Singh, J. 1982. New record of *Aspergillus candidus* Link - a potential entomogenous fungus on *Indarbela* spp. Science and Culture, 48(8):282-283.

Sah, B.N., Mehra, B.P. and Naqvi, A.H. 1972. Bionomics of *Euproctis fraterna* Moore (Lepidoptera: lymantriidae) on *Ziziphus xylopyra* Willd. Indian Journal of Agricultural Science, 42(7):630-636.

Saxena, D.K. and Rawat, R.R.1968. A study on the incidence, varietal performance and biology of the berfruitfly in Madhya Pradesh. Madras Agric. J. 55:328-333.

Sharma, A and Kushwaha, K.S. 1989. Susceptibility of different verities of *Ziziphus mauritinna* to *Larvacarus transitans*(Acari: Tenuipalpidae). In: *Progress in acarology,* Vol. 2 (eds. G.P. ChannaBasavanna and C.A Viraktamath) Oxford and IBH,New Delhi, 91-93.

Sidhu, M. and Poon C.K. 1983. Efficacy of insecticide-soaked cotton plugs against caterpillars of *Indarbela*sp. (Lepidoptera: Metarbelidae) infesting guava trees. Pesticides, 22 (9): 19-20.

Singh, D. and Mann, G.S. 1982. Behavioural studies of the mobile forms of *Syncler aunivocalis* on jujube, *Zizyphus mauritiana*. Phytoparasitica, 10(3):201-204.

Singh, M.P.1989. *Biosteresvandenboschi*Fullaway- a new braconid parasite of *Carpomyiavesuviana* Costa from the Indian desert. Entomon.14:169.

Singh, M.P. 2008. Managing menace of insect pests on ber. Indian Hort. 53(1):31-32.

Singh, P. and Bhandari.R.S. 1987.Insect, pests of *Acacia tortilis*in India. Indian Forester, 113(11):734-743.

Sonawane, B.R.1965. Study of pests of tropical ber with special reference of fruit borer *Meridarchis scyrodes* Meyr. M.Sc. (Agri) Thesis, Poona University, Poona.

Verma, A. N., Singh, R. and Khurana, A. D. 1972. Chemical control of hairy caterpillar. Indian Journal of Agricultural Science, 42(10): 928-931.

Verma, L. R., Verma, A.K. and Gautam, D.C. 2004. Pest management in Horticulture Crops. Asiatech Publishers.

Yelshetty, S.,Balikai, R.A., Biradar, B.M. and Mokashi, A.N. 1996.Management of ber fruit borer, *Meridarchis scyrodes* Meyrick; (*Carposinidae, lepidoptera*). Karnataka J. Agric. Sci. 9(3):529-532.

13

Pests of Grapes and Their Management

Romila Akoijam, Rachana R. R., Richa Varshney Sandip Patra and Narendra Prakash

Grape, *Vitis vinifera* Linn. is one of the most important fruit crops grown in India. It belongs to the family Vitaceae and is a remunerative temperate fruit crop of India. It is believed that Armenia near the Caspian and Black seas in Russia is the centre of origin of grape. By the end of 12th century, it was introduced in India from Iran and Afghanistan. Grapes are normally eaten fresh as table fruit.

Grapes are used in a wide range of food products - from raisins to wine. They are also loaded with several nutrients and antioxidants and also have high amounts of phytonutrient resveratrol. According to some studies, this phytonutrient is good for heart. It is a good source of calcium, phosphorus, iron and vitamins. In fact, while grapes are good for our overall health, it is especially lauded for its heart benefits. There are several colours of grapes, including red, black, purple or blue (concord), green (which is used to make white wine), pink and yellow. Concord grapes are usually used to make grape juice, grape jelly and grape flavouring.

Like any other crops, grape is also infested by a number of insect pests and sometimes it affects quality as well as quantity of production. Grapes are perennial in nature, the insect pests found to harbour on the crop throughout the year.

Table 1: Pests of grapes

Sl. No.	Common name	Scientific name	Family	Order
1.	Mealybugs	*Maconellicoccus hirsutus* Green. *Planococcus citri* Risso	Pseudococcidae	Hemiptera
2.	Thrips	*Scirtothrips dorsalis* Hood.	Thripidae	Thysanoptera
3.	Fleabeetle	*Scelodonta strigicollis* Motschulsky	Chrysomellidae	Coleoptera
4.	Leaf hoppers	*Arboridia viniferata* Sohi and Sandhu	Cicadellidae	Hemiptera
5.	Stem borer	*Celosterna scabrator* Fabricius	Lamiidae	Coleoptera
6.	Grape leaf roller	*Sylepta lunalis* Guenee	Pyralidae	Lepidoptera
7.	Scale insects	*Aspidiotus lataniae* Signoret	Diaspidae	Hemiptera
8.	Tobacco caterpillar	*Spodoptera litura* Fabricius	Noctuidae	Lepidoptera
9.	Stem Girdler	*Sthenias grisator* Blanchard	Cerambycidae	Coleoptera
10.	Reniform nematode	*Rotylenchulus reniformis* Linford & Oliveira	Hoplolaimidae	Tylenchida
11.	Root-knot Nematode	*Meloidogyne sp.*	Heteroderidae	Tylenchida

1. Mealybug, *Maconellicoccus hirsutus* Green. and *Planococcus citri* Risso. (Pseudococcidae: Hemiptera)

a. *Maconellicoccus hirsutus* Green.

Distribution

The centre of *Maconellicoccus* speciation appears to be Southern Asia, where many endemic species of the genus occur (Williams 1996, 2004). The species has spread to the rest of Asia, Africa and the Americas from its centre of origin. Kairo *et al.* (2000) mentioned that, spread of the pest in the Caribbean, starting in Grenada in 1994.

In Asia, *M. hirsutus* populations are widely spread in India, Bangladesh, Brunei Darussalam, Cambodia, China, Indonesia, Japan, Laos, Lebanon, Malaysia, Maldives, Myanmar, Nepal, Oman, Pakistan, Philippines, Saudi Arabia, Singapore, Sri Lanka, Taiwan, Thailand etc. (EPPO, 2005).

Hosts of commercial importance

M. hirsutus is a highly polyphagous pest and has been recorded feeding on hosts from 76 plant families and over 200 plant genera (Ben-Dov and German, 2003). It shows some preference for hosts in the families of malvaceae, leguminosae and moraceae. It attacks a wide range of predominantly woody plants, including many ornamentals. The ornamental *Hibiscus rosa-sinensis* is a typical host which is frequently attacked. It is one of the most destructive sap sucking pests of forest trees, root crops, fruit plants, vegetables, ornamental plants, cultivated and non-cultivated plantations. It infests on more than 300 species in 74 plant families (Kairo *et al.*, 2000).

Apart from *V. vinifera*, major commercially important host plants are *Mangiferaindica*, *Annona muricata*, *A. reticulate*, *A. squamosa*, *Nerium oleander*, *Allamanda* sp., *Anthurium* sp., *Colocasia* sp., *Phoenix dactylifera*, *Chrysanthemum coronarium*, *Helianthus annuus*, *Lactucasativa*, *Parthenium hysterophorus*, *Brassica oleracea*, *Opuntia* sp., *Terminalia catappa*, *Hevea brasiliensis*, *Jatropha curcas*, *Manihote sculenta*, *Tectona grandis*, *Persea americana*, *Asparagus officinalis*, *Abelmoschus esculentus*, *Morus alba*, *Musa* sp., *Psidium guajava*, *Syzygium cumini*, *Bougainvillea* sp., *Jasminum sambac*, *Averrhoa carambola*, *Passiflora edulis*, *Saccharum officinarum*, *Zea mays*, *Punica granatum*, *Prunus domestica*, *Coffea arabica*, *Citrus aurantiifolia*, *Capsicum annuum*, *Solanum lycopersicum*, *Theobroma cacao*, *Alpinia purpurata*, etc. (CABI, 2016).

Identification and biology

Field identification of *M. hirsutus* can often be achieved based on the external appearance of the mealybugs, particularly the adult females. The ovisac is elongate with loosely woven white wax filaments into which the adult female deposits several hundred oval eggs, each 0.3mm in length. Eggs are initially orange, but turn pink before hatching. First instar crawlers are pink, oval in shape with well-defined legs and antennae and lack the waxy body coating.

Later instars turn grey–pink and start to secrete white wax that covers their bodies. Crawlers average 0.37mm in length, second instars 0.70mm, third instars 1.1mm and male "pupae" 1.1mm. Adult females are wingless, oval, flattened in profile and 2–3mm in length. Body is greyish pink and covered with a thin white cotton like wax. Posterior tufts of cotton-like waxy deposits are often present. Adult males are gnat-like with a pink or orange body (1mm in length), single pair of wings and two pairs of elongated wax pencils (Aristizabal *et al.*, 2012).

b. *Planococcus citri* Risso.

Distribution

The pest is a native of Asia but is also found throughout the Americas, Africa, Europe, and Oceania (CABI, 2016).

Hosts of commercial importance

P. citri is one of the most common pests in nearly all greenhouses and nurseries, where it attacks a wide range of ornamental, citrus, and orchard crops in many temperate and tropical regions (Blumberg *et al.*, 1995). In greenhouses, it is the most common pest on bulbs, ferns, and ornamentals (Blumberg and van Driesche, 2001). It attacks new shoots and leaves of a wide range, including apple, avocado, citrus, English ivy, ficus, gardenia, jasmine, oleander, persimmon, pittosporum, rhododendron (Hill, 1983).

Host range includes commercially important plant species like *Acacia* sp, *Albizzia lebbekh*, *Annona squamosa*, *Asparagus* sp., *Bougainvillea* sp., *Brassica oleracea*, *Casuarina equisetifolia*, *Chenopodium album*, *Citrus sinensis*, *Cocos nucifera*, *Cucumis melo*, *Cucurbita* sp., *Ipomoea batatas*, *Mangifera indica*, *Musa* sp., *Psidium guajava*, *Punica granatum*, *Solanum melongena*, *S. tuberosum*, *Theobroma cacao* (Noha and Shaaban, 2010)

In the tropics, it occurs mainly on the aerial parts of crops such as cocoa, banana, tobacco and coffee and on wild trees such as *Ceibapentandra* and *Leucaena* sp. (Entwistle, 1972). In Ghana, *P. citri* has been recorded on about 54 host plants including cocoa, cola, pineapple, *Musa paradisiaca* and others within the families Bombacaceae, Euphorbiaceae, Fabaceae, Moraceae, Rubiaceae, Solanaceae, Sterculiaceae, Tiliaceae and Urticaceae (Padi *et al.,* 1999). In temperate regions, it mainly occurs on greenhouse plants such as Coleus, ferns and outdoors plants like citrus, grapes, figs etc. (Gibson and Turner, 1977).

Identification and biology

The eggs are pink in colour and hatch in 2 to 10 days. The first instar crawlers are yellowish-pink in colour and are often found congregated upon the egg sac. They pass through two additional instars before moulting to an adult female or forming a male pupa. Each instar requires 7 to 16 days to complete. Female nymphs resemble adult female in appearance, while male nymphs are more elongated. The adult female mealybug is pinkish in colour, wingless, covered with white-cottony wax and has a fringe of elongated waxy filaments that extend about the periphery of the body. Fully grown female mealybug is about 3 mm long and 1.5 mm in width. It is mobile, but lacks wings and cannot fly. Male mealybugs emerge from their pupae in 7 to 14 days and resemble elongated gnats, with tail filaments. Males are generally about 4.5 mm in length (Noha and Shaaban, 2010).

P. citri is parthenogenetic in reproduction. Each female mealybug deposits about 350-500 eggs in loose cottony ovisac of white wax during a week's time.Eggs are orange in colour (*M. hirsutus*) or yellowish white (*P. citri*). Incubation period is 5 days. Eggs hatch into nymphs which are mobile known as crawlers. After hatching they settled on the plants and start sucking the sap from tender parts. They also form colonies. Crawlers are orange in colour (*M. hirsutus*) or yellowish white (*P. citri*).

The male and female mealybugs are similar in early stages. The female and male pass through 3 and 4 nymphal instars respectively. Nymphal period for male is 19 days and for female is 21 days. The last instar of male is an inactive stage having wing buds within a cocoon (pupa) of mealy wax in the winter. Adult male and female are entirely different. Male is very rare and it has one pair of wings and one pair of halteres. Female is sedentary and responsible of damage of plant. The adult female mealybugs are yellowish white (*P. citri*) in colour and sparsely covered with white wax. Mealybug completes life cycle in about 30 days and there may be more than 15 generations in a year. In warm climates, the insects stay active and reproduce round the year.

Mode of feeding and symptoms of infestation

Nymphs and adult mealybugs suck the sap from all the tender parts of vine *viz.* the trunk, buds, spurs, leaves, nodes, shoots, panicles and bunches. Pink mealybug infestation on the growing point results in malformation of leaves and shoots tips. All the stages of mealybug excrete copious amount of honey dew that attracts ants and help in the development of sooty mould fungus on leaves, shoots and bunches. Layer of this fungal growth inhibits photosynthetic activity of the plant. Heavy infestation by this pest can cause defoliation and even death of the young vine. Sooty and sticky bunches harbouring mealybugs and their white cottony wax masses are unfit for marketing as table grapes. From such infested bunches raisins cannot be prepared. The mealybug may causeup to 100% crop loss in severe case of infestation.

Seasonal incidence

The mealybugs occur throughout the year under Indian condition. Growth of mealybug population is proportional to the vine development. After fruit pruning (September to October), they stay dormant on the trunk, cordons and stem upto the first fortnight of December. From mid-December, their population started to grow. They move out from the trunk, cordons and shoots to flower panicles and developing berries. Their populations remain low on the leaves, stem and trunk from April-September. *M. hirsutus* occurred two generations in a year (Mani and Shivaraju, 2016). The highest population (5-6 colonies/vine) was observed during the end of February to the end of March which occurred at the same time with the fruiting and harvesting season.

Economic threshold level (ETL)

ETL attained when an average of 1 mealybug per spur in mid-June or 1 % bunch infestation. Chemical control can be initiated when the number of colonies or the infestation is more than 5 % especially after fruit pruning.

Management

i.Nursery management

a. Selection of healthy planting material.

b. Soil drenching with Imidacloprid 200 SL @ 1.50 ml/l/plant in the basins around the base.

c. Foliar spray with Methomyl @ 1g/l after 30 days of soil drenching.

i. Pruning stage

a. Banding of 'Track-trap' or 'Bird Tangle Foot' on branches or on main stem before appearance of the mealybugs on bunches to prevent crawlers of mealybugs reaching the bunch. It also prevents movement of ants.

b. Pruning of the infested branches during April-May and swabbing of stem and arms with 2 ml of Dichlorvos 76 EC + 2 g of fish oil resin soap in a litre is effective in managing the pest.

c. Collection and destruction of the mealybug infested bunches and loose bark during April/May.

d. Elimination or removal of the weeds and other alternate hosts like okra, hibiscus, guava,custard apple, etc in and around the vine yards all the year round may be done.

e. Find out theant colonies and kill them by applying with Chlorpyriphos 20 EC @ 2.5 ml/lduring April-May as ants are associated with the build up of mealybug population.

f. Release the predatory Australian ladybird beetle (*Cryptolaemus montrouzieri*) @ 5000/ha. in August-September to wipe out the mealybug population.

g. Foliar sprays of two or three times *Verticillium lecanii/Beauveria bassiana* ($2x10^8$ cfu/ml/g) @ 5 g/l at 15 days interval during July-August.

iii. After fruit pruning stage

a. Monitoring and demolish the mealybug colonies as wherever found on the plants from November to February.

b. Repeat to release the Australian lady bird beetle (*Cryptolaemus montrouzieri*) @ 5000/ha. during mid-December - first fortnight of January if population persists.

c. Just before harvest, one jet spray of water can be given on the bunches if the mealybugs are still present.

2. Thrips, *Scirtothrips dorsalis* Hood (Thripidae: Thysanoptera)

Distribution

The pest is widely distributed in Bangladesh, China, India, Japan, Myanmar, Pakistan, Philippines, Sri Lanka, and Thailand etc. It is also reported from South Africa and the Ivory Coast.

Hosts of commercial importance

Not only grapes, *S. dorsalis* is known as a pest on many crops including *Actinidia chinensis*, *Allium cepa*, *Arachis hypogea*, *Camellia sinensis*, *Capsicum* spp., Citrus, *Gossypium hirsutum*, *Hevea brasiliensis*, Hydrangea, *Mangifera indica*, Nelumbo, *Ricinus communis*, Rose, *Tamarindus indica* (EPPO, 2005).

Identification and biology

Hyaline eggs are globular and laid in mass. Larvae of *S. dorsalis* are creamy white to pale in colour. Nymphs are tiny, slender, fragile and straw yellow in colour. The body of adult *S. dorsalis* is pale yellow in colour, have fringed wings and bear dark brown antecostal ridges on tergites and sternites. Dark spots forming incomplete stripes are seen on dorsal abdomen.

Female lays about 50-100 eggs. Eggs are hyaline, globular laid in mass and are inserted into the tender tissue on ventral side of the leaves. Incubation period is 8-10 days. Nymphs tiny, slender, fragile and straw yellow in colour. After feeding, nymphs move down to the soil and pupate therein. Nymphs are similar to adults but are wingless. The adults are very small, 2 mm in size, elongated and fast moving with four narrow heavily fringed wings. Adult's longevity is about 10 days. Total life cycle is completed in about 15 days. There are four stages in life cycle nymphs, pre-pupa, pupa and adult.

Mode of feeding and symptoms of infestation

Both nymphs and adults feed gregariously on under surface of the leaves. They cause damage by lacerating on lower surface of leaf with their stylets. Due to injury minute spots develop on surface thereby producing a speckled silvery effect. This appearance can be detected from a distance. In case of heavy incidence curling of the leaves is observed. During fruiting stage they also attack blossoms and developing berries which leads to poor fruit setting and yield is drastically reduced. The thrips are also responsible for scab formation on the berries. The affected berries become brown because of a corky layer development. Fruits obtained from seriously infested plants are of poor quality and fetches low market price.

Seasonal incidence

Thrips population found round the However, higher population was observed during November and December months, which corresponded with the flowering time.

Economic threshold level (ETL)

Observation of 3-4 per leaf per plant after foundation pruning or more than 2-3/ shoot or inflorescence/plant after fruit pruning, particularly during flowering and early berry setting stage prevails the ETL value. Foliar damage of 20% has also shown the ETL value (Atwal and Dhaliwal, 2005).

Management

i. Destruction of previous crop debris may be practised.

ii. Collection and destruction of weeds and alternate hosts like bhindi, hibiscus, guava etc in vineyards immediately after pruning.

iii. Deep summer ploughing or exposure/raking of soil in vineyards after April. Pruning kills the pupal stages and lower the incidence.

iv. Conservation of the predatory green lacewing, *Chrysoperla carnea* as they are usually associated with insect pests in the vineyards.

v. Application of fungal pathogens such as *Verticillium lecanii* or *Beauveria bassiana* @ 5 ml or 5 g/l helps in maintaining thrips population in cold (20-25°C) and humid (80% relative humidity) climate.

vi. Various neem formulations (EC based) depending upon the strength of botanical viz., 1% @ 2.5 ml and 5% @ 0.5 ml/l can be sprayed like insecticide @ 400 litre spray solution per acre.

vii. The insecticides like Thiamethoxam 25 WG 0.25 g/L, Lambda-cyhalothrin 5EC/CS 0.50 ml/l, Spinosad 45 SC 0.25 ml/l, Emamectin benzoate 05 SG 0.20 g/l can be applied at critical growth stages like new flush, flowering and berry developing stages (David and Ramamurthy, 2012).

3. Flea beetle, *Scelodonta strigicollis* Motschulsky (Chrysomellidae: Coleoptera)

Distribution

It is widely distributed in India (Trehan *et al.*, 1974).

Hosts of commercial importance

S. strigicollis is a specific pest of grapes.

Identification and biology

Eggs are cylindrical and yellowish to orange in colour. Newly hatched larvae are dark brown in colour but later turn light brown and whole body is covered with black shiny and rectangular plates of various sizes which gives spotted appearance to them. Pupae are bright yellow in colour with conspicuous reddish-brown eyes. Adults are oval in shape and metallic shining blue in colour with thread like antennae. The thighs of hind legs are enlarged.

Eggs are yellow cylindrical in shape. Eggs are laid singly or group of 20-40 eggs near the buds or loose cane. A single female may lay eggs from 250-500. Incubation period is 4-7 days. After hatching these larvae move towards root and start feeding there. Total larval period is 30-40 days and there are six larval instars. Pupation takes place in soil up to 6-8 cm deep. Pre-pupal and pupal period is 2-3 days and 7-10 days respectively.Adults are oval in shape, shining metallic bronze in colour, later they are reddish brown in colour with six visible spots on elytra and 4-5 mm in length. Pre-oviposition, oviposition and post-oviposition period is 30, 75 and 120 days respectively. Adults start mating after 25-30 days after emergence. Adults are nocturnal and they feed on leaves. Life cycle completes within 50-55 days.

Mode of feeding and symptoms of infestation

The adults scrap the sprouting buds after pruning. The affected buds fail to sprout successfully. The beetles also found to feed on tender shoots, leaves and tendrils causing considerable damage. Scraping on tender shoots and tendrils results in white streaks initially and later turned brown. The tender shoots may wither and dry. Heavy crop loss results when the sprouting buds are infested after forward pruning.Once shoot growth reaches 7 cm, damage caused by the grape flea beetle normally does not affect yield.

They also feed on mature leaves and give shot hole appearance. During fruiting season they also scrap unripe berries which results in scab formation. In general adults are more destructive than larvae. Larvae feed on roots leads to yellowing of leaves which in severe condition leads to death of young vine.

Seasonal incidence

Just after both foundation and fruit pruning, flea beetle population is usually found in vineyards. Peak population (5.8 flea beetle/vine) was observed during first week of November, which coincided with bud sprouting.

Economic threshold level (ETL)

ETL is estimated if 2% bud damaged are found by sampling of 10 buds on each 10 vines.

Management

i. Selection of high quality, large, vigorous seed that establishes quickly can ward off flea beetles.

ii. Wider row spacing reduces the flea beetles infestation.

iii. Elimination of the loose bark of stem at the time of pruning to impede the egg laying site. Application of Copper oxychloride paste and Carbaryl 50% WP after forward pruning give good result.

iv. Just after fourth day of pruning, spraying of insecticides like Carbaryl (0.15%) or Quinalphos (0.05%) at 3-5 days interval till the emergence of the leaves may be advocated. It will protect the sprouting bud from infestation.

v. Quiver the vines to disentangle adult beetles, collection into trays containing kerosenized water (kerosene: water, 1:9 v/v) and killing them.

vi. Placing bundles of dried shreds of banana on the pruned end of the vines in the evening. Beetles, which take shelter on these at night, can be collected in the morning and kill them.

vii. Setting up of light trap @ 1/acre.

viii. Spraying of Azadirachtin concentrations like 50000 ppm formulation @ 1ml / l, while that of 10000 ppm and 3000 ppm can be sprayed @ 2.5 ml and 5 ml per liter water respectively.

ix. Foliar spraying of Carbaryl 50 WP @ 2 g/l of water at 7-15 days after pruning.

x. Application of Imidacloprid 17.8%SL@120-160 ml in 400 l of water/acre or Lambda cyhalothrin 4.9%CS @100 g in 200-400 l of water/acre after 15-30 days of pruning.

4. Leaf hoppers, *Arboridia viniferata* Sohi and Sandhu (Cicadellidae: Hemiptera)

Distribution

It is commonly found in India, Pakisthan (Mani *et al.*, 2008).

Hosts of commercial importance

A. viniferata feeds primarily on grapes.

Identification and biology

Adult lays elongated yellow white color eggs in the midrib of the leaves. It lays approximately 15 eggs. Egg is 0.8 mm in length. Incubation period is 7-10 days. Before hatching eggs turn into greyish yellow. Nymphs are pale greenish in color. There are five instars. Nymphal period is 15-20 days. Adults are green in colour with tiny black spot on each forewing. Their size is 3.5 to 5mm. life cycle is completed in 30-35 days.

Mode of feeding and symptoms of infestation

Both nymphs and adults suck plant sap from under surface of leaf. Initially due to feeding white scattering spots develops on leaf. In severe infestation, damaged leaf curl downwards, turn yellowish brown, dried and drops off, which ultimately reduce the plant vigour.

Seasonal incidence

The pest remains active throughout the year. It is usually appear in the field from August to October in north Indian conditions while in south Indian conditions, it occurred from June to August. Increase in temperature during the vegetative growth, increases the problem of hopper infestation (Lisek, 2008).

Management

i. Keeping the nursery area free of weeds and grasses.

ii. Removal of weeds and alternate host plants viz. hibiscus, okra, custard apple, guava etc. inside and near the outside of the vineyard just after pruning.

iii. Growing of tall border crops such as maize, sorghum or pearl millet to lessen the hopper infestations.

iv. Application of Quinalphos (0.05%)as soon as the infestation of the pest is observed is effective.

vi. After pruning at 60-75 days, spraying with Imidacloprid 200 SL @ 0.3 ml/l or Thiamethoxam 25 WG @ 0.25 g/l or Lambda cyhalothrin 05 EC @ 0.5 ml/l of water are also helpful.

vii. Foliar application of Azadirachtin 1% @ 2 ml/l or 5% @ 1 ml/l of water at 90-105 days after pruning.

5. Stem borer, *Celosterna scabrator* Fab. (Lamiidae: Coleoptera)

Distribution

It is found in India as well as in Thailand (Nair, 2007).

Hosts of commercial importance

V. vinifera, Acacia nilotica, A.catechu, Cassia siamea, Casuarina equisetifolia, Dipterocarpus alatus, Eucalyptus sp, Morus alba, Prosopis cineraria, P. juliflora, Shorea robusta, Tamarix indica, Tectona grandis, Terminalia chebula, Zizyphus jijuba are the important host plants (Nair, 2007).

Identification and biology

Eggs are capsule shaped and lay singly in each of the slits and the slits are covered with a hard gummy substance. Newly hatched grubs are flat-headed and cream coloured. Pupae dark brown in colour and seen inside the tunnel made in the vine. Adult beetles are dull yellow in colour with minute spots.

Female makes slits on bark of trunk and branches and lays capsule shaped eggs singly and covered with gummy substance. It lays 12-15 eggs at a time. Eggs hatch in about 10 days. Grub is cream colored, apodous with flat head and with strong mandibles. After hatching it enters inside the trunk and branches and starts feeding there. Full grown larva measures about 75 mm in length. Larval stage lasts for 6-8 months. Full grown larva pupates inside the tunnel within a calcareous cocoon. Pupal period is about 25-30 days. An adult beetle is about 4 cm long and dull yellow with minute spots. Adult longevity is about 20-25 days. Whole life cycle completed in 10 months to one year.

Mode of feeding and symptoms of infestation

It makes slits for egg laying which may cause damage to tender shoot. Main damage is done by larva which feed inside the trunk and branches and makes tunnel upward and downwards. Gummosis and extrusion of frass through the holes are common symptoms of damage. Presence of saw dust like substance under the vine indicates damage done by the grub. Affected plan shows yellowing of leaves, drying and shedding of leaves. Damaged vines get weakened and growth gets affected which leads to poor yield. Quality of berries also get affected.

Seasonal incidence

With the onset of monsoon till September, the adult beetles start emerging and lay eggs. Soon after hatching the larvae enter into the stem and feed till April. They also pupate inside stem during May-June.

Management

i. Collection and destruction of weeds and alternate hosts like casuarinas, eucalyptus, babul, teak, etc. during July-August.

ii. Collection and destruction of borer infested branches having grubs or pupae during April-May.

iii. Collection and elimination of adult beetles or installation of light trap @2-3/ha at night time during July-August.

iv. Mechanically collection and destruction of eggs and larvae accessible in the slits on the trunks and branches by using hook or knife during July-September.

v. Injection of *Metarhizium anisopliae* @ 100 ml/l water into galleries.

vi. Removal of loose bark and pasting the branches or main trunk with 1 ml neem oil + 1ml gum+ 6 g Carbaryl + 10 g Copper oxychloride in one litre of water (David and Ramamurthy, 2012).

vii. Application of Carbarl 50WP (2 gm/l) or Chlorpyriphos 25 EC (2.5ml/l) during July-August. Pasting of carbaryl 50WP 6g + copper oxychloride 3ml + dichlorvos 3ml + neutral pH sticker soap 1ml may also give satisfactory result.

viii. Injection of petrol or Dichlorvos @ 5ml/hole or stem injection of Dichlorvos 76% EC @ 80ml/ live hole until the hole gets completely filled followed by sealing with mud or cow dung mixed with Copper oxychloride (1:3 ratio) is also effective or insertion of Aluminium phosphide tablet @ 1g/live hole has recorded hundred per cent reduction in live tunnels (Kumari and Vijaya, 2015).

6. Leaf roller, *Sylepta lunalis* Guenee (Pyralidae: Lepidoptera)

Distribution

The species is widespread in India (Sohi and Bindra, 1974).

Hosts of commercial importance

V. vinifera Ciccus quadrangularis, C. glauca, C. heyneana, C. vitiginea, C. repanda, C. pallida, Cayratia carnosa, Leeasam bucina, L. macrophylla, Tetrastigma lanceolarium, T. sulcatum are the hosts of importance for *S. lunalis* (David, 1961).

Identification and biology

Larva is pale green in colour with short hairs. Adult moths are brownish in colour with wavy lines.

Adult female lays about 100 creamy white eggs on undersurface of leaves. It hatches in 2-3 days. Larva is greenish with brown head. Larval period lasts for 15 days and there are 5 larval instars. Pupation takes place inside the rolled leaves or plant debris on ground. Pupal period is 6-7 days. Life cycle is completed in about 25 days.

Mode of feeding and symptoms of infestation

A larva rolls up leaf margin towards the midrib, exposing the under surface, the edge is held in place by bands of silk thread. Within this fold larva feeds, skeletonizing the leaf of the upper surface.

Seasonal incidence

The pest is activein the months ofAugust-November under South Indian conditions.

Management

i. During the initial stage, collection and removal of infested and rolled up leaves and destruction thereof along with larvae/pupae of the pest.

ii. Application of insecticides having contact and stomach poisonous toxicity is effective in managing the pest.

7. Scale insect, *Aspidiotus lataniae* Signoret (Diaspidae: Hemiptera)

Distribution

It is widely distributed in about 106 different countries. Important ones are Australia, Bangladesh, Brazil, China, Colombia, Cuba, Czech Republic, Egypt, France, Georgia, Ghana, Greece, Hawaiian Islands, India, Iran, Israel, Italy, Jamaica, Japan, Kenya, Madagascar, Martinique, Mauritius, Mexico, Mongolia, Morocco, Mozambique, New Zealand, Pakistan, Panama, Philippines, Poland, Portugal, Reunion, Saudi Arabia, Solomon Islands, South Africa, Spain, Sri Lanka, Sudan, Taiwan, Tanzania, Thailand, Uganda, United Kingdom, United States of America, Vietnam, Zimbabwe, etc.

Hosts of commercial importance

V. vinifera Acacia abyssinica, Albizia lebbeck , Amaranthus viridis, Annona reticulate, A. squamosal, Artocarpus heterophyllus, Asparagus officinalis, Beta vulgaris, Cajanus cajan, Capsicum frutescens, Carissa carandas, Casuarina equisetifolia, Citrus aurantiifolia, Cucurbita moschata, Cocos nucifera, Eucalyptus spp., *Ficus* spp., *Gossypium* spp., *Mangifera indica, Manihote sculenta, Melia azedarachta, Moringa oleifera, Morus alba, Musa* spp., *Nerium oleander, Passiflora edulis, Plumeria rubra, Psidium guajava, Punica granatum, Rosa indica, Solanum melongena* etc. are the hosts of commercial importance.

Identification and biology

Field identification characters are scale cover of adult female 1.5-2.0 mm diameter, circular, convex; on stems, usually tan but can be darker; on leaves, grey or almost white; exuviae subcentral, yellow to yellow-brown. If present, male scale cover elongate oval with yellow subterminal exuviae, smaller and sometimes paler than that of female. Body of adult female is bright yellow. Eggs are also yellow, elongate, each length 0.15 mm long.

Adult females lay 28-65 eggs in concentric circles under the scale cover over a period of 11-13 days, and may produce 3 or 4 consecutive batches in their lifetime. Freshly laid eggs are smooth, white later they become pale yellow. Soon after being laid, larvae (crawlers) hatch out and disperse in search of feeding site. They are free moving and recognized by presence of legs, antennae and a pair of bristles at the tip of abdomen. These are light green to yellowish brown, translucent and oblong. The legs and antennae; these are lost in the first molt. The cast skin is incorporated into the scale of the second larval stage. It takes about 16 to 18 days for the scale to molt to the second stage larva. Females molt twice, while males molt at least three times.

After settling, females remain sessile throughout their development; adult males undergo a pseudo-pupation, develop a pair of wings and can disperse by flying to find mates. Adult female has a circular or broadly oval cover that is 1.5-2.0 mm in diameter. Adult male coconut scales are small, two-winged, reddish, gnat-like insects with eyes, antennae, three pairs of legs and long appendages. Adult males do not feed and are short lived. It takes approximately 45 to 55 days for females to develop. Reproduction is mainly through parthenogenesis.

Mode of feeding and symptoms of infestation

Removal of sap from leaves, petioles, peduncles and fruits leads to discoloration, depressions and tissue distortions on leaves. It possibly introduces toxins into the plant through their saliva.

Management

i. Planting materials should be free from infestation of the pest.

ii. Removal of the loose bark at the time of pruning is effective.

iii. Scraping of encrustations of the vine and spraying with any systemic insecticides.

iv. Management of ants is important as they act as carrier of the scales.

8. Tobacco caterpillar, *Spodoptera litura* Fabricius (Noctuidae: Lepidoptera)

Distribution

The species is widely distributed throughout tropical and temperate Asia, Australasia, and the Pacific Islands (Kranz *et al.*, 1977). It is distributed in Asia including Afghanistan, Bangladesh, Brunei, Cambodia, China, Christmas Island, Cocos Islands, India, Indonesia, Iran, Japan, Korea, Laos, Lebanon, Malaysia, Maldives, Myanmar, Nepal, Oman, Pakistan, Philippines, Singapore, Sri Lanka, Syria, Thailand, and Vietnam, Europe, Russia, Africa, Australia, Hawaii, New Zealand, Solomon Islands etc.

Hosts of commercial importance

The host range of *S. litura* covers at least 120 species. Most economically important among them are *Abelmoschus esculentus, Acacia mangium, Allium cepa, Arachis hypogaea, Beta vulgaris* sub sp. *vulgaris, Boehmeria nivea, Brassica oleracea var. botrytis, Brassica oleracea var. capitata, Camellia sinensis, Capsicum frutescens, Castilla elastica, Citrus* spp., *Erythroxylum coca, Foeniculum vulgare, Gladiolus hybrids, Glycine max, Gossypium*spp., *Helianthus annuus, Hevea brasiliensis, Ipomoea batatas, Malus domestica, Manihote sculenta, Medicago sativa, Morus alba, Musa* spp., *Oryzasativa, Papaver* spp., *Paulownia tomentosa, Phaseolus* spp., *Piper nigrum, Psophocarpus tetragonolobus, Raphanus sativus, Ricinus communis, Rosa* spp., *Sesbania grandiflora, Solanum lycopersicum, S. melongena, S. tuberosum , Sorghum bicolor, Syzygium aromaticum, Tectona grandis, V. radiata*etc (Venette *et al.*, 2003).

Identification and biology

The golden yellow clustered eggs are covered by brown hairs. Caterpillars are velvety, black with yellowish green dorsal stripes and lateral white bands with incomplete ring like dark band on both anterior and posterior end of the body. Pupation takes place in soil. Adult moths are about 22 mm long and measure 40 mm across the spread wings. They are medium sized insect. Body is stout and forewings are pale grey to dark brown in colour. Wavy white criss-cross markings present on wings. Hind wings are whitish with brown patches along the margin.

Eggs are laid in clusters and fecundity is about 300. Incubation period is about 3-5 days. Fully grown larva measures about 35-40 mm in length. They pass through

6 larval instars. Larval stage lasts 15-30 days. Pupal period is about 7-15 days depending on environmental condition. The species breeds throughout the year. Adults live for about 7-10 days. Total life cycle takes 32-60 days depending on temperature and humidity.

Mode of feeding and symptoms of infestation

Early instars of the caterpillars feed gregariously. They scrap the chlorophyll content of leaf lamina giving it a papery white appearance. Later they feed on rachis and unripe berries also.

Seasonal incidence

The pests are generally occurred throughout the year and the adult moths are mainly active during August-September.

Management

i. Hand picking of well grown larvae and destruction thereof.

ii. Growing trap crop like castor all around the vineyard for oviposition is effective.

iii. Pheromone traps for *Spodoptera* @ 12-14 traps/ha keeping a distance of around 23 meters at one foot height from the plant canopy. Lures must be changed at 2-3 week interval for adult moth monitoring and capture.

iv. Spraying of NSKE 5 % against eggs and first instar larva.

v. The pest can effectively be managed by spraying Chlorpyriphos (0.08%) or Dichlorvos (0.1%).

vi. A mixture of Methomyl (0.05%) and Wettable Sulphur (0.2%) is effective to manage the early instar larvae.

9. Stem girdler, *Sthenia sgrisator* Blanchard (Cerambycidae: Coleoptera)

Distribution

They are found in Central Asia, Peninsular India and Himalayan India (Sengupta and Sengupta, 1981).

Hosts of commercial importance

They are also distributed in *Citrus spp., Excoecaria agallocha, Jatropha gossypifolia*, J. curcas, Malus domestica, Mangifera indica, etc. (Balasubramanian 1991, 1992).

Identification and biology

Adult is a medium sized, stout beetle, grey coloured with a white spot in the centre of each elytra. Grub with dark brown colour head with a pair of strong mandibles. The emerging grub tunnels into the wood and completes its life cycle within the stem. Pupation takes place within the tunnel.

Eggs are deposited in cluster of one to four in the bark. These are milky white and oval in shape. The female hollows out a small cavity in the shoot, places a single egg into it and fills the egg cavity with frass. Then it girdles the cane at two places: just below the egg cavity and several inches above. Incubation period is about 10 days. The fully-grown larva is white with a brown head, legless, and measures 10-12 mm in length. Thorax is globular with chitinous spines on the top. It burrows in the center of the shoot on either side of the egg cavity. Larval period is about 8 months.

The pupa is light- colored but becomes darker just prior to emergence. Some of the adult features such as legs and snout are already clearly visible in the pupal stage. Pupal period is about 30 days. Adult is a medium sized, stout beetle, grey coloured with a white spot in the center of each elytra. The adult beetles emerge from infested canes during August and subsequently overwinter in trash on the ground. Life cycle takes more than a year to complete.

Mode of feeding and symptoms of infestation

During night adult girdle around the main stem, young and green branches at any place from 15 cm to 3 m above the ground. Generally, branches of 1.25 to 2.50 cm thickness are preferred by beetle. The grub bores into the bark and tunnels into the dry wood. The girdling by the female causes the terminal growth of the new shoots to bend over above the upper girdle and drop to the ground. Later the whole infested shoot dies back to the lower girdle and falls from the vine. Vines 'pruned' by the grape cane girdler have a ragged appearance.

Seasonal incidence

The adult beetles normally emerge from infested canes during August and thereafter overwinter in refuse on the ground. In the next year during May, the adults leave the overwintering sites and the female starts laying eggs and girdle new vines when the grape shoots attained 30 to 50 cm long. In spring, the adults are active at night time, mate and laid eggs in clusters of 2-4 underneath the bark of girdled branches (Atwal and Dhaliwal, 2005).

Management

i. Pruning of infested shoots below the lower girdle before adult emergence in the summer and burning them.

ii. Cleaning of excess dead wood, loose bark and damaged plant parts at the time of pruning should be ensured.

iii. When noticed the presence of beetles, hand picking and destruction of the adults preferably at night is also effective.

iv. A piece of cloth soaked in a solution of Chlorpyriphos (0.1%) and then wrapping around the stem is effective.

v. Swabbing the trunk with Carbaryl 50 WP 2gm/l or Chloropyriphos 0.05% emulsion for preventing attack by the beetle may be practised.

vi. Application of Phosalone 35 EC 0.07% or Quinalphos 25 EC 0.05% or Carbaryl 50 WP 0.1%, immediately after pruning and repeatation of the same may be done 2-3 times.

In addition to the insect pests described above, some other non-insect pests like, mites, nematodes and vertebrate pests cause damage to the grape crops. They are as follows:

Nematodes

i. Reniform Nematode, *Rotylenchulus reniformis* Linford & Oliveira (Hoplolaimidae: Tylenchida)

ii. Root-knot Nematode, *Meloidogyne* sp. (Heteroderidae: Tylenchida)

Mites

Mites pose an increasing threat to grape cultivation in certain grape growing areas in India causing heavy loss of production. Six species of mites have been found to cause damage to the crop. They are, *Tetranychus urticae*, *T. cinnabarinus*, *T. neocaledonicus*, *Oligonychus angiferus*, *O. punicae* and *Eutetranychus orientalis*. Among them, two spotted spider mite and *T. urticae* cause severe loss in Maharastra and Andhra Pradesh (Mani, *et al.*, 2008).

Vertebrate pests

i. Bat, *Cynopterus sphinx* (Vahl, 1797)

ii. Jungle crow, *Corvus macrorhynchos* Wagler, 1827

iii. Bank myna, *Acridotheres ginginianus*, Indian myna, *A. tristis* (Linnaeus, 1766)

References

Aristizabal, L. F., Mannion, C., Bergh, C. and Arthurs, S. 2012. Life history of pink hibiscus mealybug, *Maconellicoccus hirsutus* (*Hemiptera: Pseudococcidae*) on three Hibiscus rosa-sinensis cultivars. Florida Entomologist, 95:89-94.

Atwal, A. S. and Dhaliwal, G. S. 2005. Integrated Pest Management. In: Agricultural pests of South Asia and their management. Kalyani publishers, Ludhiana. pp-505.

Balasubramanian, P. 1992. New nesting site of the Indian white-breasted kingfisher. The Journal of Bombay Natural History Society,89(1):124.

Balasubramanian, P. and Alagarajan, S. 1991. *Excoecaria agallocha*. An additional host to the long-horned beetle, *Sthenia sgrisator* (*Cerambicidae: Coleoptera*) from Point Calimere Sanctuary. The Journal of Bombay Natural History Society, 88(2):299-300.

Ben-Dov, Y. and German, V. 2003. Scale Net, *Maconellicoccus hirsutus*. http://www.sel.barc.usda.gov/catalogs/pseudoco/Phenacoccusmanihoti.htm.

Blumberg, D. and van Driesche, R. G. 2001. Encapsulation rates of three encyrtid parasitoids by three mealybug species (*Homoptera: Pseudococcidae*) found commonly as pests in commercial greenhouses. Biological Control, 22:191-199.

Blumberg, D., Klein, M. and Z. Mendel. 1995. Response by encapsulation of four mealybug species (*Homoptera: Pseudococcidae*) to parasitism by Anagyrus pesudococci. Phytoparasitica, 23:157-163.

CABI Compendium, 2016. Status of determined CABI editor. Wallingford, UK: CABI.

David, B. V. and Ramamurthy, V. V. 2012. Pests of field, horticultural crops and plantations. In: Elements of economic entomology. Namrutha Publications, Chennai. pp-160.

David, B. V. 1961. Some new hosts of *Sylepta lunalis* Gn in South India. Indian Journal of Entomology, 23(4):301.

Entwistle, P. F. 1972. Pests of coffee. London, UK: Longman Group Limited.

EPPO, 2005. Data sheets on quarantine pests *Maconellicoccus hirsutus.* European and Mediterranean Plant Protection Organization: EPPO bulletin, 35:413-415.

Gibson, R. W. and Turner, R. H. 1977. Insect-trapping hairs on potato plants. PANS, 23(3):272-277.

Hill, D. S. 1983. *Planococcus citri* (Rossi). In: Agricultural insect pests of the tropics and their control, 2nd Edition. Cambridge Univ. Press. pp-746.

Kairo, M. T. K., Pollard, G. V., Peterkin, D. D. and. Lopez, V. F. 2000. Biological control of the hibiscus mealybug, *Maconellicoccus hirsutus* Green (*Hemiptera: Pseudococcidae*) in the Caribbean. Integrated Pest Management Reviews, 5:241-254.

Kranz, J., Schumutterer, H. and Koch, W. 1977. Diseases pests and weeds in tropical crops. Berlin and Hamburg, Germany: Verlag Paul Parley.

Kumari, A. D. and Vijaya, D. 2015. Management of stem borer, *Coelosterna scrabrator* Fabr. in grapevine. Plant Archives, 2:1089-1091.

Lisek, J. 2008. Climatic factors affecting development and yielding of grapevine in central Poland. Journal of Fruit and Ornamental Plant Research, 16:285-293.

Mani, M. and Shivaraju, C. 2016. Mealybugs and their management in agricultural and horticultural crops. Springer, pp-655.

Mani, M., Kulkarni, N. S., Banerjee, K. and Adsule, P. G. 2008. Pest management in grapes. NRC For Grapes, Pune, India. Extension Bulletin, 2:50.

Nair, K. S. S. 2007. Tropical forest insect pest: ecology, impact and management. Cambridge University Press. pp-200.

Noha, H. A. and Shabaan, M. A. R. 2010. Host plants, geographical distribution, natural enemies and biological studies of the citrus mealybug, *Planococcus citri* (Risso) (*Hemiptera: Pseudococcidae*). Egyptian Academic Journal of Biological Science, 3(1):39-47.

Padi, B., Owusu, G. K., Kumaha, N. K. 1999. A record of *Desplatsia dewevri* (De Wild & Th. Dur) (*Tiliales: Tiliaceae*) as an alternative and potential breeding host plant for the cocoa mirid bug, *Sahlbergella singularis* Hagl. Proc. 12th Int. Cocoa Research Conf. November, 1996, Salvador, Brazil:31-35.

Sengupta, C. K. and Sengupta, T. 1981. Cerambycidae (*Coleoptera*) of Arunachal Pradesh. Records of the zoological survey of India, 78(1/4): 133-154.

Sohi, B. S. and Bindra, O. S. 1974. Comparative effectiveness of different insecticides for the control of the grapevine leaf roller (*Sylepta lunalis* Green) in the Punjab, India. *Indian Journal of Horticulture*, **31**(4):389-392.

Trehan, K. N., Wagle, K. N., Bagal, S. R. and Talegeri, G. M. 1974. Biology and control of *Sclodontastrigicollis* Mots. (*Chrysomelidae: Coleoptera*). Indian Journal of Entomology, 9:93-107.

Venette, R. C., Davis, E. E.; Zaspel, J.; Heisler, H. and Larson, M. 2003. Mini-risk assessment: Rice cutworm, *Spodoptera litura* Fabricius [Lepidoptera: Noctuidae]. http://www.aphis.usda.gov/plant_health/plant_pest_info/pest_detection/downloads/pra/sliturapra.pdf.

Williams, D. J. 1996. A brief account of the hibiscus mealybug, *Maconellicoccus hirsutus* (*Hemiptera: Pseudococcidae*), a pest of agriculture and horticulture, with descriptions of two related species from southern Asia. Bulletin of Entomological Research, 86: 617–628.

Williams, D. J. 2004. Mealybugs of Southern Asia. South dene Sdn. Bhd., Kuala Lumpur, Malaysia.

14

Pests of Apple and Their Management

Tamoghna Saha, Nithya Chandran and Tarak Nath Goswami

Apple (*Malus pumila*, though erroneously and commonly called *Malus domestica*) is a deciduous plant and considered to be the world's one of the oldest fruit crops. The tree originated from Central Asia where *M. sieversii,* wild ancestor of cultivated apple is still found today. It is one of the most important table fruits of the world and excels other fruits in having prolonged keeping quality and wide variety of flavour and taste. It is a highly remunerative crop and is grown in all temperate regions of the world. In India apple is mostly grown in Jammu and Kashmir, Himachal Pradesh, Uttarakhand and some parts of Arunachal Pradesh. Amongst the different states, Jammu and Kashmir alone account for almost sixty per cent of total production in India (Asma *et al*., 2016). In India, the total area of apple cultivation is approximately 320 thousand ha and production is almost 1885 thousand tonnes (Anonymous, 2015). However, the quality and yield of apple in our country has been severely declined as compared to that of the developed countries because of several biotic and abiotic stress factors. With regard to the biotic factors, major limitation of apple production is considered to be the insect pests and diseases.

Even though a huge number of insect pests attack apple crops throughout the world but a few of them are recognized to be very serious and require adoption of proper management options. In addition to insect pests, mites also pose a major limitation to the apple production and cause significance economic losses to commercial fruit growers. This book chapter will give an overview of major and minor insect pest of apple and their eco-friendly management.

Insect pests of Apple

Insect pests found in apple can be divided into 2 groups based on plant parts which they attack. First group of insects that feed only on apple fruits while other group of insect attacks on leaf, stem, trunk and other parts of the tree.

Table 1: Insect pests of Apple

Sl. No.	Common Name	Scientific Name	Family	Order
a. Insect pest infesting fruits				
1	Rosy Apple Aphid	*Dysaphis plantaginea* Passerini	Aphididae	Hemiptera
2	Tarnished Plant Bug Lygus lineolaris Palisot de Beauvais	*Miridae*	Hemiptera	
3	Apple Codling Moth	*Cydia pomonella* Linnaeus	Totricidae	Lepidoptera
4	Apple Maggot	*Rhagoletis pomonella* Walsh	Tephritidae	Diptera
b. Insect pest infesting Leaf, stem, trunk and other tree parts				
5	San Jose scale	*Quadraspidiotus perniciosus* Comstock	Diaspididae	Hemiptera
6	Wooly Apple Aphid	*Eriosoma lanigerum* Hausmann	Aphididae	Hemiptera
7	Tent Caterpillar	*Malacosoma indicum* Walker	Lasiocampidae	Lepidoptera
8	European Red Mite	*Panonychus ulmi* Koch.	Tetranychidae	Acarina
9	The Apple Root Borer	*Dorysthenes hugelii* Redtenbacher	Cerambycidae	Coleoptera
10	Apple Stem Borer	*Apriona cinerea* Cheverlot	Cerambycidae	Coleoptera
11	Cottony cushion scale	*Icerya purchasi* Maskell	Margarodidae	Hemiptera
12	Bark Beetle	*Scolytus nitidus* Schedl	Scolytidae	Coleoptera

Fruit infesters

1. Rosy Apple Aphid, *Dysaphis plantaginea* Passerini (Aphididae: Hemiptera)

Distribution

Rosy apple aphid is native to Europe, the Middle East and Central Asia, and has been introduced to North and South America.

Hosts of commercial importance

Apple is the preferred host for the insect species, although rosy apple aphids may also feed on pear. During the hot summer months, rosy apple aphids feed on plantain.

Identification

Newly hatched insects are dark green, and as they pass through five nymphal instars they increase in size from 0.4 to 2 mm. The colour of the aphids change with their growth, then becomes rosy brown or purple finally with a powdery white covering. Eggs are pale green when first laid then turn a shiny black when mature. Egg size is 0.5 mm and football shaped (http.//ipm.ncsu.eduapplechptr2.html).

Biology

In autumn the aphids start to produce eggs and it continues until females are killed due to cool temperature. They lay eggs on bud axils, twigs or in bark crevices. The eggs which undergo overwintering phase, produce only female aphids. Until summer the aphid population keep increasing on apple. Thereafter, winged aphids are produced which migrate to other alternative hosts, like plantain weeds. These of winged form of rosy apple aphids survive on their alternative hosts during summer and early autumn period. During late autumn these winged aphids migrate back to the apple trees and start to lay eggs to begin another cycle (http.//ipm.ncsu.eduapplechptr2.html).

Mode of feeding and symptoms of infestation

All apple varieties are attacked by rosy apple aphid although some varieties are more susceptible than others. At the time of feeding aphids inject saliva into the plant tissue. That saliva contains some toxins that cause the apple leaves to curl diagonally from tip to the base of the leaf. Toxins secreted by a closely related species the rosy leaf curling aphid (*D. devecta*) causes apple leaves to roll longitudinally and turn bright red. Leaf curling does not become obvious until petal fall. Aphids that feed on the leaves of fruit clusters cause bunching, stunting and malformation of the fruits. Honeydew is secreted by the aphids which promote the growth of black sooty mould fungi which reduces the plant vigour and the fruits give a blemished appearance and also reduce the photosynthetic activity of the plant (http://www.dpi.nsw.gov.au/__data/assets/pdf_file/0008/443294/Exotic-Pest-Alert-Rosy-apple-aphid.pdf).

Management

i. Insecticides applied with delayed dormant oil application have historically been used for control of rosy aphids.

ii. Biological control of rosy apple aphid is important since the aphid is preyed upon by many natural enemies, including ladybird beetles, syrphid flies, and aphid lions.

2. Tarnished Plant Bug, *Lygus lineolaris* Palisot de Beauvais (Miridae: Hemiptera):

The tarnished plant bug, *L. lineolaris* is a sporadic pest and occasionally becomes serious to fruit buds and fruits in early spring.

Distribution

The tarnished plant bug is found throughout North America, but it is primarily a pest in temperate non desert (humid) areas (https://ecommons.cornell.edu/bitstream/.../ tarnished-plant-bug-FS-NYSIPM).

Hosts of commercial importance

It feeds on more than fifty economically important plants, including alfalfa, cotton, strawberries, brambles, and most tree fruits that are grown in the United States (https://ecommons.cornell.edu/bitstream/.../ tarnished-plant-bug-FS-NYSIPM).

Identification

Adult tarnished plant bugs are a brassy brown in color, generally marked with variable black, white, and yellow. Adults are about 1/4 inch long, less than half as broad, flattened, and oval in outline. The nymphs and eggs are seldom seen in apple orchards (http.//ipm.ncsu.eduapplechptr2.html).

Biology

Tarnished plant bugs overwinter as adults, and remain active with the first warm days of spring. They are most abundant on apple trees just before, during, and after apple bloom. Adults feed on the developing fruit buds and young fruits. There are several generations completed for the insect annually, but comparatively little reproduction occurs on fruit trees (http.//ipm.ncsu.eduapplechptr2.html).

Mode of feeding and symptoms of infestation

Adult bugs feed on the developing fruit buds and young fruits. They inject a salivary toxin that is highly injurious to plant tissue. Feeding of plant bug causes the developing fruits to become deformed or cat faced. Fruit with sunken areas of more or less conical shaped and corky tissue at the bottom indicate specific plant bug damage (http.//ipm.ncsu.eduappl echptr2.html).

Management

i. Use of unbaited, non-reflective, white sticky boards hung low in the trees is an effective monitoring method. The best places to set the traps are in lower areas such as ditch banks and in hedge rows, which are favourable overwintering sites of the adults. White sticky traps are also available commercially.

ii. The tarnished plant bug attacked by a number of natural enemies, like other true bugs (e.g. nabids, geocorids), spiders, ladybird beetles and a number of parasitic wasps. But economical management of the pest population using these natural enemies is not possible (https://ecommons.cornell.edu/ bitstream/.../ tarnished-plant-bug-FS-NYSIPM).

iii. As the apple trees has a frequently long bloom period during which pesticides can't be applied, which make the satisfactory chemical control difficult on tree fruits and prevent optimum timing of control sprays. Also, applying pesticides during pre bloom periods may dissipate during the prolonged blooming period. Control by applying pesticides also become difficult because of the mobility of tarnished plant bug.

3. Apple codling moth, *Cydia pomonella* Linn. (Tortricidae: Lepidoptera)

It is one of the most serious fruit feeding apple pests in the world. Apple is the codling moths preferred host. The larvae are internal fruit feeders.

Distribution

The pest has been reported from Europe, USA, Canada, South Africa, Australia, New Zealand, Afghanistan and Pakistan (Sheldeshova, 1967). In India, this pest is restricted to cold arid region (Ladakh) of Kashmir state (Malik *et al.*, 1972; Wadhi and Sethi 1975). This pest is supposed to have entered into Ladakh from the North West frontier province of Pakistan where it was reported as a serious pest on the deciduous fruits (Janjua *et al,* 1943 and 1941). In Ladakh, it is widely distributed throughout the apple growing areas and has been recorded from Karkitchoo, Mangbore, Hardass, Sainikund, Batalik, Dah, Hanoo area of Kargil district and Lamayuru, Scrubachan, Khaltse, Tamisgam, Basgo, Saspol and Nimmu area of Leh district. Pawar *et al,* (1981) reported biology of codling moth and their incidence in apple growing in Ladakh region. In a study, Zaki (1999) reported that infestation of codling moth in apple was 49.7 and 42.5 per cent in Kargil and Leh district, respectively. The infestation on fallen fruit was 69.9 per cent, while on the fruits still on tree was only 27.5 percent.

Hosts of commercial importance

Apple is the major host of this pest; however, it also feed on pear, walnut, peach etc.

Identification and biology

The full grown larva of codling moth has a brown head, with pinkish white body, and is about 3/4 inch long. Adult codling moths are about 1/2 inch long with greyish, striped appearance. The male and female are similar in appearance. Chocolate brown patches at the base of the wings contain copper colored scales near the inside wing tip. Codling moth eggs are white, flattened, pancake shaped, and about

1/25 inch in diameter. Eggs are laid singly. One or 2 days prior to hatching, the dark head capsule of the larva can be seen. The adult emergence takes place in the month of June. The female moth after mating starts egg lying on fruits, leaves and twigs. The hatching of eggs take place after 7-15 days depending upon temperature and humidity. The larvae after hatching feed near the fruit surface for a period and later bore through the fruit surface from the blossom or calyx end. After that the larvae bore into deep tissues and feed on pulp and developing seeds until fully grown. Early stage larva being creamy white in colour, which later becomes light pink. From the exit hole mature larvae come out of the fruit and spin a cocoon in loose bark, cracks or on debris and may or may not pupate. The codling moth overwinters as full-grown larva within cocoon under loose bark, leaf litter, or any other sheltered place nearby. Pupation takes place during second fortnight of May, which ranges from 10-20 days depending upon temperature and other climatic condition. The pest complete one to two generations in a year depending upon the temperature and other weather parameters.

Mode of feeding and symptoms of infestation

A recent study conducted by SKUAST-K, Leh, it was observed that in Leh district alone the damage was 80 per cent. The damage is done by the caterpillars which tunnel through the fruits by eating away the pulp, fills these tunnels with frass and thus render the fruits unsuitable for human consumptions (Asma *et al.*, 2016). The symptoms are

- It is a direct pest and hence causes severe damage to the fruit.
- Through calyx, neonate larva enters into the fruit and feeds on the seed.
- Infested fruits become deformed and fall prematurely.
- 30 to 70 % fruits are rendered unmarketable.

Management

i. All the debris and weeds should be removed from old trees and orchard to prevent the hibernating larvae to find shelter.

ii. Band the trees with grass ropes or sac (Jute) cloth in 3-4 folds before the larval descend to the ground for hibernation and the larvae should be collected and destroyed.

iii. Enforce quarantine rules strictly to prevent spread of the pest.

iv. The trees should be sprayed with Chlorpyrifos (20EC) 0.02%, the spray called the calyx spray which should be applied at the time of the fall of petals and closure of calyx (Asma *et al.*, 2016).

4. Apple maggot, *Rhagoletis pomonella* Walsh (Tephritidae: Diptera)

The apple maggot also known as railroad worm

Distribution

It is an injurious apple pest in the north eastern states of India and Canada. In the Southeastern countries, it is seldom a pest in commercial orchards.

Hosts of commercial importance

Apple is the major host of this pest; however, it also feed on crab apples, hawthorn, plum, pear, and cherries.

Identification and biology

The apple maggot adult is a fly slightly smaller than a housefly. The flies are black with white bands between segments on the abdomen, four bands on the female and three on the male. The wings are conspicuously marked with four oblique black bands. Maggots are carrot shaped, white, legless, and about ¼ inch long when full grown.

Apple maggots overwinter inside a brown, 1/4 inch long puparium in the soil. In North Carolina, adult emergence begins in mid June, peaks in July, and is completed by early September. Flies begin to lay eggs 8 to 10 days after emergence. After mating is completed, the females seek out fruit. Tiny, white, elongated eggs are laid under the skin of the apple through a puncture made by the female. Maggots tunnel through the apple, causing a breakdown and discoloration of the pulp. Infested apples usually drop prematurely. The full grown maggots leave the fruit and enter the soil where they pupate. They have completed one generation in a year.

Mode of feeding and symptoms of infestation

Injury to fruit is caused by maggots boring throughout the fruit, forming irregular, winding tunnels which turn brown, often causing premature dropping of fruit. When the fruit is slightly infested, there may be no external indication of the maggots, but when the fruit ripens, the burrows show as dark, winding trails beneath the skin. Minute egg punctures and distorted, pitted areas may show on the surface. Heavily infested early varieties of fruit will be reduced to a brown rotten masses filled with the fly larvae (Varela, 2008).

Management

i. **Cultural management:** The systematic destruction of infested apples and the elimination of hawthorn and abandoned apple trees in the vicinity of orchards are considered valid control practices. Apple maggots in fruits may be killed by placing the fruit in cold storage at 32°F for a period of 40 days.

ii. **Biological management:** Hymenopterous parasites recorded from *R. pomonella,* a braconid wasp; *Opius melleus*, which attacks the larvae; and a tiny mymarid wasp, *Patasson conotracheli*, which attacks the eggs. Because the apple maggot feeds within fruit, biological control agents have not been very effective (Varela, 2008).

Leaf, stem, trunk and other parts infesters

5. San jose scale, *Quadraspidiotus perniciosus* Comstock (Diaspididae: Hemiptera)

It attacks a wide range of trees and shrubs in addition to apple. Scale insects are difficult to detect until damaging populations reaches above economic threshold. Many a times these insects are not detected until damage begins to appear on the fruits at the time of harvest.

Distribution

This insect is seen in all the countries where apple is grown. In India, this pest was introduced in 1906 from France and now this pest has been recorded to infest more than 32 different host plants (Rahman and Ansari, 1941). The records show that the insect has spread to every continent of the world except Antarctica.

Hosts of commercial importance

Apple is the major host of this pest, however, it also feeds on plum, pear, nut trees, cherry berries, peach, ornamental trees and shrubs and apricot.

Identification and biology

Mature female scales are about 1/10 inch in diameter. Color varies from nearly white for young scales to dark gray for mature females. Females have a distinct black spot in the centre of the scale. Male scales are oblong with a black spot near one end and are much smaller than the female. The active immature stage (crawler) is lemon yellow and about 1/100 inch long.

The adult female reproduces ovo-viviparously *i.e.*, the eggs developing in an ovisac inside the mother to be born as nymphs. Each female give birth to 200-400 nymphs. The first instar nymphs (crawlers) move about for 12-24 hours and fix themselves at suitable places on host tree and begin to feed by sucking the cell sap. Simultaneously, they secrete a waxy covering over themselves which give them the name scale insects. They become full grown in 3-4 days and within next 10-

14 days the female again start giving birth to new ones. The longevity of gravid mother is 50-53 days and male adult develops into a winged adult in 25-31 days and live only 24-32 hours during which it fertilizes the non winged female. The insect overwinters in the nymphal stage inside the scale covering. The insects are active from April to December. There are 6-7 overlapping generations in a year. The scale also gets disseminated by various species of birds, bats etc. (Rahman *et al.*, 1940).

Mode of feeding and symptom of damage

The insects suck the sap continuously resulting in weakness and ultimately death of young plants in the nursery. Ashy-grey scales cover the leaves, twigs, fruits and sometimes even the entire bark and it can be easily scraped off which expose the orange coloured individuals beneath. Infested fruits form pink coloured patches around the areas of scale infestation and the market value of this infested fruits are depreciated (Asma *et al.*, 2016).

Management

i. **Cultural management:** Grow attractant flowers for natural enemies: *viz.*, sunflower family, carrot family plants, buckwheat.

ii. **Mechanical management:** Pruning of infested branches and twigs, and collection and destruction of pruned infested plant material is effective.

iii. **Biological management:** Parasitoids such as *Encarsia perniciosi* and *Aphytis diaspidis* cause effective parasitization; Coccinellid predators such as *Chilocorus infernalis, Chilocorus rubidus, Pharoscymnus flexibilis* can naturally check the pest infestation to some extent.

iv. **Chemical management:** Spraying of Malathion 50% EC @ 600-800 ml in 600-800 l of water/acre or lime sulphur 22% SC @ 800-2000 ml/acre (use 2% pre-blossom and 1% post blossom in conventional sprayers) have been proved to be effective.

6. Apple wooly aphid, *Eriosoma lanigerum* (Hausmann), (Aphididae: Hemiptera)

Distribution

Wooly aphid is native to America and has cosmopolitan distribution except the hot tropical areas. In 1889 this pest was first recorded from India at Coonoor (Tamil Nadu) (Misra, 1920).

Hosts of commercial importance

The pest is active throughout the year and feed on apple, crab apple, almond, hawthorn, pear and quince. The nymphs secrete wooly filaments of wax over their bodies, making the name of the insect as wooly aphid.

Identification and biology

Young nymphs have a dark purple body covered with powdery greyish wax. Larger nymphs and adults vary from dark purple to rosy color, and are nearly covered by a wooly mass of long wax filaments. Adults and nymphs have needle like mouthparts which are inserted into the bark where they feed on the plant sap.

The aphids are viviparous (give birth to nymphs and do not lay eggs) and reproduce through both sexually and parthenogenetically after mating. The mid December to mid-February is a non-reproductive period for these insects. From March month onwards, each female produce around 30-116 nymphs parthenogenetically which can be alate (winged) or apterous (wingless). The former being present throughout the year but the apterous form mostly remains from July to October. Within 24 hours of hatching, the nymphs start secreting waxy filament to become wooly. Nymphal period having 4 instars, lasts for around 11 days during summer and 93 days during winter (Rahman Khan, 1941). In winter, sexual forms appear to mate and lay eggs while the nymphs already on the trees migrate downwards to enter the soil for hibernation. Reverse migration from soil to the aerial part takes place in May. The insects may complete 13 generations in a year.

Mode of feeding and symptoms of infestation

Activity of this pest is seen throughout the year. Primarily the underground root portion is attacked but the winged forms also attack the stems, trunk, leaf petioles, branches, twigs and fruit stalks. Downward and upward migrations are accentuated during coldest and hottest period of the year, respectively (Lal and Singh, 1947). Due to the drainage of sap by feeding, affected trees give sickly appearance; lose their vigour and their fruiting capacity also adversely affected. When infestation occur in young trees their roots get disintegrated to such an extent that moderately strong winds can blown over these trees. The fruits from heavily infested trees are of poor quality being undersized, malformed and insipid in taste.

Management

i. Use resistant root-stock like Golden Delicious, Northern Spy and Morton Stock 778, 779, 789 or 793 has been found to be effective in preventing the damage by aphids (Atwal and Dhaliwal 1999).

ii. The aphid population can be effectively checked by an exotic parasitoid *Aphelinus mali* Haldeman.

iii. Spray 800 ml Malathion 50 EC in 500 liters of water per ha during summer months. Avoid insecticidal sprays where parasitoid *Aphelinus mali* is present.

iv. Conserve predators such as *Coccinella septempunctata, Chrysoperla zastrowi sillemi, Menochilus sexmaculatus, Syrphus confrator*.

7. Tent Caterpillar, *Malacosoma indicum* Walker (Lasiocampidae: Lepidoptera):

Distribution

It is an important pest of apple in north-western India, being more serious in the Kashmir valley and Shimla hills.

Hosts of commercial importance

This pest is found on a number of fruit and forest trees including ornamentals; but apple being its preferred host. Besides the apple almond, apricot, cherry, gooseberry peach, pear, walnut are also occasionally attacked.

Identification and biology

The pest is active from mid of March to May. A female lays around 200-400 eggs on branches as broad bands or rings, and pass the remaining 9 months in egg stage. The hatching of eggs coincides with the appearance of buds on the plants. They hatch in around 9-10 months between mid March to mid May. The larvae live gregariously and soon after emerging spin a nest at a convenient and sheltered place on the tree. Larvae rest in the nests during day time and feed voraciously on the leaves at night. The larval stage lasts 39-68 days and later spin oval, white and compact cocoons, each about 25 mm in length. The pupal stage lasts for about 8-22 days which is passed in side these cocoons. Sometime, moths emerge in the 3rd week of May and continue till the beginning of June and are short lived. The female may survive for 3-5 days. They completed only one generation in a year (Atwal and Dhaliwal, 1999).

Mode of feeding and symptoms of infestation

The caterpillar web a tent like nest at the forking of twigs and hide in this nest during day. At night the caterpillars congregate on leaf lamina and feed voraciously leaving behind only the midribs and portions of hard veins. In case of serious infestations, around 40-50 % of apple plants in an orchard may be defoliated. The tents and skeletal remains of the leaves are conspicuous symptoms of the pest attack (Asma *et al*., 2016).

Management

i. To check the infestation of this pest, all the egg bands are need to be destroyed at the time of pruning in December-January.

ii. The caterpillar can be killed by mopping up the tents with a pole and some rags dipped in kerosene tied on its end. Place open containers with kerosenized water below the infested tree so that the larvae that fall may also get killed (Asma *et al.*, 2016).

iii. Spray 500 ml of Dimethoate 30 EC or 2.0 kg of Carbaryl 50 WP in 500 liters of water per hectare.

8. European red mite, *Panonychus ulmi* Koch (Tetranychidae: Acarina)

Distribution

Red mite has been introduced from Europe in around 1910. This spider mite has become one of the most important pests of all fruit growing belts of the world and considered by many growers to be their most important pest which is sometimes difficult to control.

Hosts of commercial importance

It is having a wide host range and includes deciduous bushes and trees belonging to the family Rosaceae, but it is found commonly in association with fruit trees such as apple, walnut, almond, vines, mulberry, blackberry, peach, pear, plum, prune and cherry that it reaches economic importance (Asma *et al.*, 2016).

Identification and biology

Adult female European red mites are bright to brownish red with an elliptical body. Females have four rows of curved spines on their backs, each spine borne on a whitish tubercle. These spines can be seen with a hand lens. The overwintering eggs are bright red, spherically shaped and have a distinct white stalk about as long as the diameter of the egg. This white stalk can be seen only under magnification. Summer eggs are translucent.

P. ulmi overwinter as diapausing eggs laid on the bark of trees or smaller branches and spurs. During heavy infestation, areas of bark may even appear red due to the presence of many eggs. Eggs hatch from mid April to end June depending upon climatic conditions. On hatching, juvenile mites move to the underside of leaves and begin feeding. In about 3 weeks they become mature after third moult. Mostly 8-10 generations are completed in a year. *P. ulmi* lays two types eggs, *i.e.*, summer eggs and winter eggs. Summer eggs are laid on the leaves of host plant and are of the non-diapause type and develop without interruption. The winter eggs are deposited predominately on the bark. Summer females exposed to cool conditions lay diapausing eggs. Overwintering eggs are deposited in groups on roughened bark, especially around buds and fruit spurs.

Mode of feeding and symptoms of infestation

European red mite injure the trees by feeding on leaves destroying chlorophyll, and increasing respiration. This is accomplished by insertion of the mite mouth parts into the leaf cells to withdraw the contents. All motile stages feed on the foliage, preferably on lower surface of leaves, but both leaf surfaces are attached when populations are high. A characteristic of brown foliage, starting as a subtle cast to the green leaf which becomes bronze in severe cases, results from heavy mite feeding. Damage is more severe when mite infestation is followed by *Alternaria* fungus. High population of mites during late season can cause further indirect downgrading of fruit by depositing overwintering eggs (Asma *et al*., 2016).

Management

i. **Biological management:** Conserve the predators such as *Chrysoperla zastrowi sillemi*, anthocorid bug, predatory mite (*Amblyseius fallacis*), coccinellid (*Stethorus punctum*) etc.

ii. **Chemical management:** Spray Fenazaquin 10% EC @ 160 ml in 400 l of water/acre or Propargite 57% EC @ 0.5-1 ml/l of water and use spray fluid of 10 l/tree or Spiromesifen 22.9% SC @ 120 ml in 400 l of water/ acre.

9. Apple root borer, *Dorysthenes hugelii* Redtenbacher (Cerambycidae: Coleoptera)

Distribution

The main host of this borer is living and dead roots of apple trees including other host like apricot, cherry, peach, pear, walnut and a few forest trees.

Hosts of commercial importance

It is confined to foot hills of Himalayan range and is a serious pest of apple in Kumaon hills (Asma *et al*., 2016).

Identification and biology

The adult beetles start mating immediately after emergence sometime around the end of June to mid of July. The males die soon and females live for about 10-12 days and lay around 200 ovoid shaped yellow white eggs, below the soil surface. Eggs hatch within 30-40 days and the grubs go down in the soil about 10-25 cm deep and feed on the root system of host plants. The duration of grub lasts for 3 years 6 months and without food can survive for 24-90 days. Pupation take place in earthen cocoons inside the soil and the pupal period lasts for 3 months. The beetle emerges from the soil after the first shower of monsoon (Asma *et al.*, 2016).

Mode of feeding and symptom of damage

The grubs feed on the roots and girdle the roots and feed on the internal tissues. As a result the main roots are severely damaged from the base and the trees, if young. Die away while the older ones become weak and fall down by strong winds (Asma *et al.*, 2016).

Management

i. To check the incidence of this pest, avoid dry sandy soils for planting apple orchards.
ii. Inter culturing in the orchard helps in killing of grubs.
iii. Use well rotted FYM and mix thoroughly with soil around the tree.

10. Apple stem borer, *Apriona cinerea* Cheverlot (Cerambycidae: Coleoptera)

Distribution

It is reported from India, Pakistan and Afghanistan. In India it is more commonly found in Kashmir, Himachal and Uttar Pradesh.

Hosts of commercial importance

The major host plant of apple stem borer is apple, peach, fig, pear etc.

Identification and biology

Grubs are creamy yellow with dark brown and flat head, while adults are ashy grey beetle with numerous black tubercles at the base of elytra.

The beetles appear in July-August with the onset of monsoon and after mating female excavate an oval patch on the shoot and lays eggs inside the cavity. The incubation period is 7-8 days and after hatching grubs start feeding in the tree trunk. The feeding activity starts in March reaching tree trunk by autumn and activity gets slowed and go in hibernation during winter. As the summer season comes, it starts activity again and when full fed it pupates inside the tunnel in the woody tissue that lasts for more than a month. The life cycle is completed in about two years (Atwal and Dhaliwal, 1999).

Mode of feeding and symptoms of infestation

The grubs bore through the shoots and make circular galleries downward leading to main stem and trunk of the trees. Due to feeding by the grubs, the affected trees may not die for many years but their vitality and productivity is reduced greatly. Though the adult beetles feed on bark but does not cause considerable loss (Atwal and Dhaliwal, 1999).

Management

i. Collection and destruction of beetle and grub is very effective method to reduce the pest population

ii. Prune the branches containing grubs before they entered the tree trunk

11. Cottony cushion scale, *Icerya purchasi* Maskell (Margarodidae: Hemiptera)

After its introduction into California from Australia in 1868 or 1869, cottony cushion scale devastated citrus there. In 1888, a lady beetle called the vedalia lady beetle (*Rodolia cardinalis* (Mulsant) was introduced from Australia, and within 18 months the cottony cushion scale was reduced to non-pest status. This was the first successful biological control of an insect pest (Baker and Frank, 2011).

Identification and biology

Female cottony cushion scales are rusty red with black legs and antennae. They are about 4.5 mm long. The mouthparts are of piercing/ sucking type. The body is often obscured by wax. Male cottony cushion scales are small (3 mm), slender, reddish-purple insects with 2 metallic blue wings (Baker and Frank, 2011).

Each female lays a large number of eggs in a dense, fluted ovisac from which this insect got its name. This ovisac is made of wax secreted from the lower side of the scale. Three weeks (summer) to 8 weeks (winter) after being laid, the eggs hatch to first nymphs. In most scales, this stage is called the crawler but most stages of the cottony cushion scale can crawl. The first nymphs can crawl up to 1 yard in 10 minutes, and they tend to crawl up to the leaves and twigs to feed. After 2 or 3 weeks (much longer in winter), they moult into second nymphs which also soon secrete a layer of wax. After 2 or 3 weeks, the second nymphs moult to become third nymphs. In many cottony cushion scale infestations, males are rare or nonexistent in some cases (Baker and Frank, 2011).

Mode of feeding and symptoms of infestation

Cottony-cushion scale extracts plant sap from leaves, twigs and branches, reducing the plant vigour. If infestations are heavy, leaf and fruit drop can occur along with twig dieback. It also secretes honeydew, which promotes the growth of sooty mold that may discolor the fruit and block photosynthesis.

Management

i. Select healthy and pest free rootstock
ii. Collection and destruction of the infested plant parts
iii. Spray application of neem oil 2%, NSKE 5%
iv. Spray application of Chlorpyriphos 20 EC 0.04% with sticking agent
v. Field release of some coccinelid predators and *Chilocorus nigritus*.

12. Bark Beetle, *Scolytus nitidus* Schedl (Scolytidae: Coleoptera)

It is an important shot-hole borer which has caused severe economic losses to apple fruit trees in the Kashmir valley and its population has witnessed manifold increase during the last few years due to conducive environmental conditions, mainly due to drought.

Distribution

It is distributed in Himachal Pradesh, Kashmir and Uttar Pradesh. On an average 5-10 per cent apple trees get damaged annually by the attack of *S. nitidus*, which can increase up to 44 per cent in the mismanaged orchards during dry and hot weather conditions (Buhroo and Lakatos, 2007).

Hosts of commercial importance

Apart from apple, this pest causes severe damage to apricot, peach, plum and other forest plantation.

Identification and biology

This bark beetle overwinters in larval stage on apple trees in Kashmir. They resume their activity from the first fortnight of March. Pupation start from the first fortnight of April and the adult emergence start from the fourth week of April. The insect completes three generations in a year. The first generation lasted from the second fortnight of April to July having a total life span of 97-120 days. The second generation emerges from the first fortnight of July to the 15th of October with a total life span of 108-120 days while the third overwintering generation takes 245-284 days. After emergence the adults fly to host trees and undergo maturation feeding for 5-8 days. The copulation takes place at the entrance hole. The females lay about 60 eggs which hatch in 4 to 8 days. The larvae attain 5 instars and accomplish their development in 40 to 55 days constructing larval galleries. The larvae pupate for 8-20 days. The adults survive for 48-65 days and the total life-span of this insect ranges from 94 to 128 days (Asma *et al.*, 2016).

Mode of feeding and symptoms of infestation

Galleries made by the insects under the bark, severely hamper the translocation of food and water. During the early phase of infestation the tree does not show symptoms but their growth is arrested. Attacked trees also show reduction in yield and foliage density. Heavily infested trees attract fungal diseases (Atwal and Dhaliwal, 1999).

Management

i. Infested branches or tree should be removed from the orchards.

ii. New orchard should not be planned near infested forest trees.

iii. Spraying of Dimethoate 30 EC 100 ml per 100 liters of water should be done for controlling the insect pest.

References

Anonymous, 2015. Horticultural Statistics at a Glance. Department of Agriculture, Cooperation & Farmers Welfare. Ministry of Agriculture & Farmers Welfare Government of India.

Asma, S., Malik, M. and Ashraf, A. W. 2016. Insect Pests of Apple and their Management. In: Insect Pests Management of Fruit Crops: 295-306. Biotech Books® 4762-63/23, Ansari Road, Darya Ganj, New Delhi - 110 002.

Atwal, A. S. and Dhaliwal, G. S. 1999. Agricultural Pests of South Asia and their management. Pp-274-286.

Baker, J. R. and Frank, S. 2010. Cottony Cushion Scale, *Icerya purchasi* (Maskell), (Margarodidae:Hemiptera).

Buhroo, A. A. and Lakatos, F. 2007. On the Biology of the Bark Beetle *Scolytus nitidus* Schedl (Coleoptera:Scolytidae) attacking apple orchards. Acta Silvatica et Lignaria Hungarica. 3:65-74.

Janjua, N. A. and Samuel, C. K. 1941. Fruit pests of Baluchistan. ICAR Miscellaneous Bulletin, 42:1-41.

Janjua, N. A., Mustafa, A. M. and Samual, C. K. 1943. On the biology and control of codling moth *Cydia pomonella* L., in Baluchistan. Indian J. Agri. Sci., 13:112-128.

Lal, K. B. and Singh, R. N. 1947. Seasonal history and field ecology of the wooly aphid in the Kamaon hills. Indian J. Agric. Sci., 17(4):211-218.

Malik, R. A., Punjabi, A. A. and Bhat, A. A. 1972. Survey study of insect and non insect pests in Kashmir. *Horticulturist* (J&K), 13 (3):29-44.

Misra, C. S. 1920. Index to Indian fruit pests. Report. *Proc.* 3rd Entmol. Management, Pusa (Bihar), February 1919, 2: 564-595.

Pawar, A. D., Tuhan, N. C., Balsubramanian, S. and Parry, M. 1981. Distribution, damage and biology of codling moth, *Cydia pomonella* (L). Indian J. Plant Prot. 10:111-114.

Rahman Khan, A. 1941. Occurrences of the gypsy moth, *Lymantria obfuscate* Walk. In Shimla hills. Indian J. Ent. 31(2):338.

Rahman, K. A. and Ansari, R. 1941. Scale insects of the Punjab and northwest frontier province usually mistaken for San Jose (with description of the new species). Indian J. Agric. Sci. 14(4):308-310.

Rahman, K.A. Asa, N. and Kalia, 1940. Volant animals which act as carriers of San jose scale. Current Sci. 9(5): 235.

Sheldeshova, G.G. 1967. Ecological factors determining distribution of codling moth, *Laspeyria pomonella* L (*Lepidoptera: Torticidae*) in northern and southern hemispheres. Entomological Review, 46:349-361.

Varela, L. G. 2008. Apple maggot, *Rhagoletis pomonella*. UC Pest Management Guidelines (no longer available online).

Wadhi, S. R. and Sethi, G. R. 1975. Eradication of codling moth- a suggestion. J.Nuclear Agri. and Biol., 4:14-19.

Zaki, F. A. 1999. Incidence and biology of codling moth, *Cydia pomonella* L., in Ladakh (Jammu and Kashmir). Appli. Biol. Res. 1:75-78.

Web resources cited

http://agritech.tnau.ac.in/crop_protection /crop_prot_crop_insect pest_apple_main.html

http://ipm.ncsu.edu/apple/chptr2.html

http://www.dpi.nsw.gov.au/__data/assets/pdf_file/0008/443294/Exotic-Pest-Alert-Rosy-apple -aphid .pdf

https://ecommons.cornell.edu/bitstream/.../ tarnished-plant-bug-FS-NYSIPM

15

Pests of Cashew Nut and Their Management

Amitava Banerjee and Biswanath Bandyopadhyay

Cashew (*Anacardium occidentale* L., Family: Anacardiaceae) is a very important crop in India for earning foreign exchange. Portuguese travelers introduced this plant into India from Brazil during sixteenth century and the purpose was to check soil erosion on the coastal area. Though initially this crop is considered as a remedy for soil conservation, wasteland development and afforestation, but it gradually became a commercial important crop. Cashew is now widely grown in tropical regions and well adapted to the climatic condition of India. Now India becomes the largest producer, consumer, processor and exporter of cashew in the world. In India, cashew is mainly cultivated in Goa, Maharashtra, Kerala and Karnataka along the West coast and Andhra Pradesh, Tamil Nadu, Orissa, Pondicherry and West Bengal along the East coast. To a limited extent it is also grown in some non-traditional areas like Kolar region of Karnataka and Bastar region of Chhattisgarh.

Among the various factors that reduce in the production, insect pests are one of the important limiting factors that cause a crops loss of about 30 to 40 % (Satapathy, 1993). A number of insect pests infest cashew during its different growth and developmental stages. Rai (1984) gave an exhaustive list of 133 species of insect pests that infest cashew. In India more than 50 species have been reported to infest cashew with a varying intensity (Devasahayam and Nair, 1986). However, only four pests are reported to cause major damage to the cashew crop, *i.e.* cashew stem and root borer (CSRB), tea mosquito bug (TMB), nut borer, leaf miner and other minor pests including leaf and blossom webber, thrips and mealy bug (Maruthadurai *et al.*, 2012). National Institute of Plant Health Management (NIPHM), Hyderabad, India identified six insect pests having national significance viz. mosquito bug, stem and root borer, leaf miner, leaf and blossom webber, flower thrips and three species of foliage thrips along with some pests having regional significance viz. apple and nut borer, mealy bug, hairy caterpillar, leaf folder, aphids, leaf twisting weevil, cashew scale etc. (Anon., 2014).

Different reports depicted that during 1997-2000, incidence of cashew stem and root borer, shoot and blossom webber, flower thrips, leaf miner and shoot tip caterpillar was significantly higher in different cashew growing regions of West Bengal and interestingly the population of tea mosquito bug was found low (Anon., 1998; Anon., 2000). Bandyopadhyay and Banerjee (2007) reported that cashew stem and root borer, flower thrips and mealy bugs causes severe damages to the crop. Kar *et al.* (2016) stated that among different insect pests infesting cashew, cashew stem and root bore, tea mosquito bug, leaf miner, leaf and blossom webber, thrips and apple nut borer are relatively more important in terms of damage potentiality in West Bengal.

A list of insect pests occurring on cashew comprising both major and minor pests is being presented here in tabular form.

Table 1: Insect pests of cashew nut.

Sl. No.	Common name	Scientific name	Family	Order
1.	Mosquito bugs	*Helopeltis antonii* Signoret, *H. anacardia* Miller, *H. schoutedeni* Reuter	Miridae	Hemiptera
2.	Stem and root borer	*Plocaederus ferrugineus* Linn.	Cerambycidae	Coleoptera
3.	Leaf miner	*Acrocercops syngramma* Meyk.	Gracillariidae	Lepidoptera
4.	Shoot and blossom webber	*Lamida moncusalis* Wlk.	Pyraustidae	Lepidoptera
5.	Mealy bug	*Ferrisia virgata* Cockerell	Pseudococcidae	Hemiptera
6.	Shoot tip and inflo-rescence caterpillar	*Chelaria heligramma* M.	Gelechiidae	Lepidoptera
7.	Thrips a) Foliage thrips	*Selenothrips rubrocinctus* (Giard)	Thripidae	Thysanoptera
	b) Foliage thrips	*Rhipiphorothrips cruentatus* Hood	Thripidae	Thysanoptera
	c) Foliage thrips	*Retithrips syriacus* Mayet.	Thripidae	Thysanoptera
8.	c) Flower thrips	*Rhynchothrips raoensis* G.	Thripidae	Thysanoptera
9.	Apple and nut borer	*Thylacoptila paurosema* Meyrick	Pyralidae	Lepidoptera
10.	Leaf folder	*Sylepta aurantiacalis* Fisch.	Pyralidae	Lepidoptera
11.	Hairy caterpillar	*Metanastria hyrtaca* Cramer.	Lasiocampidae	Lepidoptera
12.	Mango aphid	*Aphis (Toxoptera) odinae* van der Goot.	Aphididae	Hemiptera
13.	Hard scale (Ar-moured scale)	*Pseudaonidia trilobitiform-is* Green.	Diaspididae	Hemiptera

(*Source*: Modified from Rai (1984), Devasahayam and Nair (1986) and Maruthadurai *et al.*, 2012)

The distribution, alternate host plants, biology, mode of feeding and symptoms of infestation to the crop and possible management aspects of the insect pests causing significant damage to cashew are discussed in brief.

1. Tea mosquito bug, *Helopeltis antonii* Signoret. (Miridae: Hemiptera)

Tea mosquito bug (TMB) is one of the most serious pests of cashew in India. It causes significant crop loss. It is estimated that this pest alone caused 30-50 per cent yield loss (Ambika and Abraham, 1984, Pillai *et al.*, 1984) and loss may reach up to 100 % in outbreak situations (Sundararaju and Sundarababu, 1999). The pest causes inflorescence blight and immature nut fall which result in an economic loss of more than 30%. In West Bengal, the damage of 25-30% has been reported by Chatterjee (1989). Cashew trees are also infested by another 2 species of TMB viz., *H. theivora* and *H. bradyi*.

Distribution

Helopeltis antonii is distributed from India through to New Guinea (Entwistle, 1972).

Hosts of commercial importance

The pest also found to infest on guava, cocoa (Puttarudriah and Appanna, 1955), neem, cinchona, mahogany, cotton, grapes, apples, tea (Pillai and Abraham, 1974), pepper, cinnamon, camphor (Pruthi, 1969), avocado, tamarind (Butani, 1979) and red gum (Nair *et al.*, 1979).

Identification and biology

The adult bugs are about 6-8 mm long, reddish-brown in colour with a black coloured head, red coloured thorax, black and white coloured abdomen. Female bug prefer to lay eggs on the soft tissues of inflorescence branches and tender tissues of newly formed shoots. Eggs are kidney shaped and creamy white in colour. The occurrence of chorionic threads outside the tissues indicates the presence of eggs inside. An average of 50 eggs is laid by a single female bug. The eggs have an incubation period of 5-7 days. The adult usually feeds during 6-10 AM. They make about 150 feeding punctures in a day. After hatching young nymphs feed on the tender leaves and because of feeding the leaves turn necrotic. Nymphs don't have any wing and smaller in size, but they resemble to the adults on other aspects. The young nymphs are orange in colour and look like ants having long legs.

They have 5 nymphal instars and nymphal period lasts for 10-15 days. Life span of female adult is about 7 days, while incase of male life span is of 9-10 days. Total life-cycle is completed within 25-32 days. They can be seen to damage the young plants during the monsoon period.

Mode of feeding and symptoms of infestation

Both nymphs and adults suck sap from young shoots, tender flushes, panicles, inflorescence, growing young nuts and apples. Dark brown patches develop on the green tender stem of young shoots and inflorescence rachis. As they suck the sap from the tender leaves, it causes crinkling and curling of the leaves. Long black lesions develop on the affected shoots. When the pest infests the immature nuts, characteristic eruptive spots develop on nuts, become shriveled and finally fall off.

Heavily infested trees give a scorched appearance that lead into the death of shoots and growing tips. Secondary fungal infection (*Botrydiplodia theobromae*) is also seen in severely affected branches that lead to die back disease. Each nymph/adult during its entire life damages at least 3 panicles/ tender shoots. From December onwards, when the cashew trees are in their most active growth stage, the situation becomes quite congenial for feeding and multiplication by both the nymphs and adults. The nymphs and adults are highly attracted by the new flush and tender inflorescence of the trees.

Seasonal incidence

During February-March, the pest reaches its peak population. In West Bengal, the maximum incidence has been reported during winter months (Kar *et al.*, 2016). During these periods, trees are in the full blossom stage. Young trees are more affected, because of their succulent growth throughout the year. Low temperature is favourable for increasing the population densities of TMB (Chakraborti and Majumder, 2007).

Management

i. "Bhaskara", a variety developed at ICAR-Directorate of Cashew Research, Puttur, escapes TMB damage due to non-overlapping of the cropping period with that of peak pest population. In the varieties like Ullal-1 and Madakkathara-2 major flowering /fruiting is delayed *i.e.*, in the month of February onwards. As these months are not favourable for the multiplication and establishment of TMB, these varieties suffer less due TMB.

ii. The volunteer neem plants should be removed in and around cashew plantations. Proper monitoring of the pest situation is very important.

iii. *Telenoums* sp., *Trichogramma* sp., *Chaetostricha* sp. and *Erythmelus helopeltidis* parasitize the tea mosquito bug eggs. Predatory spiders like *Oxyopes schireta, Phildippus patch* and *Hyllus* spp. feed on nymphs and adult tea mosquito bugs. There are other different predators e.g., red ant, dragon fly, ladybird beetles, spiders, praying mantis, black ant and anthocorid bug. Among them red ant (*Oecophylla smaragdina* Fab.) was the most effective predator against this pest (Manjanaik and Chakravarthy,

2013) and because of their higher efficiency to repel the tea mosquito bug, their colonies should be encouraged in cashew plantations.

iv. Application of pongamia seed kernel extract (2%) has a definite impact for controlling the pest (Thirumalaraju, 2002, Manjanaik and Chakravarthy, 2013).

v. For pesticidal management, following schedules of three sprays may be followed:

a. 1st spray with Lambda cyhalothrin @ 0.003% during new flushing stage (November- December), 2nd spray with Carbaryl @ 0.1% or Chlorpyriphos @ 0.05% or Profenophos @ 0.05% at flowering stage (December- January) and for 3rd spray, the first spray should be repeated at the initial fruiting stage (February-March) (Manjanaik and Chakravarthy, 2013, Navik and Godase, 2014).

b. However, for pesticidal management, the spray should be well in advance before the crop is attacked by the insect and thorough foliar coverage is necessary. Repetition of the same insecticide should be avoided in the 2nd round. As synthetic pyrethroids cause flare-up of sucking pests, so indiscriminate use of this insecticides should be avoided.

Sprayings of insecticides should be done before 9 am or after 4 pm to avoid harming of non-target pollinators. When out-break of TMB occur, the management programme should be introduced on large scale community basis.

2. Stem and root borer, *Plocaederus ferrugineus* Linn. (Cerambycidae: Coleoptera)

The cashew stem and root borer (CSRB) is the most serious threat to cashew plantation as its damage can cause death of plants. It bores into the internal tissue. It can cause up to 40% infestation but it varies in different periods. When the trees are attacked severely, they die within a period of two years causing a substantial tree loss (Faleiro and Desai, 2007).

Distribution

The pest is seen in all the cashew growing regions of India and some parts of Nigeria (Asogwa*et al.*, 2009a).

Hosts of commercial importance

Bombax malabaricum, Boswelia serrata, Buchanania latifolia, Diospyros melanoxylon, Hardwickia binata, Lannea grandis, Moringa are the alternate host plants of the pest (Pillai *et al.*, 1976).

Identification and biology

The adult is a reddish-brown long horned beetle, size ranges from 25-40 mm. Their head and prothorax are dark brown. In the males, the antenna is longer than the body and in females, it is much shorter than the body length (Gahan, 1906). The preoviposition period ranges from 3 to15 days. Males live for a mean period of 18 days and females for 15 days. They prefer to lay eggs on trees which are more than 4-5 years old having rough bark with more number of cracks which are either damaged by stem borers in the preceding season or by physical activities like heavy pruning.

They prefer to lay eggs into the live tissues in crevices of loose bark in the trunk portion or portion of the roots which are exposed above the soil. The eggs are 3 mm in length having an ovoid shape, whitish in colour with smooth surface (looks like rice grain). Adult females lay about 60-90 eggs. The incubation period varies from 7-9 days. The first instar larvae are dirty white in colour with a brownish head, 5 mm long and 1 mm broad at the prothoracic region. The grown-up grubs are 7-10 cm in length and off-white in colour. The larval period ranges from 138-150 days. The larvae construct an oval chamber in the heartwood for pupation. The cocoons are found up to 1 meter height on the trunk and also inside the unexposed roots. The cocoon is ovoid and flattish, about 5 cm long and 3 cm broad. It is whitish but covered with chewed up fibrous tissues (Asogwa *et al.*, 2009b). The pupal period ranges from 108-166 days. The insect completes its life cycle in about 338 days. The borer has one generation per year.

Mode of feeding and symptoms of infestation

The pest is generally noticed in older plantations of 15 years over. The larvae bore into the stem and roots and kill the trees outright. Presence of small bore holes at the collar region of the tree indicates the stem borer infection. The young grubs after boring into the bark, feed on the sub-epidermal and vascular tissues. The bored holes are plugged with reddish mass of chewed fibre and excreta. A resinous material oozes out from the infested portion (Sathiamma, 1977). Later stages cause extensive irregular tunneling in the basal stem and root regions. As a result of tunneling, supply of water and nutrients is hampered because of the damaged cambial tissues. Because of this yellowing and shedding of leaves take place. The branches start drying up, the tree gets weakened and results in the death of the tree. If the injury is severe on anchoring roots, affected trees become tilted on one direction.

The damage by the pest in the above ground portion has been estimated as 61.62% and it was 38.38% in below ground portion of the plant (Anon., 2000).

Seasonal incidence

Even though the infestation is noticed throughout the year, the peak period of infestation was observed during summer months. But in West Bengal, the occurrence is mostly confined to summer as well as in rainy season (Kar *et al.*, 2016). The incidence of stem and root borer is not significantly influenced by the weather parameters (Chakraborti and Majumder, 2007).

Management

i. Planting of CSRB host trees like silk cotton and drumstick should be prevented inside or in the vicinity of new cashew plantations (Haripriya, 2014).

ii. Severely infested tree and dead tree are to be uprooted before and after monsoon season.

iii. Causing injury to the plants by sickle and other garden tools should be avoided, which otherwise will attract the adult for egg laying.

iv. The trees, which are high yielding, are to be examined at their base at monthly interval especially during harvest time (January-May) for eggs and grubs of CSRB. The affected bark should be removed along with the grubs during initial stages of infestation mechanically after examining critically the sickly looked trees including stout root region.

vi. Neem oil (50 ml neem oil + 1 litre water + 0.5 ml teepol/ 5 gm of soap) (Anon., 2014) or Chloropyriphos 0.2 per cent solution may be swabbed on the trunk up to a height of 1 meter (Raviprasad, 2009).

vii. The pest can be controlled biologically with the application of *Metarhizium anisopliae* spawn @ 250 g per tree along with neem cake @ 500 g per tree (Sahu and Sharma, 2008). The efficacy of other entomopathogenic fungi like *Beauveria bassiana* and *B. brongniartii* besides *M. anisopliae* has been reported to control the CSRB (Sundararaju, 2002, Saminathan *et al.*, 2004).

viii. Application of Phorate granules into the bore holes and swabbing the affected area with 0.1 per cent Carbaryl, Dimethoate or Fenitrothion after removal of grubs have been suggested as control measures (Sathiamma, 1978). Synthetic pesticide like Carbaryl is also effective against this pest (Mohapatra, 2004).

3. Leaf miner, *Acrocercops syngramma* Meyk. (Gracillariidae: Lepidoptera)

It is one of the serious pests infesting cashew crop causing serious damage to the tender foliage. The incidence of this pest is severe along the west coast of

Karnataka. In a severely attacked plot up to 60% leaf damage is recorded (Rai and Vasantha, 1980). However, the damage of 18-20% by leaf miner has been reported in West Bengal condition (Chatterjee, 1989).

Distribution

It is known from Hong Kong, India, Indonesia and Thailand (Bai *et al.*, 2009).

Hosts of commercial importance

This pest is also recorded on jamun and mango (Butani, 1979).

Identification and biology

The adults are small moths with a wing span of 4-5 mm. The antennae are long and filiform. Wings are fringed and shining silvery grey in colour. It lays eggs on the tender leaves. Larvae after hatching are pale white in colour whereas fully grown caterpillars are pinkish or reddish brown in colour. Full grown larvae measure about 5 mm in length.

The caterpillars make their way to come out from mined areas and fall to the ground where they pupate, though in some cases pupation can also take place in leaf-folds in a thin membranous cocoon. The total larval period lasts 10 to 15 days. Pupal period ranges for 7-9 days and their entire life cycle is completed in 20 to 25 days.

Mode of feeding and symptoms of infestation

After hatching the larvae begin to mine the epidermal layer on the dorsal surface of the tender leaves. Blistered patches of grayish white colour are formed in the affected areas because of their feeding. As the infested leaves become mature, the damage is identified as big holes. Nursery seedlings and young plantations are more prone to the infestation of this pest than the older ones. Occurrence of this pest is common during post-monsoon flushes and young plantations. In West Bengal, the leaf miner usually becomes serious in September to November and moderate in July, August and December (Chatterjee, 1988), though Kar *et al.* (2016) observed the pest throughout the year. However, cool temperature conditions enhance the building up of the pest population densities (Chakraborti and Majumder, 2007).

Management

i. Management of this pest is required only during severe incidence especially on young plants (Vanitha *et al.*, 2015). *Chelonus* sp. and *Sympiesis* sp. are the parasitoids while, lace wings, robber fly, Coccinellids, spiders, red ants, dragon fly, praying mantis etc. are the predators of leaf miner.

ii. The population is taken care by three eulophid larval parasitoids viz., *Chrysocharis* sp., *Closterocerus* sp. and *Aprostocetus* sp. that brings down the pest population even up to 50% under field conditions (Vanitha, 2015).

iii. Need based application of Triazophos @ 0.05% or Cypermethrin were found effective against these pests (Athalye and Patil, 1999).

4. Shoot and blossom webber, *Lamida moncusalis* Wlk. (Pyraustidae: Lepidotera)

Distribution

The pest is found in India (Rao *et al.*, 2002) particularly in the coastal district of Andhra Pradesh and gradually spreading in the cashew plantations of Tamil Nadu and Odisha.

Hosts of commercial importance

It is solely recorded in cashew. However, the tree species *Terminalia catappa* and *T. arjuna* are reported as new hosts of *Lamida moncusalis* from Tamil Nadu, India.

Identification and biology

Adults are medium sized moths with dark brown forewings having markings and dirty white hind wings with brownish margin. On the dorsal side of the head of males there is a hornlike projection which when disturbed appears like a brush. The females possess two such projections. The adults live for 3-6 days.

They lay eggs singly or in small groups of 3-5 on the underside of the tender leaves mostly during night time. Freshly laid eggs are yellowish green in colour and change to pinkish before hatching. A single moth lays 60-90 eggs and the eggs hatch within 5-6 days. Larvae on hatching are about 1.2 mm in length, greenish brown and sparsely covered with hairs. Full grown larvae are brownish, 2.4-2.6 mm in length with brownish head. The ventral side is green with a yellow longitudinal band and a black line on the dorsal side. Larval period occupies 16-21 days with 5 larval instars. Pupation occurs in the leaf webs or in the soil within a cocoon. The pupal case is formed by silken threads. Pupa is reddish brown and the pupal period ranges from 8-11 days. The life cycle is completed in 30-40 days (Murthy *et al.*, 1974).

Mode of feeding and symptoms of infestation

The caterpillars cause serious damage to the cashew crop by feeding on the tender leaves and shoots at new flush and the flowers at blossoming time. Infestation initially occurs as webbing on the terminal portions of new shoots and blossom.

The caterpillars remain inside the web and feed on them. Symptoms of attack include the presence of silken webs reinforced with pieces of plant parts, chewed material on the terminal portion of tender shoots and blossoms and drying up of webbed shoots. The incidence is found severe mostly on young trees. It is reported to cause significant damage in Andhra Pradesh (Pillai *et al.*, 1976).

Seasonal incidence

The population reaches its peak in December-February, it is moderate in November-March and low in April-October in West Bengal condition (Chatterjee, 1988).Kar *et al.* (2016) reported this insect as a winter season pest.

Management

i. Two species of *Apanteles* and one Braconid have been reported as natural parasites (Nair *et al.*, 1979). Among the parasitoids, *Tetrastichus* sp., *Trichogramma* sp., *Bracon* sp., *Goniozus* sp. and *Trichospilus pupivora* and among the predators, King crow, common mynah, wasp, reduviid bug, big eyed bugs, pentatomid bug, earwigs, ground beetle, rove beetle, lacewing, ladybird beetle, spiders, praying mantis, dragon fly, robber fly, red ants etc. have been identified.

ii. Spray of Carbaryl or Malathion @ 0.15% (Ayyanna *et al.*, 1977) or Chlorpyriphos @ 0.05% at flushing period can control this pest effectively.

5. Mealy bug, *Ferrisia virgata* Cockerell. (Pseudococcidae: Hemiptera)

The mealy bug, *Ferrisia virgata* is a major pest of cashew which is found in all areas where cashew is grown. Other mealy bug species that infest cashew includes *Planococcus citri* and *Planococcus lilacinus.*

Distribution

The pest is distributed worldwide (Williams, 2004).

Hosts of commercial importance

Apple, custard apple, crotons, Jatropa (Ayyar, 1963), banana, guava, citrus, grapevine and jack fruit (Butani, 1979) are some of the alternate host plants of this pest. Other host plants reported are Cocoa, Coconut, Mango, Millet, Papaya, Passion fruit, Pineapple, Sugarcane, Yam etc.

Identification and biology

They are soft bodied insect with milky white waxy covering on their body. They have reproductive potential of laying 100-300 eggs in masses covered with cottony white mealy filaments. The eggs are amber coloured and they hatch into crawlers within a day. Nymphal period continues for 26-45 days. Adult longevity is 15-20 days. The total life cycle is completed within 45-65 days.

Mode of feeding and symptoms of infestation

The nymphs and adults, both stages of mealy bugs suck profuse amount of sap from the tender parts of the plant that results in withering of inflorescence, growing shoots and developing fruits. They are generally present on the ventral surfaces of tender leaves, inflorescence, twigs, panicles and fruit peduncles. The bugs also excrete a huge amount of honey dew which enhance the growth of sooty mould fungus that hamper the normal photosynthetic activity of the plant. Under severe outbreak of mealy bug, heavy loss of nut yield is observed.

Management

i. To manage this pest continuous monitoring and detection of infestation at early stage is essential. Weeds and alternate hosts of mealy bugs should be removed from the cashew plantation and neighboring areas.

ii. The infested plant parts such as inflorescence, leaves and twigs along with mealy bug colonies should be pruned and destroyed. Mealy bug infested fallen leaves should be collected and destroyed to avoid further spread.

iii. Predators are very effective in controlling the mealy bug. Mirid bug, dragonfly, spiders, robber fly, praying mantis, red ants, lacewings, big-eyed bugs (*Geocoris* sp.)., Coccinelids such as *Cryptolaemus montrouzieri, Cheilomenes sexmaculata, Rodolia fumida, Scymnus coccivora* and *Nephus regularis* are the predators of mealy bug.

iv. Some commonly found parasitoids are *Aenasius advena, Blepyrus insularis* and *Anagyrus* sp. Building up of the predator population e.g. *Chrysoperla carnea, Menochilus sexmaculatus, Coccinella septempunctata* and *Scymnus coccivora* should be encouraged in and around the cashew plantations.

v. Spraying with Dichlorvos @ 0.2% or Dimethoate @ 0.05% along with fish oil resin soap @ 20 g per litre of water can reduce the incidence of mealy bug effectively. But spraying should be done in such a way so that the entire lower surface of leaves, twigs and branches should be covered because in these areas the crawlers are present in large numbers. In is also necessary not to spray same insecticides repeatedly as in increase the chance of resistant development.

6. Shoot tip and inflorescence caterpillar, *Chelaria heligramma* M. (Gelechiidae: Lepidoptera)

Distribution

The pest is distributed in India and Thailand (Ponomarenko, 1997).

Hosts of commercial importance

The pest is reported to be occurred in cashew only.

Identification and biology

The adults are tiny dark brown moths. The larvae measure about 12 mm in length, yellowish brown and move about very briskly. Pupation takes place within the folded leaves of shoot tips or on floral branches. The pupa is reddish and the pupal stage lasts 7-8 days.

Mode of feeding and symptoms of infestation

The larvae of this moth feed on the shoot tips and inflorescence and hence known as shoot tip and inflorescence caterpillar. Occasionally they bore into the tender shoot tips up to a depth of about 2.5 cm. as a result of injury the growing shoot sometimes withers and dries up. Shoot tip damage is mostly noticed at the time of new flush. The caterpillars fold the fresh leaves and feed on them. Damage up to 60% of shoots by this caterpillar is being recorded (Abraham, 1959).

Seasonal incidence

Incidence of the caterpillar is more in the 2nd fortnight of February and 1st fortnight of March because this period coincides with the blossoming and inflorescence period of the tree. Incidence gradually decreases with no infestation during 2nd fortnight of May to 1st fortnight of October in Andhra Pradesh condition (Rao *et al.*, 2006). However, in West Bengal condition, the pest remains active from August to April with two peaks during February and September (Anon., 2000).

Management

Insecticidal sprays given at the time of new flush for the control of tea mosquito bug controls this pest also. However, this pest was effectively controlled by Thiodicarb (0.075%) or Spinosad (0.015%) or Profenophos (0.05%) or Chlorpyriphos (0.05%) or Cartap hydrochloride (0.05%) (Rao *et al.*, 2006).

7. Foliage thrips

a) Foliage thrips, *Selenothrips rubrocinctus* (Giard) (Thripidae: Thysanoptera)

Distribution

This thrips is a tropical and subtropical species thought to have originated in northern South America and is found in Asia, Africa, Australia and Pacific Islands, North America, Central America, West Indies and South America (Wang, 1984).

Hosts of commercial importance

The pest also found to feed on avocado, guava, pear, mango (Butani, 1979) and arecanut (Pillai *et al.*, 1976).

Identification and biology

They are minute, soft bodied, slender, fragile, fast moving insects. Wings of adults are fringed with hairs. They have rasping sucking type of mouthparts. The female starts to oviposit soon after emergence and can lay up to 50 eggs in her life time.

Adult females insert their eggs singly into the epidermis of leaf and then cover them with a drop of excremental fluid which later dries and form a black disc like pellicle. After about 12 days, the nymphs emerge out and wander freely. At the apex of the abdomen, they carry a drop of excremental fluid. The nymphs have greenish yellow colour and across their first and last abdominal segments presence of red bands is the distinguishable characteristics of the nymph. Incase of adults they possess a highly polygonally reticulate body with terminal antennal joints which look like a needle and broad wings having dark strong stiff setae. The nymphs feed for 10 days, involving a mono-feeding pre-pupal instar that lasts for one day and non-feeding pupal instar that lasts for further 2-3 days (Panda, 2013).

Mode of feeding and symptoms of infestation

Adults and nymphs both found on the lower surface of leaves. They suck the sap from inflorescence, leaves, apples and nuts. Because of their rasping and sucking activity the leaves turn into pale brown in colour. Due to thrips damage scab develop on the floral branches, nuts and apples and incase of severe infestation corky layers develop on the affected plant parts along with shedding of leaves and stunting of trees occur.

Seasonal incidence

Their population starts to increase during the dry periods (December-January) and reaches its peak in April-May and then there is a rapid decline in the population during wet season. The population of thrips varies from tree to tree in the same area and the insects do not feed indiscriminately, some trees even remain uninfested, while others are heavily infested. Those trees produce leaves after the onset of monsoon, are found free from attack. Cool temperature conditions favour the population build-up of leaf thrips (Chakraborti and Majumder, 2007).

Management

i. The natural enemies of foliage thrips are Predatory mite (*Amblyseius swirskii*), predatory thrips (*Aeolothrips* sp.), insidious flower bugs (*Orius insidiosus*), ant lion, lygaeids, ladybird beetle, anthocorids etc.

ii. Spraying with Fenitrothion 0.05% during the emergence of panicles and tender shoots is effective for managing the pest (Nair *et al.*, 1979).

b) Foliage thrips, *Rhipiphorothrips cruentatus* Hood. (Thripidae: Thysanoptera)

Distribution

It is reported from India and Taiwan (Hood, 1919 and Chiu, 1984).

Hosts of commercial importance

Some other important host plants are rose, grapevine (Ayyar, 1963), country almond, custard apple, pomegranate, mango (Butani, 1979), jamun (Atwal, 1976), cocoa and arecanut (Pillai *et al.,* 1976).

Identification and biology

The adults are minute, about 1.4 mm long, blackish-brown with yellowish wings. The eggs are laid inside the leaf tissue by making a slit. Eggs are dirty white and bean shaped. A female lays up to 50 eggs. The eggs hatch in 3-8 days (Atwal, 1976).

The nymphs are yellowish brown and the nymphal period ranges from 9-20 days. Pupation takes place on the leaves.

Mode of feeding and symptoms of infestation

The nymphs and adults cause damage by sucking the sap from the leaves. Infested leaves show whitish spots, wither and turn brown. This thrips can cause severe damage to young plantations during summer months (Abraham, 1958).

Management

The thrips can be controlled by spraying with Dimethoate @ 0.05% (Ayyanna *et al.*, 1979).

c) Foliage thrips, *Retithrips syriacus* Mayet. (Thripidae: Thysanoptera)

Distribution

This thrips species is distributed in the Middle East, Africa, India, North and Central America (Ben-Yakir, 2012).

Hosts of commercial importance

Grapevine, custard apple, pomegranate are some of its alternate host plants (Butani, 1979).

Identification and biology

The adults have a flattened dark brown body and the fore-wings are dark grey. The life cycle occupies 11-14 days.

Mode of feeding and symptoms of infestation

The nymphs and adults of this thrips species lacerate the leaf tissues and suck the exuding sap. Severe infestations often result in leaves getting bleached, filling up with excreta and gradual drying up.

Management

Spraying with any systemic insecticide is effective to manage this thrips.

8. Flower thrips, *Rhynchothrips raoensis* G. (Thripidae: Thysanoptera)

Distribution

It is described from India (Ramakrishna, 1928).

Hosts of commercial importance

Mallotus philippinensis (Kumkum) is one of the alternate host plants of this pest (Ananthakrishnan and Sen, 1980).

Identification and biology

The adults are minute, soft bodied and slender insects having wings fringed with hairs. They insert their eggs singly in the epidermis of the leaf. The nymphs emerge out within 10 days. The nymphs are pale yellow and wingless. They moult 2-3 times passing through 3-4 instars in 12-18 days according to the prevailing temperature.

Mode of feeding and symptoms of infestation

The flower thrips infest the inflorescence and tender leaves (Abraham, 1958). The feeding causes scabs on the floral branches, nuts and apples. In severe cases, malformation of nuts has also been reported. Infestation on developing nuts results in the formation of corky layers on the affected fruits and even immature fruits drop. It is reported that the infestation of flower thrips alone accounts for 16.39% fruit drop (1.91% in mustard stage, 12.57% in peanut stage and 1.91% in later stages) in cashew (Panda, 2013).

Seasonal incidence

Low temperature along with moderate relative humidity enhance the population build-up of the flower thrips (Chakraborti and Majumder, 2007).

Management

i. Predatory mite, predatory thrips, insidious flower bugs, ant lion, lygaeids, ladybird beetle, anthocorids etc. are the natural enemies of flower thrips.

ii. Control measures taken against tea mosquito bug manage this pest also. However, this thrips can be controlled by Profenophos @ 0.05% or Ethofenprox @ 0.015% (Mohapatra *et al.*, 2000) or Dimethoate or Quinalphos @ 0.05% (Nair, 2009).

9. Apple and nut borer, *Thylacoptila paurosema* Meyrik. (Pyralidae: Lepidoptera)

Distribution

It has a wide range and is found on the Canary Islands, Saudi Arabia, Namibia, South Africa, Madagascar, Indonesia, Burma, India and Ascension Island (Asselbergs, 1999).

Hosts of commercial importance

This pest also occurs on mango.

Identification and biology

Adults are a medium sized moth with a wing expansion of 18-24 mm. The forewings are dark and hind wings are pale white. Eggs are laid on the fruits and eggs hatch within 3-5 days. The caterpillars are 20-25 mm in length, very active, cylindrical in shape and dark pinkish in colour. There are 5 larval instars taking 15-33 days to go for pupation. Pupation takes place in soil in earthen cocoons as the fully grown larvae drop to the ground. The pupal period lasts for 8-10 days (Butani, 1981).

Mode of feeding and symptoms of infestation

The caterpillars can attack all the stages of fruits resulting in shriveling and premature fall of nuts. At early stages, the young larvae move to the joints of apple and nut, they scrape the epidermis and bore into them and plug the entry holes with excreta. At later stages, larvae bore into the tender apples and nuts and feed on them. When the nuts are affected by the borer they do not develop rather they get shriveled and dried up and fall down pre-maturely (Butani, 1981). Sometimes the infested apples are completely hollowed. Generally a single caterpillar is seen either in the nut or apple, but according to some reports up to 5 larvae can be seen in apples and three in the nuts. Damage up to 10% of the apples and nuts is being reported (Nair *et al.*, 1979). This pest also attacks stored cashew nuts (Ayyar, 1963).

Management

i. *Trichogramma* sp. and *Bracon* sp. are parasitoids of the pest. Some predators are *Chrysoperla carnea*, ladybird beetle, King crow, common mynah, wasp, dragonfly, spiders, robber fly, reduviid bug, praying mantis, red ants, big eyed bugs (*Geocoris* sp.), pentatomid bug (*Eocanthecona furcellata*), earwigs, ground beetles, rove beetles, shield bug, anthocorids etc.

ii. An effective cultural method to control apple and nut borer population is to remove and destroy the dead and dried inflorescence completely during the pre-flowering season.

iii. Spraying of Carbaryl @ 0.1% at the time of fruit setting or spraying of Dichlorvos @ 0.2% during the off-season may be followed.

10. Leaf folder, *Sylepta aurantiacalis* Fisch. (Pyralidae: Lepidoptera)

Distribution

It is found across southern Europe, Africa, Asia, Turkey, Ukraine, Hawaii as well as New South Wales and Queensland.

Hosts of commercial importance

It also feeds on *Quercus serrata* and *Castanea* sp.

Identification and biology

The adults are yellowish brown coloured medium sized moths with a wing expanse of 32-36 mm having brownish wavy lines running across both the wings. Female moths lay several eggs singly on emerging flush leaves. The eggs are oval in shape and creamy white in colour. Eggs hatch in 3-4 days. The neonate larvae are pale green and full grown larvae are glistening green with brownish head and measure about 30-35 mm in length. Total larval period lasts for 18 to 30 days. Pupation occurs in the leaf fold itself and pupal period lasts for 8-12 days.

Mode of feeding and symptoms of infestation

The caterpillars during early stages roll up the tender leaves from tip downwards or longitudinally towards dorsal side throughout the breadth or length to form a tubular niche. The leaves are tightly fastened with silken thread of salivary secretion of the caterpillars. Normally one caterpillar is present in each roll and feed on the leaves by scraping the green matter. As a result of the damage the terminal and tender leaves dry up. This affects the terminal growth in root stock and young grafts. On regular flushes, the panicle emergence is arrested (Abraham, 1958). Its activity is seen from September to January synchronizing with flushing and flowering seasons of cashew. Cool temperature favours the building up of the population densities of this pest (Chakraborti and Majumder, 2007).

Management

i. *Cotesia flavipes, Chrysoperla carnea*, coccinellids, King crow, common mynah, wasp, dragonfly, spiders, robber fly, reduviid bugs, praying mantis, red ants, *Geocoris* sp., *Eocanthecona furcellata*, earwigs, ground beetles and rove beetles are common natural enemies.

ii. Since stray incidences are noticed, removal and destruction of rolled up leaves will help in controlling this pest. Spraying of Chloropyriphos or Trizaphos @ 0.05% during emergence of new flushes in October-November will effectively control this pest (Ambethgar *et al.*, 2001).

11. Hairy caterpillar, *Metanastria hyrtaca* Cramer. (Lasiocampidae: Lepidoptera)

Distribution

The hairy caterpillar is distributed in India, Nepal, Myanmar, Thailand, Malaysia, Indonesia, Vietnam, China and Taiwan (Muniappan, 2012).

Hosts of commercial importance

The pest also infests almond, jamun and sapota (Butani, 1979).

Identification and biology

The moths are stout and of moderate size; the wing expanse in males varies from 5-6 cm. The males have a deeper colour than females and have a dark brown, irregular patch with a white spot on the forewing. The antennae are short and bipectinate. The wings are grayish brown in colour, cryptic in design and densely covered with scales. The moths lay eggs in clusters on the lower surface of the leaves. A female lays about 250 eggs. Eggs are spherical, dull white with two brownish markings.

The eggs hatch in 8-12 days. Early instar caterpillars are gregarious. There are 5 larval instars which occupy from 21-47 days. Full-grown caterpillars are grayish in colour, about 6-7 cm long, stout and densely clothed with black hairs. The hairs cause irritation when touched. Tufts of long hairs project forward on either side of the head, metathorax and first abdominal segment. There is a velvety black patch on the mesonotum, a reddish dorsal line and yellow ventral patches on the body. The larvae pupate on the lower surface of the leaf in a long cocoon. The pupa is brownish red and about 25 mm in length. The pupal period ranges from 8 to 12 days. The life cycle is completed in 54-76 days (Nair *et al.*, 1974).

Mode of feeding and symptoms of infestation

These caterpillars feed voraciously on the leaves and cause severe defoliation. Early instar caterpillars feed on the tender foliage while, full-grown caterpillars feed on mature leaves. During daytime the caterpillars rest in groups on the basal portion of tree trunk and feed on the foliage during night. Presence of excretory pellets at the base of trees is an indication of the presence of these caterpillars.

Management

Coccinellid beetles, spiders, praying mantis, dragonfly etc. predate upon *Metanastria hyrtaca. Perilampus microgastrii* is a parasite of this pest (Pillai *et al.*, 1976). The larvae resting on tree trunks in groups if detected can be easily destroyed by burning them. Otherwise, Parathion @ 0.025% is effective against this pest.

12. Mango aphid, *Aphis (Toxoptera) odinae* van der Goot. (Aphididae: Hemiptera)

Distribution

This aphid is distributed widely in Asia and sub-saharan Africa (Blackman and Eastop, 1994).

Hosts of commercial importance

This species is rather polyphagous and is found on shrubs and some trees of the families Anacardiaceae (*Anacardium*, *Mangifera*, *Rhus*), Araliaceae (*Aralia*, *Polyscias*, *Kalopanax*), Caprifoliaceae (*Viburnum*), Ericaceae (*Rhododendron*),

Pittosporaceae (*Pittosporum*), Rubiaceae (*Coffea*, *Mussaenda*) and Rutaceae (*Citrus*) (Butani, 1979, Blackman and Eastop, 2006).

Identification and biology

The aphids are brownish in colour with prominent cornicles and found in colonies. Wingless females are usually grey-brown to reddish-brown. The colonies are ant-attended. It can be distinguished from other aphids colonizing these plants by its short dark siphunculi, only about half the length of the black, finger-like cauda. In almost all regions this species is anholocyclic, that is, it reproduces parthenogenetically throughout the year (Margaritopoulos *et al.*, 2013). However, it has recently been found to have a sexual phase on various hosts in Japan (Blackman *et al.*, 2011). Particularly, males have been reported from *Viburnum* and *Rhus*, sexual females (oviparae) from *Kalopanax*, *Viburnum* and *Rhus*, and fundatrices (stem mothers) from *Juglans* and *Rhus*. A laboratory study revealed that the cashew aphid has four nymphal instars lasting for 2.25, 2.26, 2.32 and 2.81 days, respectively (Satapathy and Urs, 1998).

Mode of feeding and symptoms of infestation

The aphids are seen on cashew inflorescence and tender shoots in groups. Wingless females feed on the undersides of leaves along the main veins, and also on young shoots. Though they cause serious damage on those plant parts, they attack tender apples and green nuts also. Sucking of the sap by the pest causes shedding of flowers. Honeydew secreted by the aphids attracts ants.

Management

i. Coccinellids like *Menochilus sexmaculatus* and *Scymnus* sp. and syrphids like *Paragensyer buriensis, Ischiodon scutellaris* and *Dideopsis aegrota* are the common predators of cashew aphid (Satapathy, 1993).

ii. Spraying with Dimethoate @ 0.05% gives good control against aphids.

13. Hard scale (Armoured scale), *Pseudaonidia trilobitiformis* Green. (Diaspididae: Hemiptera)

Distribution

It is distributed throughout the world and a common pest of cashew in Brazil and East Africa (Ohler, 1979; Feng and Wei, 2011).

Hosts of commercial importance

Some of the recorded host plants of this pest are *Ixora, Echitis* (Ayyar, 1963), guava and mango (Butani, 1979).

Identification and biology

The eggs are protected by the waxy scale of the female. The nymphs are pale green and later turn to purple brown at the adult stage. The adults are almost circular and about 3 mm in diameter. After death they turn to grey white colour.

Mode of feeding and symptoms of infestation

The nymphs of the scale insect settle along the midrib and veins of cashew leaves on both the surfaces and suck the sap. Due to draining of the plant sap the leaves turn yellow, dry up and drop off.

Management

Spraying with Malathion or Diazinon in recommended doses effectively reduces pest incidence.

14. Leaf twisting weevil, *Apoderus tranquebaricus* Fabricius. (Curculionidae: Coleoptera)

Distribution

This pest in distributed in Europe and Asia (Legalov, 2007).

Hosts of commercial importance

The weevil is also noticed on mango (Pillai *et al.*, 1976) and *Terminalia catappa* (Butani, 1981).

Identification and biology

The adult weevil is a medium sized reddish brown insect with long neck and snout. Eggs are oval, orange coloured and about 2 mm in length. The eggs are laid on the leaf tips. The larvae feed on the rolled up leaf tissues. They have a brownish head and yellowish body. The grubs are legless and possess few short hairs on the body. Pupa is about 3 mm long, bright yellow in colour and found inside the folded leaf.

Mode of feeding and symptoms of infestation

The weevil is commonly noticed on cashew during new flush (Abraham, 1958). The weevils fold the leaf along the midrib, lay eggs on the tip and twist the leaf in a compact mass.

Management

i. All adult weevils should be removed from tree prior to destruction and also after removal of bark all larvae and pupae may be killed (Anon., 2014).

ii. Insecticidal sprays given for the control of tea mosquito bug controls this pest also. However, Dichlorvos @0.05% may also be applied.

15. Cashew weevil, *Mecocorynus loripes* Chevrolat. (Curculionidae: Coleoptera)

It is usually a minor pest of cashew but neglected orchards are liable to be under severe attacks.

The cashew weevil is a large weevil, dark grey-brown in colour about 20 mm long, and knobbed appearance. The female weevil lay eggs singly in small holes in the bark of the trunk or branches. The larvae are legless grubs, whitish in colour with a brown head. Fully grown larvae pupate in a chamber of about 2 cm below the bark.

The larvae bore through the bark and moves downwards by tunneling under the bark and feeding on the sapwood of tree. Brown-black gummy frass is seen on the trunk and main branches. At certain intervals it makes frass ejection holes to the exterior. Heavily infested tree becomes ringed by damaged sap wood and eventually die.

Lightly infested tree can be treated by killing all evident adults, cutting of bark to expose the larvae and pupal galleries and then removing and killing the larvae and pupae. This treatment should be repeated for further six months if required. Very severely infested tree should be destroyed. Different stages of the weevils may be collected and destroyed. The affected trees should be felled and debarked to expose all the larval galleries.

16. Cashew stem girdler, *Paranaleptes reticulate* Thomson. (Cerambycidae: Coleoptera)

It is also a minor pest of cashew. Adults are large sized long horn beetles, with body length of about 25-35 mm and antennae longer than the body. Head and the thorax of the beetle are dark brown in colour, elytra are orange with large black blotches giving them a reticulate appearance.

Female beetles lay elongated eggs in transverse slits made in the bark of the girdled branch at points above the girdle. Larvae are yellow in colour and reach a length of 45 mm when fully grown. They mine in dead wood of the girdled branches. Pupation takes place in the dead wood. Adults girdle branches from 3-8 mm in diameter leaving a V-section cut. Only a narrow, central pillar round the pith zone is left behind which eventually breaks off. The lifecycle takes about one year.

For management of the pest, once a year (in November or December) all girdled branches should be collected and destroyed.

Other minor pests of cashew are bark-eating caterpillar, *Indarbela tetraonis* Moore, leaf webber, *Orthaga exvinacea* (Hampson), coreid bug, *Pseudotheraptus devastans*, nut crinkler, *Paradasynus* sp., hairy caterpillar, *Amsacta albistriga*, cricula silkmoth, *Cricula trifenestrata* (Helfer), leaf roller *Dudua aprobola* (Meyrick), red cotton bug / cotton stainer, *Dysdercus koenigii* Fabricius., bagworm *Eumeta variegata* (Snellen), common baron *Euthalia aconthea* (Cramer), defoliating looper, *Achaea* sp. etc.

References

Abraham, E.V. 1958. Pests of cashew (*Anacardium occidentale*) in South India. Indian Journal of Agriculture Science, 28(4):531-544.

Abraham, E.V.1959. Killer insects of cashew. Indian Farming, 5(10):26-27.

Ambethgar V., Swamiappan M. and Rabindra R.J.2001. Evaluation of certain synthetic insecticides and neem products against cashew leaf folder, *Sylepta aurantiacalis* Fisch. (Pyralidae: Lepidoptera). Indian Journal of Plant Protection, 29(1&2):106-109.

Ambika, B. and Abraham, C.C. 1984. Effect of tropical application of JH analogue on the development and survival of the cashew mired bug (*Helopeltis antonii* Sign.) (Miridae: Heteroptera). In: Cashew Research and Development. Indian Society for Plantation Crops, CPCRI, Kasaragod, India. pp-111-115.

Ananthakrishnan, T.N. and Sen, S. 1980. Taxonomy of Indian Thysanoptera. Handbook Series No. I. Zoological Survey of India, Calcutta. Pp-234.

Anon. 1998. Annual Report of AICRP on Cashew (1998). Regional Research Station, Bidhan Chandra Krishi Viswavidyalaya, Jhargram, Midnapore, West Bengal, India.

Anon. 2000. Annual Report of AICRP on Cashew (2000). Regional Research Station, Bidhan Chandra Krishi Viswavidyalaya, Jhargram, Midnapore, West Bengal, India.

Anon. 2014. AESA based IPM–Cashewnut. Department of Agriculture and Cooperation, Ministry of Agriculture, Government of India.

Asogwa, E.U., Anikwe, J.C., Ndubuaku, T.C.N. and Okelana, F.A. 2009a. Distribution and damage characteristics of an emerging insect pest of cashew, *Plocaederus ferrugineus* L. (Coleoptera: Cerambycidae) in Nigeria: A preliminary report. African Journal of Biotechnology, 8(1):53-58.

Asogwa, E.U., Anikwe, J.C., Ndubuaku, T.C.N., Okelana, F.A. and Hammed, L.A. 2009b. Host plant range and morphometric descriptions of an emerging pest of cashew, *Plocaederus ferrugineus* L. (Coleoptera: Crambycidae) in Nigeria: a preliminary report. International Journal of Sustainable Crop Production, 4(3):27-32.

Asselbergs, J.E.F. 1999. *Arenipses sabella* Hampson, 1901, from South Spain: new to the European fauna; *Thylacoptila paurosema* Meyrick, 1885, new to the fauna of the Canary Islands; *Ancylosis (Ancylosis) albidella* Ragonot, 1888, new to the Spanish fauna (Lepidoptera: Pyralidae, Galleriinae, Phycitinae). SHILAP Revista de Lepidopterologia, 27(105): 125-129.

Athalye, S.S. and Patil, R.S. 1999. Bionomics, seasonal incidence and chemical control of cashew leaf miner. Journal of Maharashtra Agricultural University, 23(1): 29-33.

Atwal, A.S. 1976. Agricultural pests of India and South-East Asia. Kalyani Publishers, Ludhiana. 502 p.

Ayyanna, T., Arjuna Rao, P., Narayana, K.L. and Rao, B.H.K. 1979. Chemical control of cashew leaf thrips, *Rhipiphorothrips cruentatus* H. Abstracts published during International Cashew Symposium, CPCRI, Kasaragod. pp. 47.

Ayyanna, T., Subbaratnam, G.V. and Rao, B.H.K. 1977. Chemical control of *Macalla moncusalis* Walker and *Nephopteryx* sp. on cashew. Cashew Bulletin, 14(9): 8-10.

Ayyar, T.V.R. 1963. Hand Book of Economic Entomology for South India. Government Press, Madras. 516 p.

Bai, H.Y., Li, H.H. and Kendrick, R.C. 2009. Microlepidoptera of Hong Kong: Checklist of Gracillariidae (Lepidoptera: Gracillarioideae)-SHILAP, Revista de Lepidopterologia, 37: 495-509.

Bandyopadhyay, B. and Banerjee, A. 2007. Comparative study of cashew pest complex in different agro-climatic zones of West Bengal. In: Coastal Resource Management (edtd. S. Maity). pp. 51-55.

Ben-Yakir, D. 2012. The black vine thrips *Retithrips syriacus* (Mayet) as a pest of fruit trees and grape vine. Alon HaNotea, 66(11):40-41.

Blackman, R.L. and Eastop, V.F. 1994. Aphids on the World's Trees. CAB International with the Natural History Museum, London. Pp-987.

Blackman, R.L. and Eastop, V.F. 2006. Aphids on the World's Herbaceous Plants and Shrubs. Volume 2 The Aphids. John Wiley & Sons with the Natural History Museum, London. pp. 1025-1439.

Blackman, R.L., Sorin, M. and Miyazaki, M. 2011. Sexual morphs and colour variants of *Aphis* (formerly *Toxoptera*) *odinae* (Hemiptera, Aphididae) in Japan. Zootaxa,3110: 53-60.

Butani, D.K. 1979. Insects and Fruits. International Book Distributors, Dehradun.pp- 415.

Butani, D.K. 1981. Insect pests of cashewnuts and their control. Pestology, 5(4):9-11.

Chakraborti, S. and Majumder, A. 2007. Meteorological determinants of the incidence of insect pests and their damage on cashew (*Annacardium occidentale*) in coastal region of West Bengal. Indian Journal of Agricultural Sciences, 77(4): 264-266.

Chatterjee, M.L. 1988. Seasonal occurrence of pest complex on cashew (*Anacardium occidentale* L.) in West Bengal. *Cashew Bulletin*, 25(12): 11-14.

Chatterjee, M.L. 1989. Insect pest of cashew in West Bengal and status of some important Pests. The Cashew, 3(3):19-20.

Chiu, H.T. 1984. The ecology and chemical control of grape-vine thrip (*Rhipiphorothrip scruentatus* Hood) on wax apple (Chinese). Plant Protection Bulletin Taiwan, 26(4): 365-377.

Devasahayam, M. and Nair, C.P.R. 1986. The tea mosquito bug, *Helopeltis antonii* Signoret on cashew in India. Journal of Plantation Crops, 14(1):1-10.

Entwistle, P.F. 1972. Pests of Cocoa. Longmans, London. Pp-779.

Faleiro, J.R. and Desai, A.R. 2007. Chemical controls of cashew stem and root borer, *Plocaederus ferrugineus* L.: a case study. Souvenir and Extended Summarises published in National Seminar on "Research, Development and Marketing of Cashew" held at ICAR Research Complex for Goa, Old Goa, India during November 20-21, 2007. Pp-72-73.

Feng,J.N. and Wei, J.F. 2011. A new species of revision of the genus *Pseudaonidia* Cockerell, 1897 (Hemiptera: Coccoidea: Diaspididae: Aspidiotinae), with description of one new species and a chicklist (sic.) of the species from the Oriental Region. Transactions of the American Entomological Society, 137(1&2):173-178.

Haripriya, S. 2014. Sustainable cashew production in Cuddalore district – a case study. Tamil Nadu Agricultural University, Coimbatore.

Hood, J.D. 1919. On some new Thysanoptera from southern India. *Insecutorinscitiaemenstruus,* 7:90-103.

Kar, A., Poduval, M. and Jash, S. 2016. Influence of abiotic factors on seasonal incidence of insect pests in cashew grown under Red and Lateritic Zone of West Bengal. Abstracts of

International Conference on "Agriculture, Food Science, Natural Resource Management and Environmental Dynamics: The Technology, People and Sustainable Development" held at BCKV, West Bengal during August 13-14, 2016. pp. 25.

Legalov, A.A. 2007. Leaf rolling weevils (Coeloptera: Rhynchitidae, Attelabidae) of the world fauna. Novosibirsk, Agro-Siberia. Pp-523.

Manjanaik, C. and Chakravarthy, A.K. 2013. Sustainable management practices for tea mosquito bug *Helopeltis antonii* Signoret (Miridae: Hemiptera) on cashew. Karnataka Journal of Agricultural Science, 26(1):54-57.

Margaritopoulos, J.T., Papapanagiotou, A.P., Voudouris, C.C.H., Kati, A. and Blackman, R.L. 2013. Two aphid species newly introduced in Greece. Entomologia Hellenica, 22:23-28.

Maruthadurai, R., Desai, A.R., Prabhu, H.R.C. and Singh, N.P. 2012. Insect Pests of Cashew and their Management. Technical Bulletin No. 28, ICAR Research Complex for Goa, Old Goa.

Mohapatra, L.N. 2004. Management of cashew stem and root borer *Plocaederus ferrugineus* L. Indian Journal of Plant Protection, 32(1):149-150.

Mohapatra, R., Rath, L.K., Mohapatra, L.N. and Dash, D.D. 2000. Evaluation of some new chemicals against insect pests of cashew in Orissa. The Cashew, 14(2):26-28.

Muniappan, R. 2012. Arthropod Pests of Horticultural Crops in Tropical Asia. CABI, 2012. Pp-168.

Murthy, M.M.K., Sayi, I.V., Rao, S.V. and Rao, B.H.K. 1974. Note on the biology of shoot and blossom webber, *Macalla moncusalis* Walker (Lepidoptera: Pyralidae) of cashew. Indian Journal of Entomology, 36(1):76-77.

Nair, C.P.R., Abraham, V.A. and Pillai, G.B. 1974. Biology of *Metanastria hyrtaca* Cram., A defoliator of cashew. Journal of Plantation Crops, 2:32-33.

Nair, K.P.P. 2009. The agronomy and economy of some important industrial crops. In: Advances in Agronomy-Vol. 101 (edtd. D.L. Sparks). Academic Press. Pp-183-314.

Nair, M.K., Bhaskara Rao, E.V.V., Nambiar, K.K.N. and Nambiar, M.C. 1979. Pests. In: Cashew – Monograph of Plantation Crops-I. Central Plantation Crops Research Institute, Kasaragod, India. pp-60.

Navik, O.S. and Godase, S.K. 2014. Tea mosquito bug and their management in cashew ecosystem. Popular Kheti, 2(3):94-99.

Ohler, J.G. 1979. Cashew. Department of Agricultural Research, Royal Tropical Institute, Amsterdam. Pp-260.

Panda, H. 2013. Pests of cashew. In: The Complete Book on Cashew (Cultivation, Processing & By-Products). Asia Pacific Business Press Inc., Delhi. Pp-74-90.

Pillai, G.B. and Abraham, V.A. 1974. Tea-mosquito – A serious menace to cashew. Indian Cashew Journal, 10(1):5-7.

Pillai, G.B., Dubey, O.P. and Singh, Vijaya 1976. Pests of cashew and their control in India – A review of current status. Journal of Plantation Crops, 4(2):37-50.

Pillai, G.B., Singh, V., Dubey, O.P. and Abraham, V.A. 1984. Seasonal abundance of tea mosquito bug, *Helopeltis antonii* on cashew in relation to meteorological factors. In: Cashew Research and Development. Indian Society for Plantation Crops, CPCRI, Kasaragod, India. pp-103-110.

Ponomarenko, M.G. 1997. Catalogue of the Subfamily Dichomeridinae (Lepidoptera, Gelechiidae) of the Asia. Far East. Ent., 50:1-67.

Pruthi, H.S. 1969. Text Book on Agricultural Entomology. Indian Council of Agricultural Research, New Delhi. Pp-977.

Puttarudriah, M. and Appanna, M. 1955. Two new hosts of *Helopeltis antonii* Signoret in Mysore. Indian Journal of Entomology, 17:391-392.

Rai, P.S. 1984. Hand book on cashew pests. Researchco Publications, Delhi. Pp-117.

Rai, P.S. and Vasantha, S.T. 1980. The cashew leaf miner, *Acrocercops syngramma* M. (Lepidoptera: Gracillaridae) and its control. Cashew Causerie, 2(1):7.

Rao A.R., Naidu V.G., Prasad P. R. 2002. Studies on the Biology of Cashew Shoot and Blossom Webber (*Lamida moncusalis* Walker). Indian Journal of Plant Protection, 30(2):167-171.

Rao, Y.S., Rajasekhar, P., Rao, V.R.S. and Rao, V.S. 2006. Seasonal incidence and control of shoot tip and inflorescence caterpillar on cashew. Annals of Plant Protection Sciences,14(1):218-273.

Raviprasad, T.N. 2009. Integrated pest and disease management in cashew. In: Proceedings of 7th National Seminar on "Cashew Development in India – Enhancement of Production and Productivity" held at Bhubaneswar, Orissa during November 2-3, 2009. Pp-33-39.

Saminathan, V.R., Mahalingam, C.A. and Ambethgar, V. 2004. Virulence of entomopathogenic fungi to cashew stem and root borer, *Plocaederus ferrugineus* Linnaeus (Coleoptera: Cerambycidae) under laboratory conditions. Journal of Biological Control, 18(2):215-218.

Satapathy, C.R. 1993. Bioecology of major insect pests of cashew (*Anacardium occidentale* Linn.) and evaluation of certain pest management practices. Ph.D. Thesis submitted to University of Agricultural Sciences, Bangalore. Pp-224.

Satapathy, C.R. and Urs, K.C.D. 1998. Biology and morphometrics of *Toxoptera odinae* infesting cashew. Indian Journal of Entomology, 60(1):8-12.

Sathiamma, B. 1977. Nature and extent of damage by *Helopeltis antonii* S., the tea mosquito on cashew. Journal of Plantation Crops, 5:58-62.

Sathiamma, B. 1978. Occurrence of insect pests of cashew. Cashew Bulletin, 15(4):9-10.

Sundararaju, D. 2002. Pest and disease management of cashew in India. *The Cashew*,(4): 32-38.

Sundararaju, D. and Sundarababu, P.C. 1999. *Helopeltis* spp. (Heteroptera: Miridae) and their management in plantation and horticultural crops of India. Journal of Plantation Crops, 27:155-174.

Thirumalaraju, G.T. 2002. Ecology and management of tea mosquito bug *Helopeltis antonii* Signoret (Hemiptera: Miridae) on cashew. Ph.D. Thesis submitted to University of Agricultural Sciences, Bangalore. Pp-115.

Vanitha, K. 2015. A report on the occurrence of eulophid parasitoids on the cashew leaf miner, *Acrocercops syngramma* Meyrick. Journal of Threatened Taxa, **7**(12):7933-7936.

Vanitha, K., Bhat, P.S. and Raviprasad, T.N. 2015. Pest status of leaf miner, *Acrocercops syngramma* M. on common varieties of cashew in Puttur region of Karnataka. Pest Management in Horticultural Ecosystems, 21(1):55-59.

Wang, W.X. 1984. Bionomics and control of *Selenothrips rubrocinctus*. Acta EntomologicaSinica, 27: 81-86.

Williams, D.J. 2004. Mealybugs of southern Asia. Kuala Lumpur, Malaysia: Southdene SDN. BHD. Pp-896.

16

Pests of Strawberry and Their Management

Rajesh Kumar and S.J. Prasanthi

Strawberry (*Fragaria vesca*) is an important fruit crop in India and its commercial production is possible in temperate and sub-tropical areas of the country. According to the Ministry Agriculture, Government of India estimates, production of the fruit in India has dropped by 40% from 8,000 metric tonnes in 2014-15 to around 5,000 metric tonnes in both 2015-16 and 2016-17. Strawberries mainly consist of water (91%) and carbohydrates (7.7%). They contain only minor amounts of fat (0.3%) and protein (0.7%). The nutrients in 100 gm of raw strawberries are, calories 32, 91% water, 0.7gm protein, 7.7 gm carbohydrate, 4.9 gm sugar, 2 gm fiber and 0.3 gm fat. Strawberries should not lead to big spikes in blood sugar levels and are considered safe for people with diabetes. It is also useful for weight loss and can help prevent many diseases.The most abundant vitamins and minerals in strawberries are, vitamin C, vitamin B9, potassium and manganese. To a lesser extent, strawberries also provide iron, copper, magnesium, phosphorus and vitamins B6, K and E.

Among the various biotic and abiotic factors for low production of fruits, insect-pest forms one of the most important restraining factors. Strawberry is not an exception and several insect pests cause damage to strawberry out of which leaf rollers, white grubs, aphids and red spider mitesare of national significance in India.

Table 1: Important pests of strawberry

Sl. No.	Common name	Scientific name	Family	Order
1.	Black cutworm or Greasy cutworm	*Agrotis ipsilon* Huf.	Noctuidae	Lepidoptera
2.	White grub	*Phyllophaga* spp.	Scarabaeidae	Coleoptera
3.	Leaf roller	*Ancylis comptana* Frolich.	Totricidae	Lepidoptera
4.	Aphids	*Chaetosiphon fragaefolii* Cock., *Aphis forbesi* Weed	Aphididae	Hemiptera
5.	Fruit fly	*Drosophila melanogaster* Meigen	Drosophilidae	Diptera
6.	Tarnished plant bug	*Lygus hesperus* Knight.	Miridae	Hemiptera
7.	Spittle bug	Philaenus spumarius Linn.	Cercopidae	Hemiptera
8.	Strawberry bud weevil	*Anthonomus signatus* Say	Curculionidae	Coleoptera
9.	Red spider mite	*Tetranychus urticae* (Koch)	Tetranychidae	Acarina

1. Black cutworm or Greasy cutworm, *Agrotis ipsilon* Huf. (Noctuidae: Lepidoptera)

Distribution

This pest is widely distributed in India, China, northern Europe, Canada, Japan, South America and New Zealand (Capinera, 2006).

Hosts of commercial importance

It is a polyphagous pest feeding on extensive range of various host plants of economic importance.

Identification and biology

It is the most common cutworm infesting strawberries in most strawberry growing areas. Other species such as *Peridroma saucia* is also found sporadically (Zalom *et al,* 2008). Adult is a fairly large moth (1.5 inches) having pale brown forewings with dark purplish brown along distal costal margin and white hind wing with brown tinges. Sexual dimorphism in antenna has been found where male exhibit bi-pectinate antenna and female has filiform antenna. After a pre-oviposition period of 10 days, moth searches low-growing broad leaves for oviposition and in their absence, it ends up laying eggs on dead strawberry leaves during July-August (Ayre, 1980 and Capinera, 2006). Soil is an unsuitable oviposition site for this pest (Capinera, 2006).

After an incubation period of 3-6 days the eggs hatch into larvae which move into the soil. These remain in the soil during the day time and move to the surface at night and feed on young plants. They molt for five to six times which is dependent on temperature and adequacy of diet. After 25-35 days of feeding, larva pupates in the soil pupal stage lasts about 12-15 days. Cutworms can have two to four generations per year, and overwinter as pupae or mature caterpillars (Anonymous, 2018).

Mode of feeding and symptoms of infestation

The damage by early instars of cutworms generally appears as small, webless perforations in the newly expanding crown leaves. Later, they begin their characteristic stem cutting along with chewing larger, irregular holes in the foliage. Leaf feeding results in irregular-shaped holes and usually does not result in yield reduction. However, stem feeding can significantly reduce the size of the strawberry crown and the plant's subsequent yield. During harvest, cutworms can cause rather pronounced holes in the fruit (Zalom *et al.*, 2008).

Management

i. Delay in planting help plants escape the damage by cutworm.

ii. Deep summer ploughing help reducing pest population. Pruning also reduces overwintering larvae.

iii. As weedy fields tend to attract more moths for egg laying, good weed management will reduce the infestation.

iv. Like all caterpillars, cutworms are killed by a particular strain of the bacteria *Bt* (*Bacillus thuringiensis*). Cutworm bait may be prepared by mixing *Bacillus thuringiensis* var. *kurstaki* with moist bran and molasses. Spreading the bait over the soil surface has been found to reduce infestation by the pest.

v. Encircling of young plants with a cutworm collar that extends 2″ below and 2″ above the soil surface. This can be an effective low-tech solution.

vi. Installation of bird percher is effective in reducing pest population.

2. White grub, *Phyllophaga* spp. (Scarabaeidae: Coleoptera)

Distribution

Phyllophaga spp. and related insects are distributed throughout the Asia, United States and Canada. However, the distribution of individual species usually is more restricted.

Hosts of commercial importance

White grub, a Scarabaeid larval form of June beetles is a polyphagous pest feeding on roots of wide range of agriculturally important crops.

Identification and biology

Adults are the large, hard-shelled, block-shaped beetles May and June beetles that range in length from to 1 inch and vary in color from tan to dark brown. They fly at night and are seldom seen in strawberry fields. Females deposit eggs in soil at a depth of one to several inches during late spring or early summer, preferably on grassy sod that has not been disturbed for years.

Eggs hatch in 3-4 weeks.The grub is with a tan or brown head capsule and six prominent spiny legs.When dug from the ground, the larvae always lie in a curved position, forming the letter "C". The newly hatched larvae feed on the roots and often completely cut off strawberry plants just below the crown throughout the summer, then burrow deep in the soil to overwinter. During the following year they again migrate to the root zone to feed. These larger larvae cause much greater damage than they did the year before. After overwintering again well below the soil surface, white grubs pupate early in the following summer, and adults emerge from pupal cells in the spring three years after the cycle began.

Mode of feeding and symptoms of infestation

Symptoms of white grub injury on strawberry plants include stunted growth and pruned roots. Damage is most common in lighter soil types when strawberries are grown after pasture, sod or occasionally after a grain crop (Omafra, 2006 and Anonymous, 2004).

Management

i. Avoid planting strawberries on newly ploughed grassland.
ii. Insect damage can be reduced by rotating crops and cleanly cultivating the crop that precedes strawberries.
iii. Collection of destruction of adult beetles in soapy water may be practised particularly in case of small cultivation.
iv. Good cultivation before planting will injure grubs and expose them to birds and the sun.
v. Installation of light trap help reducing pest population.
vi. Leave one or two rows of strawberry plants to serve as a trap before ploughing infested plantings under shortly after harvest

v. Deep ploughing after summer showers would expose the pupae and beetles to hot sun or bird predation.

vi. Utilisation of fungal pathogens like *Metarhizium anisopliae*, *Beauveria brongniartii* is now under consideration.

3. Leaf roller, *Ancylis comptana* Frolich (Totricidae: Lepidoptera)

These leaf rollers are the caterpillars of Totricid moths known for rolling up leaves to protect themselves while they grow. This is considered as a minor or occasional pest of strawberry.

Distribution

Strawberry leafroller is found in eastern North America, occurring west to Illinois, Iowa, Oklahoma and Texas. It is also recorded from California, Colorado, Nevada, Alaska, the Pacific Northwest and several Canadian provinces. The pest is also found to infest strawberry in India.

Hosts of commercial importance

The species is found primarily on plants in the rose family (Rosaceae), including cultivated strawberry (*Fragaria* spp.), blackberry and raspberry (*Rubus* spp.), as well as roses (*Rosa* spp.) and *Dryas octopetala*. It has also been recorded on small burnet (*Sanguisorba minor*) and thyme (*Thymus* spp.).

Identification and biology

The caterpillar is slender, green to brown in colour, about 12 mm long with dark-brown or black head. They are very active and moves backwards when disturbed. Moths have reddish brown forewings (wing span of 1/2 inch), with distinct white areas streaked with darker brown, particularly on distal ends. Hindwings are gray in colour.

Females deposit about 20-120 translucent eggs that look similar to fish scales on the lower surface of strawberry leaves. After an incubation period of about 5-17 days, eggs hatch and the newly hatched larvae crawl in search of suitable leaves to roll. As larvae feed, they secrete silken threads to fold strawberry leaves together. Larva remains inthese folded leaves and feed on the epidermis of each leaf. Pupae are formed within the leaf roll. The average duration of the entire larval period is 24 days and the pupal period is 6-18 days. There may be two generations per year.

Mode of feeding and symptoms of infestation

The infestation of strawberry leafroller is confined to leaves. Due to their exhaustive feeding, the fruit may become small and distorted. Larvae feed on and fold or roll leaves, causing shriveling and withering so that fields appear scorched or burnt when populations are high (Renkamma, 2016, Anonymous, 2014). Damage by the pest is most common on new plantings late in the season.

Seasonal incidence

Strawberry leaf rollers overwinter as larvae or pupae in folded leaves or leaf litter, emerge during April-May and starts feeding on the plants.

Management

Low levels of strawberry leafroller infestation (about 10-20% of leaves infested) can be sustained by strawberries without causing economic losses and populations rarely reach high densities, therefore control is not often warranted in minimal level of infestation. However, taking following precautions help protecting the crop better.

i. Maintenance good field sanitation measures.

ii. Pinching to kill larvae in rolled leaves also may be available option in small scale cultivation.

iii. There are several natural predators and parasites that do keep the populations in check. Two parasites, *Macrocentrus ancilovorous* and *Cremastes cookii*, often kill a high percentage of strawberry leafroller larvae, especially in summer generations.

iv. A decision to apply insecticides should be based on scouting for damage or results of temperature models used to predict occurrence of egg hatch in the spring.

4. Aphids, *Chaetosiphon fragaefolii* Cock., *Aphis forbesi* Weed (Aphididae: Homoptera)

Though strawberry hosts many species of aphids, *Chaetosiphon fragaefolii* is considered to be a true strawberry aphid. It is an important destructive pest of cultivated strawberries worldwide, primarily because it transmits strawberry crinkle virus (SCV). *Aphis forbesi* is also sometimes found to infest strawberry crop.

Identification and biology

Strawberry aphid is pale green to yellowish in color. Both adults and nymphs appear to have transverse striations across the abdomen and are covered with knobbed hairs that are readily seen with a hand lens. These striations and hairs are not found on any of the other aphid species infesting strawberry.

The life cycle of the strawberry aphid consists of egg, nymph, and adult (apterous and alate forms). Eggs are white-yellowish in color when deposited, but hours later they become shiny and black. Nymphs are small and similar to adults. Nymphs vary in color from light green to pale yellow. Adults are 1.3-1.5 mm long, pale to yellowish green, and have short setae (hairs) over the body. The antennae are as long as or longer than the length of the body. The cornicles are long, pale, and slender, about 1/4 of the body length, and the legs are pale green and almost translucent. Wingless adults are small (0.9-1.3 mm long), elongate oval, translucent, and yellowish-white or pale green. Winged adults are medium size and have a pale greenish abdomen with a brown dorsal patch. Aphid females can reproduce throughout the year without males; however, oviparae (winged reproductive females and wingless males) may occur in greenhouse and laboratory culture. Oviparae forms have not been reported from the field.

Mode of feeding and symptoms of infestation

Aphid nymphs and adults feed on plant sap, preventing plant growth. Typical symptoms of aphid damage include curled leaves, yellowish spots, and the presence of sticky honeydew. A black sooty mold may develop on the leaves, affecting photosynthesis and possibly reducing plant yields (Zalom *et al.*, 2008 and Rondon *et al.*, 2005).

Management

i. Predators such as syrphid fly or green lacewing larvae often provide a greater level of control.
ii. Cultural and biological controls and sprays of insecticidal soap, azadirachtin, neem oil, and pyrethrin are acceptable for use on organically certified strawberries.
iii. Row covers (plastic tunnels enclosures) have reduced aphid populations to below economic levels.

5. Fruit fly, *Drosophila melanogaster* Meigen (Drosophilidae: Diptera)

Fruit flies, also known as fruit or pomace flies. It is a small, yellowish flies and are commonly attracted to all kinds fermenting fruits. *Drosophila* may feed on fruit that has been previously damaged. *Drososphila* species are unable to penetrate the surface of sound fruits and must wait until the fruit rot or are damaged by other causes, such as pathogens, mechanical injuryor other insects (Burrack, 2014).

Distribution

This is cosmopolitan in its distribution. But, not being found at extremes of altitude and latitude.

Identification, biology and symptoms of infestation

Fruitflies lay eggs on very ripe or damaged fruits. This is primarily a problem in strawberries picked for freezing. Because these fruits are allowed to ripen in the field in order to allow easy removal of the strawberry calyx. Here, harvest interval is increased and the fruit becomes more susceptible to infestation. Fruit fly eggs and larvae are primarily a contamination problem (Kuhar and Pfeiffer, 2000).

The larvae are about 6 mm long, can be found in very ripe and damaged fruit in the fields. Affected fruits are readily recognized since rotting develop rapidly and the skin around the sting marks becomes discolored. Ideal temperatures for development of this insect are in the low 80°F (27° to 30°C). The flies do not lay eggs at temperatures below 54°F or above 91°F (Zalom *et al.*, 2008).

Management

i. Maintenance of field sanitation by removing over-ripen fruits from the field and around packing sheds.

iii. Protein bait for the control of female flies is recommended for use in spots or bands in the field.

iii. Male Annihilation Technique (MAT) is also applicable in managing the pest.

iv. Treat portions of fields or obvious sources of flies with synthetic pyrethroids to control adult flies.

6. Tarnished plant bug, *Lygus hesperus* Knight, *L. lineolaris* (Miridae: Hemiptera)

Tarnished plant bug also known as lygus bug is a common sap-feeding insect of strawberry in many countries.

Distribution

The pest well distributed in several strawberry growing regions in Asia, Africa, North America and Oceania.

Hosts of commercial importance

Lygus bug has a wide host range that includes more than 100 species of cultivated crops and wild host plants (Fye, 1982 and Mueller *et al.*, 2005) that include cultivated crops such as alfalfa, broccoli, celery, cauliflower, grapes and strawberries.

Identification and biology

Adults are about 6 mm long, oval and flattened. They are greenish or brownish in colour and have reddish brown markings on their wings. In the centre of their back is a distinct but small, yellow or pale green triangle that helps distinguish them from other insects (Zalom *et al.*, 2008). The insect overwinters as an adult in protected areas such as organic debris, bark cracks etc. (DeFrancesco *et al.*, 2018). In the spring, females begin laying eggs into stems and leaf midribs of host plants (Burrack and Toennison, 2014a). Egg hatch takes place about a week later and the newly hatched nymph, is smaller and bright green, resembling an aphid, but much more active (Schloemann, 2016). Nymphs develop through 5 instars, reaching the adult stage in approximately 30 days (Brust, 2013).

Mode of feeding and symptoms of infestation

Both adults and nymphs feed on the developing flowers and fruit, sucking out plant juices with causing cell death, distorted growth, and seed damage all of which result in mis-shapen fruit. This deformed fruit, typically "cat-faced" berries, also called nubbins or button berries are generally unmarketable (Brust, 2013). Poor pollination might also be another reason for small under-developed seeds. Straw brown seeds that are large and hollow area good indication of lygus bug damage (Zalom *et al.*, 2008). Lygus bug damage is more of a problem in strawberry-growing areas where continuous fruit production occurs.

Management

i. Elimination of weed hosts that serve as protection and early season food for the insect.
ii. There are several predators that feed on the immature stages of TPB. These include big-eyed bugs, damsel bugs, minute pirate bugs and several species of spiders.
iii. Once adults are present on weeds, they will migrate into strawberries when the weeds are removed.
iv. Spraying of insecticides like Fipronil/Thiamethoxam will reduce the population of these bugs.

7. Spittle bug, *Philaenus spumarius* Linn. (Cercopidae: Homoptera)

The meadow spittlebug is an annoying pest on strawberries that can stunt plants and reduce berry size. They often are seen in meadows and on plants in uncultivated fields. Although the spittle is harmless, pickers object to being wetted by the insect excretion (Bessin, 1993).

Distribution

The species is widely distributed, covering most of the Palearctic regions and extending to Nearctic, as well as most of the temperate regions of earth and oceanic islands (Drosopoulos and Remane, 2000). It has been reported for North Africa, several parts of the former Soviet Union, Afghanistan, Japan, USA, Canada, Azores, Hawaii and New Zealand (Yurtsever, 2000). Karban and Strauss (2004) suggested that the species northward shift in California since 1988 is related to variations in humidity and temperature.

Hosts of commercial importance

The species is highly polyphagous and occurs in most of the terrestrial habitats (Stewart and Lees, 1996).

Identification and biology

Adult spittlebugs are inconspicuous, often greenish or brownish insects, about 1/4 inch in length. Adult spittlebugs look like leafhoppers and readily jump or fly when disturbed.

Egg laying occurs primarily during September and October. Eggs are inserted into the lower parts hidden parts of the strawberry, such as the sheath between leaves and stems. The insect overwinters as an egg. Nymphs appear in May or June and complete their development in 5 to 8 weeks. Immature spittlebugs are recognized by the frothy white mass that nymphs surround themselves with on plant tissue where they feed (Penn state extension, 2017).

Mode of feeding and symptoms of infestation

Both nymphs and adults feed by sucking plant juices out of their host plants with their straw-like mouthparts (Burrack and Toennison, 2014b). Initially the nymphs feed at the base of the plants, but later move up to the tender foliage (Bessin, 1993). Feeding activities of large numbers of these insects cause plants to become stunted and berries do not attain full size. However, significant yield loss seldom occurs.

Management

They do not usually cause significant damage. Handpicking or washing off with water may be practiced to lower down its population build up. Cutting of weeds or washing spittlebugs off these alternate hosts before the insects mature may be helpful to protect the crop.

8. Strawberry bud weevil, *Anthonomus signatus* Say (Curculionidae: Coleoptera)

This species often is called the "strawberry clipper" or simply "clipper" because clipped buds are characteristic symptom of infestation.

Distribution

The species is well distributed in the northeastern United States and Canada.

Hosts of commercial importance

It is a pest of several small fruits grown during spring.

Identification and biology

Adult is a snout bearing beetle *i.e.*, weevil which is about 3-4 mm long, chestnut brown in color and possessing two black spots on its elytra. Larvae are very small, creamy white-colored grubs found inside un-opened flower buds.

Ovipositing females puncture unopened buds with their snouts and deposit egg into the bud singly. Damage results when females sever the strawberry bud from the pedicel following oviposition, causing it to hang by part of the stem or fall to the ground (Annonymus, 2014). Thus, infestation by this pest prevents fruit formation. Larvae develop in the severed buds and reach maturity in 3-4 weeks. Strawberry bud weevil has one generation per year.

Mode of feeding and symptoms of infestation

After feeding on the pollen of various flowers for a short period of time, the new adults seek hibernating sites in mid-summer and remain inactive for the rest of the season.Weevils remain in these sites until the next spring. Injury is most likely to occur when strawberries are grown adjacent to woodlots or other suitable hibernating sites.

Management

i. Attention to be given in fields near woods and hedgerows. Treatment of field borders may be sufficient in some instances.

ii. Mulches and full canopy beds can encourage newly emerged adults to remain in the planting, causing damage in succeeding years. Ploughing under old beds immediately after harvest.

iii. Removing the foliage and mulch to reduce the suitability of overwintering sites help lessen the chances of clipper damage.

iv. Pistillate varieties of strawberries are relatively immune from attack since only varieties with staminate flowers seem to provide adequate food for developing larvae.

v. Early-fruiting varieties are more susceptible to attack than later-fruiting varieties. Topping of plants and removal of foliage and mulch immediately after harvest and then applying a follow up chemical spray to kill overwintering adults is beneficial.

9. Red spider mite, *Tetranychus urticae* (Koch) (Tetranychidae: Acarina)

The most common mite pest of strawberry crops is two spotted mites (*Tetranychus urticae*). It is also known as spider mite or redspider mite.

Identification and biology

Two-spotted spider mite lays eggs, initially spherical, clear, and colorless when laid but become pearly white, on the undersides of leaves. Nymphs, adult males, and reproductive adult females are oval and generally yellow or greenish with one or more dark spots on each side of their bodies, and the top of the abdomen is free of spots. Adult female two-spotted spider mites may stop reproduction during the coldest winter months in production areas of colder inland valleys. Diapause is indicated by a change in color to bright orange. In coastal growing areas it is rare to have a significant proportion of the population undergo diapause. Mating and egg laying typically occur year-round in all coastal strawberry-growing regions.

Mode of feeding and symptoms of infestation

Two spotted spider mite damage is particularly severe to strawberries during the first 2 to 5 months following transplanting in late summer or fall. Mite feeding during this critical period of plant growth substantially reduces berry number per plant and overall plantation yield. Typical symptoms include stippling, scarring, and bronzing of the leaves and calyx. Plants that sustain infestations may become severely weakened and appear stunted, dry, and turn red (Zalom *et al.,* 2008).

Management

i. When transplanted in summer, short duration cultivars are relatively tolerant. It is to be ensured that transplants have received adequate chilling and receive proper irrigation and fertilization.

ii. Conservation of natural enemies like *Phytoseilus persimilis, Neoseiulus californicus* etc.

iii. Entomopathogenic fungi such as *Beauveria bassiana* is acceptable for use on organically certified strawberries.

v. Applying a shortresidual miticide to reduce spider mite numbers before a predator release may improve biological control under some conditions.

References

Anonymous. 2004. Retrieved from http://ipm.illinois.edu/fruits/insects/white_grubs/index.html (Accesssed date: 31.03.2018).

Anonymous. 2014. Retrieved from https://content.ces.ncsu.edu/leafrollers-in-strawberries (Accessed date: 2.4.2018).

Anonymous. 2018. Retrieved from https://www.cabi.org/isc/datasheet/3801 (Accessed date: 30.3.2018).

Ayre, G. L. 1980. The biology and life history of the cutworm *Amphipoea interoceanica* (Lepidoptera: Noctuidae), a new pest of strawberry in manitoba. The Canadian Entomologist,112(2):127-130.

Bessin, R. 1993. Strawberry pests. Retrieved from https://entomology.ca.uky.edu/ef207 (Accessed date: 5.5.2018).

Brust, G. 11th February, 2013.Tarnished Plant Bug (Lygus) Management in Strawberries. Retrieved from https://extension.umd.edu/TarnishedPlantBug_LygusInStrawberries (Accessed date: 1.5.2018)

Burrack, H. 2014. https://content.ces.ncsu.edu/other-drosophila-species-in-strawberries (Accessed on 30.04.2018)

Burrack, H. and Toennison, A. 2014a. Tarnished plant bug in strawberries. Retrieved from https://content.ces.ncsu.edu/tarnished-plant-bug-in-strawberries (Accessed date: 1.5.2018).

Burrack, H. and Toennison, A. 2014b. Spittle bugs in strawberries. Retrieved from https://content.ces.ncsu.edu/spittlebug-in-strawberries (Accessed date: 5.5.2018).

Capinera, J. L. 2006. Retrieved from http://entnemdept.ufl.edu/creatures/veg/black_cutworm.htm (Accessed on 30.03.2018).

DeFrancesco, J., Edmunds, B. and Bell, N. 2018. Small fruit crops: Blueberry pests. PNW Insect Management Handbook. A Pacific Northwest Extension Publication, pp-12.

Drosopoulos, S., Remane, R. 2000. Biogeographic studies on the spittlebug *Philaenus signatus* Melichar, 1896 species group (Hemiptera: Aphrophoridae) with the description of two new allopatric species. Ann. Soc. Entomol. Fr., 36:269-277.

Fye, R. E. 1982. Weed hosts of the lygus (Heteroptera: Miridae) bug complex in Central Washington. *J. Econ. Entomol.*, 75:724-727.

Karban, R., Strauss, S. Y. 2004. Physiological tolerance, climate change, and a northward range shift in the spittlebug, Philaenus spumarius. Ecol. Entomol, 29:251-254.

Kuhar, T. P. and Pfeiffer, D. G. 2000. Insect pests of strawberries and their management. http://www.virginiafruit.ento.vt.edu/StrawMaster.html (Accessed on 30.04.2018).

Mueller, S. C., Summers, C. G. and Goodell, P. B. 2005. Composition of *Lygus* species found in selected agronomic crops and weeds in the San Joaquin Valley, California. Southwest. *Entomlo.*, 30: 121-127.

Omafra. 2006. Notes on Strawberry Insects: White grubs. Retrieved from http://www.omafra.gov.on.ca/english/crops/pub360/notes/strawwhgrub.htm (Accessed date: 31.03.2018).

Penn State Extension. 2017. Meadow spittle bugs in strawberries in home fruit plantings. Retrieved from https://extension.psu.edu/meadow-spittlebug-on-strawberries-in-home-fruit-plantings (Accessed date: 5.5.2018).

Renkamma, J. 2016. Retrieved from http://entnemdept.ufl.edu/creatures/FRUIT/strawberry_leafroller.htm (Accessed date: 2.4.2018).

Rondon, S. I., Cantliffe, D. J. and Krey, K. L. 2005. Retrieved from http://edis.ifas.ufl.edu/hs253 (Accessed date: 4.4.2018).

Schloemann, S. G. 2nd May, 2016. Strawberry IPM: Tarnished Plant Bug. https://ag.umass.edu/fruit/fact-sheets/strawberry-ipm-tarnished-plant-bug (Accessed on: 1.5.2018).

Stewart, A. J., Lees, D. R. 1996. The colour/pattern polymorphism of *Philaenus spumarius* (Linn.) (Homoptera: Cercopidae) in England and Wales. Philos. Trans. R. Soc. B., 351:69–89.

Yurtsever, S. 2000. On the polymorphic meadow spittlebug, *Philaenus spumarius* (Linn.) (Homoptera: Cercopidae). Turk. J. Zool., 24:447-460.

Zalom, F. A., Bolda, M. P., Dara, S. K. and Joseph, S. 2008. UC IPM Pest Management Guidelines: Strawberry. Retrieved from http://ipm.ucanr.edu/PMG/r734300511.html (Accessed on 30.03.2018).

17

Pests of Temperate Fruit Crops and Their Management

G. Mahendiran, Shahid Ali Akbar and Mudasir Ahmad Dar

Resource management has become a formidable task with loss of cultivable land to urbanization. Coercion to increase production has resulted in adoption of intense farming. This, however, has resulted in significant increase in the pest problem. Despite ten-fold increase in insecticide use during the last 80 years, losses inflicted by pests have nearly doubled. Estimates of yield losses caused by various pests attack range from 10-35% in various crops during different seasons which account for about 2000 crore revenue loss to the horticulture industry and is further projected at 5000 crore or more in future due to changing climate and intense farming (Reddy, 2010 and Rather, *et al*, 2013).

The temperate fruit crops are prone to attack by insect pests which are one of the major limiting factors in their successful production. Integrated pest management (IPM) is an eco-friendly approach for managing insect pest problems, encompassing available methods and techniques such as cultural, mechanical, biological and chemical in compatible and scientific manner.

Effective management of pest depends on understanding of its biology, alternate host plants, nature of damage and seasonality of occurrence. Some of the important pests of temperate fruits viz., almond, apricot, peach, plum, pear, cherry, walnut and various strategies for their successful management are discussed in this chapter. For the sake of convenience, the major pest species of above mentioned crops are grouped into three main categories; sucking pests, borers, and defoliators.

Table 1: Major insect pests of some dominant temperate fruit crops.

S.No.	Common name	Scientific name	Family	Order
A.	**Sucking pests**			
1.	San Jose Scale	*Quadraspidiotus perniciosus* (Comstock)	Diaspididae	Hemiptera
2.	Peach leaf curl aphid	*Brachycaudus helichrysi* (Keltenbach)	Aphididae	Hemiptera
3.	Large peach aphid	*Pterochloroides persicae* Cholodkovsky	Aphididae	Hemiptera
4.	Small walnut aphid	*Chromaphis juglandicola* (Kaltenbach)	Aphididae	Hemiptera
5.	Blossom thrips	*Frankliniella schultzei* (Try-bom)	Thripidae	Thysanoptera
6.	Pear psyllid	*Cacopsylla pyri* Linn.	Psyllidae	Hemiptera
7.	European red mite	*Panonychus ulmi* (Koch)	Tetranychidae	Acarina
8.	Two spotted mite	*Tetranychus urticae* Koch	Tetranychidae	Acarina
B.	**Borers**			
1.	Stem borer	*Apriona cinerea* Chev.	Cerambycidae	Coleoptera
2.	Shot or pine hole borer	*Scolytus* spp. *Xyleborus* spp.	Scolytidae	Coleoptera
3.	Flat headed borer	*Sphenoptera lafertei* Thom.	Buprestidae	Coleoptera
4.	Root borer	*Dorysthenes huegelii* Redt.	Cerambycidae	Coleoptera
5.	Codling moth	*Cydia pomonella* (Linn.)	Tortricidae	Lepidoptera
6.	Cherry fruit fly	*Rhagoletis indifferens* Curran.	Tephritidae	Diptera
7.	Anar butterfly	*Deudorix epijarbas* Moore	Lycaenidae	Lepidoptera
C.	**Defoliators**			
1.	Indian gypsy moth	*Lymantria obfuscate* (Walker)	Lymantridae	Lepidoptera
2.	Tent caterpillar	*Malacosoma indicum* (Walk-er)	Lasiocampidae	Lepidoptera
3.	Chafer beetle/ white grub	*Holotrichia* spp., *Anomala* spp.	Scarabaeidae	Coleoptera

A. Sucking pests

Sucking pests include aphids, leafhoppers, thrips, scale insects, mites, whitefly and bugs. These pests suck nutrients out of plants, inject toxic saliva, cause virus transmission, sooty mould and other fungal developments.

1. San jose scale, *Quadraspidiotus perniciosus* (Comstock) (Diaspididae: Homoptera)

Distribution

Widely distributed in all the apple growing regions of the world. The native home of this insect is believed to be China where from it spread to all temperate region of the world. This is the most injurious insect in North-West India, particularly Kashmir and Kallu valley in Himachal Pradesh.

Hosts of commercial importance

It has been recorded on more than 150 host plants, including almond, apple, apricot, cherry, chestnut, citrus, crab apple, grapevine, gooseberry, mulberry, peach, pear, plum, quince, raspberry, strawberry and ornamental trees and shrubs.

Identification and biology

Adult females of the San jose scale are yellow, circular, sac-like, legless insects. They secrete and live beneath a protective covering. The covering is round, grey brown and made up of concentric rings surrounding a raised nipple near the center. The circular female covering is about 1.5 mm in diameter. Adult males are tiny, golden-brown, two-winged insects, about 1 mm long with a narrow, dark band across the abdomen. Nymphs or crawlers have six legs, are yellow, somewhat oval, about 0.3 mm long.

The nymphs (black cap stage) hibernate from December to March and resume their activity around the end of March and mature in about 4 weeks. The males (winged) are short lived, they just fertilize the females (wingless) and die. The females produce young ones (crawlers) about 300 - 400 per female. Emerged crawlers move to new site of infestation and settle, then insert their stylet into host and begin to feed. Nymphal period varies between 40-50 days. The number of generations in a year depends mainly on elevation and climatic conditions. There are six to seven overlapping generations in a year.

Mode of feeding and symptoms of infestation

San jose scale is a highly destructive and polyphagous pest. Both nymphs and adults suck the sap from all the tender parts of tree and fruits. Infested trees show a general decrease in vigour, growth as well as productivity. The affected fruits present pink-coloured areas around the scales. Early season fruits infestation may result in small, deformed and poorly coloured fruit. It is the most destructive pest in nurseries and to the young trees.

Management

i. Cultural

a. Orchard sanitation greatly reduces infestation by this pest. Even simple removal of alternate host, weed hosts, volunteer plants and crop residue greatly reduce the infestation potentiality.

b. Heavily infested branches and twigs should be pruned and burnt.

ii. Biological

a. Release endoparasitoid *Encarsia perniciosi* (Tower) @ 2000/ infested tree and repeating it if necessary (Singh, 1997).

b. Avoiding summer spray if biocontrol agents like coccinellid predator *Chilocorus bijugus* and parasitoids, *Encarsia perniciosi* and *Aphytis diaspidis* are active and found in good number. Rawat and Pawar (1992) recorded that repeated release of *Chilocorus bijugus*, provided pest reduction averaged 8-17% (as a measure per tree) or 11-44% (as measure per twig).

iii. Chemical

a. Avoidance of infested field for raising nursery, selecting healthy plant from nursery and treating them with Chlorpyriphos 0.05% before planting in the field (Dhiman and Arif, 2006).

b. Spray the dormant miscible spray oil (2% of HP spray oil E, Servo oil, IPOL, Loven-30 or Wonder TSO) during the winter season when the trees are in dormant stage.

c. Bhardwaj (1988) reported that a single application of Fenitrothion (0.05%), Chlorpyriphos (0.02%) at first crawler emergence provides effective control of the pest.

d. The best treatment for controlling San jose scale is 2% miscible oil followed by a spray of 0.4% Chlorpyriphos in May (Khajuria and Sharma, 1997).

e. Spray with Chlorpyriphos (0.02%) or Dimethoate (0.03%) to kill crawlers and newly settled scale insects in May. If necessary, spray again in September-October (postharvest spray) with Chlorpyriphos (0.02%) or Fenitrothion (0.05%).

2. Peach leaf curl aphid, *Brachycaudus helichrysi* (Kalt.) (Aphididae: Homoptera)

Distribution

The peach leaf curl aphid is a highly destructive and polyphagous pest. It is prevalent both in the plains and in the mountainous areas of India, up to an altitude of 2200 meters.

Hosts of commercial importance

Peach, almond, apricot, plum etc.

Identification and biology

Both wingless and winged forms are present. The body of wingless adult female is green to brownish yellow with a waxy bloom. The size of aphid is about 0.9-2.0 mm in length, oval in shape.

The aphid is heterocyclic, passing summer and rainy season on alternate or secondary hosts, namely *Trifolium pretense*, Cineraria, *Ageratum conyzoides* etc. The eggs hatch on *Prunus* during spring, each giving rise to a morph called a fundatrix, which is apterous and reproduce parthenogenetically when temperature is above 7°C. These individuals produce 3-8 generations of apterous aphids. At the end of spring, winged forms are produced for dispersal to secondary hosts.

Mode of feeding and symptoms of infestation

Nymphs and adults suck the cell sap from tender leaves, flower buds and newly formed fruits. Affected leaves turn pale and curl up, blossoms wither and fruits do not develop into normal size and drop prematurely.

Management

i. Cultural

a. For effective management of aphid, secondary host plants (weeds) should be removed from the field.

b. Pruning of current years vegetative growth (which carried eggs) during Dec-Jan also reduces infestation.

c. The peach cultivars Stark Early Giant, Early white Giant and Flavour Crest are found moderately susceptible (10-30% leaf whorl infestation) against leaf curl aphid (Gupta *et al.*, 1988) and hence, in aphid prone areas, cultivation of those cultivars should be avoided.

ii. Biological

a. Release of biocontrol agents such as predatory coccinellids, syrphid flies, anthocorids and chrysopids should be done at appropriate time.

iii. Chemical

a. Two applications of Imidacloprid or Thiamethoxam (each 0.006%) or Acetamiprid (0.008%) at pre-flowering and pea stage of crop found effective (Sharma, 2010).

b. Post bloom spray should not be given to encourage biotic agents like coccinellids, syrphids flies and anthocorids bugs which are very active in the Month of March-April (Singh, *et al.*, 2010).

3. Large peach aphid, *Pterochloroides persicae* Cholodkovsky (Aphididae: Homoptera)

Distribution

The pest is well documented with distributional records available for Cyprus, Greece, Italy, Romania, Sicily, Spain, Turkey, Yugoslavia, Egypt, Tunisia, Armenia, Republic of Georgia, Kazakhstan, Latvia, Tadzhikistan, Turkmenistan, Uzbekistan, Bhutan, India, Indonesia, Java, Iran, Iraq, Israel, Lebanon, Pakistan, Saudi Arabia, Syria, Yemen Republic (Blackman and Eastrop, 1994).

Hosts of commercial importance

Peach, plum, almond etc.

Identification and biology

Pterochloroides persicae also called as brown peach aphid, woody aphid, clouded peach bark aphid, cloudy-winged peach aphid. Apterous viviparous are the most conspicuous form in the colony. When fully grown, these are large in size, having morphometric ranges of body length and width as 3.75-4.00 mm and 2.61-2.70 mm, color is dark brown to black with some white patches, dorsum of abdomen with a double row of large tubercles. Forewing with large distinctive dark patches and long rostrum easily demarcates the species from other con-generic members.

The life cycle of the pest is complex, having alteration of parthenogenetic and sexual generations, apterous and alate forms exhibited and persistently overlapping. Holocyclic populations with sexual phase in autumn have only been observed in cooler regions with oviparous females very similar to apterous viviparous females, except for their larger size (Wieczorek, *et al.*, 2013). Environmental conditions predominately influence population dynamics of the species with 18 overlapping generations reported in a year (Darwish, *et al.*, 1989). However, fewer progenies are produced per female in the winter than in summer (Velimirovic, 1977). It takes 66-82 hrs, 64-80 hrs and 144-152 hrs respectively for the first, second and third instar to complete development, 16-48 hrs, 122-280 hrs and 8-24 hrs for the pre-oviposition, oviposition and post-oviposition periods. About 17-55 nymphs/female are raised in the entire reproductive life with an estimated reproductive rate of 4.51 nymphs/female per day (Darwish, *et al.*, 1989). Khan, *et al.* (1998) reported 4-instars with 8 days nymphal period and affirmed that average life span of the pest is 41 days.

Mode of feeding and symptoms of infestation

The pest mainly infests branch cortex (phloem). Apart from general weakening of the young fruit trees, withered branches and reduced yield are observed (Stoetzel,

1994). The pest is not listed as transmitting a virus (Chan, *et al.*, 1991) but severe infestation results in premature fruit drop, leaf curling, irregular curvature of twigs, stunted growth and sooty mould fungal growth development due to excessive honey dew production (Hondru, *et al.*,1986).

Management

i. Cultural

a. Pruning of current years vegetative growth (which carried eggs) during Dec-Jan reduces infestation by this pest.

ii. Biological

a. Very little is known about the natural enemies of the pest. However, green lacewing, *Chrysoperla carnea*, *Coccinella undecimpunctata*, *Pauesia antennata* and *Coccinella algerica* considered as most promising bio-control agents of the pest (Mdellel and Kamel, 2012).

b. *Metarhizium anisopliae* (entomopathogenic fungi), yeast, vegetable oil, mineral oil, potassium sulphate + detergent have been trialed successfully against control of the pest (El-Salam, 2001).

iii. Chemical

a. Several chemicals at different concentrations and varying compositions have also been successfully used against the pest like, Imidaclorpid, combined preparation of Chlorpyrifos and Cypermethrin (Ateyyat, 2008); Diflubenzuron) and Cyromazine as biorational insecticides (ElSalam, 2001).

b. High volume sprays of Chlorpyrifos, Dimethoate and Quinalphos (Sandhu and Sohi, 1978), non-specific insecticide like Acephate along with addition of some wetting agent (Ciampolini and Martelli, 1980) have been advocated for its effective management.

4. Small walnut aphid, *Chromaphis juglandicola* (Kaltenbach) (Aphididae: Homoptera)

Distribution

The small walnut aphid is a serious pest known in all the walnut growing regions of the world.

Identification and biology

These are small aphids typically found scattered on the lower side of leaves. It is easily identified and differentiated from the large dusky walnut aphids (*Panaphis juglandis*) which are having row of dark bands along the dorsal surface and are much larger.

The aphid overwinters on trunks as egg masses which hatch into nymphs, move towards buds at the end of February. Eggs hatch as soon as leaf buds begin to open and the young aphids settle on the leaflets, they mature into larger, yellowish aphids. The females give birth to several nymphs without mating resulting in formation of several overlapping generations during the progress of season. Alate forms appear throughout the season and eventually mate with males to deposit overwintering egg masses on main trunks of walnut tree.

Seasonal incidence

The population mostly peaks during July.

Mode of feeding and symptoms of infestation

Aphid feeding reduces tree vigour, nut size, yield, and quality of the fruit. These aphids excrete honey dew which promotes sooty mold growth causing husk surface to turn black leaf drop, exposing more nuts to sunburn and darkens or shrivels the kernels.

Management

i. Cultural

a. Pruning of current years vegetative growth (which carried eggs) during Dec-Jan reduces infestation of the pest.

ii. Biological

a. The parasitic wasp, *Trioxys pallidus* is a dominant natural enemy of the pest along with few beetles.

iii. Chemical

a. Some of the recommended chemicals for the aphid control are, Imidacloprid, Acetamiprid, Chlorpyriphos and several narrow range oils. For these chemicals the recommended dosage varies mostly with region and severity. However, not not more than 4 applications/season to be done.

Some of the other prominent aphid pests of above mentioned temperate fruit crops are, Dusky-veined large walnut aphid (*Panaphis juglandis*); Green apple aphid

(*Aphis pomi*); Mealy plum aphid (*Hyalopterus pruni*); Pomegranate aphid (*Aphis punicae*) etc.

5. Blossom thrips, *Frankliniella schultzei* (Trybom) (Thripidae: Thysanoptera)

This is one of the most important pests of temperate fruit crops. In addition to the aforementioned species some other species of the pests are, *Frankliniella dampfi*, *Taeniothrips* spp., *T. rhopalantennalis*, *Thrips flavus*, *T. florum*, *T. carthami*, *Aeolothrips collaris* and *Haplothrips tenuipennis* etc.

Distribution

The common blossom thrips has a very wide distribution and is mainly found in tropical and sub-tropical areas throughout the world. Reported worldwide distribution is as follows: Bangladesh, India, Indonesia, Iran, Iraq, Israel, Java, Malaysia, Pakistan, Sri Lanka.

Hosts of commercial importance

Apple, pear, peach, plum, cherry, apricot etc.

Identification and biology

There are two larval instars and two inactive and non-feeding stages in life cycle of the species. The latter two stages are known as pre-pupa and pupa. Females of *Frankliniella schultzei* insert their eggs in flower tissue. The embryonic stage lasts for four days and the 1st and 2nd larval instars, prepupa and pupa take on an average of 2.5, 2.5, 1.2, and 2.1 days respectively. The adult female and male longevity is approximately 13 days.

Mode of feeding and symptoms of infestation

These are associated with the blossom and their population increases with the onset of spring and flowering. Both nymphs and adults lacerate all floral parts and unfolding leaves of vegetative buds. Consequently, brown spots develop. The lesions on androecia and gynoecia deteriorate the fruit quality and reduce fruit set and yield. The set fruits may also drop. Heavily infested flowers bear sticky and faded appearance with indication of early senescence (Reddy, 2009).

Management

i. Coccinellids, anthocorid (*Orius* sp.) and lygaeid bugs (*Geocoris* spp.) feed on blossom thrips (Kumar and Ananthakrishnan, 1984) and check its population build up.

ii. If the population of blossom thrips is too high, spraying 0.05% Fenitrothion or Imidachloprid or Thiomethoxam could be done at pink bud stages when predator activity is very low.

iii. Singh and Singh (1987) reported that Phosalone at 0.035% was the most effective compound, causing an 85% reduction in abundance.

6. Pear psyllid, *Cacopsylla pyri* Linn. (Psyllidae: Hemiptera)

Distribution

The pest is one of the dominant psyllids in many South-Eastern countries including India, Iran, USA and others.

Identification and biology

The eggs are yellowish orange and about 0.3 mm long. They are deposited in the creases of the bark, in old leaf scars and around buds. They are pear-shaped, elongated and have a smooth, shiny surface. Small, yellow, wingless nymphs are about 3 mm long when newly hatched. They have red eyes and turn green, then brown as they mature. The nymphs feed and develop primarily on the newer, tender growth. The adults are dark reddish brown and have four wings and resemble a tiny cicada.

Life cycle of the pest completes in 55-61 days with 5 developmental stages (instars). Hatching starts after incubation period of 8-10 days and undergoes 5 developmental stages before adult formation. The freshly laid eggs are whitish in color, as the development precedes color of the eggs change to deep yellow and at completion to eclosion red eye spots become most prominent feature of the eggs. First instar nymphs are whitish soft bodied, as they feed on tree sap these secrete large quantities of transparent honey dew which soon hardens on exposure to air. After the first molt and in all preceding instars external wing pads are prominent feature. The succeeding molting involves attainment of larger size and deeper colouration. The fifth and last nymphal stage is distinct in its appearance; dark brown with prominent wing pads also known as "hard shell" stage.

Mode of feeding and symptoms of infestation

The pest is known to cause heavy economical losses, sooty mold growth and possible transmission of phytoplasms. The pest secretes large amounts of honeydew, which runs down over foliage and fruit and in which a sooty fungus grows. This causes the skin of the fruit to become blackened and scarred and the leaves to develop brown spots. Excessive feeding and the injection of toxic saliva by large populations may cause partial to complete defoliation of trees, reducing vitality and preventing the formation of fruit buds. Return bloom and fruit yield are often reduced the following season.

Management

i. Cultural

a. During dormancy, trees should be pruned regularly and destroy pruned woods by burning.

ii. Biological

Coccinella septempunctata, *Adalia tetraspilota*, *Calvia flaveola*, *Sphaerophoria* sp., *Anthocoris* sp. and *Eristalinus* sp. are foremost natural enemies of the pest.

7. European red mite, *Panonychus ulmi* (Koch.) (Acarina: Tetranychidae)

Distribution

It was recorded in North America during 1911. In India, it was reported in 1974 from Jammu and Kashmir, Himachal Pradesh and Uttarakhand. This pest has now been established as a major pest in almost all the apple growing areas of Jammu and Kashmir and Himachal Pradesh (Khajuria and Sharma, 1996).

Hosts of commercial importance

It is a polyphagous species and cause damage to apple, pear, peach, apricot, plum, walnut, quince, jack fruit, rose and *Hibiscus*.

Identification and biology

The adult is red in colour, very small (0.4 mm long) in size and just visible to naked eye due to its colour. Eggs are red with a slender stripe arising from the center. Newly hatched mites are green but with feeding turn them red.

European red mite overwinters in the egg stage on the bark of tree or smaller branches and spurs. Eggs start hatching in early April on rising of temperature. After hatching, juvenile mites move to the underside of leaves and begin feeding. They reach maturity in approximately three weeks after three moults. The number of generations per year varies depending upon weather condition. Dry and hot weather is highly favourable for their reproduction and multiplication.

Mode of feeding and symptoms of infestation

Both nymph and adult cause damage by sucking the sap from the leaves and tender plant parts. The initial symptom of mite infestation appears in the form of a minute speckling of the leaves. Increase in mite population render the leaves lose their bright green colour which become dull green and finally turn brownish green or bronze colour. They also become brittle and fall off prematurely. Continuous feeding by mite causes production of unripe, sour, undersized and poorly coloured fruits which are prone to early dropping. The economic losses caused by this mite have estimated at more than 30% of the harvest (Bhardwaj, 2004).

8. Two spotted mite, *Tetranychus urticae* (Koch) (Acarina: Tetranychidae)

Distribution

It has a worldwide distribution.

Hosts of commercial importance

A polyphagous pest having a very wide range of host plants.

Identification and biology

Adult mite body has two large dark spots, often visible through the transparent body wall. Since these spots are accumulation of body wastes, newly molted mites lack them. Larvae usually are light green in colour and develop these distinct spots in the middle of the body at protonymph stage. The body colouration becomes darker and darker with developmental stages. The male is elliptical with the caudal end tapering and smaller than the female.

The two spotted mite, is normally pale yellow, pale green or straw coloured. The adult has two dark green or black spots on the body. The development is highly temperature dependent. Life cycle is completed in 6 days at 35°C, 12 days at 20°C and 55 days at 10°C. They overwinter as adults in debris, fallen leaves and loose bark just below soil surface in cracks and dead leaves of other plants and resume activity in spring. Eggs hatch in about 5-14 days depending upon temperature. Before becoming adult the mite passes through the larva, protonymph and deutonymph stages in about a week time. In contrast to the summer form the hibernating mites produced during autumn are uniformly pink without spots on their body. There have 6-7 overlapping generations during the season (Bhardwaj, 2004).

Mode of feeding and symptoms of infestation

The nymphs and adults are found feeding on ventral surface of leaves under protective cover of fine silken webs. As a result of their feeding innumerable yellow spots appear on the dorsal surface of leaves and the affected leaves gradually start curling and finally get wrinkled and crumpled. This in turn affects the growth and fruit formation capacity of the plants.

Management of mite pests

i. Cultural

a. Keep the tree basin free from weeds to avoid mite multiplication.

b. The winter sprouts should be removed from the trees.

c. During dormancy, trees should be pruned regularly and destroy pruned woods by burning.

ii. Biological

a. Conserve natural enemies like, predatory mites, *Typhlodromus* sp., *Amblyseius fallacis* A-H, *Zetzellia mali* (Ewing), coccinellid beetle *Stethorus punctum* Le Conte, green lace wing *Chrysoperla* spp., anthocorid *Orius* spp. and predatory thrips *Leptothrips mali* (Khajuria, 2009).

iii. Chemical

a. Delayed dormant spray (Green tip stage) with Horticultural mineral oil (HMO) (2%) + Ethion 50 EC @ 1 ml/l found effective.

b. The suitable Acaricides namely Fenpyroximate 5 SC @ 1 ml/l, Fenazaquin 10 EC @ 0.4 ml/l, Propergite 57 EC@ 1 ml/l and Ethion 50 EC @ 1 ml/l are effective against the European red mite.

c. New chemicals like, Abamectin 1.8 EC @ 0.5 ml/l, Spiromesifen 240 SC @ 1 ml/l and Milbemectin 1 EC @ 1 ml/l are also effective in pome and stone fruits.

d. Neem formulations are effective to some extent and safe to natural enemies.

e. Avoid spray if natural enemies like predatory mites, coccinellid beetles, green lace wing, anthocorid bugs and predatory thrips are active and found in good number. Avoid repetition of same chemical for the control of mites in subsequent sprays.

B. Borer pests

These insects encompass many species and families of beetles whose larval or adult forms eat and destroy wood. Insect borers are now emerging as major pests of horticultural crops and affect almost all types of fruit tree like cherry, apple, peach, apricot, walnut, grapes, pomegranate and plum. These can be grouped into three prominent categories viz., stem borers, root borers and fruit borers.

1. Stem borer, *Apriona cinerea* Chev. (Cerambycidae: Coleoptera)

Distribution

Common in the North-West Himalayas and the adjoining plain regions.

Hosts of commercial importance

Apple, peach, poplar, fig, mulberry etc.

Identification and biology

It is a polyphagous pest. The grubs are creamy yellow with dark brown, flat head. Adult beetles are ash grey to greyish yellow in colour. The most characteristic feature of adult is their long antennae, which in males can be more than twice the body length.

Adults are ashy grey to greyish yellow with numerous black tubercles at the base of elytra. During July-August, females excavate oval patches on the shoots and lay eggs inside. Incubation period lasts for about 7-9 days. After hatching, grubs bore inside the stem. Grubs are creamy white colour and elongate in shape. Each grub makes 8-10 circular holes on the stem or branches. The grubs continue to feed till October, diapause till March and again resume feeding and reach tree trunk by September-October. Then, they again undergo to diapause during winter and become pupae. Adults emerge out in summer. The life cycle is completed in about 2 years (Dhiman and Arif, 2006 and Reddy, 2009).

Mode of feeding symptoms of infestation

Larvae bore into shoot, feed under the bark and cut circular holes on bark at 10-15 cm distance on branches. Finally, these reach the trunk. Frass comes out of the holes at the entrance. The affected trees are fragile and often get broken even with moderate wind.

Seasonal incidence

The pest remains active from August-November.

Management

i. Cultural

a. The stem borer can be controlled by pruning and destruction of all attacked branches and killing the older larvae by inserting flexible wire into the live tunnel.

b. Heavily infested trees should be uprooted and removed from the orchard and burnt.

ii. Chemical

a. To avoid serious damage by stem borer clean the hole and then 0.5g of Paradichlorobenzene in the hole and plug it with peddling material or insert cotton wick soaked in petrol or Kerosene or Dichlorvos (0.15 ml/l) or Methyl parathion (1ml/l) and plug the holes with mud.

2. Shot or pin hole borer, *Scolytus* spp., *Xyleborus* spp. (Scolytidae: Coleoptera)

Distribution

Western Himalayas.

Hosts of commercial importance

Apple, almond, peach, pear, plum, cherry, walnut and apricot.

Identification and biology

The adult beetles are tiny, brown or black in colour, bark feeders and often attack weak and poorly managed trees. Adult females bore tiny holes in the bark and lay eggs in the cambium layer of the tree. When the eggs hatch, young larvae feed and excavate secondary galleries at right angles to the egg gallery. The outline of the gallery system resembles a centipede. There is one to three generations each year.

A female lays about 50-75. Larvae become full fed in about a month, pupate at the end of larval gallery and adults emerge within two weeks of pupation. There may be up to four overlapping generations, the last over wintering as larva or sometimes as adult, if adult emergence occurs by November (Reddy, 2009).

Mode of feeding symptoms of infestation

Adults are bark feeders and often attack weak and poorly managed trees. Adults and grubs tunnel into the sapwood and hardwood of the plant making galleries and pin holes and often reduce it to powder. Surface of infested branches get perforated followed by yellowing and wilting of leaves. A serious infestation may kill the whole tree.

Management

i. Cultural

a. Pruning and burning of borer infested branches during autumn.

b. The trees by application of balanced dose of fertilizer shall help to reduce the attack of shot hole borer.

c. Leave scattered fallen tree trunks and branches in the orchard for trapping egg laying scolytids during the month of June and burn them at the end of June or first week of July.

ii. Chemical

a. Installation of light traps with killing agents to collect and kill the beetles.

b. When severe attack appears, spraying of the trunk with Fenitrothion (0.05%) gives protection against this pest.

3. Flat headed borer, *Sphenoptera lafertei* Thom. (Buprestidae: Coleoptera)

Distribution

India, Kashmir, Pakistan, Iran.

Hosts of commercial importance

Cherry, plum, apple, peach etc.

Identification and biology

Adult beetles are flattened, hard, dark greenish bodied and boat shaped with short antennae. Larvae are cream-colored and legless with widened, flattened body segments just behind the heads.

Overwintering adult beetles or pre-pupae emerge, lay eggs singly on crevices of the trunk and main branches. Hatched larvae bore into the bark and resinous gum oozes out from the entry hole. Grown up larvae penetrate deep into the sapwood. Larval duration varies from 2 months in summer to 6 months in winter. Pupal period lasts for about 8-12 days. Depending on altitude, there may be 2-4 overlapping generations in a year.

Mode of feeding and symptoms of infestation

The larvae feed beneath bark or into the sapwood and they produce oval or flattened tunnels in cross section. Galleries are often winding and packed with frass.

Management

i. Proper training and pruning of the tree to avoid direct exposure of the stem to the Sun and covering the trunk with craft paper or muslin cloth or dry grass provide protection from the borer.

ii. Maintaining tree vigour is very important. In case of localized infestation, spray the infested tree with Fenitrothion (0.05%).

4. Root borer, *Dorysthenes hugelii* Redt. (Cerambycidae: Coleoptera)

Distribution

Jammu and Kashmir, Uttarakhand and some other states.

Hosts of commercial importance

Apple, apricot, cherry, peach, pear, plum, walnut and some forest trees.

Identification and biology

Adults are of chestnut red colour with head and thorax darker than elytra. The pest is a serious problem in fruits and forest trees in Jammu and Kashmir, Uttarakhand and some other states.

Adult lay on an average 300 eggs singly or in small cluster at about 8-12 mm depth in soil during June –July. Eggs hatches in 3-7 weeks. Larvae bore and feed on the roots and become full grown in about three and half years. Pupation takes place in earthen cocoon inside the soil and the pupal period lasts for about three months. The life cycle may complete in about 4 years.

Mode of feeding and symptoms of infestation

It attacks on the living and dead roots of fruit tree. The grubs either bore into or girdle around the roots, branches wither, tree become shaky and death may occur within three years. Young and small trees suffer the most, lower leaves usually turn yellow and cracks may result in the stem.

Management

i. Cultural

a. Avoiding dry sandy soils for plantation of new orchard.

b. Use of well rotten farm yard manures.

c. Marking the infested trees in July and digging their root system in winters to collect and destroy the grubs manually.

ii. Chemical

a. Drenching the tree basin with 0.04% Chlorpyriphos in July against neonate grubs. The roots of infested trees can be treated with Chlorpyriphos 0.06% during November-February (Singh, *et al*, 2010).

b. Soil application of *Metarhizium anisopliae* was found effective against the root borer that reduced infestation by 70% in Himachal Pradesh.

5. Codling moth, *Cydia pomonella* (Linnaeus) (Tortricidae: Lepidoptera)

Distribution

Widely distributed throughout Europe, North America and Australia, Baluchistan (Pakistan). In India, it is confined to Ladakh region (Leh and Kargil district) of Jammu and Kashmir.

Hosts of commercial importance

Apple, apricot, peach, pear, quince, walnut, citrus etc.

Identification and biology

Cydia pomonella is a small greyish moth with chocolate brown patches near the tip of forewings. The larvae are pinkish-white caterpillars with black or mottled black head.

The first adults begin to emerge in late April/early May and can be seen flying around trees from May through July. They lay eggs immediately, usually on leaves. Once these larvae hatch, mature, and pupate, a new generation of adults can emerge in late summer (mid-July to early September). There are typically two generations of adults per year.

Mode of feeding and symptoms of infestation

Codling moth is a serious pest and about 30-70% fruits are rendered unmarketable by this pest. Eggs are laid singly on leaves, blossoms and fruits. The freshly hatched caterpillars feed on leaves for a while, then burrow inside the fruits and feed on the pulp. The entry holes become quite conspicuous as these are filled with dry brown frass and are surrounded by a dark reddish ring. The infested apples become brighter in colour than those that are not infested and also ripe prematurely.

Management

i. Cultural

a. Strict domestic quarantine has to be followed by screening of consignments of fruits to prevent the spread of the insect from Ladakh to other apple growing regions.

b. All loose bark of apple tree should be scraped off to remove overwintering sites for the caterpillars.

c. On a small scale, corrugated card board 10 cm wide to be placed around the tree trunk in July will enable fully grown larva to be collected and destroyed.

d. Mechanical collection and destruction of cocoon and deep burying of fallen fruits will reduce the infestation.

ii. Chemical

a. Spraying Chlorpyriphos (0.02%) or Ethion (0.05) at time of complete petal fall stage when the moths start emerging. Repeat the spray after 2-3 weeks if require.

b. Application of 0.2% Pyrethrum extract is also helpful in checking the pest infestation. The protective treatment may be applied about ten days before ripening of the fruits.

c. Use sticky traps (Delta traps) baited with the synthetic codling moth pheromone/ lure to monitor the flight of male moth as an aid to the timing of chemical sprays.

iii. Biological

a. Weekly release of egg parasitoids, *Trichogramma embryophagum* and *T. caeoeciae pallidum* at the rate of 20,000 adults/50 trees/ week should be undertaken from first fortnight of June to end of August.

b. Birds, especially tits feed on overwintering caterpillars also important for causing natural mortality.

6. Cherry fruit fly, *Rhagoletis indifferens* Curran. (Tephritidae: Diptera)

Distribution

India, United States, Switzerland, Germany and also in South-eastern and South-central Canada.

Hosts of commercial importance

Cherry

Identification and biology

The adult cherry fruit fly is slightly smaller than house fly and about 1/5 inch long, black body with yellow markings near base of wings and white strips across abdomen. Wings have black markings that differentiate it from the other related fruit flies. The maggots are white, legless with no distinct head and about 5-6 mm long when mature.

Adult flies mate shortly after they emerge and begin laying eggs about 5 to 10 days later. One female can lay between 50 to 200 eggs. After hatching the legless maggots feed for 1-2 weeks before cutting exit holes and dropping to the ground to pupate. These insects spend nearly 85% of their life (nearly 10 months) buried in the soil waiting for spring to arrive. Only one generation is produced each year.

Mode of feeding and symptoms of infestation

The larvae (maggots) feed in the flesh near the tip of cherries rendering them unmarketable. These flies also feed on sticky deposits of honeydew, plant pollen and sometimes enjoy those little delicacies in life like bird droppings.

Management

i. Cultural

a. Destruction of infected cherries before the larvae emerge.

b. Removal of all cherries at harvest will eliminate breeding sources and reduce fruit fly numbers for next season.

c. Destroy any unmanaged trees where possible.

ii. Biological

a. Place Methyl eugenol sex lure trap in the orchard @10 per ha to capture male flies.

iii. Chemical

a. Use yellow sticky cards to detect when the first fruit fly emerges and apply the first spray within a week of the trap catch.

b. Prepare bait with Methyl eugenol 1% mixed with Dichlorvos 0.1% and take 10 ml of this mixture per trap and keep these in 10 different places in one hectare.

c. Racking up soil below the tree and drenching with Chlorpyriphos 20 EC @ 2.5 ml/L to kill the pupa.

d. Bait spray combination Dimethoate or Spinosad with protein hydrolysate or molasses or jaggery at two weeks interval is recommended.

7. Anar butterfly, *Deudorix epijarbas* Moore (Lycaenidae: Lepidoptera)

Distribution

It is a serious pest in North India especially Himachal Pradesh and Jammu & Kashmir.

Hosts of commercial importance

Besides pomegranate, the pest also damages guava, amla, annona, apple, ber, citrus, litchi, loquat, sapota, mulberry, peach, pear, plum etc.

Identification and biology

The males are slightly smaller, glossy, bluish violet and the females are brownish violet and have conspicuous patches on fore wings (Butani, 1976). Females are dark brown and lay eggs singly on the calyx of flower/fruits. The larvae bore into the fruit of pomegranate and can destroy up to 50% of fruits. The female butterfly lays eggs singly on calyx of flowers or small fruits. On hatching, the caterpillars bore inside the developing fruits and feed on pulp and seeds. This will allow the entry of fungi and bacteria causing fruit rot.

Female butterflies lay eggs in the day time single on the calyx of flower or on fruits which are shiny white and round in shape and hatch in 7-10 days depending upon the season. Caterpillars feed on the seed and fill the fruit with their excreta. Full grown caterpillars are dark brown with short hairs and white patches all over the body and measure 17-20 mm in length. Larval period extends from 18-47 days. Pupation takes place outside the fruits and lasts 7-34 days. The pest breeds throughout the year on various host plants. Four generations have been reported in a year. Adult butterflies have 40-50mm wingspan.

Mode of feeding and symptoms of infestation

The conspicuous symptoms of damage are offensive smell and excreta of caterpillars coming out of the entry holes. The affected fruits rot, get dried and ultimately fall down.

Management

i. Removal and destruction of all the affected fruits (fruits with exit holes).

ii. Spray Decamethrin 2.8 EC at 1ml/l at the time when more than 50% of fruits have set. Repeat after two weeks with Carbaryl 50 WP at 4g/l or Fenvalerate 20 EC at 0.25 ml/l or Quinalphos 25EC at 2ml/l in non-rainy season.

iii. Kakar, *et al.* (1987) found that Cypermethrin at 50g a.i./ha, Fenvalerate at 50 g a.i./ha and Deltamethrin at 7.5 g a.i./ha are effective.

iv. Two sprays of Cypermethrin or Deltamethrin in July followed by a cover spray after 45 days gave effective and profitable control. Remove flowering weeds especially of Compositae family.

v. After fruit setting, cover the fruit with paper or cloth bags. The release of egg parasitoid, *T. chilonis* at 2.5 lakh/ha 4 times at 10 days interval reduces the fruit borer infestation up to 50% (Kakar and Sharma, 1988).

vi. The egg parasitoids namely, *Telenomus* sp., *Ooencyrtus papilionis* and *Trichogramma chilotraeae* are known to cause up to 60% parasitism on *Deudorix isocrates* in Peninsular India (Mani and Krishnamoorthy, 1999).

vii. Weekly spraying of *B. thuringiensis* at 1g/l is effective. Spray application of 5% NSKE gave satisfactory control of fruit borer.

C. Defoliators

These pests belong to various insect orders and are mostly responsible for defoliating the tree by eating leaves. Different fruit trees vary in their tolerance to defoliation. Healthy and vigorously growing trees generally survive impacts of defoliators. Some of the most important defoliators of temperate horticultural crops are described in brief here.

1. Indian gypsy moth, *Lymantria obfuscate* Walker (Lymantriidae: Lepidoptera)

Distribution

It is a pest of temperate fruits and deciduous forest trees in Western Himalayas including Kashmir.

Hosts of commercial importance

Apple, apricot, cherry, walnut, peach, plum etc.

Identification and biology

The female has vestigial wings and cannot fly but crawl whereas the male has well developed wings and are swift fliers. The caterpillars are clothed by the tuft of hairs.

Female moth lay eggs on the bark of the tree. Eggs are round, shining and light greyish brown. Larval period varies from 37 to 60 days. Pupal period ranges from 12 to 14 days. Adult longevity of male and female is about 3.5 to 6 days and 7 to 9 days respectively. Thus, the life cycle of *Lymentria obfuscata* is univoltine, having one generation in a year.

Mode of feeding and symptoms of infestation

Defoliation of leaves occurs due to gregarious larval feeding during night time, which results in failure of fruit formation.

Management

i. Cultural

a. Collection and destruction of egg masses and larvae trapped in gunny burlaps tied on tree trunks near ground level.

ii. Behavioural

a. Masoodi *et al.* (1990) reported that the removal of a substantial proportion of males by pheromone traps (disparlure) resulted in a significant suppression of mating.

iii. Chemical

a. Spraying of 0.03% Fenvalerate at time of hatching of caterpillars from the eggs gives effective control (Gupta, 2004).

2. Tent caterpillar, *Malacosoma indicum* Walker (Lasiocampidae: Lepidoptera)

Distribution

India, North America and Eurasia.

Hosts of commercial importance

Apple, pear, apricot, walnut etc.

Identification and biology

Adults are medium sized insects. They are often considered pests due to their habit of defoliating trees. The larvae of this species are notorious and do not make tents, but rather, weave a silky sheet where they lie together during molting. Tent caterpillars are readily recognized because they are social, colorful, diurnal and build conspicuous silk tents in the branches of host trees. Some species, such as the eastern tent caterpillar, *Malacosoma americanum,* build a single large tent which is typically occupied through the whole of larval stage, while others build a series of small tents that are sequentially abandoned. Whereas tent caterpillars make their tents in the nodes and branches of a tree›s limbs, webworms enclose leaves and small branches at the ends of the limbs.

The adult tent caterpillar moth lays its eggs in mid-summer and early autumn. The eggs are deposited in dark brown saddle-like cases which are 2 cm long containing 150 to 350 eggs. In the spring, the eggs hatch into young caterpillar larvae that make communal tent webs, usually in tree or branch forks. The caterpillars mature in four to six weeks, reaching a length of about 2 to 3 cm. In June to early July, the caterpillars enter the pupal stage of development, encasing themselves in cocoons. After about 10 days, the adult moth emerges and mates within 24 hours. The female immediately begins to lay eggs for the next spring, producing only one generation of tent caterpillars every year.

Mode of feeding and symptoms of infestation

The caterpillar feed gregariously on foliage, leaving only the midrib and other harder veins. In case of severe infestation, the entire plant may be defoliated and subsequently the caterpillar may feed even on the soft bark of twigs. The pest is active from mid-March to May and passes the remaining nine months of year as diapausing larva (Joshi and Aggarwal, 1987).

Management

i. The infestation can be minimized to some extent by collection and destruction of conspicuous egg clusters during pruning in December-January.

ii. The menace of tent caterpillar can be checked by dipping up the tents with a pole and some rags dipped in kerosene tied on its end. The best results are obtained when the operation is carried out from 12 noon to 3 p.m. on clear sunny days.

iii. Kerosenized water in an open vessel should be placed below the tree so that larvae that fall may also be killed readily (Atwal and Dhaliwal, 1997).

iv. During severe infestation, Nimbecidine 0.5% or *Bacillus thuringiensis* based bio insecticides can be sprayed (Singh and Pandey, 2004).

3. Chaffer beetle/ White grub, *Holotrichia* spp., *Anomola* spp. (Coleoptera: Scarabaeidae)

Distribution

India (Andhra Pradesh, Karnataka, Tamil Nadu, Western Ghats, Kashmir) Pakistan, United States and Canada.

Hosts of commercial importance

Grape, raspberry and blackberry over many other plants, including blueberry.

Identification and biology

Adults are shiny smooth beetles, which are adapted to feed on leaves of many different plant species. Adult beetles feed on the foliage and fruits of about 300 plant species, including fruits, field and garden crops, ornamentals and many wild plants.

Eggs are laid in soil and larvae live inside soil feeding and undergoing development for 3-4 months. Pupation too will take place in soil. Total life cycle takes about 150 – 180 days.

Mode of feeding and symptoms of infestation

Adult beetles skeletonize the foliage, consuming the soft mesophyll tissues between the veins, leaving a lacelike skeleton. Beetles cluster at the tops of plants, so the upper canopy is usually defoliated more severely. In many fruit crops, foliar damage is unlikely to affect yield. Once fruit becomes ripe, beetles may feed and aggregate on ripening fruit. The beetles feed during night and remain hidden during day time. In nursery it may cause severe damage, eating away the foliage and thus resulting in drying and death of nursery plants. The grubs of chaffer beetles also cause damage by feeding on the roots of the plants inside the soil. Grubs cause more damage in the nursery fields where complete wilting and drying of the plants may result.

Seasonal incidence

The pest remains active mostly from April-September.

Management

i. Cultural

a. Control measures should be taken on community basis. Install a light trap near the orchard to collect and kill the beetles in kerosenized water.

b. Autumn ploughing helps in reduction of the pest by exposing the grub stages of this pest to predators like birds etc.

ii. Biological and behavioural

a. Local isolates of entomopathogenic fungi namely *Beauveria bassiana* and *Metarhizium anisopliae* @ 10^8 spores/ml was found effective.

b. Pheromone dispenser be placed (3-4 dispensers/ tree) on the host tree in the evening, continuously for three evenings at the time of beetle emergence. Well rotten FYM should be used.

iii. Chemical

a. Spraying of Imidacloprid 17.8 SL 0.027% or Carbaryl 0.2% or Quinalphos 0.05% within three days of first pre-monsoon rain, with the objective to kill the beetle before they lay the eggs in the soil.

b. Insecticides like Fipronil 0.3G 12 kg/ha should be applied in the soil for the grub stages of this pest in the orchard as well as in the nursery.

References

Atwal, A. S. and Dhaliwl, G. S. 1997. Pests of temperate fruits *In: Agricultural pests of south Asia and their management.* Kalyani publishers, Ludhiana, pp-274-286.

Bhardwaj, S. P. 1988. Effect of summer application of insecticides on San Jose scale, *Quadraspidiotus perniciosus*, in orchards of apple (*Malus pumila*). Indian Journal of Agricultural Sciences, 58(8):655-656.

Bhardwaj, S. P. 2004. Red spider mite: A major threat to apple production. *In: Recent trends in horticulture in the Himalayas* (K.K. Jindal and R.C. Sharma (eds.)), Indus publishing Co., new Delhi, pp-293-299.

Blackman, R. L. and Eastrop, V. F. 1994. Aphids on the World›s Trees. An Identification and Information Guide. Wallingford: CAB International. ISBN 0-85198-877-6.

Butani, D. K. 1976. Insect pests of fruit and their control-21: Pomegranate. Pesticides 10(6):23-26.

Chan, C. K., Forbes, A. R. and Raworth, A. 1991. Aphid-transmitted viruses and their vectors of the world. Agric. Canada Res. Branch Tech. Bull., pp-216.

Ciampolini, M. and Martelli, M. 1980. Appearance in Italy of the peach trunk aphid *Pterochloroides persicae* (Cholodk.). Bollettino di Zoologia Agraria e di Bachicoltura,14:189196.

Darwish, E. T. E., Attia, M. B. and Kolaib, M. O. 1989. Biology and seasonal activity of giant brown bark aphid *Pterochloroides persicae* (Cholodk.) on peach trees in Egypt. Journal of Applied Entomology, 107:530-533.

Dhiman, S. C and Arif, M. 2006. Insect pests of temperate fruit crops. *In: Insect pests of fruit crops, vegetables, spice and condiments and their management, Vol. 3* (Ananad Prakash, Jagadiswari Rao and V. Nandagopal (eds.)), AZRA, CRRI, Cuttack, India.

ElSalam, S. A. A. 2001. Toxicity and biochemical effects of some safe alternative materials against *Pterochloroides persicae* Chlod. Individuals (Order Homoptera: Fam. Aphididae) on peach trees. Egyptian Journal of Agricultural Research, 79:13871397.

Gupta, P. R. 2004. Integrated pest management in temperate fruits. *In: Recent trends in horticulture in the Himalayas* (K. K. Jindal and R. C. Sharma (eds.)), Indus publishing Co., New Delhi, pp-280-292.

Gupta, P. R., Thakur, J. R. and Dogra, G. S. 1988. Evaluation of some dormant miscible oils applied during the late dormancy against San Jose scale, *Quadraspidiotus perniciosus* (Comstock) on apple. (abstr.). National Symp. on Tree protection, Solan.

Hondru, N., Margarit, G. and Popa, I. 1986. A new aphid pest of fruit orchards *Pterochloroides persicae* Cholodkovsky (*Homoptera Aphidina Lachnidae*). Analele Institutului de Cercetări pentru Protecția Plantelor, 19:151154.

Joshi, K. C and Agarwal, S. B. 1987. Bionomics of the tent caterpillar, *Malacosoma indica* Wlk. Bulletin of Entomology, 28(1):1-11.

Kakar, D. K. and Sharma, J. P. 1988. Feasibility of *Trichogramma* spp. for the biological suppression of fruit borers of pomegranate. *National Symp. on Tree protection*, Dr. Y. S. Parmar Univ. Hort. & Forestry, Solan, Himachal Pradesh, India.

Kakar, K. L., Dogra, G. S and Nath, A. 1987. Incidence and control of pomegranate fruit borer, *Virachola isocrates* (Fabr.) and *Deudorix epijarbas* (Moore). Indian Journal of Agricultural Sciences, 57:749-752.

Khajuria, D. R and Sharma, J. P. 1996. Outbreak of phytophagous mites on apple in Kullu Valley of Himachal Pradesh. Indian Journal of Plant Protection, 24(1-2):134-138.

Khajuria, D. R. 2009. Predatory complex of phytophagous mites and their role in integrated pest management in apple orchard. Journal of Biopesticides, 2(2):141-144.

Khajuria, D. R and Sharma, H. K. 1997. Use of miscible oil and insecticidal combinations in management of San Jose scale (*Quadraspidiotus perniciosus*) on apple (*Malus domestica*). Indian Journal of Agricultural Sciences, 67(10):488-489.

Khan, A. N., Khan, I. A. and Poswal, M. A. and Ahmad, S. 1998. Mass rearing/culturing of brown peach aphid, *Pterochloroides persicae* (Kholodkvoskii) (Lachninae:Aphididae) under laboratory condition. Sarhad Journal of Agriculture, 14:355368.

Kumar, N. S and Ananthakrishnan, T. N. 1984. Predators-thrips interaction with reference to *Orius maxidentex* Ghauri and *Caryonocoris indicus* Muraleedharan (Anthocoridae: Heteroptera). Proc. Indian Nat. Sci. Acad. 50B:139-145.

Mani, M. and Krishnamoorthy, A. 1999. First record of *Trichogramma chilotraeae* Nagaraja and Nagarakatti on pomegranate butterfly, *Deudorix isocrates* (Fabr.) Entomon 24:254-259.

Masoodi, M. A, Trale, A.R and Bhat. A. M. 1990. Suppression of *Lymantria obfuscata* Walker by sex pheromone trapping of males. Indian Journal of Entomology, 52(3): 414-417.

Mdellel, L. and Kamel, B. M. 2012. Prey consumption efficiency and fecundity of the ladybird beetle,*Coccinella algerica* Kovàr (*Coleoptera: Coccinellidae*) feeding on the giant brown bark aphid, *Pterochloroides persicae* (Cholodkovsky) (*Hemiptera: Lachninae*). African Entomology, 20:292299.

Rather, N. A., Lone, P. A., Reshi, A. A. and Mir, M. M. 2013. An Analytical Study on Production and Export of Fresh and Dry Fruits in Jammu and Kashmir. International Journal of Scientific and Research Publications, 3(2):1-7.

Rawat, U. S and Pawar, A. D. 1992. Biocontrol of San Jose scale, *Quadraspidiotus perniciosus* (Comstock) by predatory beetle, *Chilocorus bijugus* Mulsant in Himachal Pradesh. Plant Protection Bulletin, Faridabad, 44(4):7-10.

Reddy, P. P. 2009. Temperate fruits. *In: Advances in Integrated pest and disease management in Horticultural crops Vol.1 Fruit crops*. Studium press (India) Pvt. Ltd. New Delhi, pp-1-347.

Reddy, P. P. 2010. Plant protection in Horticulture, Vol. 1. Scientific Publishers, New Delhi, India.

Sandhu, G. S. and Sohi, A. S. 1978. High volume and ultra low volume spraying for the control of peach stem aphid in Punjab, Pesticides, 12:1517.

Sharma, D. R. 2010. Bio efficacy of insecticides against peach leaf curl aphid, *Brachycaudus helichrysi* (Kaltenbach) in Punjab. Indian journal of entomology, 72(3):217-222.

Singh, M and Singh, M. 1987. Chemical control of blossom thrips in apple. *Indian Journal of Entomology*, 49(4):580-582.

Singh, S. P. 1997. Biological control in India. *In: IPM System in Agriculture* (Rajeev K Upadhyay, K.G. Mukerji and R.L. Rajak (eds.)), Aditya Books (P) Ltd., New Delhi.

Singh, S. S. and Pandey, J. C. 2004. Integrated pest management in temperate fruits. In: Apple Souvenir, Lal, H., Singh, S.S and Singh, V.P. (eds.) published by Dept. of Horticulture on apple show held at Nanital from 27-28 August, pp-46-50.

Singh, S. S., Gupta, N. N and Rai, M. K. 2010. Pest management in temperate fruits: opportunities and challenges. Progressive Horticulture, 42(1):76-83.

Stoetzel, M. B. 1994. Aphids (Homoptera: Aphididae) of potential importance on Citrus In the United States with illustrated keys to species. Proc. Entomol. Soc. Washington, 96:74-90.

Velimirovic, V. A. 1977. Contribution to the study of peach aphid *Pterochloroides persicae*. Zastita Bilja, 28(1):37.

Wieczorek, K., Kanturski, M. and Junkiert, L. 2013. The sexuales of giant black bark aphid, *Pterochloroides persicae* (*Hemiptera, Aphidoidea: Lachninae*). Zootaxa, 3626(1):94-98.

Bhardwaj, S. P. 2004. Red spider mite: A major threat to apple production. *In: Recent trends in horticulture in the Himalayas* (K.K. Jindal and R.C. Sharma (eds.)), Indus publishing Co., new Delhi, pp-293-299.

Dhiman, S. C and Arif, M. 2006. Insect pests of temperate fruit crops. *In: Insect pests of fruit crops, vegetables, spice and condiments and their management, Vol. 3* (Anand Prakash, Jagadiswari Rao and V. Nandagopal (eds.)), AZRA, CRRI, Cuttack, India.

Reddy, P. P. 2009. Temperate fruits. *In: Advances in Integrated pest and disease management in Horticultural crops Vol.1 Fruit crops*. Studium Press (India) Pvt. Ltd., New Delhi, pp-1-347.

18

Mite Pests of Fruit Crops and Their Management

Pijush Kanti Sarkar and Debashis Roy

Perennial nature of horticultural fruit crops provides a stable environment for the buildup and development of mite populations, as against the ephemeral habitat of short-lived plants. The phytophagous mites are represented by 169 species whereas 148 species belong to the predatory group. Mites (subclass Acari) are ecologically and morphologically very diverse group of tiny invertebrates that belong to the class Arachnida (along with scorpions and spiders), subphylum Chelicerata under Arthropods. As arthropods also include insects, but mites are differ because they have four pair of legs (insects are hexapods), lack of true head and conspicuous segmentation of the body. Though about 50,000 mite species has been identified till date, but it is estimated that the true number may be almost 20 times higher. Some mites act as agricultural pests, some are natural enemies (predators), some have medical and veterinary importance (scabies mites, house dust mites, ticks), while the species that lives in water and soil are important indicators of environment.

In India 12 species of phytophagous mites are considered as major pests and 12 as minor pests on different crops (Gupta, 1991 and Singh *et al.*, 2000). Out of different groups of mites, the members represented in four families, *i.e.*, Tetranychidae (spider mites, >75 species), Tenuipalpidae (false spider mites, >29 species), Tarsonemidae (yellow mites, >4 species) and Eriophyidae (gall mites, >64 species) are of major phytophagous importance. Adaptation of advanced agricultural practices, including exhaustive usage of fertilizers, pesticides, irrigation and mono cropping and decreased genetic diversity has further aggravated the problem.

The plant feeding mites cause injuries by scrapping of the leaf chlorophyll resulting in bronzy appearance of the foliage, premature defoliation and stunted growth of the plant. Certain species of mites also cause various types of feeding deformities like galls, crinkling, erenium and curling which finally reduce the yield drastically. Apart from direct damages some species are also known to transmit certain viral diseases of plants. The major mite species attacking the fruit crops are listed below-

Table 1: Important mite pests of fruit crops inIndia

Sl. No.	Mite Species(Common Name)	Major host plants
Family- Tetranychidae		
1.	*Bryobia praetiosa* Koch (Brown clover mite)	Pear
2.	*Eutetranychus africanus* Tucker	Pear, Peach,
3.	*Eutetranychus anneckei* Meyer (Lowveld citrus mite)	Pear, Peach, Citrus, Orange, Jackfruit, Mulberry
4.	*Eutetranychus orientalis* Klein (Oriental citrus mite)	Pear, Peach,
5.	*Eotetranychus hirsti* Pritchard and Baker (Fig spider mite)	Fig
6.	*Oligonychus mangiferus* Rahman and Sapra (Mango red spider mite)	Mango
7.	*Panonychus citri* Mcgregor (Citrus red mite)	Apple, Pear, citrus
8.	*Panonychus ulmi* Koch (European red mite)	Pear, Apple
9.	*Tetranychus urticae* Koch (Two spotted spider mite)	Pear, Peach, Apple
Family- Tenuipalpidae		
10.	*Brevipalpus californicus* Banks (Citrus flat mite)	Citrus
11.	*Larvacarus transitans* Ewing (Ber.gall mite)	Ber (*Zyziphus mauritiana*)
12.	*Dolichotetranychus floridanus* Banks (Pineapple flat mite)	Pineapple
13.	*Raoiella indica* Hirst (Coconut red mite)	Coconut
14.	*Tenuipalpus ludhianaenis* Sadana and Chhabra	Pear
15.	*Tenuipalpus pruni* Maninder and Ghai	Pear, Peach
16.	*Tenuipalpus pyrusae* Maninder and Ghai	Pear, Peach, Guava
17.	*Brevipalpus phoenicis* Geijskes (Scarlet mite/ Reddish black flat mite/Leprosis mite)	Guava
Family- Eriophyidae		
18.	*Aceria mangiferae* Sayed (Mango bud mite)	Mango
19.	*Aceria litchi* Kiefer (Litchi erineum mite)	Litchi
20.	*Aceria guerreronis* Keifer (Coconut eriophyid mite)	Coconut, Palmyra palm
21.	*Eriophyes cernuus* Massee [Jujube (Ber) gall mite]	Jujube (ber)
22.	*Epitrimerus pyri* Nalepa (Pear bud mite)	Pear
23.	*Phytoptus pyri* Pagenstecher (Pear leaf blister mite)	Pear
24.	*Aculus cornutus* Banks (Peach silver mite)	Pear, Peach, Almonds
25.	*Aceria mangiferae* Sayed (Mango bud mite)	Mango

A. Spider Mites (Tetranychidae)

Tetranychid mites or spider mites feed on the plant sap causing considerable economic losses to the crops. They attack many fruit crops of economic importance, along with vegetables and grain crops. Some of these mites have acquired the destructive pest status in India and many other parts of the world, and became great concern to Indian agriculture. These mites generally prefer to feed on leaves but with their increasing population and overcrowding, they migrate to other infested plant parts. Generally, they prefer undersides of leaves as feeding site, but in severe infestation they may occur on both leaf surfaces as well as on the stems and fruits. During feeding they injure the epidermis with their needle like chelicerae to suck cell sap. Their feeding activity deprives the leaves of their chlorophyll content and this leads to stippling, scars, blotches or bronzing of leaves. Thus the mite attack reduces the luster of leaves along with speckles followed by drying, loss of turgour pressure followed by wilting and subsequently death of leaves and defoliation. Severe infestation may lead to reduction of plant vigour which in turn results in to adverse effect on flowering, fruit setting vis-à-vis the quality of fruit.

Further, the dust entangled in the web and silken strand formed by tetranychid mites on leaves prevent the sunlight reaching the leaf surface and as result photosynthesis is impaired, the leaves turn pale and drop down. This affects the vitality of plant. Hot, dry conditions are often associated with population build-up of spider mites. Under optimal conditions (approximately 27°C), the two-spotted spider mite can hatch in as little as 3 days, and become sexually mature in 5 days. One female can lay up to 20 eggs per day and can live for 2 to 4 weeks, laying hundreds of eggs. Fertilized eggs produce diploid females. Unmated as well as unfertilized females lay eggs which will originate exclusively haploid males. The accelerated reproductive rate allows spider mite populations to adapt quickly to resist pesticidal toxicity, which results in rendering the chemical control methods somewhat ineffectual when the same chemical is used over a prolonged period of time.

1. Two spotted spider mite, *Tetranychus urticae* Koch

The two-spotted spider mite, *Tetranychus urticae* Koch is a cosmopolitan polyphagous species. They spin a canopy of webs, which can be present in large quantities in dense colonies. All life stages live beneath this canopy as a protection against a number of natural enemies. There are several other species of *Tetranychus*, very similar in appearance and habits, which are difficult to distinguish from two-spotted spider mite. The summer form of the adult female is approximately 0.65 mm long, rotund oval in shape, and covered in fine hairs. The body is pale yellow, greenish or straw coloured, with two characteristic lateral dark green or black spots, easily distinguishable from the European red mite and *Bryobia* mite, which are red.

The adult male is slightly smaller than the female and has a narrower, more pointed abdomen, which makes the legs appear longer. In autumn, females become an orange or brick red colour and lose their characteristic dark body spots, although they retain the bright red eyespots. Overwintering females usually occur in clusters under the bark of the tree, under dry leaves and in other protected sites. The minute, spherical, translucent eggs are about 0.14 mm in diameter which turn pale yellow as they reach in advance age and the red eyespots of the developing mite are visible just before hatching.

The larvae grow and moult through two nymphal stages, the protonymph and deutonymph, which are successively larger and have four pairs of legs. The nymphs are also pale yellow, pale green or straw coloured and have two dark spots on the body which is major characteristic of this species. Between each stage the nymphs enter a quiescent phase which proceeds moulting, and during this time they can appear silvery in colour. As winter approaches, the normal colouring of these stages can include pale orange if they are developing into over wintering females. Peak injury to trees typically occurs in late February through mid-June when weather conditions are hot and dry. Mite infestations often begin as "hot spots" within an orchard that spread to neighboring trees if weather conditions are favorable and predatory mites are unable to suppress the infestation.

Two spotted spider mites damage plants by piercing and sucking out the cell sap, which causes collapsing and dying of cells. Pale coloured spots appear on the upper surface of the leaves. In case of heavy infestations, the mites remove nearly all the chlorophyll, giving leaves a silvered appearance, which turns to brown as the leaves deteriorate. The feeding by two spotted spider mites results into collapsing of the mesophyll tissue, followed by formation of small chlorotic spots in the feeding site. They can destroy 18 to 22 cells per minute. Continuous feeding results into stippled-bleached effect and afterwards, the leaves turn yellow followed by greyish or bronzy in appearance. Complete defoliation may be encountered in case of severe uncontrolled infestation. The silk like webbing is seen over the damages and dying leaves and, as they dry out and falls, the mites move away to feed on healthy foliage. Females can disperse from heavily infested trees on threads of webbing and may drift or are blown on to other plants. Defoliation of heavily infested trees decreases their vigour and leads to a reduced or damaged crop. Overwintering orange females may move onto apples, particularly at the calyx and stem ends. This can reduce the market acceptability, especially when large numbers of mites are present.

2. Brown clover mite, *Bryobia praetiosa* Koch

Brown clover mite or Brown mite, is phytophagous in nature which had been described during 1950's as pest on different horticultural plants. This species is dark red-brown in colour and is one of the largest plant-infesting mites extending up to 0.75 mm in length. The front legs are longer than the others and extend in front of the body, appearing similar to antennae. Feather like plates are present on the body. Brown mites become active in the spring when temperature ranges from 14–24°C. They rupture the cells of the host leaf tissue with their needle-like stylets and feed on the sap. Feeding injury appears as fine flecking or stippling on the foliage.

3. Fig Spider Mite, *Eotetranychus hirsti* Pritchard and Baker

Eotetranychus hirsti is a serious pest of fig and reported from India and Pakistan (Baradaran *et al.* 2002).This mite has been recorded from Delhi, West Bengal,

Punjab and Uttar Pradesh. They are host specific, and infest exclusively fig. It is greenish yellow in colour with black blotches all over dorsum. Due to their feeding, green transparent patches can be seen on the under surface of leaves across light. Afterwards, the patches turn yellowish green, later becoming brown, rough and dry. Infested leaves and the fruits drop prematurely, resulting in complete defoliation of plants. The mite normally colonizes under surface of the leaves but in case of heavy infestation, they may move to the upper surface of the leaves and fruits as well. The infested fruits get shrivelled, fail to ripe and ultimately fall off. During November, the mites migrate to the terminal buds where they hibernate till February or early March.

4. Mango Red Spider Mite, *Oligonychus mangiferus* Rahman and Sapra

It is an important pest of mango distributed widely in tropics and subtropics. It can also attack cotton and is the second most serious pest of pomegranate in Egypt (Jeppson *et al.,* 1975). This mite is dark red dorsolaterally and pale red medio dorsally and infests the upper surface of the leaves. The females are dark red with darker spots on the (dorsam) hysteronotum, measuring around 0.35-0.45 mm. The body of the male is dark red and almost 0.35 mm in length. A female produces around 20-35 eggs, life cycle is completed in a fortnight, and requires 190 day degrees for development. This mite lives and feeds mostly on the upper leaf surface, located at bottom to middle of trees, where it spins delicate silk threads in which dust may gather. Mite feeding on mango leaves causes wilting, followed by reddening and finally dropping, thereby reducing the yield. Sucking of sap from leaves generally produces the appearance of characteristic yellowish spots. In case of mango the middle area of lamina becomes much yellowish with less browning, whereas in case of black berry, the attack is more near the margin.

5. Citrus red mite, *Panonychus citri* McGregor

Citrus red mite has a worldwide distribution on citrus (Meyer, 1981). This polyphagous pest feeds on variety of crops and plants throughout the world, and in India, another species *P. ulmi* has been found to infest peach, plum, apple, wheat, fig, hibiscus, tomato, apricot and ivy from Jammu and Kashmir region. The mite is dispersed by "ballooning", floating on air streams by holding on to a silken thread. The eggs, each with a prominent stalk, are placed along the main vein, usually on the upper side of young leaves. Pest populations are more numerous during early winter, least abundant in summer, but their reproduction is continuous, with no summer diapause. The mite is very susceptible to high temperatures and low humidities, causing a population decline during summer.

Feeding causes stippling spots that combine to produce yellow or silvery areas on the leaves. Due to heavy infestations, symptoms like leaf dropping, twig dieback, low-quality fruits are produced (Jeppson *et al.,* 1975). The expression of damage is enhanced by dry, hot winds. Further, this mite has also been reported as a serious pest of mulberry in the Terai zone of West Bengal, India. Their damage impairs the quality of mulberry leaves which in turn cause detrimental effects on the biological and economic parameters of silkworm. A recent investigation has suggested that in citrus-cultivating farmers *P. citri* is the most common causative allergen with symptoms like asthma or allergic rhinitis (Lee *et al.*, 2000).

6. Oriental spider mite, *Eutetranychus orientalis* Klein

The oriental spider mite can be found in eastern Mediterranean countries, from India to China and in almost all parts of Africa. This mite has been recorded from more than two hundred dicotyledonous host plants. The adult female is about 0.4-0.5 mm in length, brownish, with a oval body and legs that are about as long as the body. The male is reddish, smaller, its hysteronotum is pointed and the legs are longer than the body. The short dorsal setae are spatulate in shape and placed on small tubercles. The rounded eggs are pale-brown. The mites colonize the upper side of leaves where on their eggs are deposited along the midrib. 10°C is favourable for development and it requires approximately 15 days to complete the life cycle at around 25°C. Each female produces 30-40 eggs during two-three weeks, and almost 20 overlapping generations can be found.

Oriental mite is an important pest of citrus in certain dry parts of the world, as it produces important infestations on leaves and fruits, eventually causing a lack of pigmentation. In stressed trees, leaves can drop, and the branches can dry up. Higher densities are concentrated in areas that are most exposed to the sun. They attack the abaxial surface of leaf and feed voraciously, and reflection of their damage appears as yellow-grey stippled spots followed by leaf wilting and drop. Heavy infestations of this mites lead to fruit dropping followed by decaying of upper branches, which affects the blossoming in the next season. Reduced irrigation increases the damage.

B. Eriophyid Mites (Eriophyidae)

In the groups of phytophagous mites, the eriophyid mites are peculiar in several ways unlike other mites, they are elongate, vermiform and having only two pairs of legs both in the immature and adult stages. Most eriophyids average around 200 microns in length, worm like and with the body differentiated into an anterior short cephalothorax and annulated posterior abdomen. An anteriorly projecting or down curved rostrum with needle like chelicerae is characteristic of this group of mites. Body setation is reduced to the minimum. Eriophyid mites are exclusively plant parasites. Only perennial plants are considered to act as hosts to these mites. A few species, however, may infest annuals, but even in such cases the species concerned has other hosts which provide for its continuous feeding and breeding. Many species induce different kinds of malformation; some of which may appear like symptoms caused by fungi, bacteria and viruses. The common symptoms produced by this group of mites are the outgrowth of felt-like hairs (erineum) on the infested surface, outgrowth of closely-packed papillae on leaf surface, production of various kinds of galls, distortion of flower buds, induction of witches broom effect and others. Brooming of leaves and fruits may be caused by some of the vagrant species. Besides the direct damage caused by the feeding of the mites, at least six viruses are known to transmit by different species of eriophyid.

1. Mango bud mite, *Aceria mangiferae* Sayed

Eighteen species of eriophyid mango mites from different parts of the world are known, and mango bud mite, *Aceria mangiferae* Sayed and mango rust mite, *Metaculus mangiferae* Attiah are prevalent in mango orchards having detrimental effect on its yield. The most familiar symptoms caused by these mites are rusting, bud blasting, impedance of new growth, bud distortion and leaf chlorosis. If such type of infestation continues for few successive years, young trees become stunted and cannot develop normally. The mango bud mite, *Aceria mangiferae* Sayed, remains within the tightly apposed axillary buds moving into terminal buds. Reproduction occurs through arrhenotoky and life cycle (egg to egg) requires 2-3 weeks in summer, twice that much in winter.

It lives in the buds and inflorescences of the host, causing "cauliflower"-like growths. Mite feeding causes stunting and malformation of the buds, resulting in leaf drop and arrest the plant growth. Young trees are more heavily attacked than older ones. The mango malformation which is a serious malady of mango has been found associated with this mite. This malady has assumed an alarming magnitude in northern India and it is threatening the very existence of mango industry. It is widespread in Punjab, Delhi, Uttar Pradesh and also to some extent in Gujarat and Maharashtra. It causes two types of malformations *i.e.* the vegetative and floral malformation. Nursery seedling and young plants greatly suffer the vegetative malformation. The floral malformation is characterized by a condensed mass of flower buds. Both types of malformation may also occur in the same branch.

2. Coconut mite, *Aceria guerreronis* Keifer

The eriophyid mite, *Aceria guerreronis* Keifer was not known in India till 1984, and was first recorded from Srivilliputhur area of Tamil Nadu in 1989. In India, the mite attained a major pest status in the three peninsular states of India viz., Kerala, Karnataka and Tamil Nadu and is spreading throughout the country, and has drawn national attention as a threat to the coconut plantation.

On coconut trees, the mite populations develop on the meristematic zone of the fruits covered by the perianth (bracts) which is composed with several tepals. Feeding in this zone causes physical damage that leads to severe necrosis. The feeding injury by large number of mites results in development of brownish patches. Besides, damage to young nuts results in poor development of the nut, reduction in kernel content and poor quality husk. As the nut grows, the injury on the nuts leads to warting and longitudinal fissures on the nut surface. Mites can be found on coconut inflorescences during the dispersion process. Although the pest remains present throughout the year, the infestation becomes more severe in relatively dry climates or during the dry periods of wetter climates. It has been determined that the mite species can walk at a rate of about 22.5 cm in 30 min being negatively geotactic, and tends to move from older to younger inflorescences or upwards. It has been observed that dispersal involves mostly movement of inseminated females. Coconut trees provide a large target for aerially dispersing organisms, but the mortality associated with aerial dispersal is probably high. Coconut mite is an elongate wormlike, yellowish white eriophyid. Eggs are small, white, round to oval. Except for the sizes and presence of genital openings in adults, all developmental stages look similar. Plant nutrients exert pronounced effect on resistance to pests through host plant.

3. Litchi erineum mite, *Aceria litchi* Keifer

Aceria litchi Keifer is known as litchi erineum mite and is one of the most important pests of litchi in India, Pakistan, Bangladesh and Hawaii. The mite remains active on litchi leaves throughout the year (Thakur and Sharma,1990) and the population may increase according to favorable climate. However, lowest pest population may be recorded during December and highest in the month of May. These mites develop erineum on under side of the leaves. In the beginning the leaf looks light green than normal leaf but later turns light brown to deep or chocolate brown with velvety growth of tuft of hairs. The death of erineum is indicated by deep brown colour and then mites begin to abandon the leaf for newer plant parts to infect. The upper surface of affected leaves develops tumour like bumpy galls. The affected leaves eventually become distorted or curled which ultimately dry and fall off. Mite feeding also affects the growth of flower buds due to damage which is more pronounced during flush growth periods. Previously it was known that *A. litchi* is responsible for the formation of litchi erineum. But later several research works established that the hair like erineum forms due to infestation of an opportunist parasitic alga, *Cephaleuros virescens*, which takes entry though the puncture made by the mites during feeding and proliferate. It grooms in small yellow/green spots usually on ventral surface of the leaves, which turns reddish brown to chocolate brown at maturity.

4. Jujube (Ber) gall mite, *Eriophyes cernuus* Massee

This mite produces woody galls usually on one side of jujube stem, ringing the base of the developing buds or shoots. The mite population commences from September and reaches its peak in June, and size of the galls show similar trend as that of population load. Maximum population of the mite can be noticed on young galls. The galls are irregular, globose, lobed or tuberculate and hard in nature. The galls present on axillary branches often have the appearance of mass of cauliflower like surface, composed of undifferentiated parenchyma. The cell lining of the gall cavities form a nutritive layer, and the cells of this zone contain dense cytoplasm and big nuclei. Hypertrophy and hyperplasia of cells play an important role in the development of gall.

5. Pear rust mite, *Epitrimerus pyri* Nalepa

The mite causes injury to foliage and russeting of the fruits, and is found only on pear where it is almost invisible with the naked eyes. The dull white adults are pale brown with wedge shaped body which is larger at the head end than the posterior end. The two nymphal stages look similar in shape but vary in size. The round shaped eggs are colorless when first laid and extremely small (Howitt, 1999).The mite species overwinters as adult females beneath bud scales of leaf spurs and under loose bark of 1 to 2 year old twigs. With increment in ambient temperature in early April onwards, usually before buds break, mites move to developing clusters and start feeding on the succulent parts of buds. Eggs are produced shortly after mites become active.

As buds open, adults and immature move to the expanding leaf tissue and eventually move to fruits as the leaves mature and harden. Immature mites develop quickly through two instars, each followed by a resting stage. There are 4 or 5 generations completed in each growing season and each cycle take only 10-14 days under warm summer conditions. During June and July, all stages are commonly found together on the lower leaf surface and around the calyx end of fruit. By late summer, only females are produced and they soon move to protective overwintering phase. Cosmetic damage to the surface of fruit is especially important on clear-skinned cultivars of pear destined for the fresh fruit market. Depending upon the extent of russet, fruits can be downgraded or culled from fresh market sale, although the tolerance for cosmetic damage is dependent upon local market expectations and conditions

6. Pear leaf blister mite, *Phytoptus pyri* Pagenstecher

Pear leaf blister mite is most commonly encountered in the unsprayed trees in the orchard. The infestation of this mite is restricted to pear, apple and it causes up to 25% reduction in production. The mites are elongated worm like having average body length of 0.2 mm. While in the bud, mites are white; later in the season they assume a pinkish color, and overwinter under the buds. The mites begin to feed and oviposit as buds swell in the spring and blisters begin to form at the cluster bud stage, on leaves. After 1-2 weeks, blisters begin to form small holes through which mites enter into the holes and lay the eggs. Usually only one female enters in each blister. Each female lay around 7-21 eggs that require 18 days to complete the embryonic development. However, in warmer weather it requires only 5-8 days. There are 2-3 generations per year. Their activity ceases during June to August, and resumes with arrival of cooler climate. Mites return to buds from September to November. This complex causes blisters on the undersides of pear and apple leaves, especially on younger foliage, usually in a row along the mid vein. The blisters are tiny green swellings at first, later expanding and turning red. These blisters eventually turn necrotic and brown. The leaf epidermis becomes loose and wrinkled on the underside. Sometimes, the blisters may coalesce forming larger blistered areas along the mid vein. The leaf blade may turn yellow, leaving a dark band along the mid vein, dark green on the top and brown on the bottom of leaf. Small blisters may also occur on stems and around the fruit calyx. This may cause fruit drop, but this is usually less common than the foliar injury.

7. Peach silver mite, *Aculus cornutus* Banks

Aculus cornutus is a robust, light amber or yellow colored spindle shaped mite, measuring around 185-205 microns in length. The males and protogynes are distinguished by having strong dorsal micro tubercles and pair of small anterior spines on the dorsal shield. The overwintering females (deutogynes) lack anterior spines on the dorsal shield. The life history involves alternation of generation. The mites feed on both surfaces of the leaf blade. Besides peach, *A. cornutus* also attacks nectarine, almond and pear. Feeding of this mite produces the well-known silvering symptoms of peach foliage that develops late in the season just before the leaves drop. The mite damage also causes "yellow spot" characterized by yellow spotting and chlorosis alongside the leaf veins accompanied by upward longitudinal folding of the leaf margin. The size of spots ranges from point to more than a millimeter in diameter and are usual circular. In severe infestations, spots coalesce producing a mottled effect. Transverse sections of affected areas of the leaf blade show disorganization of cell contents and hypertrophy of parenchyma cells. Spots and chlorosis alongside veins appear to be the result of mite infestation. Mature leaves are not generally affected by this mite species.

C. False Spider Mites (Tenuipalpidae)

Mites of the family Tenuipalpidae have a worldwide distribution and can be seen on many cultivated plants. Several species have been reported, especially by growers of greenhouse ornamentals. These mites look similar as spider mites (family Tetranychidae) but can't spin web, and their body is rather flat compared to other mites. Mites belonging to the genera *Brevipalpus* and *Tenuipalpus* are severely detrimental for the cultivated plants. Most false spider mites are rust-red to brownish yellow in colour having similar body shape as spider mites but are usually smaller. They move sluggishly and are mostly found along the veins and veinlets on the underside of leaves. The eggs are clear red in colour and elliptical in shape. They are laid in a fold in the leaf or along the midvein, often in dense clusters of several hundreds.

1. Scarlet mite/reddish black flat mite, *Brevipalpus phoenicis* Geijskes

The adult male is flat, reddish and more wedge-shaped than the female. Eggs are deposited singly, in cracks, crevices and other protected areas on the plant surface. These clusters of bright reddish-orange eggs are more easily seen with the naked eye than any other life stage. Eggs have a stipe (tail-like projection) that extends from the slightly pointed end that comes out of the female mite last. This stipe often breaks off if the egg is handled. A day before hatching, the eggs become opaque white and the red eyes of the larva are visible within. There are two nymphal stages: the protonymph and the deutonymph. The deutonymph is similar in appearance to the protonymph except for having an extra pair of legs and two additional setae (hairs), and being slightly larger. This mite completes its life cycle in about 3 weeks. Larval, protonymph and deutonymph stages have active periods during which they feed, grow and disperse followed by an inactive stage in which they transform into the next stage of development. Feeding, growth and dispersal activity are similar in all stages becoming more pronounced with the passing of each stage (Haramoto, 1969).

This mite has an extensive host range and may cause economic damage depending on the host. *Brevipalpus phoenicis* is the most common false spider mite, and was reported for attacking over 65 host species. Feeding by this mite weakens the plant and causes mesophyll tissues to collapse. Damage is visible on both sides of the leaves. Damage is mostly characterized by browning of the damaged area, and some hosts may exhibit deformed leaves. Symptoms are more prevalent in the spring, summer, and fall. It is a serious pest on guava and heavy infestation has been recorded on guava in Punjab. Commonly this mite is found in the field with mixed population of *B. californicus* and *B. obovatus* and become very difficult to manage. A large population of this mite is found on under surface of leaves along the midrib. In case of guava brownish patches appear on fruits. In citrus, *Brevipalpus phoenicis* transmits the Citrus leprosis virus (CiLV), which causes yield reduction and eventual death of the affected trees.

2. Citrus flat mite, *Brevipalpus californicus* Banks

The mites are difficult to find or to see with naked eyes because they lie flat against the surfaces of leaf and fruit and move very slowly. The life cycle completes in within 3 weeks at normal condition. Mites prefer green to ripe fruit to feed. Feeding injury symptoms on selected plants include: chlorosis, blistering, bronzing, or necrotic areas on leaves (Childers *et al.* 2003). A toxin or virus is transmitted to the leaves and fruit by the feeding of this mite and causes "pinhead" rust symptoms/nail head rust. In USA this symptom on twigs and branches is called

"Florida scaly bark". Both types of symptoms are known by the term "Leprosis" in Argentina. They begin as small scabby lesions that are brown to bronze and corky. They may occur on leaves or fruits. As the damages increase, the lesions become proportionately larger, and glandular areas develop on the bark of twigs. The mite attack affects the quality, quantity and size of fruits. Feeding causes pitting, scarring and splitting of orange skins and mandarin fruits which downgrade their quality. On lemons, silver scars can occur during severe infestations.

3. Ber Gall Mite, *Larvacarus transitans* Ewing

Presently many Ber orchards are facing severe pest problems and one such alarming problem is the false spider mite, *Larvacarus transitans.* The unique character of this mite is the presence of 3 pairs of legs in both stages, *i.e.*, nymphal and adult. Unlike eriophyids, these mites approach the gall forming habit which shows only at early stages of adaptation to such life. The adoptions elongate the body of mites and reduce the rear pair of legs. The female can be differentiated from male having round opisthosoma, whereas the opisthosoma of male consists of a pair of terminal stylet like rods known as genital stylets. During monsoon period the galls rupture and the mites emerge out from the galls and hide themselves in cracks and crevices of the tree trunk, and start fresh infestation when new flushes come after rainy season in September. The twigs bearing blister galls bear less fruit in comparison to healthy twig.

4. Flat mite, *Dolichotetranychus* spp.

More than 22 species of *Dolichotetranychus* have been reported, out of which two species *viz. D. floridanus* and *D. cocos* have been reported to cause serious damages in South India. *D. cocos* inhabit beneath the perianth of coconut and infest the coconut by causing the infested area to become cracked and dried. However, *D. floridanus* is monospecific and once it is established its eradication become extremely difficult. This small mite is also reported as a pest of pineapples in India, Puerto Rico, Florida, Mexico, Hawaii, Cuba, Japan, Java, Australia, Philippine Islands etc. The adult mite has a bright orange colour and it measures about 0.3 mm in length and approximately one third in width. *D. floridanus* exhibits no specificity towards a particular pineapple variety and affects all pineapple varieties and clones equally. It damages tissue of both the leaf and fruit components as well as the planting material. As the epidermal tissues get damaged, it causes drying and cracking of affected region that often permits the entry of bacterial and fungal phytopathogens and leads rotting of tissues. Damage caused by this mite severely alter the normal crop cycle of the infested plants and underdevelopment of the crop thereby leading to uneven establishment of the crop and extended harvesting periods. This is further exacerbated in ratoon crops and ultimately increases production costs.

Pineapple flat mite, *Dolichotetranychus floridanus* Banks was first recorded from south India (Karnataka) in 1971. The females have an elongated, oval body and are red to orange in colour. The body of the male is pointed posteriorly. The eggs are oval and orange while six-legged larvae are oval and amber in colour, with distinct dark red eyes. The pineapple flat mite attacks plants of all ages, however, they cause most of the damage to young plants. The mite feeds on the soft white tissues at the plant base. They produce rust-like spots which can be infected by fungi and bacteria, so that bud rotting may occur. When the plants are severely infested, they may remain small without fruits (Meyer, 1981 and Singh *et al.*, 2000). This mite was also recorded on coconut inflorescence in Sri Lanka. But the studies of *D. floridanus* on arecapalms are limited.

5. Coconut red mite, *Raoiella indica* Hirst

This mite is easily distributed by wind currents and movement of infested plants through nursery stock and cut branches of plants. It is likely to establish throughout tropical and subtropical areas throughout the Western Hemisphere of the world. Females of *R. indica* are are oval and reddish in color. Females develop dark markings on the dorsum of the body after feeding. The dorsum is smooth, except for the presence of punctae (sculptured depressions). The male is smaller, but similar to the female in shape except for having a tapering of the posterior end of the body. Adult females are larger than males and less active. Males and females are sexually mature when they emerge and males actively seek out females, suggesting there is a sex pheromone involved.

The population reaches at peak twice, in June and September in India (Sarkar and Somchoudhury, 1989 and Singh *et al.*, 2000). Palms affected by *R. indica* show scattered yellow spots on both leaflet surfaces to a strong yellowish coloration on the entire leaflet, with most of the leaflets affected located in the middle area of the leaf. Coconut palms severely affected by the mite showed entirely yellow leaves, particularly on the lower third region of the plant. It is unknown if this condition is solely the result of mite feeding in combination with the dry season or the presence of a plant pathogen transmitted by the mite; there is no information as to whether this mite is a disease vector. Very young coconut palms to very old palms (more than 50 feet tall) can be severely affected, with most of their lower leaves yellow. Young coconut plants may be the most affected by the feeding of the mite. The yellow color of the leaflets is followed by the abortion of the flowers or small nuts in coconut palms.

6. *Tenuipalpus ludhianaensis* Sadana and Chhabra

This species of phytophagous mite was first reported on pear plant from Punjab during 1980. The economic importance of this mite is not known.

7. *Tenuipalpus pruni* Maninder and Ghai

This tanuipalpid mite was observed for the first time from Himachal Pradesh on peach plant by in 1978 with no economic importance

8. *Tenuipalpus punicae* Pritchard and Baker

This phytophagous mite was first time described and reported on pomegranate and peach from Spain in 1958. In India, this species of mite recorded on peach and guava from Punjab. This mite occurs on the under-surface of leaves as well as on twigs. The small colonies are usually seen near the midrib. The feeding on the leaves causes the appearance of brownish patches. Such leaves ultimately fall of. There is no information available on economic losses caused by *T. punicae.*

Management of mites

Chemical management

Over the past half of the 20th century there is a substantial increase in the global consumption of acaricides. Till date several generations of structurally diverse synthetic acaricides have been commercialized which posses ability against various biochemical and physiological targets. Besides specific acaricides, a number of insecticides are reported to have acaricidal activity (*i.e.* avermectins, pyrethroids, thiourea compounds, benzoylureas), while for the control of phytophagous mites, some older neuroactive compounds are still available. Some acaricides of the modern generation exert their effects by disrupting the respiratory processes. Another group act on growth and development. Phytophagous mites resistant to acaricide, especially the spider mites have become a menace in management programme of mites. According to APRD (*Arthropod Pesticide Resistance Database*) there are more than 700 cases of phytophagous mites resistance to acaricides have been reported, of which, 93% of these refer to two spotted spider mite (*Tetranychus urticae*), most important pests throughout the world. Therefore, there was an urge for development of new generation acaricides having novel biochemical modes of action and it is also necessary to apply those acaricides judiciously to prevent or delay the evolution of resistance and prolong their life span.

Four different compounds belonging to different chemical classes were commercialized as acaricides that acts on the respiratory system, namely Fenpyroximate (pyrazole), Fenazaquin (quinazoline), Pyridaben (pyridazinone) and Tebufenpyrad (pyrazole carboxamide), that act through inhibit the MET (mitochondrial electron transport) at complex I. These compounds, also called as METI (mitochondrial electron transport inhibitor) acaricides and show higher efficacy against both eriophyid and tetranychid mites through quick knockdown effect and have a long-lasting impact. The Complex I inhibitors also include Pyrimidifen (pyrimidinamine), as well as, the insecto-acaricide tolfenpyrad (pyrazole carboxamide), and Flufenerim, derivative of pyrimidifen. Complex II inhibitor is recently introduced as insecto-acaricide is Cyenopyrafen, (acrylonitrile class of chemistry). Complex III inhibition acaricides are Acequinocyl (naphthoquinone compound) which is a pro-acaricide and effective against all stages of spider mites, Bifenazate (carbazate compound) is highly effective against both immature and adult stages of spider mites, with quick knockdown effect and also a pro-acaricide, and Fluacrypyrim, effective against all stages of tetranychids. The Insecto-acaricide includes Diafenthiuron (novel thiourea compound; a pro-acaricide effective against all motile stages of spider mites and also effective against eriophyoid mites), and Chlorfenapyr (pyrrole compound; a pro-acaricide effective against all stages of spider mites and eriophyoid mites).

The acaricides that influence growth and development include Etoxazole (oxazoline compound), a mite growth inhibitors which is highly effective against eggs and immature stages of spider mites. Though it is non-toxic to adults, but it can reduces fertility of treated females Spirodiclofen and Spiromesifen (tetronic acid derivatives) are highly toxic to egg and immature stages of spider mites, while their effects on adult females are slower with fertility and fecundity reduction. They have long-lasting and stable acaricidal effect. These two new acaricides are only used to control eriophyoid mites. Spirotetramat, which was initially developed for controlling whiteflies and aphids, has also been proved to be an effective acaricide.

Natural acaricides and other alternative solutions

Probably triterpenoid azadirachtin is the most studied botanical insecticide in the last 20 years. It is derived from *Azadirachta indica.* Over 40 countries have registered neem based products to suppress arthropod pests. Milbemectin (a mixture of milbemycin A3 and milbemycin A4), which is a fermentation product, isolated through the fermentations of *Streptomyces hygroscopicus* subsp. *aureolacrimosus,* introduced several years after introduction of abamectin. This acaricide has neuroactive properties (chloride channel activator) which is effective against eriophyoid and tetranychid mites and is a relatively safe compound due to rapid uptake into treated plants. Moreover, different essential oils from basil, caraway, citronella, clove, lemon eucalyptus, mint, peppermint, rosemary, thyme, and other plants had shown a significant acaricidal activity. Studies have shown that, the constituents of essential oils, carvone, carvacrol, cineole, cinnamaldehyde, cuminaldehyde, eugenol, geraniol, limonene, linalool, menthol, thymol are effective against spider mites.

Biological management

The management of the phytophagous mites by using natural enemies is yet to prove success in our country. Although many developed countries are successfully employing Phytoseiid predators to manage plant mites specially the tetranychid in protected cultivation and different fruit orchards. Laboratory oriented mass rearing techniques for these predators have also been standardized. These predators have established themselves as efficient and effective in checking the population buildup of tenuipalpid, tetranychid and eriophyid mites both in the laboratory as well as in the protective field experiments. There is immense scope to utilize these predators in field level. There are some more species mainly Cheyletidae, Stigmaeidae, Anaystidae etc. are also found effective for biological control of mite pests. Coccinellid beetle *Stethorus pauperculus* is also found effective for utilization in the management programme of phytophagous mites. Certain specialist ladybirds of the genus *Stethorus* are potential biocontrol agents of tetranychid mites, especially at high density level. The feeding potential of various *Stethorus* sp. has been studied in depth by many researchers and they found, that, the prey was detected by contact.

The inner contents of the chorion of the eggs sucked out by the grub and discarded the empty shells. The body of motile stages of mite was first punctured and then their inner contents were sucked. Some other predatory coccinellids with immense potentiality against mites are *Menochilus sexmaculatus*, *S. auperculus*, *Coccinella septempunctata, Chilochorus nigratus, Brumus suturalis* etc. The most active Neuropteran predators of spider mites families are Chrysopidae and Coniopterygidae. *Chrysoparla carnea* has been reported to consume 1000 to 1500 citrus red spider mites daily but fails to complete its life cycle on a mite diet. Another Chrysopid, *Chrysopa vulgaris* is known to have better searching ability than *Stethorus* and devours 30–50 European red mite larvae per hour.

The entomopathogenic fungi are also an efficient alternative to synthetic chemical acaricides. Only one mycoacaricide was available at the beginning of the 1980s (based on *Hirsutella thompsonii*) and used to suppress citrus rust mite. Quarter of a century later, about 30 products are commercialized which act against eriophyoid, tetranychid and tarsonemid mites. Relatively few viruses are found to be effective against mites. First recorded evidence of viral disease of spider mite was made in 1955 (Muma, 1955). It was reported that the infected mites showed diarrhea symptom and the cadavers adhered to the surface of leaf by a black resinous material which was excreted from their anus. Isolates of *Bacillus thuringiensis* was found to show toxicity towards spider mites and *B. thuringiensis* strain isolated from dead two spotted spider mites, *T. urticae* (Jung *et al.*,2007).

Predatory mites include families like Anystidae, Ascidae, Bdellidae, Cheyletidae, Cunaxidae, Erythraeidae, Phytoseiidae, Stigmaeidae, and Tydeidae. Out of these predatory mite families, members of Phytoseiidae are effective predators because of their specific nature, ability to feed on alternate food sources and they can survive even their prey is absent. However, life cycle of many phytoseiid mites are comparatively shorter, have equivalent reproductive potentials of their prey, have good host searching capacity and also have capacity to survive when prey density is relatively less and thus they are comparatively more effective in predation and better in management of a number of phytophagous mites in both field and greenhouse conditions (Dhooria, 2016). Introduction, conservation, and release of predatory mites began few years back (Hoy, 2011).

Integrated management

A satisfactory and long lasting management strategy of mite pests is the integration of biological and chemical methods. In areas where chemical control against mite pests is not effective because of resistance development, it becomes necessary to go for biological control. Integration of these two methods can be achieved by two ways. First method is that acaricides may be used to suppress the higher mite population into lower levels and then manage this low population of mite through predators. Second method is that, acaricides which do not cause mortality to predators can be used. It is necessary to initiate and intensify studies in this direction. It may be mentioned that in tackling mite pest problems, stable habitat like the habitat of perennial tree is more suitable for an integrated approach.

Environmentally benign biorational and microbial pesticides have been gaining attention worldwide as an important tool for and safe integrated pest management. Biorational or "reduced risk" insecticides are synthetic or natural compounds that effectively control insect pests, and they are especially valuable because their toxicity to non-target animals and humans is usually low. Hence, these compounds are considered important components of IPM programs for controlling mites and other agricultural pests. The primary obstacle to using microbial and biologically derived biopesticides is that they work comparatively slowly and also may decompose rapidly in sunlight. Laboratory evaluation of the compatibility of various entomopathogenic fungi viz., *Beauveria bassiana*, *Metarhizium anisopliae*, and *Paecilomyces* sp. with various neonicotinoids including acetamiprid, imidacloprid, and thiamethoxam are found very effective for rapid and prolonged control of mite populations.

References

Baradaran, P., Arbabi, M. andRanjbar, V. A. 2002. Comparative population fluctuation of 1391 fig spider mite (*Eotetranychus hirsti*) on fig varieties in Saveh region. Journal of 1392 Entomological Society of Iran, 22 (1):49-61.

Childers, C. C., McCoy, C. W., Nigg, H. N., Stansly, P. A., Rogers, M. E. 2013. 2014 Florida citrus pest management guide: Rust mites, spider mites, and other phytophagous mites. EDIS. (9 October 2018).

Dhooria, M.S. 2016. Fundamentals of appl.acarol.. Springer, Singapore. Pp-20.

Gupta, S. K. 1991. The mites of agricultural importance in India with remark on their economic status. In: Modern Acarology Vol. 1 (eds. F. Dusbabek and V. Bukva). Academia, Prague, 509-522.

Haramoto, F. H. 1968. Biology and Control of *Brevipalpus phoenicis* (Geijskes) (Acarina: Tenuipalpidae). Hawaii Agric. Exp. Sta. Tech. Bull. No. 68:1-63.

Howitt, A. H. 1999. Fruit IPM fact sheet, Pear Rust Mite. *Extention bullatine*, Michigan State University, Van Buren County, 3.

Hoy, M.A. 2011. The Phytoseiidae: effective natural enemies. In: Agricultural acarology: 1415 introduction to Integrated Mite Management. Taylor and Francis Group, Boca Raton. pp 1416 159-184.

Jeppson, L. R., Keifer, H. H., Baker, E. W. 1975. Mites for injurious to economic plants. Berkeley, University of California Press, pp-614.

Jung, Y.C., Mizuki, E., Akao, T., and Cote, J. C. 2007. Isolation and characterization of a 1420 novel *Bacillus thuringiensis* strain expressing a novel crystal protein with cytocidal 1421 activity against human cancer cells. Jr Appl.Microbiol.,103: 65–79.

Lee, M.H.; Cho,S.H.; Park, H.S., Bahn, L.W.; Lee, B.J.; Son, J.W.; Kim, Y.K.; Koh, Y.Y.; 1425 Min, K. U, and Kim, Y. Y. 2000. Citrus red mite (*Panonychus citri*) is a common 1426 sensitizing allergen among children living around citrus orchards. Ann.Allergy Asthma 1427 Immunol. 85(3):200-204.

Meyer, M.K. and Smith,P. 1981. Mite pests of crops in southern Africa. Sci. Bull. Dept.Agric. 1435 Fish Repub. S. Afr. No. 397,92.

Muma, M.H. 1955. Factors contributing to the natural control of citrus insects and mites in 1440 Florida. Journal of Economic Entomology, 48: 432–438.

Sarkar, P.K and Somchoudhary, A. K. 1989. Influence of major abiotic factors on the 1471 seasonal incidence of *Raoiella indica* and *Tetranychus tijiensis*on coconut. In: *Progress in* 1472 *acarology,* Vol. 2 (eds. G.P. Channa Basavanna and C.A Viraktamath) Oxford and IBH, 1473 New Delhi, 67-72.

Singh, J., Singh, R. N. and Rai, S. N. 2000. Expanding Pest Status of Phytophagous 1475 Mites and Integrated Pest Management. In: *IPM System in Agriculture* (eds: R.K. 1476 Upadhayay, K.G. Mukherjee and O. P. Dubey), New Delhi, India. 7:1-29.

Thakur, A.P. and Sharma, D.D. 1990. Influence of weather factors and predators on the populations of *Aceria litchi* Keifer. Indian J. Plant Protection 18: 104-112.

19

Nematode Pests of Fruit Crops and Their Management

Tushar Kanti Dutta

Most of the plant parasitic-nematodes (PPNs)are worm-shaped microscopic animals. They are virtually invisible through naked eyes when present in soil or plant materials. It has been estimated that of the known nematode species, approximately 50% are free-living found in soil and fresh water, 25% are marine (found in sea water), 15% are animal parasites, and 10% are plant-parasites (Walia and Bajaj, 2004; Sasser and Freckman, 1987). Free living nematodes contribute to maintain the soil bio-dynamic system, by influencing microbial colonization of substrates and mineralization of nutrients (Neher, 2001), whereas PPNs feed on plants and reduce crop growth and yield efficiency. Among the PPNs, a few nematode species feed on aerial plant parts (leaves, stems, flowers and seeds), but the majority of feed on underground plant parts (roots, bulbs, and tubers). They feed on plant cells using a spear-like structure called stylet, which is distinct in all the PPN species. They use this sharp, hollow needle like structure to withdraw plant cell juice by sucking and thereby weaken the host plants. PPNs can be endoparasitic (remaining and feeding inside the cell), semi-endo or semi-ecto parasitic (half of the body inside and half outside plant surface), and ectoparasitic (remaining and feeding from outside the plant cell) (Luc *et al.*, 2005; Jones *et al.*, 2013).

When large population of nematodes feed on a plant root system, they interfere with the roots' ability to take up water and minerals and to transport them to the shoots. This restricts the root growth, reduces plant vitality, and retards shoot growth, the combination of which results in decreased yield and quality of crops. In addition to the direct injury to the plants resulting from feeding activities, nematodes inject various effector molecules into the plants while feeding. These effector molecules help PPNs to overcome the structural barrier of host plant (by secreting call-wall degrading enzymes), counteract host defence responses and modify the host cell at their will to make permanent nutrient source for them(for example Giant cells for root-knot nematodes, Syncitium for cyst nematodes etc.). Nematode injury to plant roots also aggravates invasion by other disease-causing fungi and bacteria. Certain nematode species are also known to transmit plant-pathogenic viruses

(Perry and Moens, 2006). Overall average annual loss of the world's major crops due to damage by the PPNs was estimated to be 12.3%. For the 20 major crops which serves as the human's primary food source, annual yield loss of 10.7% was estimated, while for another group of 20 crops mainly of commercial importance, a 14% annual yield loss was assessed. Developing countries had a crop loss of 14.6% compared to 8.8% in developed countries (Sasser, 1989).

Symptoms caused by most of the plant-parasitic nematodes are difficult to distinguish from those caused by soil related problems, like nutrient deficiency. Thus, identification of nematode damage based on specific symptomology is difficult, and the types of generalised symptoms (Fig. 1) are caused by PPNs are as follows:

a. Non-uniform plant growth, mostly in patches and poor plant establishment giving unhealthy appearance.
b. Plants weakened over time and dies when combined with other problems.
c. Chlorosis and stunting (resembling symptoms of some virus infections).
d. Knots (galls), lesions on roots, stubby roots or bifurcation of roots.
e. Excessive branching of roots and hairy root symptoms.
f. Poor root health, growth and establishment

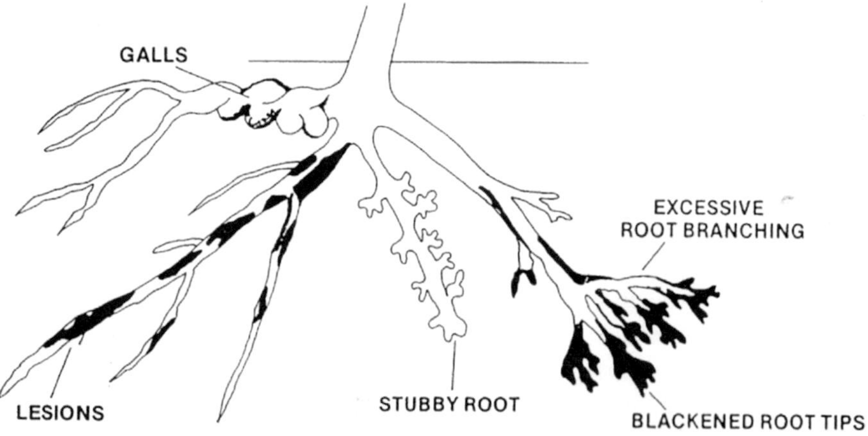

Fig. 1: Different types of symptoms caused by the PPNs.

While concerning the fruit crops, injury by the PPNs predispose them to cold injury, high soil pH susceptibility and micro nutritional deficiency. In addition, poor plant nutrition, inefficient orchard floor management, poor soil health status, and secondary disease infections may also predispose the plants to nematode damage. The most common PPN species associated with fruit crops are *Pratylenchus* spp., *Xiphinema* spp., *Meloidogyne* spp., *Helicotylenchus* spp., *Criconemoides* spp., *Tylenchulus* spp.etc. These nematodes alone or in combination with other factors reduce the productivity of the trees. The PPN species such as *Xiphinema* spp., *Longidous* spp. transmit plant viruses. These viruses in combination with nematodes weaken the trees to such extent that they are more easily killed by secondary infection, like cold injury, or any other stress factor (Swarup and Dasgupta, 1986; Swarup *et al.*, 1989).

Although hundreds of different nematode species are associated with plants, many of them may not cause significant economic damage. Further, in many cases, PPN populations occur in polyspecific communities. Nematodes from soil samples or infected plant parts must be extracted, identified, and counted in order to determine if and which nematode species is causing poor plant growth. If the population level is high enough to cause economic damage (*i.e.*, at or above the "economic threshold level" for particular species), then the application of specific control strategies are recommended.

Important Plant-Parasitic Nematode Genera in Fruit Crops

The root-knot nematodes (*Meloidogyne* spp.), burrowing nematode (*Radopholus similis*) and citrus nematode (*Tylenchulus semipentrans*) are the major threat to fruit crop cultivation.

Root-knot nematode (*Meloidogyne* spp.)

Root-knot nematodes are basically sedentary endoparasites. After penetrating the root tissue, normally behind the root tip, the second stage juveniles (J2s) start feeding on cortical cells followed by making a 'U' turn to progress into the vascular cylinder behind the root epidermis. After reaching vascular bundle they select a suitable cell, become sedentary and convert that cell into a feeding site (giant cell) which serves as permanent nutrient sink for nematodes. The J2s then develop to non-feeding 3rd (J3) and 4th(J4) stage juveniles. Third and fourth moults occur in quick succession, ultimately leading to the development of saccate, pear-shaped, white adult females. The non-feeding males are vermiform. The reproduction of *Meloidogyne* spp. Occurs primarily through parthenogenesis and rarely through amphimixis (in few species only). Each adult female lays about 400-500 eggs in the gelatinous matrix which can survive in soil for long time and hatching of eggs occur when favourable environment prevails. The total time taken for completion of one life cycle under optimum temperature (27-30°C) is 3-4 weeks. Hence, the nematodes can complete several generations in a year, and plant root cells around the female nematode are hypertrophied and enlarged to produce galls (Perry *et al.*, 2009).

Emerging root-knot nematode problems in fruit crops

Meloidogyne enterolobii in Guava

Due to the *M. enterolobii* infestation a unique type of dieback symptom occurs followed by complete death of guava trees. Yellowing of plants is the earliest symptom leading to withering and barren appearance of plants at later stage with dirty root appearance comprising beaded knots (Fig2). In addition, galled roots often show rotting aggravated by secondary infection of microorganisms. Plants become flaccid and when broken reveal the presence of hollow twigs. Multiple galls may merge together to form compound galls, and number of females varies from 50-200 per gall. Egg masses are laid outside the roots. Each egg mass contains around 150-200 eggs. Rootstocks bought from nurseries are mostly prone to nematode infestation (Fig-3).

Fig. 2: *Meloidogyne enterolobii* infection progression in guava trees. Photographs obtained from AICRP (Nematodes).

Fig. 3: Rootstock of guava showing yellowing of leaves and profuse galling. Photographs obtained from AICRP (Nematodes).

Meloidogyne incognita in Pomegranate

Meloidogyne incognita is becoming a serious bottleneck for pomegranate cultivation in Maharashtra and Tamil Nadu. Orchards having 4-5 year old plants show symptoms like yellowing and shedding of leaves, stunted growth with lesser number of flowers and fruits and more number of galls compared to healthy plants (Fig. 4). Although presence of no other nematodes was detected but the association of wilt fungus, *Fusarium oxysporum* cannot be ruled out.

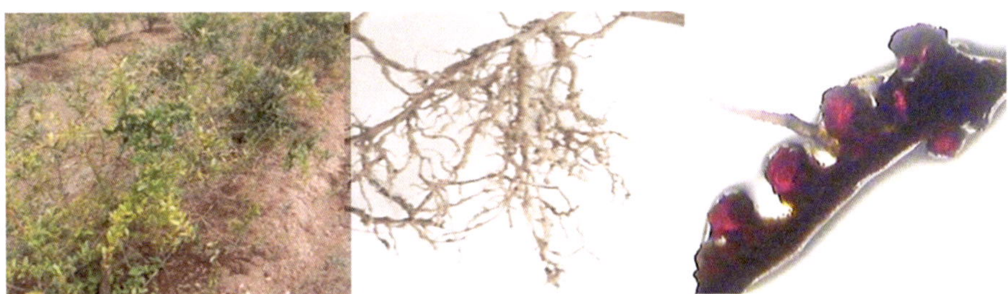

Fig. 4: *Meloidogyne incognita* infection in pomegranate. Egg masses are stained with acid fuchsin. Photographs obtained from AICRP (Nematodes).

In addition, *Meloidogyne indica* has the potential to cause yield loss in citrus plantation by inflicting galls in the citrus roots (Fig5). *Meloidogyne incognita*, *M. javanica* and *Rotylenchulus reniformis* (reniform nematode) also cause substantial yield loss in papaya under field condition. *Meloidogyne incognita* is considered as a minor pest for banana plantation. Although they do not cause serious damage in mature plants, they can be a menace in the young tissue culture seedlings (www.iihr.ernet.in). They cause swellings or galls mainly on root tips of smaller roots, sometimes producing necrotic spots on root tissues when developing deep inside roots (Fig-6).

Meloidogyne mali, parasitic on apple, is known only from Japan. *Meloidogyne hapla* and *M. incognita*, which have world-wide distributions, have also been reported from apple. All the four of the common *Meloidogyne* species, *i.e.*, *M. incognita, M. javanica, M. arenaria* and *M. hapla* have been reported from peach (Evans *et al.*, 1993; Ferrez *et al.*, 2002).

Fig. 5: *Meloidogyne indica* infection in citrus.
Photographs obtained from AICRP (Nematodes).

Fig. 6: Banana root galls due to *M. incognita* infection. Mature females and second stage juveniles of *M. incognita***.** Photographs borrowed from a book by Bridge & Starr, 2007.

Burrowing nematode (*Radopholus similis*)

The burrowing nematodes are migratory endoparasites in nature. Although the stages remain vermiform throughout the life cycle, sexual dimorphism is apparent with adult males being degenerated and non-feeding. Eggs are normally laid by gravid females in infested tissues over 7-8 days at the rate of about four eggs per day. The life cycle from egg to egg extends over 20 to 25 days with eggs taking 8-10 days to hatch and the larvae 10-13 days to mature. Wounding of roots by the nematodes usually induce reddish-brown cortical lesions which are diagnostic symptom of the disease in case of banana (Fig. 7). These lesions are clearly seen when an affected root is split open longitudinally and examined immediately. Root and rhizome necrosis is manifested by varying degrees of retarded growth, leaf yellowing and falling of mature plants. With the increase in nematode population, feeder roots are invaded and destroyed as fast as they are formed. The resulting setback in the uptake of plant nutrients leads to debility of the plants and production of smaller fruits. The lesions of the primary roots together with the girdling and death of these anchor roots make the plant prone to 'tip over or topple' by wind action in case of banana (Fig-8).

Two species of *Radopholus* are known to damage citrus seriously, but both are much localized. *R. similis* citrus race (previously named *R. citrophilus*) is restricted to the centre of Florida in sandy soils. It causes 'spreading decline' showing reduced fruit and leaf size and quantity. Twig die-back is normally associated with the nematode, and wilting can occur during dry spells. The *R. similis* populations on citrus are considered as a biotype separate from those populations that occur on banana and other crops. The other species is *R. citri*, which has only been found in Sumatra, Indonesia. *R. citri* also causes stunted growth. *Radopholus* spp. cause cortical lesions and extensive root damage in citrus (Swarup *et al*., 1989). Root systems invaded by the nematode are reduced and become dark brown resulting in poor growth (Fig-9).

Fig 7: Rhizome necrosis in banana due to *R. similis* infestation. Photographs obtained from IIHR, Bengaluru.

Fig. 8: Delayed and distorted flowering followed by toppling of banana due to *R. similis* infection. Photographs obtained from IIHR, Bengaluru.

Fig. 9: Citrus seedling infested with Radopholus citri (left) showing brown, necrotic roots and

reduced growth compared to healthy seedling of the same age (right). Photographs borrowed from a book by Bridge & Starr, 2007.

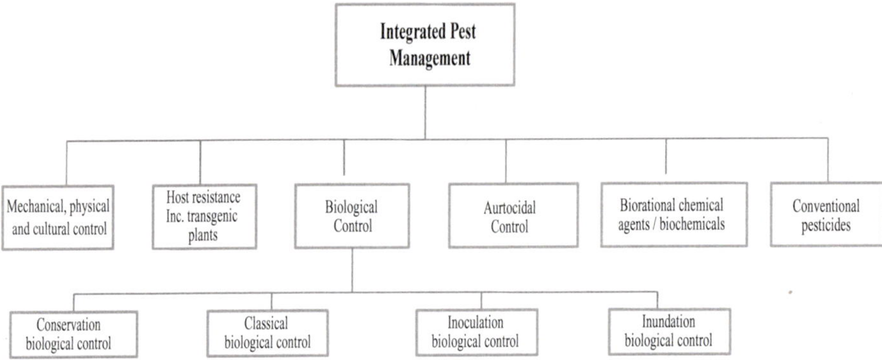

Fig. 10: Twig dieback on citrus is associated with severe infestation by *T. semipenetrans* (left). Dirty root of citrus due to nematode infestation (middle). Female nematode is embedded in the root cortex (right). Photographs borrowed from a book by Bridge & Starr, 2007.

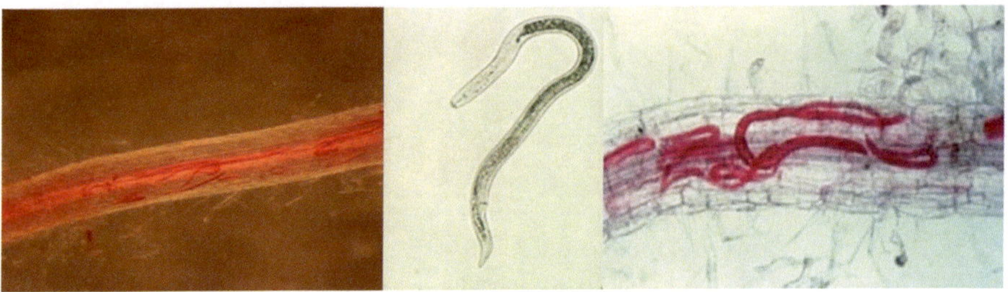

Fig. 11: Pattern of lesion nematode (*Pratylenchus* sp.) infestation in cortical tissue. Photographs obtained from AICRP (Nematodes).

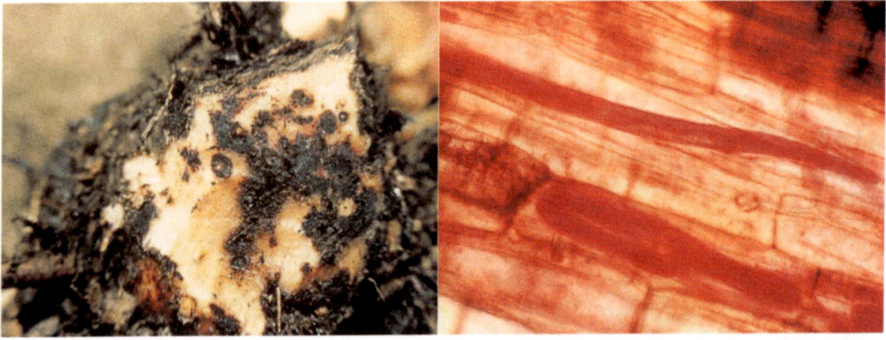

Fig. 12: Necrotic lesions in banana corm at root attachment points caused by *P. coffeae.* Stained nematodes in root cortex. Photographs borrowed from a book by Bridge & Starr, 2007.

Citrus nematode (*Tylenchulus semipenetrans*)

This is a semi-endoparasitic nematode that causes severe damage to citrus crops, but can also cause moderate damage in grapes, olive and apple. *Tylenchulus semipenetrans* exhibits sexual dimorphism, reproduces sexually and occasionally by facultative parthenogenesis. From the eggs, second-stage juveniles (J2s) emerge and search for host roots. The motile and vermiform J2s molt into vermiform J3s and J4s, and finally develop to sedentary preadult. The immature female penetrates the deep cortical layers of the root, and establishes a permanent feeding site consisting of specialized cells called 'nurse cells' which serve as main source of nutrients. Upon maturation, the posterior portion of the nematode body swells and protrudes from the root surface while its elongate neck and head remains embedded into the cortex. Mature females produce eggs that are embedded in a gelatinous matrix. The length of the female life cycle from egg to egg ranges from four to eight weeks. The development of the J2s into preadults is completed in seven days and does not require feeding. Citrus nematode infected roots are thicker, darker, decayed and show a dirty appearance because of large number of females sticking on the infected roots. The infected root systems due to the nematode damage lose the ability to absorb enough water and nutrients for normal growth (Fig-10). It causes slow decline in citrus and in soils with high pH the damage due to *Tylenchulus* is generally higher (Swarup *et al.*, 1989).

Root lesion nematode (*Pratylenchus* spp.)

Prevalent in well-drained sandy soil, lesion nematodes can be a major cause of orchard replant failures. Lesion nematodes enter the roots and burrow tunnels through the root cortex. All life stages except the egg stage are motile and can invade roots. Eggs are laid inside root tissues or in soil. The eggs hatch, and juveniles then enter the roots and contribute to root injury. Root lesion nematodes are migratory (Fig-11) and therefore are capable of repeatedly entering and exiting from root tissue, although several generations can occur inside the roots without the nematodes migrating into the surrounding soil. These nematodes cause small brown lesions on the white lateral roots and kill the fine feeder roots. The entire root system appears discoloured when these lesions merge. Such root lesions are frequently invaded by other root-rotting pathogens. Severely affected trees may lose all feeder roots. Ultimately young replant trees may die and existing trees lack uniformity. If not dead, the top part of the infected tree often exhibit stunting, chlorosis, and twig dieback. They can cause a decline in vigor of existing peach and cherry orchards. Tip die-back may also be observed in some fruit crops (Swarup *et al.*, 1989; Swarup and Dasgupta, 1986;Perry and Moens, 2006).

A large number of lesion nematodes in a root frequently cause the roots to turn brown and die. The presence of brown lesions composed of dead root tissues is a common symptom of lesion nematode infection; however, roots of some plants are not sensitive and do not turn brown. In addition to the death of root cells caused directly by the nematode movement and feeding, wounds resulting from nematode activity allow other soil microorganisms to enter the root tissues and contribute to root decay. The most common root-lesion nematode in fruits that causes damage in apple, peach, cherry, grapes and so many other crops is *Pratylenchus vulnus* (in warmer climate). In cooler climate *P. penetrans* is encountered most frequently in fruit crops. Nematode parasitism contributes substantially to complex disease syndromes, especially the peach tree short life syndrome and cherry decline (Evans *et al.*, 1993).

Fig. 13: Roots of toppled banana, broken and necrotic due to *P. goodeyi* infection (left).Necrotic lesions in the outer cortex of banana root due to *H. multicinctus* infection (right). Photographs borrowed from a book by Bridge & Starr, 2007.

Pratylenchus coffeae is important on bananas, plantains, and abaca (*Musa textilis*) in the Pacific Island countries, Central and South America, and Cuba. It is also present as an economically important pest on dessert bananas in South Africa causing crop losses as high as 80% where it occurs. Due to the nematode attack toppling and uprooting occurs, root lesions are purple to black extending throughout the root cortex and extensively on corms with necrotic patches at root attachment points (Fig-12).

Pratylenchus goodeyi is only found as a pest of commercial dessert bananas in the Canary Islands and Crete, otherwise it is a major pest exclusively of small-scale cultivation of banana and plantain food crops in Africa. Above-ground symptoms and root damage are very similar to those caused by *Radopholus similis*. Growth and bunch size are severely reduced, toppling/uprooting is common (Fig-13); reddish to purple necrosis of roots extends through the cortex to stele and necrosis of basal portion of corms occurs (Bridge & Starr, 2007).

Dagger Nematode (*Xiphinema* spp.)

There are seven common *Xiphinema* spp., known as dagger nematodes, that are commonly associated with deciduous tree crops, including *X. basiri*, *X.*

diversicaudatum, *X. vuittenezi*, and the *X. americanum* group (which also includes *X. brevicolle*, *X. californicum* and *X. rivesi*). The combined distribution of these species on deciduous crops is world-wide. They are most commonly associated with apple, cherry, peach, and plum (Bridge and Starr, 2007).

This nematodes cause direct damage to a wide variety of plants by their feeding activities, but several dagger nematode species also transmit soil-borne plant viruses. Females lay eggs singly in the soil near plants, and these eggs hatch to produce first stage juveniles. They may have 3-4 juvenile stages and may require 6 to 12 months to complete life cycle. Males are rare, and this nematode may live up to 3 years under favourable soil conditions. Ectoparasitic dagger nematodes do not burrow into roots, but insert their long stylet deep into root tips where they feed on root tip cells. Dagger nematode feeding causes some necrosis and stunting and swelling of root tips. Several lateral roots may appear above the damaged root tips. Root-tip swelling may be confused with the galls of root-knot nematodes. Shoots show symptoms of general decline, early leaf and fruit drop, and fruits may be reduced in size, and lack typical flavor. Parasitism by *Xiphinema* spp. is often accompanied by symptoms of the NEPO viruses these nematodes transmit (tomato ringspot virus, cherry rasp leaf virus, peach rosette mosaic virus and strawberry latent ringspot virus). In addition to foliar symptoms of viral disease, there may be stem pitting of peach and, if the root stock is susceptible and the scion is resistant to the virus, there may be necrosis of the graft union such that girdling of the tree occurs, followed by tree death (Swarup *et al.*, 1989; Swarup & Dasgupta, 1986; Perry & Moens, 2006).

Grapevine Fan Leaf Virus (GFLV) is transmitted by *X. index*. Cherry Rasp Leaf Virus (CRLV) is transmitted by the *X. americanum* species complex. Another virus vectored by *X. rivesi* and *X. americanum* is Tomato Ring Spot Virus (TmRSV), which causes Prunus stem pitting (PSP). Some other important plant viruses transmitted by dagger nematodes are: grapevine yellow vein virus, tobacco ring spot virus, peach rosette mosaic virus, and three strains of TmRSV causing peach yellow bud mosaic, prune brown line, and Prunus stem pitting. The viruses are acquired within 24 hours of the initiation of feeding and are carried in the esophagus lumen for up to 12 months. Virus particles are transmitted by both adults and juveniles of nematode (Chen *et al.*, 2004a).

Spiral nematode (*Helicotylenchus* spp.)

Spiral nematodes feed at or near the root tips cause severe root injury, and induce stubbiness to primary and lateral roots and coarse root systems which lacks feeder roots. They may contribute to stress on fruit trees, especially when present high in number. Spiral nematodes generally are ectoparasites, but some are semi-endoparasites and a few are endoparasites. Ecto and semi endoparasites lay their eggs in soil near to the roots whereas endoparasites lay eggs inside the roots.

They have a wide host range including lawn grasses and other ornamentals, such as corn, soybeans, and tobacco. *Helicotylenchus multicinctus* cause substantial yield loss in banana. Reduced plant growth and bunch size are seen as a gradual decline in production over a number of years, sometimes accompanied by nutrient deficiency symptoms.

Damage symptoms are more apparent on bananas grown in fringed areas outside the main tropical banana growing areas. On roots, small, rounded black lesions are first seen on the root surface leading to complete blackening of the outer parts of the roots. Internally, roots have shallow, black to brown necrosis in the epidermal and outer cortical tissues but not penetrating to the stele (Fig. 13). *H. multicinctus* is considered to be the main damaging nematode pest in areas which are sub-optimal for banana production where temperature and rainfall are limiting factors such as the Mediterranean and Middle East regions. Due to the prevailing lower temperatures, these areas also lack the more damaging nematode species such as *Radopholus* and *Pratylenchus* spp. (Swarup *et al.*, 1989; Chen *et al.*, 2004a;Bridge & Starr, 2007).

Ring nematode (*Criconemoides* spp.)

Several species of *Criconemoides* and related genera are found to affect fruit crops. They are small, cigar-shaped strictly ectoparasitic nematodes that feed on the surface of the root and cause browning on tender tissues. The life cycle takes 25-35 days. After feeding for several days on roots, females deposit single eggs in every two to four days. Second stage juveniles (J2s) hatch from the egg in 11-15 days, moult to J3s in three-five days, subsequently moult to J4s in 4-7 days, and become adults after 5-6 days. Males are very rare. Ring nematodes incidence is seen more in coarse soils than in fine textured soils. Optimum conditions for reproduction include a pH near 7 and soil temperatures between 75 and 80°F. Ring nematodes tend to be sensitive to low soil moisture and low pH (Swarup and Dasgupta, 1986 and Swarup *et al.*, 1989).

Criconemoides curvatum and *Criconemella xenoplax* cause the greatest losses to peach, almond and walnut. The typical symptoms of parasitism of roots of deciduous crops by the ring nematodes are a general decline in apparent root health, based on root biomass and colour. The abundance of feeder roots is diminished and the roots are discoloured, often due to the presence of necrotic tissues. Longitudinal cracks of older almond roots and necrosis of phloem tissues have been associated with parasitism by *Criconemella xenoplax*. The shoots show various symptoms of decline and poor vigour; including some chlorosis, tip die-back, and thinner crowns. Nematode parasitism is frequently associated with complex disease syndromes, especially the peach tree short life syndrome. Bacterial canker caused by *Pseudomonas syringae* and increased winter freeze damage are components of the peach tree short life syndrome, and ring nematodes increase the susceptibility of peach and walnut to these maladies (Ferraz *et al.*, 2002; Chen *et al.*, 2004a; Bridge & Starr, 2007).

Reniform nematode (*Rotylenchulus reniformis*)

It is obligate, sedentary semi-endoparasite species attacking a wide range of fruit and plantation crops. The common name 'reniform' was derived from the kidney-shaped mature female. It has worldwide distribution and is receiving importance as national pest of crops. At present, several species of reniform nematodes known but *R. reniformis* is the most widespread. The symptoms of damage to crops are non-specific on the above ground or even in the below ground parts necessitating a close observation to confirm their presence and damage. It feeds on cortical tissue, phloem and pericycles and its infection may cause formation of necrosis on roots of certain crops. Symptoms appear as root discolouration, shedding of the leaves and formation of malformed fruits and seeds in papaya. In addition to causing direct damage to plants roots, the nematode in concert with other pathogens like *Fusarium* spp., *Verticillium* spp., *Sclerotium rolfsii* and *Rhizoctonia solani* develop disease complexes. It has also been reported to parasitize the bacterial nodules. The first moult occurs within eggs and eggs are hatched in water without the influence of root exudates. Juveniles develop to pre-adult stage without any feeding of host tissue and quickly completing three superimposed moulting. The young female is the only infective stage. After infection to the roots, young female orient themselves perpendicularly to the longitudinal axis of roots with the posterior part remain outside the root. After establishing the feeding site, it develops into kidney shaped female with posterior portion protrude outside the root. Egg laying starts within 7-10 days after invasion and eggs are laid into a gelatinous matrix secreted by six specialised cells around vulva. Each egg mass contains 30-200 eggs. Total life cycle is completed within 3-4 weeks depending upon temperature and host suitability (Swarup & Dasgupta, 1986; Swarup *et al.*, 1989).

Management of nematodes in fruit crops

Some of the generalized methods for managing nematodes in fruit crops below the economic threshold level include (Swarup & Dasgupta, 1986; Swarup *et al.*, 1989; Whitehead, 1998; Chen *et al.*, 2004b; Bridge & Starr, 2007):

i. Provide enough organic matter to increase the free-living nematodes as they suppress the PPNs in the soil due to increased competition for space.

ii. Marigold, Sudan grass and *Brassica* spp. can be used as green manure crops to manage PPNs. Alpha-terthinyl of marigold and Glucosinolate or isothiocyanate of *Brassica* species are known to be allellopathic in nature.

iii. Soil solarization is very effective to control PPNs and other soil borne pathogens. For soil solarisation purpose, plough the field to ensure looseness, ensure adequate moisture, cover with thin transparent plastic mulch, seal the mulch to make it air tight and maintain that for at least 45 days in warmer months like June and July.

iv. Pre-plant applications of a fumigant nematicide is recommended when replanting into an established orchard site known or suspected to be infested with root-knot nematodes. Apply soil fumigants, such as Vapam (Metam sodium)when the soil temperature is high at 6 inch soil depth with appropriate field moisture. These conditions are critical to get effective results. Covering soil after application of the soil fumigant increases efficacy of the treatments. Post-plant application of non fumigant nematicides such as Phenamiphos provides better suppression of populations of *C. xenoplax*.

v. Host resistance: Many rootstocks are reported to have tolerance or resistance to PPNs. Notably, grapevine cultivars like Khalili, Kishmish, Beli, Banquabad, Cardinal, Early Muscat, Loose Perlett etc. are known to be resistant to *M. incognita* and *M. javanica* infection.In banana, cultivars like Kadali, Pedalimoongil, Kunnan, Pey Kunnan, Pisang Seriby etc. are known to be resistant to *R. similis* infection.In papaya, cultivars like Solo, Washington, Coorg Honey Dew etc. are known to be resistant to *R. reniformis* infection (Starr *et al*., 2002).

vi. Alternate wetting and drying of soil results in the sharp decline in *R. reniformis* population. The nematode is capable of surviving in air dried soil for a long period of time (Gaur & Perry, 1991). The retention of moulted cuticles of previous stages is a unique adaptation for survival of *R. reniformis* in soil. Individual young females, males and fourth stage juveniles could survive in a coiled anhydrobiotic state with encrusted cuticles in soil. Survival of this nematode is inversely related to the rate of moisture loss in soil which can be exploited to manage them in papaya. Several peach rootstocks are also known to be resistant to *M. incognita*.

vi. For use of nematode free planting material, hot water treatment of bulbs, corms, tubers and fleshy roots can be adopted which can kill the dormant nematodes inside the root.

vii. The NEPO viruses have alternate broad-leaf weed hosts, such as dandelion, and are spread over long distances by the seeds of these weeds. Thus management of broad-leaf weeds in the orchard by cultivation or herbicides is important to prevent introduction of the viruses. Nematode transmission is the only mechanism by which the virus can move from the weeds to the orchard trees.

Specific case studies

- Integrated management of root-knot nematode on pomegranate. Soil application of *Trichoderma* + @ 20 kg/ha + Castor cake @ 2 t/ha reduced nematode population (44%), root galling (38%), increased yield by 29.7%.
- Paring (trimming of suckers) + Hot Water Treatment (55°C), neem cake @ 1 kg/plant was most effective in reducing burrowing nematode (48%), and increasing banana yield by 26.3%.
- Neem cake @ 100 g/m^2 as soil application enhanced grapevine yield by 40.7% and reduced root-knot nematode population by 10.8%.
- Soil application of egg parasitic fungi, *Paecilomyces lilacinus* @ 20kg/ hawas most effective in reducing citrus nematode population by 43% and increasing yield of sweet orange by 19.5%.

To manage the root-knot nematode problem in pomegranate, guava and citrus major focus should be to check the spread of nematode disease through infected saplings, more basic studies on the role of nematodes and other pathogens in disease complex should be conducted and network projects should be formulated based on specific problems (Khan *et al.*, 2014).

Fig. 14: Nematode free banana seedling using IIHR technology. Left panel-nematode infested seedling; Right panel-healthy seedling.

Fig. 15: Nematode free papaya seedling using IIHR technology.

Production of healthy seedlings or grafts at IIHR, Bengaluru

Securing healthy seedlings or grafts is essential to ensure optimum plant population stand, good crop growth and higher yields. Infestation by heavy populations of nematodes and other pathogens will result in very weak seedlings or grafts with poor root growth. Seedlings or grafts with stunted root system cannot establish well after transplanting. Nematode attack on the root system makes the seedlings or grafts weak and also vulnerable for the infection by secondary pathogens (soil borne fungi and bacteria). Nematode damage also results into the breakdown of the resistance against pathogenic fungi. Further, nematode infected seedlings facilitate the spread of the nematodes in the main fields making the problem more difficult to manage in a larger area. Because of all these reasons it was inevitable to produce the seedlings or grafts without infestation of nematodes or other pathogens.

The IIHR, Bengaluru has developed a bio-pesticide formulation to manage nematode problem in banana seedlings. This is an organic formulation, consists of *Pseudomonas fluorescens* and *Trichoderma harzianum*. One ton of soil mixture or any substrate used should be prepared by mixing 2 kg of each of *Pseudomonas fluorescens* 1% WP, *Trichoderma harzianum 1%* WP and *Paecilomyces lilacinus* 1% WP + 25 kg of neem cake or pongamia cake. After mixing of all these materials the soil mixture or any substrate can be used for hardening of banana and papaya seedlings (Fig-14 and 15) or producing grafts. Alternatively in field condition, apply two tons of FYM or 500 kg of neem cake / pongamia cake or one ton of vermicompost enriched with *Pseudomonas fluorescens* + *Trichoderma harzianum* + *Paecilomyces lilacinus* during the land preparation (http://www.iihr.ernet.in).

RNAi-based management of plant nematodes

Existing management approaches such as use of nematicides are posing a threat on the environment and are costly. So, the adoption of novel strategies can combat these hidden enemies effectively and can be proved as environment friendly and cost effective. The discovery of RNAinterference (RNAi) in *Caenorhabditis elegans*, in which double stranded RNA (dsRNA) induces the degradation of cognate endogenous mRNA and so prevents synthesis of the encoded protein, has provided a significant new tool to study gene manipulation and analysis of gene function. Therefore, RNAi can be utilized as a powerful reverse genetic tool for developing RNAi-based transgenic plants to combat plant-parasitic nematodes. Plants can be modified to express dsRNA molecules by cloning the sense and antisense cDNA sequences of the PPN target gene, separated by an intron or spacer, into an RNAi vector under the direction of a plant promoter. While feeding on the host tissue PPNs can ingest dsRNA molecules on its own. Transgenic hairy roots of grape expressing siRNAs of *16D10* gene (responsible for nematode parasitism) showed less susceptibility to *M. incognita* infection even after 5 weeks post infection. Transgenic walnuts expressing dsRNAs of *Pv010* gene (housekeeping gene) showed reduced multiplication of *P. vulnus* with no visible lesions even after 2 months post infection. These results suggest the immense potential of RNAi-based fruit crop transgenic to manage plant nematodes (Dutta *et al.*, 2015).

References

Bridge, J., Starr J. L. 2007. Plant nematodes of agricultural importance – a colour handbook. Manson Publishing, London, U.K. pp-152.

Chen, Z. X., Chen, S. Y., Dickson, D. W. (eds) (2004a). Nematology: Advances and Perspectives. Vol. I. Nematode Morphology, Physiology and Ecology.Tsing University Press, China; CABI Publishing, Wallingford, UK.

Chen, Z. X., Chen, S. Y., Dickson, D. W. (eds) 2004. Nematology: Advances and Perspectives. Vol. II. Nematode Management and Utilization.Tsing University Press, China; CABI Publishing, Wallingford, UK.

Dutta, T. K., Banakar, P., Rao, U. 2015. The status of RNAi-based transgenic research in plant nematology.5:760.doi: 10.3389/fmicb.2014.00760.

Evans, K., Trudgill, D. L., Webster, J. M. (eds) 1993. Plant Parasitic Nematodes in Temperate Agriculture. CABI Publishing, Wallingford, UK.

Ferraz, L. C. C. B. and Brown, D. J. F. 2002. An Introduction to Nematodes: Plant Nematology. Pensoft Publishers, Sofia, Moscow.

Gaur, H. S., Perry, R. N. 1991.The role of the moulted cuticles in the desiccation survival of adults of Rotylenchulus reniformis. Revue De Nematology 14: 491-496.

Jones, J. T., Haegeman, A., Danchin, E. G. J., Gaur, H. S., Helder, J., Jones, M. G. K., Kikuchi, T., Manzanilla-López, R., Palomares-Rius, J. E., Wesemael, W. M. L., Perry, R. N. 2013.Top 10 plant-parasitic nematodes in molecular plant pathology. Molecular Plant Pathology 14:946-961.

Khan, M. R., Jain, R. K., Ghule, T. M., Pal, S.2014. Root knot Nematodes in India-a comprehensive monograph. All India Coordinated Research Project on Plant Parasitic nematodes with Integrated approach for their Control, Indian Agricultural Research Institute, New Delhi. Pp-78.

Luc, M., Sikora, R. A., Bridge, J. (eds) 2005. Plant Parasitic Nematodes in Subtropical and Tropical Agriculture. 2nd edn.CABI Publishing, Wallingford, UK.

Neher, D. A. 2001. Role of nematodes in soil health and their use as indicators. Journal of Nematology 33: 161-168.

Perry, R. N., Moens, M. 2006. Plant Nematology. CABI Publishing, Wallingford, U.K. 447p.

Perry, R. N., Moens, M., Starr J.L.2009. Root-knot nematodes.CABI Publishing, Wallingford, U.K. Pp-488.

Sasser, J. N. 1989. Plant parasitic nematodes, the farmer's hidden enemy. Coop Publication, North Carolina State University, Raleigh, USA. 115p.

Sasser, J. N., Freckman, D. W.1987. A world prospective on Nematology: The role of the society In: Vistas on Nematology: A Commemoration of the twenty-fifth anniversary of the Society of Nematologists. Eds. Veech & Dickson.Pp-7-14.

Starr, J. L., Cook, R., and Bridge, J. 2002. Plant Resistance to Parasitic Nematodes. CABI Publishing, Wallingford, UK.

Swarup, G., Dasgupta, D.R. 1986. Plant Parasitic Nematodes of India. ICAR Publications, New Delhi.Pp-497.

Swarup, G., Dasgupta, D. R., Koshy, P. K. 1989. Plant diseases. AnmolPublishing Pvt. Ltd., New Delhi.Pp-396.

Walia, R. K., Bajaj, H. K. 2004. Textbook onIntroductory Plant Nematology, ICAR Publications, New Delhi.Pp-227.

Whitehead, A. G. 1998. Plant Nematode Control.CABI Publishing, Wallingford, UK.

Web sources cited

http://www.iihr.ernet.in/content/iihr-publications.

20

Conservation Biological Control Approaches in Orchard Ecosystems

Suprakash Pal and Biwash Gurung

Chemical pest management is neither economically nor ecologically sustainable. As for example, the USA spent an estimated $11 billion on pesticides in 2008, applying more than 480 million pounds of these chemicals to its agricultural fields (Fernandez-Cornejo, *et al*, 2009). Despite this widespread use of pesticides, which cause severe harm to the mankind as well as the ecosystem, estimate suggest that 37% of the crop yield of the USA are lost due to pests (Pimentel, *et al*, 1992). This type of situation is witnessed all throughout the world. Under these circumstances, the natural enemies of agricultural pests may offer a sustainable solution to pest problems and the management they offer is worth of $13 billion per year in the USA (Losey and Vaughan, 2006). So, conservation of these biological control agents in the ecosystem is very much important for the sustainability as well as the health of the environment. Conservation and augmentation of natural enemies are among the main task in Conservation Biological Control (CBC).

These objectives should rely on in-depth knowledge of potential predators dwelling the agroecosystem of interest. Conservation of natural enemies involves environmental modification to benefit (or minimize harm to) natural enemies.

Conservation Biological Control (CBC)

Conservation biological control (CBC) has been defined as 'modification of the environment or existing practices to protect and enhance specific natural enemies of other organisms to reduce the effect of pests' (Eilenberg, *et al*, 2001).In practice, CBC is effected by either

(1) reducing the pesticide-induced mortality of natural enemies through better targeting in time and space, reducing rates of application or using compounds with a narrower spectrum efficacy, or

(2) by habitat manipulation to improve natural enemy fitness and effectiveness.

The second approach often involves increasing the species diversity and structural complexity of agro-ecosystems. In the context of CBC, habitat manipulation aims to improve natural enemies with resources such as nectar, pollen, physical refugia, alternative prey, alternative hosts and lekking sites. However, the improvement in the number of natural enemies does not always result in better yield. This is due to complex tritrophic interactions in the ecosystems and the importance of appraising the complex relationship between natural enemy diversity and effective biological control has been highlighted by many authors (Tscharntke *et al.*, 2007; Jonsson *et al.*, 2008; Martin *et al.*, 2013).

One of the conclusions from these studies is that increased biodiversity in the ecosystems is no guarantee of successful pest control (Landis *et al.*, 2000) and this is attributed to the negative impact of enhanced intraguild predation on biological control (Jonsson *et al.*, 2008). Prasad and Snyder (2006) found, for example, that creating beetle banks increased the activity density of ground living beetles in the crop but this did not lead to increased fly egg predation. The authors conclude that the deployment of beetle banks increased intraguild predation within the ground beetle community. Floral resources provided to improve fitness of certain natural enemies may also affect the pest (Lavandero *et al.*, 2006) or antagonists of biological control agents (Araj *et al.*, 2006) with negative consequences for crop protection. Generally, the basis for successful biological control combines the use of tactics that increase the relative abundance of the most effective natural enemies within the community (Straub and Snyder, 2006).

Few important terminologies on habitat manipulation and ecological engineering

a. Habitat manipulation: It is an intervention in an agro-ecosystem's vegetation with the intended consequence of suppressing pest densities.

b. Bottom-up effects (resource concentration effects): It is the action of vegetation (first trophic level) on herbivore pests (second trophic level) is called as bottom-up-effects.

c. Top-down effects (enemies hypothesis): It is the action of natural enemies (third trophic level) on herbivore pests (second trophic level) is known as top-down effects.

d. Ecological engineering: It is defined as a refinement of habitat management whereby the intervention is explicitly supported by evidence to maximize impact.

Conservation biological control approaches in orchard ecosystems

CBC is especially useful in perennial cropping systems where both pests and natural enemies are active and abundant all year round (Barbosa, 1998 and Landis *et al.*, 2000).Thus, resident populations of natural enemies may persist from year to year in perennial crops due to lower levels of disturbances. Citrus, as a permanent evergreen crop, is a system where CBC has many applications (Jacas and Urbaneja, 2010).One of the most popular CBC practices implemented in citrus orchards has been the application of selective pesticides to reduce harmful impact on parasites and predatory arthropods (Ruberson *et al.*, 1998; Jacas and Urbaneja, 2010).

On the other hand, the use of banker plants, cover and reservoir plants provide citrus growers new avenues to augment the activity of natural enemies (Jacas and Urbaneja, 2010). These beneficial plants provide refuge, hibernation or aestivation sites, alternative hosts or prey, as well as plant resources, *i.e.*, pollen or nectar for the parasites and predators as food source (Landis *et al.*, 2000 and Jonsson *et al.*, 2008).

Diversity of natural enemies in orchard ecosystems

Coccinellids are important predators of aphids and other soft bodied insects in citrus orchards. In a survey conducted in the sub-Himalayan terai region of West Bengal, a total of six species of coccinellid beetles was found mostly feeding on citrus aphids, *Toxoptera* spp. The species were *Cryptogonus quadriguttatus* Weise, *Cheilomenes sexmaculata* (Fabricius), *Jauravia pallidula* (Motschulsky), *Micraspis discolor* (Fabricius), *Cryptogonus bimaculatus* Kapur, *Synonychimorpha chittagoni* Vazarani. Among them *Cryptogonus quadriguttatus* Weise was the most abundant species (Gurung, 2017).

In a similar study on Assam lemon in Jorhat district of Assam a total of twelve species of coccinellids was found. The species were *Coccinella transversalis* (Fabricius), *Coelophora bowringii* (Crotch), *Coelophora saucia* (Mulsant), *Cryptogonus bimaculatus* (Kapur), *Cryptogonus* spp., *Cryptolaemus montrouzieri* (Mulsant), *Harmonia conglobata* (Linnaeus), *Harmonia dimidiata* (Fabricius), *Illeis confusa* (Timberlake), *Propylea* spp., *Platynaspis kapuri* (Chakraborty and Biswas) and *Scymnus* spp. Among them Coccinellids, *Coccinella transversalis* and *Harmonia dimidiata* were found to be the most abundant and *Cryptogonus* spp. was observed to be the lowest (Ramya and Thangjam, 2016).

In Jharkhand (India) 16 species of coccinellid beetles were recorded in the mango orchard of which *Anegleis cardoni*, *Cheilomenes sexmaculata* and *Illeis indica* were found to be the most frequent and abundant (Choudhary *et al.*, 2014).

Cryptogonus quadriguttatus *Cryptogonus bimaculatus* *Jauravia pallidula*

Cheilomenes sexmaculata *Micraspis discolor* *Synonychimorpha chittagoni*

Predaceous ladybird beetle diversity in the citrus orchard in sub-Himalayan West Bengal (*Source*: Gurung and Pal (2017), National Conference ZooCon 2017, OP 06.)

Shantibala and Singh (1987) studied the coccinellids preying on 9 aphid species infesting 11 different fruit plants in Manipur, India. They found 6 coccinellid species on *Aphis gossypii* Glover infesting *Artocarpus heterophyllus* (=*integrifolia*) Lam, *Psidium guajava* and *Pyrus malus* at altitudes 700-2000 m. *Hyalopterus pruni* (Geoffroy) that infests *Prunus dulcis* (Mill.) and *Prunus persica* (L.) has 9 coccinellid predators but with only one common species viz., *Cryptogonus bimaculatus* Kapur. *Rhopalosiphum nymphae* (L.) which infests *Prunus dulcis* (Mill.) from March to May has two coccinellid predators. *Citrus grandis* Merr. is infested by *Toxoptera citricidus* (Kirkaldy) which has only two coccinellid species. On the other hand, two other species of this genus, *T. odinae* (v.d.G) infesting *Rhussemia lata* and *T. aurantii* (B.d.F.) infesting *Citrus aurantium* Linn. each has only one coccinellid species. Shantibala *et al*. (1995) while studying the population dynamics of *Cervaphis quercus* Takahashi on *Quercus serrata* Thunberg observed

6 coccinellid species during the month from March to June. Singh *et al.* (1995) found 9 species of coccinellids feeding on *Tuberculatus nervatus* Chakrabarti and Raychaudhuri infesting *Quercus serrata* Thunberg at two elevations, 780m and 1750m.

Albertini *et al.* (2016) studied the role of carabids (Coleoptera: Carabidae) as a potential natural enemy of third instar larvae and pupae of *Bactrocera oleae* (Diptera: Tephritidae) in the olive orchard agroecosystem. Olive orchard supported a well-structured carabid assemblage, whose species phenologies revealed a temporal overlapping within the pest cycle. The assemblage of adjacent woody semi-natural habitats is more of conservation interest, but plays a weaker role in *B. oleae* control provisioning. Further, they suggested the identification of carabids main traits for *B. oleae* conservation biological control as a cost-effective strategy for addressing future attention and resources only to those predators that satisfy basic requirements.

Role of flowering plants in conservation of natural enemies

Natural enemies rely on flowering plants for nectar and pollen or shelter. Floral and extrafloral nectars are important food sources that can increase longevity, fecundity, searching and realized parasitism and predation as well as the female ratio of natural enemies (Gurr *et al.*, 2017). Plants commonly known to provide these resources are crops like buckwheat, swe*et al*yssum, faba bean, dill and coriander. To stop the decline of agrobiodiversity and the loss of biological control functions in intensively used landscapes, a minimum of 10% of agricultural land is recommended for ‘ecological compensation area’. Ecological compensation areas include non-crop habitats (such as hedges, set-asides, woodlots and weedy strips) and low-input agricultural habitats, for example low-input meadows such as *Mesobrometum*, a dry, unfertilised meadow (Pfiffner and Wyss, 2004). However, most of the intensively used agricultural habitats have less than 2-3% of area devoted to such ecological compensation areas, therefore additional non-crop habitats are needed. One option is sowing wildflower strips, which may also fulfill many goals of agroecology and nature conservation. Though artificial, these strips are a good supplement to natural and semi-natural habitats including other field margins. By the provision of these non-crop habitats various environmental requisites for natural enemies are supplied, like supplementary foods (alternate host or prey, or in some cases pollen); complementary foods (nectar, pollen, honeydew); modified microclimate; and shelter, for example after agricultural practices, and wintering or nesting habitat.

Crop stubbles and weeds serving as refugia for predators like *Micraspis* spp. and *Paederus* spp.

Flowering weeds on bunds serve as good source of pollen and nectar for natural enemies (*Source*: Gurung, 2017)

Use of cover crops to promote natural enemies

The results of the studies conducted in eastern Spain demonstrate that the use of the plant *Festuca arundinacea* Schreb (Poaceae) as cover crop in citrus orchards can significantly reduce red spider mite, *Tetranychus urticae* population on associated citrus trees and, consequently, the need for chemical applications (Jaques *et al.*, 2015). On the contrary, conventional alternatives as bare soil and wild cover might lead to increased *T. urticae* populations in citrus trees and hence to higher pesticide use, with the consequent enhanced production and environmental costs (Aguilar-Fenollosa *et al.*, 2011). Implementation of *F. arundinacea* as cover crop may result in better biological control of other citrus key pests, such as the Mediterranean fruit fly, *Ceratitis capitata* (Wiedemann), by increasing soil dwelling predators (Monzo *et al.*, 2009, 2010, 2011), and aphids, by promoting the early arrival of natural enemies (Gomez-Marco *et al.*, 2012).

Modifying landscape composition and configuration to promote natural enemies

Biological control is affected by composition and configuration of landscapes surrounding the agricultural fields. Natural enemy communities are typically more diverse, and effective at providing biological control services, in complex compared to simple landscapes (Cohen and Crowder, 2017).The most common method to classify landscape complexity is a binary system where all habitat patches are classified as either 'semi-natural/natural' or 'agricultural/developed'. While there is no definitive standard for what defines a 'simple' landscape, a common approach is to define landscapes with less than 20% non-crop habitat as 'simple' and those with greater than 20% non-crop habitat as 'complex' (Tschartnke *et al.*, 2005).Agricultural landscapes may also differ in compositional heterogeneity (*i.e.*, the number of different cover types) and configurational heterogeneity (*i.e.*, the spatial arrangement of cover types), being more heterogeneous when having a higher number of cover types or more smaller and irregular patches. As both compositional and configurational heterogeneity increase, so does natural enemy diversity in landscape, particularly when more natural habitats remain (Fahrig *et al.*, 2011).

Conclusion and future prospects

Studies have proved that the conservation approaches in the orchards can increase the diversity of natural enemies and improve the natural enemy fitness in the ecosystem. But, the question is whether this enhancement in natural enemy population translates in increased crop yield and more profit for the growers. Future research should be directed to divulge the utility of CBC in the yield or profit level. In addition to pest management the conservation of natural enemies leads to multiple ecosystem services also. The farmers manage complex agricultural business system and are not focused on pests in isolation. So, in future research should be carried out with active participation of farmers to develop locally appropriate form of implementation of CBC measures in a more holistic way.

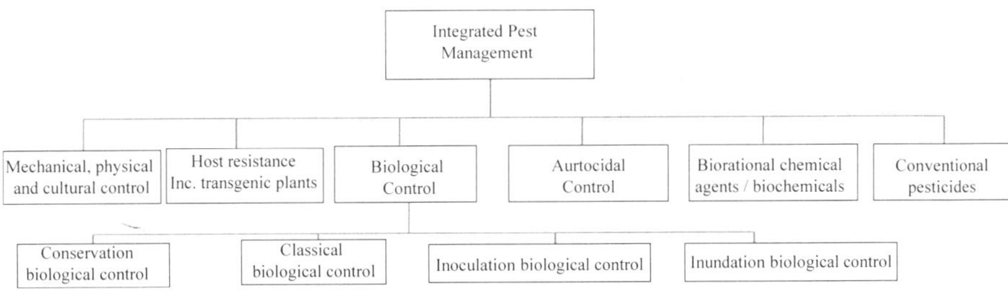

Fig. 1: Biological approaches in relation to other tactis available to integrated pest management (*Source*: Eilenberg *et al.*, 2001)

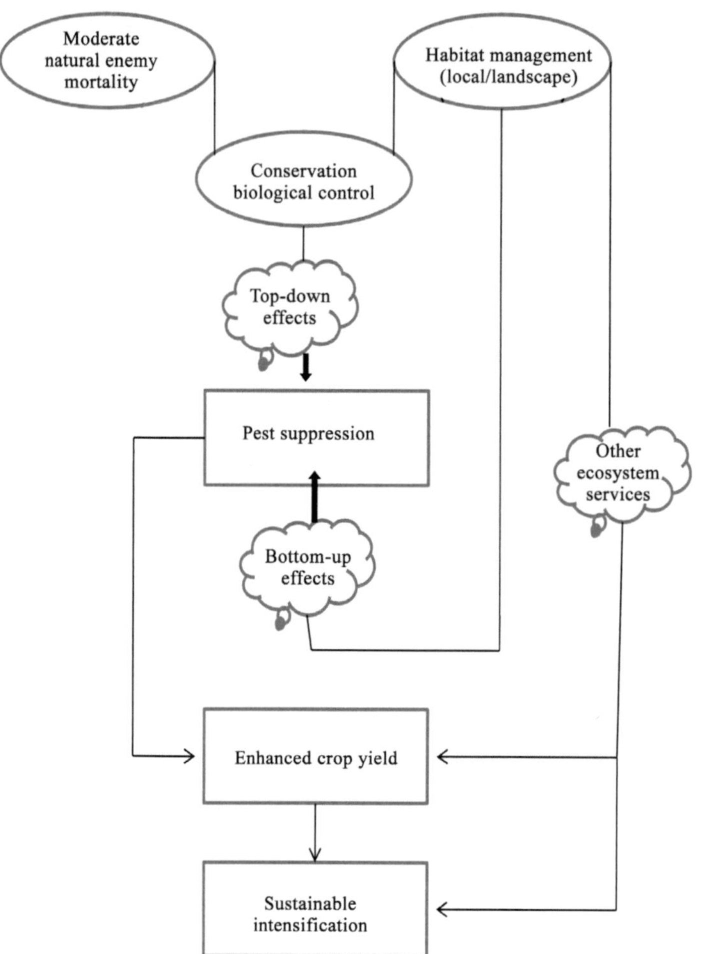

Fig. 2: Habitat manipulation in the context of conservation biological control (*Source*: Gurr *et al*., 2017)

References

Araj, S. A., Wratten, S. D.; Lister, A. J. and Buckley, H. L. 2006. Floral nectar affects longevity of the parasitoid *Aphidius ervi* and its hyperparasitoid *Dendrocerus aphidum*. New Zealand Plant Protection, 59:178-183.

Aguilar-Fenollosa, E., Pascual-Ruiz, S.; Hurtado, M. and Jacas, J. A. 2011. Efficacy and economics of ground cover management as a conservation biological control strategy against Tetranychusurticae in clementine mandarin orchards. Crop Prot, 30: 1328-1333.

Albertini, A., Pizzolotto, R. and Petacchi, R. 2016. Carabid patterns in olive orchards and woody semi-natural habitats: first implications for conservation biological control against Bactrocera oleae. Bio Control, DOI 10.1007/s10526-016-9780-x

Barbosa, P. (ed) 1998. Conservation biological control. Academic, San Diego

Choudhary, J. S.,Naaz, N.; Mukherjee, D.; Prabhakar, C. S.; Maurya, S. and DAS, B; Kumar, S. 2014. Biodiversity and seasonality of predaceous coccinellids (coleopteran: coccinellidae) in mango agro-ecosystem of Jharkhand. The Ecoscan, 8(1-2):53-57.

Cohen, A. L., and Crowder, D. W. 2017. The impacts of spatial and temporal complexity across landscapes on biological control: a review. Current opinion in insect science, 20:13-18.

Eilenberg, J., Hajek, A. E. and Lomer, C. 2001. Suggestions for unifying the terminology in biological control. BioControl, 46: 387-400.

Fahrig, L., Baudry, J. and Brotons, L. 2011. Functional landscape heterogeneity and animal biodiversity in agricultural landscapes. Ecology Letters, 14: 101-112.

Farnandez-Cornejo, J.,Nehring, R.; Sinha, E. N.; Grube, E. and Vialou, A. 2009. Assessing recent trends in pesticide use in US agriculture. Presented at the Annual Meeting of the Agricultural and Applied Economics Association (AAEA), 26-28 July, 2009. AAEA, Milwaukee, WIAvailable at Web site: http://purl.umn.edu/49271.

Gomez-Marco, F., Tena, A.; Jacas, J. A. and Urbaneja, A. 2012. Ground cover management in citrus affects the biological control of aphids. Paper presented at the XII international citrus congress, Valencia, 18-23 Nov 2012.

Gurr, G. M., Wratten, S. D.; Landis, D. A. and You, M. 2017. Habitat management to suppress pest populations: progress and prospects. Annu. Rev. Entomol., 62: 91-109.

Gurung, B. 2017. Biodiversity and bioecology of predaceous coccinellids. Thesis submitted to Uttar Banga Krishi Viswavidyalaya, Pundibari, West Bengal, India.

Gurung, B. and Pal, S. 2017. Biodiversity of Predaceous Coccinellid beetles in different crop ecosystems. In: Book of Abstracts, National Conference ZooCon 2017: Animal Science in 21st Century, February 11-12, 2017, Department of Zoology, University of North Bengal, Siliguri, Darjeeling, West Bengal, India, OP 06.

Jacas, J. A. and Urbaneja, A. 2010. Biological control in citrus in Spain: from classical to conservation biological control. In: Ciancio, A. and Mukerji, K. G. (eds) Integrated management of arthropod pests and insect borne diseases, vol 5. Springer, The Netherlands, pp. 57-68.

Jaques, J. A., Aguilar-Fenollosa, E.; Hurtado-Ruiz, M. A. and Pina, T. 2015. Food web engineering to enhance biological control of *Tetranychus urticae* by phytoseiid mites (Tetranychidae: Phytoseiidae) in citrus. In: Prospects for biological control of plant feeding mites and other harmful organisms, Progress in Biological Control 19, DOI 10.1007/978-3-319-15042-0_10

Jonsson, M., Wratten, S. D., Landis, D. A. and Gurr, G. M. 2008. Recent advances in conservation biological control of arthropods by arthropods. Biol Control, 45: 172-175.

Landis, D. A., Wratten, S. D. and Gurr, G. M. 2000. Habitat management to conserve natural enemies of arthropod pests in agriculture. Annual Review of Entomology, 45: 175-201.

Lavandero, B. I., Wratten, S. D.; Didham, R. K. andGurr, G. M. 2006. Increasing floral diversity for selective enhancement of biological control agents: a double-edged sword? Basic and Applied Ecology, 7:236-243.

Losey, J. E. and Vaughan, M. 2006. The economic value of ecological services provided by insects. Bioscience, 56:311.

Martin, E. A.,Reineking, B.; Seo, B. andSteffan-Dewenter, I. 2013. Natural enemy interactions constrain pest control in complex agricultural landscapes. Proc Natl Acad Sci USA 110:5534-5539.

Monzo, C., Molla, O.; Castanera, P. and Urbaneja, A. 2009. Activity-density of *Pardosa cribata* in Spanish citrus orchards and its predatory capacity of *Ceratitis capitata* and *Myzus persicae*. BioControl, 54:393-402.

Monzo, C., Sabater-Munoz, B.; Urbaneja, A. and Castanera, P. 2010. Tracking medfly predation by the wolf spider, Pardosacribata Simon, in citrus orchards using PCR-based gut-content analysis. Bull Entomol Res, 100:145-152.

Monzo, C., Sabater-Munoz, B.; Urbaneja, A. and Castanera, P. 2011. The ground beetle *Pseudophonus rufipes* revealed as predator of *Ceratitis capitata* in citrus orchards. Biol Control, 56:17-21.

Pfiffner, L. and Wyss, E. 2004. Use of sown wildflower strips to enhance natural enemies of agricultural pests. In: Ecological engineering for pest management: advances in habitat manipulation for arthropods. CSIRO Publishing, Collingwood, Australia, pp-165-186.

Pimentel, D., Acquay, H.; Biltonen, M.; Rice, P.; Silva, M. and Nelson, J. 1992. Environmental and economic costs of pesticide use. Bioscience, 42:750-760.

Prasad, R. P. and Snyder, W. E. 2006. Polyphagy complicates conservation biological control that targets generalist predators. Journal of Applied Ecology, 43:343-352.

Ramya, H. R. and Thangjam, R. 2016. Predatory coccinellids of insect pests of Assam lemon (*Citrus limon* L. Burmf) in Jorhat district of Assam. *Journal of Biological control*, **30**(2): 121-123.

Ruberson, J. R., Nemoto, H. and Hirose, Y. 1998. Pesticides and conservation of natural enemies in pest management. In: Barbosa, P. (ed) Conservation biological control. Academic, San Diego, pp. 207-220.

Shantibala, K., Varatrajan, R. and Singh, T. K. 1995. Population dynamics of *Cervaphisquercus* Takahashi (Homoptera: Aphididae) on *Quercusserrata* Thunberg (Fagaceae) in relation to density dependent and independent factors. *Sericologia*, *35*(4):737-742.

Shantibala, S. and Singh, T. K. 1987. Aphids and their coccinellid predators on fruit trees in Manipur, India. Journal of Aphidology, 1(1/2): 78-79.

Singh, L. S., Singh, T. K. and Varatrajan, R. 1995. Field density of the oak aphid, *Tuberculatus nervatus* Chakrabarti and Raychaudhuri (Homoptera: Aphididae) on *Quercus serrata* Thunberg (Fagaceae) in relation to predators and abiotic factors. Phytophaga, 7: 33-40.

Straub, C. S. and Snyder, W. E. 2006. Species identity dominates the relationship between predator biodiversity and herbivore suppression. Ecology, 87: 277-282.

Tscharntke, T., Bommarcob, R.; Clougha, Y.; Cristc, T.O. and Kleijnd, D. 2007. Conservation biological control and enemy diversity on a landscape scale. Biol Control, 43: 294-309.Tscharntke, T.; Klein, A. M.; Kruess, A.; Steffan-Dewenter, I.; Theis, C. 2005. Landscape perspectives on agricultural intensification and biodiversity ecosystem service management. Ecological Letters, 8:857-874.

21

Pesticide Residue in Fruit Crops and Its Management

Samrat Saha, Nripendra Laskar and Victor Phani

India is an agrarian country where around 70% of the total population is engaged in farming activities. Agriculture and allied sectors are undoubtedly the largest livelihood provider of India, along with significant contribution towards the Gross Domestic Product (GDP) of the country. Agricultural production has been raised dramatically in the recent past to feed our burgeoning population with the help of modern inputs like fertilizers, pesticides, mechanical devices for tilling, seeding and irrigating the crops. With the advancement of agro-inputs, incidence of various kinds of pests also increases that hugely threaten our food security. Additionally, the technological advancement and changing climatic condition are joining hands to shape this new dreaded scenario. Beyond the threshold level pests become noxious, destructive and troublesome to human and his interests. As a result, cultivated crops experience challenges due to increasing pressure of insects, weeds, fungi, viruses, nematodes, protozoa, MLOs, rodents, mollusks etc. Amongst different categories of agricultural pests insects are one of the most important that cause a huge loss every year by affecting the product qualitatively as well as quantitatively. Owing to combat the problem, the farmers largely depend on insecticides (pesticides) as they are cheap, easy to use and offer quick visible impact in protecting the crops.

Pesticides are mixture of substances used for preventing, destroying, controlling any pest including the vectors of human being or animal diseases, unwanted species of plants or animals causing harm during or otherwise interfering with the production, processing, storage, transport or marketing of food and agriculture commodities (FAO, 2002). Thus, pesticides play pivotal role in increasing the yield of agricultural products. Pesticides can be classified in different ways according to the target pests, chemical structure of the compounds, mode of action, and degree and type of health hazard involved. Currently, about 4 million tonnes of pesticides are used per year in the world, most of which are herbicides (56%), followed by the fungicides (25%), insecticides (19%), and others like nematicides and rodenticides (FAO, 2018).

Historically, the use of chemical pesticides in agriculture started with the use of elemental sulfur by the Sumerians during 2500 B.C. for management of plant diseases. By the 15th century different toxic chemicals like arsenic, mercury and lead were used to kill the pests of agricultural crops. The growth of synthetic pesticides was accelerated with the discovery of insecticidal properties of DDT by Paul Muller in 1939. Post introduction of DDT a number of synthetic organic insecticides like organophosphates, carbamates, synthetic pyrethroids were discovered propelling huge food production; and the agricultural industry breathed a short-term sigh of relief by sorting out the pest problem.

Pesticide in Indian agriculture

India ranks 12th in the world and 3rd in Asia following China and Turkey in terms of pesticide consumption, which is 1% of the total global pesticide use. According to the data of FAO (2018) India utilized around 58160 ton of pesticides, with application rate of 0.31 kg per hectare in 2017. However, the application rate in developed countries like China, Japan and America was near about 13.07, 11.76 and 3.57 kg per ha, respectively. So it is clear that pesticide use per unit area is very less in India, but non-judicious use of pesticide is responsible for the presence of high residues of pesticides in both physical and natural environments. Only around 0.1 percent of the applied pesticides are believed to reach the targeted organisms and the rest pollute the environment causing environmental hazards (Carriger *et al.*, 2007; Gill and Garg, 2014).

Presently India follows a pesticide use pattern highest with insecticides, followed by herbicides, bactericides, fungicides, and others; whereas in other parts of the world herbicides rank the first followed by bactericides, fungicides, insecticides, and others. According to the data obtained from Research and Markets, the Indian pesticide market worth Rs. 214 billion in 2019; and by 2024, it is expected to reach Rs. 316 billion with 8.1 percent compound annual growth rate (TAAS, 2020). Consumption wise, Maharashtra ranks first, followed by Uttar Pradesh, Punjab and Haryana. On the other hand, Punjab (0.74 kg) ranked first in consumption of pesticides per hectare followed by Haryana (0.62 kg) and Maharashtra (0.57 kg) during 2016-17(Pesticide Management Bill, 2020).

Table 1: Year-wise chemical pesticide consumption in India

Year	2014-15	2015-16	2016-17	2017-18	2018-19
Grand total of chemical pesticide consumption in India (Unit: Quantity in MT Tech. Grade)	56121	58221	52755	62183	53453

(*Source*: Government of India, Ministry of Agriculture & Farmers Welfare, Department of Agriculture, Cooperation & Farmers Welfare, Directorate of Plant Protection, Quarantine & Storage)

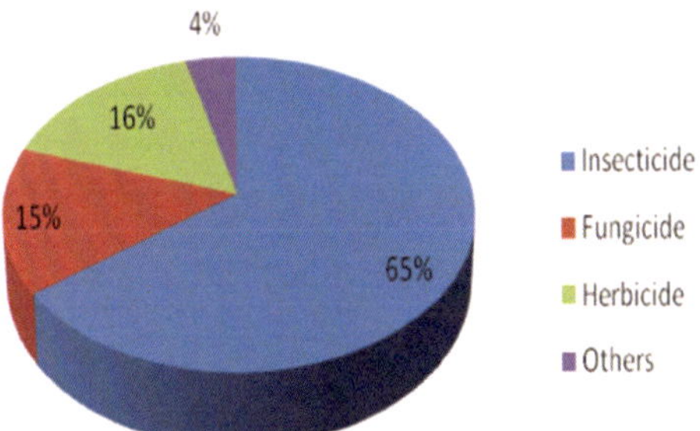

Fig. 1: Share of different pesticides in India, FY-2012 (Source: Industry reports, Analysis by Tata Strategic)

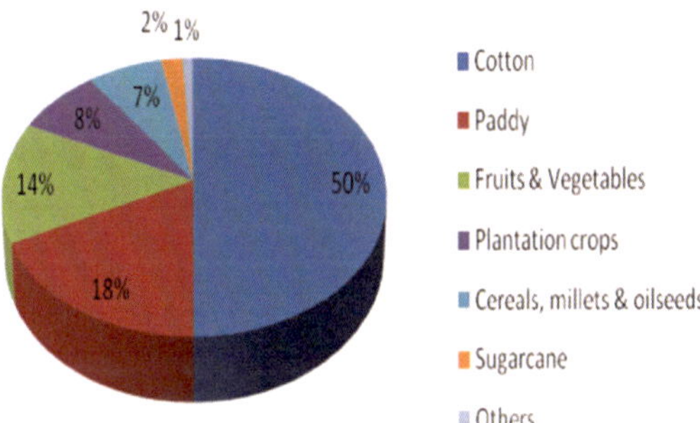

Fig. 2: Crop-wise pesticide consumption in India, FY-2012 (Source: Industry reports, Analysis by Tata Strategic)

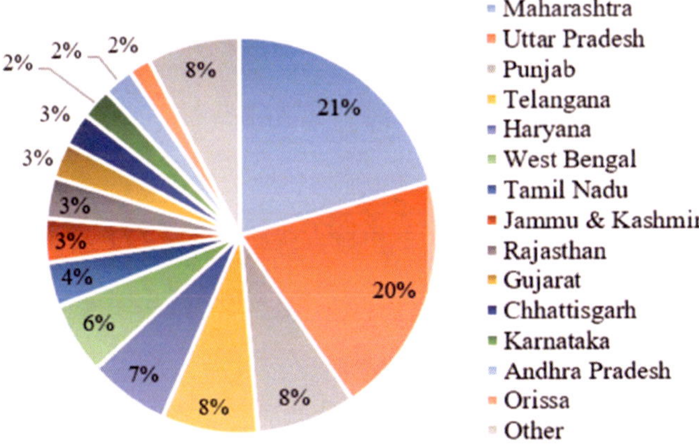

Fig. 3: State wise pesticide consumption in India (2019-2020) (Source: GOI, 2020)

Table 2: Insecticides and formulations registered for use in the country under the Insecticide Act, 1968.

S. No.	Name of the insecticide	Formulation registered
1.	Acetamiprid	20 SP
2.	Acephate	75% SP
3.	Allethrin	0.5% Coil, 4% Mat, 0.5% Aer., 3.6% L, 0.2% & 0.02% Coil
4.	Alphacypermethrin	10% EC, 5% WP, 0.5% Chalk, 10% SC, 0.1%RTU
5.	Azadirachtin (neem products)	25%, 10%, 0.03% EC 0.1 EC, 0.15 EC, 5 EC, 0.3% 15% extract concentrate, 1% EC, 0.1% Gr
6.	Bacillus thuringiensis (Bt)	Liquid & WP formulations, 5% AS
7.	Barium Carbonate	1% P
8.	Beta cyfluthrin	2.45% SC
9.	Bendiocarb	80% WP
10.	Benfuracarb	40%EC, 3.0% GR
11.	Bifenthrin	10% EC, 2.5% EC, 23.4%, MUP (Imp), 8% SC(FI), 0.05% MC(11 Hrs.), 10% WP
12.	Bifenozates	22.6% SC
13.	Buprofenzin	25% SC
14.	Carbaryl	5% DP, 10% DP, 50% WP, 85% WP, 4% Gr., 40% LV, 42% Flow
15.	Carbofuran	3% CG, 50% SP for Government use.
16.	Carbosulfan	25% DS, 25% EC, 6% Gr.
17.	Carpropamid	27.8% SC
18.	Cartap Hydrochloride	4% Gr., 50% SP
19.	Chlorfenapyr	10% SC (FI)
20.	Chlorfluazuron	5% EC W/V, 5% w/v, 5.4% w/w EC
21.	Chlorofenviphos	10% Gr.
22.	Chromafenozide	80% WP
23.	Chlorpyriphos	20% EC, 10% Gr., 1.5% DP, 50% EC, 2% RTU
24.	Chlorpyriphos Methyl	40% EC
25.	Chlorantraniliprole	18.5% SC, 0.4% Gr.
26.	Clothianidin	50%WG(FI), 50% WG formulation
27.	Cyantraniliprole	-
28.	Cyfluthrin	10% WP, 5% EW, Cyfluthrin + Propoxur (0.5%) (0.015%)
29.	Cypermethrin	10% EC, 25% EC, 1% Chalk, 0.1% Aquous (HH), 0.25 DP, 3% Smoke Generator
30.	Cyphenothrin	5% EC, 0.15% in combination as Aer. , 7.2% VP
31.	Decamethrin (Deltamethrin)	2.5% Flow, 2.5% WP, 2.8% EC, 0.5% Chalk, 1.25% ULV, 25% Tab., 11% EC, 0.5% Tablet bait
32.	Diafenthiuron	50% WP
33.	Diazinon	20% EC, 10% Gr, 2% DP, 40% WP, 5% Gr., 25% Micro Encapsulation

S. No.	Name of the insecticide	Formulation registered
34.	Dichloro Diphenyl Trichloroethane (DDT)	50% WP, 75% WP
35.	Dichloropropene and Dichloropropanes mixture (DD Mixture)	1:1
36.	Diclorvos (DDVP)	76% EC
37.	Diclofop-methyl	28% EC
38.	Dicofol	18.5% EC
39.	Diflubenzuron	25% WP, 2% Tab, 2% GR. (FI)
40.	Dimethoate	30% EC
41.	Dinetofuran	20% WG (F.I.)
42.	Dithianon	75% WP
43.	D-trans allethrin	2% Mat, 0.1% coil, 0.1% coil (12 hr.)
44.	Emamectin Benzoate	5% SG (FI) & (FIM) , 1.9% EC
45.	Endosulfan	2% DP, 4% DP, 35% EC, 4% Gr.
46.	Ethephon	39% SL, 10% Paste
47.	Ethion	50% EC
48.	Ethofenprox (Etofenprox)	10% EC
49.	Ethylene Dichloride and Carbon Tetrachloride mixture (EDCT mixture 3: 1)	3:1
50.	Fenazaquin	10 EC
51.	Fenitrothion	5% DP, 40% WP, 50% EC, 82.5% EC, 2% Spray, 20% OL
52.	Fenobucarb (BPMC)	50% EC
53.	Fenpropathrin	10% EC, 30% EC
54.	Fenthion	82.5% EC, 2% Gr., 2% Spray
55.	Fenvalerate	0.4% DP, 20%EC
56.	Fenpyroximate	5% SC,
57.	Fipronil	0.3% Gr., 5% SC, 0.05% Gel (Import) & FIM, 80% WG, 2.92% EC
58.	Flonicamide	50% WG
59.	Flubendiamide	39.35%SC,20%WG
60.	Flumite	20% SC
61.	Flufenoxuron	10% DC
62.	Fluvalinate	25% EC
63.	Fosetyl-Al	80% WP
64.	Hexythiazox	5.45% EC
65.	Hydrogen cyanamid	49% age, 50% SC
66.	Imidacloprid	17.8% SL, 70% WS, 48% FS, 30.5% SC, 2.5% Gel, 70% WG, 0.3% SG
67.	Imiprothrin	50% MUP

S. No.	Name of the insecticide	Formulation registered
68.	Indoxacarb	14.5% SC,15.8%EC
69.	Kresoxim-methyl	44.3%
70.	Lambda cyhalothrin	5% EC, 10% WP, 2.5% EC, 0.5% Chalk, 22.9% CS (FI), 4.9%
71.	Lime Sulphur	22% SC
72.	Lufenuron	5.4% EC
73.	Malathion	5% DP, 25% WP, 50% EC, 0 .25% Spray and 96% ULV, 2% Spray, 5% Spray
74.	Milbemectin	1%EC
75.	Metaflumizone	22% SC (FI)
76.	Methomyl	40% SP
77.	Methabenzthiazuron	70% WP
78.	Methyl Parathion	2% DP, 50% EC
79.	Monocrotophos	36% SL, 15% w/w SG
80.	Myclobutanil	36% SL, 15%SG,10%WP
81.	Novaluron	10% EC (FI), 8.8% SC, 10% EC
82.	NPV of *H. armigera*	0.43% AS, 2.0% AS
83.	NPV of *Spodoptera litura*	0.5% AS
84.	Oxydemeton-methyl	25% EC
85.	Pencycuron	22.9% SC
86.	Permethrin	25% EC, 5% SG., 2% EC
87.	Phenthoate	2% DP, 50% EC
88.	Phorate	10% CG
89.	Phosalone	4% DP, 35% EC
90.	Phosphamidon	40% SL,
91.	Primiphos-methyl	25% WP, 50% EC, 1% Spray
92.	Prallethrin	0.8% mat for 12 hours, 1% Mat, 0.8% L, 1.6% L, 0.5% mosquito coil, 0.04% Mosquito coil, 1.2% mat, 19% w/w VP, 0.6% mat
93.	Profenophos	50% EC
94.	Propanil	35% EC
95.	Propergite	57% EC
96.	Propetamphos	20% EC, 1% Spray
97.	Propineb	70% WP
98.	Propoxur	20% EC, 1% Aer., 2% Aer. 1% HH Spray, 2% Bait
99.	Pyrethrins (Pyrethrum)	0.2% DP, 2.5% EC, 0.05% Spray, 0.2% PH, 2.0% EC
100.	Pyriproxifen	0.5% Gr
101.	Quinalphos	1.5% DP, 25% EC, 20% AF
102.	S-Bioallethrin	2.4% mat
103.	Sodium Cyanide	Used as Tech., 96% a.i. min
104.	Spinosad	45% SC, 2.5% SC

S. No.	Name of the insecticide	Formulation registered
105.	Spiromesifen	22.9% SC
106.	Sulphur	85% DP, 80% WP, 40% SC, 80% WG/WDG , 55.16 SC (800 gm / L) 40% WP, 52% SC
107.	Temephos	50% EC, 1% Sand Granules
108.	Thiodicarb	75% WP
109.	Thiomethoxiam	25% WG, 70% WS, 30%FS
110.	Thiometon	25% EC
111.	Thiacloprid	21.7% SC
112.	Thifluzamide	24% SC
113.	Transfluthrin	0.88% Liquid Vaporiser, 0.03% Mos. Coil, 20% MV Gel(30 days mat tray), 1% FU, 1.2% LV, 1.6% LV
114.	*Trichoderma viride*	1% WP (2×106 CFU/g), 0.5% WP, 5% WP
115.	Triazophos	40% EC, 20% EC
116.	Trichlorfon	5% DP, 50% EC, 5% G
117.	*Tricoderma harzianum*	0.5% WS

(Source: Directorate of Plant Protection, Quarantine & Storage, Central Insecticides Board & Registration Committee, Faridabad, 121001, Department of Agriculture & Cooperation, Ministry of Agriculture, GOI)

Table 3: Approved formulation of combination insecticides.

Sl.No.	Combination Product	Company
1.	Acephate 25% + Fenvalerate 3% EC	M/s Rallis India Ltd., Bangalore
2.	Acephate 5% + Imidacloprid 1.1%	M/s United Phosphorus Ltd.
3.	Acephate 50% + Imidacloprid 1.8% SP	M/s United Phosphorus Ltd.
4.	Acetamiprid 0.4% + Chlorpyriphos 20% EC	M/s Gharda Chemicals Ltd, Mumbai
5.	Beta cyfluthrin 8.49% + Imidacloprid 19.81% OD	
6.	Carbaryl 4% + Gamma BHC 4% Gr.	M/s Avantis Crop Science India Ltd., Mumbai
7.	Chloropyriphos 16% + Alphacypermethrin 1% EC	M/s Acco Industries Ltd., Mumbai
8.	Chlorpyriphos 50% + Cypermethrin 5% EC	M/s De-Nocil, Mumbai
9.	Cyfluthrin 0.025% + Tranfluthrin 0.04%	M/s Bayer India
10.	Cymoxanil 8% + Mencozeb 64% WP	EI Dupont
11.	Cypermethrin 3% + Quinalphos 20% EC	M/s United Phosphorus Ltd., Mumbai
12.	Deltamethrin 0.75% +Endosulfan 29.75% EC	
13.	Deltamethrin 1% + Triazophos 35% EC	
14.	Deltamethrin 0.05% + Allethrin .04% L	
15.	Endosulfan 35% + Cypermethrin 5% EC	M/s Excel Industries Ltd., Mumbai
16.	Ethion 40% + Cypermethrin 5% EC	M/s Rallis India Ltd., Bangalore.

Sl.No.	Combination Product	Company
17.	Ethiprole 40% + Imidacloprid 40% (80% WG)	M/s Bayer Crop Science Ltd, Mumbai
18.	Fipronil 40% + Imidacloprid 40% WG	
19.	Indoxacarb 14.5% + Acetamiprid7.7% SC	M/s Rallis India Ltd., Bangalore.
20.	Imiprothrin 0.1% + Cyphenothrin 0.15%	
21.	Imiprothrin 0.05% + Cypermethrin 1.0% CL	M/s Godrej Consumer Products Ltd., Mumbai
22.	Methyl bromide 98% + chlorpicrin 2%	
23.	Novaluron 5.25% + Indoxacarb 4.5% SC	M/s Makhteshim Agan India Pvt Ltd, Hyderabad
24.	Phosalone 24% + Cypermethrin 5% EC	M/s Aventis Cropscience Ltd..
25.	Phosphamidon 40% + Imidacloprid 2% SP	M/s United Phosphorus Ltd.
26.	Profenofos 40% + Cypermethrin 4% EC	M/s Novartis Crop Protection Ltd., Mumbai
27.	Propoxur 0.25% + cyfluthrin 0.025% Aerosol	M/s Bayer India
28.	Propoxur 0.5% + Cyfluthrin 0.025% Spray, Propoxur 0.5% + Cyfluthrin 0.015% Spray	
29.	Pyriproxyfen 5% + Fenpropathrin 15% EC	M/s Sumitomo Chemical India Pvt. Ltd.

(Source: Directorate of Plant Protection, Quarantine & Storage Central Insecticides Board & Registration Committee, Faridabad, 121001, Department of Agriculture & Cooperation, Ministry of Agriculture, GOI)

Role of pesticides in growth of agricultural industry

Tremendous success was achieved by use of pesticides in agricultural activities and public health (Aktar *et al.*, 2009). Nowadays, pesticides are an indispensable part of agricultural industry being used to control a wide range of pests. Since the beginning of the 20th century, the agricultural yield has increased significantly to cope up with the demographic growth. The world population increased from 1.5 billion in 1900 to about 6.1 billion in 2000, corresponding to a three times more growth rate than during the entire history of humanity. Since 2003, there was an increase of another billion in population, and it is projected to reach 9.4–10 billion by 2050 according to the current growth statistics (Carvalho, 2017). Survival of this humongous population is not possible without a parallel surge in food production. Notably, if pesticides are not used there will be an estimated loss of 78, 54 and 32 per cent in fruit, vegetable and cereal production, respectively (Lamichhane, 2017). Thus, pesticides become a significant contributor to alleviate hunger by providing sufficient supply of food.

Pesticides also provide a quick knock down effect to the obnoxious pests and pathogens with long-term sustainability in farm and agribusiness revenues, food safety, nutrition and health improvement, quality of life, life expectancy, maintenance costs, international spread of diseases, export revenues, workforce productivity etc. (Cooper and Dobson, 2007). Further, higher agricultural productivity using proper and judicious pesticides significantly increase income of the farming families (Miller, 2016).

Insect pest problems and insecticides in fruit production

"An apple a day, keeps the doctor away", every person has heard that phrase at least once in life. This is quite true as fruits constitute an important part of our life. During ancient times, human often relied on nature's wonderful gifts like fruits and vegetables to survive. Fruits play an important role in building health and supplying nutrition containing a variety of vitamins, (especially vitamins A and C), minerals (electrolytes), essential phytochemicals, antioxidants etc. Additionally, fruits are also recommended as a good source of dietary fiber (Slavin and Lloyd, 2012). Most nutritional and global recommendations stated that an adult should consume at least 2 servings of fruits per day (WHO, 2002). Eating a diet high in fruits can reduce the risk of developing cancer, heart disease, diabetes and inflammation.

India has a diverse climatic condition that ensures availability of all varieties of fresh fruits. According to National Horticulture Database (Second Advance Estimates) published by National Horticulture Board, India produced around 99.07 million metric ton of fruits during 2019-20 and around 6.66 million hectares area is under cultivation of fruit crops (APEDA, 2022). India also exports huge quantity of fruits in the world. During the year 2020-21 India has exported 609,612.91 MT of fresh fruits other than Grapes and Mango which worth of USD 302.00 Million (APEDA, 2022). With increase in production, pest problem pose a major hindrance in quality food production. As per 1996 estimates, globally insects cause 6% fruit crop losses even after the use of insecticides and when insecticides were not used for protection, these losses may reach up to 23% (Krattiger 1997). Insects can reduce the yield of fruit crops through direct damage (by affecting different parts of the fruit crops, viz. flowers, fruits, foliage, twigs, stem) and also act as vectors of various disease-causing pathogens. Besides the existing insect pests, invasion of exotic insect pests are also becoming a threat to the fruit industry. With import of foreign commodities, a number of exotic insect species has already entered in our country. India is also rich in cultural heritage and attracts numerous tourists every year which accidentally introduce insects or other organisms (Kiritani and Yamamura, 2003). Thus, the problems of pests in fruit production are increasing day by day creating multifarious hindrances in fruit production. To overcome these problems, the farmers are compelled to use several hazardous toxic chemicals on the fruit crops. Indiscriminate use of toxic chemicals in the name of crop protection are polluting the environment, killing non-target organisms, destabilizing ecosystem leaving residual toxicity in the produce. This is obviously a matter of great concern to the consumers, as well as the growers.

Post application fate of insecticides

Fate of an insecticide refers to the pattern of distribution of it, its derivatives or metabolites in an organism, system, compartment or population of concern as a result of transport, partitioning, transformation or degradation (OECD, 2003). Through different transfer processes such as adsorption, leaching, volatilization, spray drift, and runoff, insecticides relocate from the target site to other non-target sites (Robinson *et al*., 1999). The different groups of chemicals differ in their environmental behavior. For example, organochlorine compounds like DDT have low acute toxicity but have a higher tendency to accumulate in tissues and can persist in nature for longer period and cause long-term damage. While organophosphate insecticides have low persistence but have appreciable acute toxicity in mammals (Kim *et al*., 2017).

Insecticide Migration

Sorption

Post application of insecticides, only a small portion offer protective role in fighting against the insects. In contrast, a large amount of insecticides reach the soil and causes severe soil pollution (Qin *et al*., 2014). The sorption is a process in which insecticides bind to soil particles due to attraction between chemical and soil particles (Qin *et al*., 2014). The soil absorption is influenced by factors like organic matter content, pH and soil amendment. Soils which are rich in organic matter or clay shows much adsorptive potential to insecticides because they have a greater surface area of particle or more sites for binding of insecticides whereas sorption is low in coarse, sandy soils (Boskovic *et al*., 2020). Dry soils have higher potential of absorbing more insecticides compared to wet soils as in wet soils there is a competition for binding sites between water molecules and insecticides. The colloid of humic acid influence the adsorption of DDT in sediments (Gao *et al*., 2014).

Leaching

Worldwide large amount of insecticides leach to the groundwater causing ground water pollution. Several factors are responsible for leaching (Singh, 2012). Singh (2002) indicates that solubility of insecticides is an important factor for leaching. Water soluble insecticides can easily dissolve and mix with soil water. Another crucial factor is soil permeability that influence insecticide leaching (Fontana *et al.*, 2010). Furthermore, it also depends on persistence of insecticide in the environment. If an insecticide has low persistence it is less likely to leach because of its short time of presence in the environment (Geng *et al.*, 2017). For example, imidacloprid has higher persistence (DT50 in soil = 187 days), thus it has a higher environmental fate. Moreover, meteorological factors like annual average temperature and rainfall also influence the leaching process (Singh, 2012). Precipitation influence the flux of downward leaching to groundwater along with insecticide solutes and temperature acts on the evapotranspiration of soil which, in turn, affects the leaching process.

Spray drift

Spray drift is the transportation of spray droplets through wind activity from site of treatment during pesticide application and cause environmental pollution and food contamination (Singh, 2012). For example, aquatic ecosystems are contaminated by various insecticide residues, like chlorpyrifos, because of spray drift and agricultural runoff from adjacent crop field and cause toxicity in aquatic organisms. Thus the sub-lethal concentrations of chlorpyrifos can induce oxidative stress and histological alterations in the tilapia tissue (Farhan *et al.*, 2021). In modern agriculture unmanned aerial vehicle (UAV) have been adopted in several commercial sectors and application of insecticides at low volume using fine and very fine droplets, causes spray drift hazards which could raise the environmental and regulatory concerns.

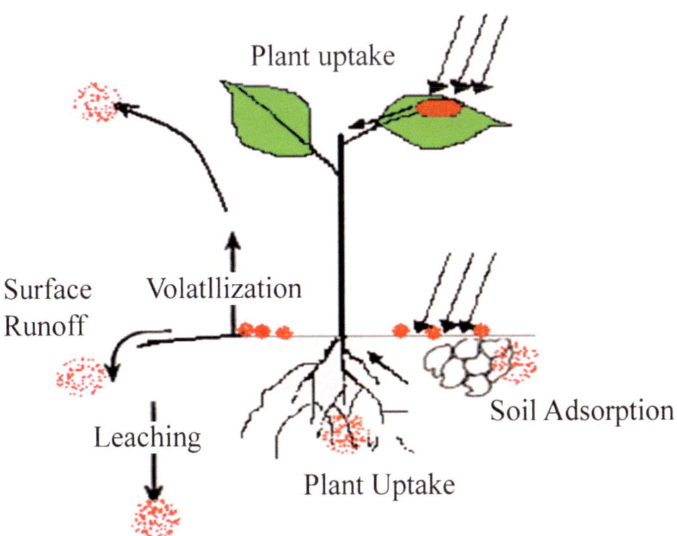

Fig. 4: Transportation of the applied insecticide

Volatilization

Volatilization refers to conversion of liquid or solid into a gas. Once insecticides have been volatilized, air currents carry the particles away from the treated surface (Singh, 2012). Factors that determine the volatilization of insecticide are vapor pressure, air movement, temperature, humidity, and soil factors like texture, moisture and organic matter content. Higher vapor pressure makes the insecticides more volatile. In addition, low relative humidity, air movement and high temperatures tend to increase volatilization (Chakraborty *et al.*, 2015). An insecticide which is tightly adsorbed to soil particles shows less tendency of volatilization.

Surface runoff

Runoff is the movement of insecticides in water through a sloping surface. Runoff is related with factors like slope or grade of an area, texture and moisture content of the soil, irrigation and amount and timing of rainfall. Runoff occurs when the speed of water in the field is fast that soil is unable to absorb it. Over-irrigation also causes insecticide runoff. Insecticide runoff from the agricultural lands eventually pollutes the water bodies affecting the in-built ecosystem.

Insecticide degradation

Post application of insecticides to the target organism, they get degraded by various chemical reactions, light or microbes (Abian *et al.*, 2002). Depending on the conditions of environment and chemical characteristics of the insecticides, degradation of insecticides may take from hours to several days or even years (Tcaciuc *et al.*, 2018). Insecticide degradation processes control persistence of pesticide in soils and form different metabolites (Tariq and Nisar, 2018). It also gives the half-life concept of the insecticides in environment. For example, the major metabolite produced from chlorpyrifos is 3,5,6-trichloro-2-pyridinol (TCP),which is much mobile and toxic than its parent insecticidal compound (Zhao *et al.*, 2016). These compounds were frequently detected in soils, groundwater and sediments having potential endocrine-disruption properties

There are 3 types of insecticide degradation, microbial degradation, chemical degradation and photochemical degradation. Microbial degradation is the process of insecticides degradation by the activity of microorganisms such as bacteria and fungi. For example, biodegradation is the main path of degradation of niclosamide by aerobatic and anaerobic microorganisms (Luo *et al.*, 2018). There are several factors that influence microbial degradation of insecticides such as soil pH, soil moisture, oxygen, temperature, and soil porosity. Insecticides being organic molecules can serve as the substrates for microbial growth and energy. These insecticides support microbial growth if it is metabolized to supply C, N, S and energy. If these insecticides are added to the soil, the microbial population increases. On the other hand, some insecticides may be modified by microorganisms, but do not serve as a nutrient source. The microorganisms do not obtain energy from the transformation reaction and require another substrate for growth. This normally results in simple modification of that insecticide. This type of metabolism is called co-metabolism. Insecticides that are co-metabolised are usually relatively stable. But the rate of co-metabolism of an insecticide can be enhanced by adding the substrate analog, which can be metabolized by the organisms. In many tropical areas characterized by intermittent heavy rain and dry seasons, soils are subjected to alternate periods of flooding and drying with concomitant increases in the activities of anaerobic and aerobic microorganisms, respectively. Such alternate reduction and oxidation cycles in the soil could provide a favorable environment for more extensive destruction of organic compounds than in either system alone (Sethunathan *et al.*,1982). For example, diazinon was readily cleaved via hydrolysis in flooded soils, but complete mineralization of the resulting aromatic-ring metabolite only occurred under aerobic conditions following anaerobiosis (Walker and Welch, 1992).

Pesticides also can be degraded through chemical reactions in the soil, called as chemical degradation. The rate and type of chemical degradation influenced by factors like soil temperature, moisture, pH levels and binding of insecticides to soil. The most important reaction is hydrolysis, a pH-influenced reaction that many organic molecules undergo in the presence of water. Even very dry field soil has enough moisture for some hydrolysis to occur. Many insecticides, especially the organophosphate and other ester type, are prone to degradation by hydrolysis in high pH soils. Moreover, degradation of molecules on soil surfaces may occur because of the chemical reaction by solar radiation (Quan *et al.*, 2015). All insecticides are photo-degradable to some extent, and the factors influencing the rate of degradation include intensity of light, properties of the insecticide and length of exposure (Singh, 2012). For example, niclosamide is photo degraded to 5-chlorosalicylic acid and 2-chloro-4-nitroaniline (Luo, 2018).

Pesticide residue

The promising results in protecting agricultural produce from pesticide application as pest control agent simultaneously led to indiscriminate use of pesticides. The misuse and abuse of pesticides ultimately resulted in three sad 'R's. The sad 'R's are as follows:

1. Resistance
2. Resurgence
3. Residue

Pesticide residue is the substances detected in any matrix resulting from the use of pesticide including any specified derivatives such as degradation and conversion products, metabolites, reaction products and impurities known to cause adverse effects on living organisms. Indiscriminate use of pesticides destabilizes ecological balance causing environmental pollution. Additionally, in view of non-biodegradable nature of some of the pesticides, the residues are bio-accumulated and bio-magnified in the food chain resulting in hazardous effects on human health. The amount of residues primarily depend on nature of pesticides, environmental conditions, good agricultural practices (GAP), waiting periods and storage conditions.

The possible reasons for pesticide residues are-

a. Indiscriminate use of pesticides

b. Non-observance of prescribed waiting periods

c. Use of counterfeit pesticides

d. Wrong advice and supply of pesticides to the farmers by pesticide dealers

e. Continuance of DDT and other use of pesticides in Public Health Programmes

f. Effluents from pesticide manufacturing units

g. Wrong disposal of left over pesticides and cleaning of plant protection equipment

h. Pre-marketing pesticides and

i. Post-harvest treatment of fruits and vegetables.

Health hazards associated with pesticide contaminated fruits

As compared to other food materials, health hazard associated with pesticide residue in fruits are more detrimental. As the cereals, pulses, vegetables, spices are consumed with good cook and boil, several residue molecules get degraded or converted into less toxic forms. But in case of fruits, we consume it raw. So there is higher chance of consuming more pesticides that increases intake of more pesticide residue in our daily life. There remain a number of signs which indicate that the person is affected by pesticides; however the symptoms are similar to that of other illnesses. Health hazards associated with pesticide consumption includes headache, sweating, irritation of nose and throat, fatigue, diarrhea, insomnia, nausea, weakness, eye irritation, restlessness, loss of appetite, skin irritation, dizziness, nervousness, sore joints, thirst, vomiting, inability to breathe, loss of reflexes, unconsciousness, increased rate of breathing, fever, damage to nerves, thirst, convulsions, reproductive problems, damage to various organs, constriction of eye pupils, hormonal effects, some cancers, muscle twitching, impaired metabolism, reduced ability of blood to clot etc. In long term, there remains every chance for some other chronic or acute health problems with no preliminary outward symptoms.

To combat the problems of pesticide residue in food commodity, the Government of India initiated All India Coordinated Research Project on Pesticide Residue (AICRP on PR)during the year 1984-85 under the aegis of Indian Council of Agricultural Research. Its objective is to overcome the problem of widespread contamination of pesticides in various food commodities and environmental factors such as water and soil in India. Later on, the project was re-designed as the All India Network Project on Pesticide Residues (AINP-PR) and is still continuing. The project presently holds 19 coordinated centres, with project coordinating cell (Nodal Centre) at ICAR-IARI, New Delhi. A total of 14 laboratories are accredited by the National Accreditation Board for Testing and Calibration of Laboratories (NABL) in the field of chemical testing as per ISO/IEC 17025 to ensure quality data generation. The project has been generating persistence or residue data of pesticides on crops in different agro-climates for fixing the Pre-Harvest Interval (PHI), approval of label claim and fixation of maximum residue limits (MRLs) for the safety of the consumers. Department of Agriculture, Cooperation & Farmers' Welfare, Ministry of Agriculture & Farmers' Welfare sponsored central sector scheme on "Monitoring of Pesticide Residues at National Level" in 2017-18, in which 27 NABL accredited laboratories participated from different parts of India, collected and analysed the samples of rice, wheat, pulses, fruits, vegetables, spices, tea, meat, egg, curry leaves, milk, fish/marine, red chilli powder and water from retail outlets, APMC markets, mother dairy, organic outlets and farm gate for examining the presence of possible pesticide residues. In this scheme a total of 2,274 fruit samples were collected and analysed by 17 participating laboratories for the presence of pesticide residues. Samples include apple, mango, orange, banana, pear, pomegranate, grapes, guava and sapota. It was found that 1780 (78.3 %) fruit samples did not contain any residue and remaining in 494 (21.7 %) samples were detected with residue out of which only 25 (1.1 %) samples were found to contain residue exceeding FSSAI MRL. Among the fruit samples 277 (12.8%) samples were detected to contain non-approved pesticides.

Table 4: Center-wise monitoring data of pesticide residue in fruit samples.

Centers	Sample analysed	Samples with no detected residues	Samples with detected residues	No of samples with detection of non-approved pesticides	Samples Above FSSAI MRL
AAU, Anand	149	113	36	17	3
BCKV, Kalyani	120	112	8	0	0
CIARI, Port Blair	6	6	0	0	0
Dr. YSPUH&F, Solan	220	192	28	7	0
IIHR, Bangalore	104	47	57	35	4
IITR, Lucknow	82	81	1	0	0
IPFT, Gurgaon	108	101	7	2	0
KAU, Vellayani	117	92	25	17	9
MPKV, Rahuri	60	58	2	1	0
NIOH, Ahmedabad	98	97	1	1	0
NIPHM, Hyderabad	202	127	75	49	1
NPQS, Delhi	275	271	4	1	0
PC Cell, Delhi	199	123	76	58	4
PJTSAU, Hyderabad	128	55	73	42	2
RPQS, Chennai	156	99	57	33	1
RPQS, Mumbai	138	114	24	3	1
TNAU, Coimbatore	112	92	20	11	0
Grand Total	2274	1780 (78.3%)	494 (21.7%)	277 (12.8%)	25 (1.1%)

(*Source*: AINP on PR, ICAR, Govt. of India)

Table 5: State-wise prevalence of pesticide residues.

Sl.No.	State	Predominant pesticide residues found in food commodities above PFA MRL
1.	Gujarat	Chlorpyriphos & Monocrotophos, HCH-D, HCH-B & HCH-A
2.	West Bengal	Cypermethrin, Chlorpyriphos, Endosulfan-T, Fenitrothion & Fenpropathrin
3.	Kerala	Chlorpyriphos, Quinalphos, Phorate
4.	Punjab	Chlorpyriphos
5.	Tamil Nadu	Chlorpyriphos, Dichlorvos, Aldrin, Phorate & HCH-B
6.	Maharashtra	Monocrotophos, Endosulfan-T, Lindane, Cypermethrin, Phorate & Dimethoate
7.	Himachal Pradesh	Chlorpyriphos & Cypermethrin
8.	Karnataka	Cypermetnrin, Triazophos, Quinalphos & Fenitrothion
9.	Haryana	Cypermethrin, Cyfluthrin-beta & Monocrotophos
10.	Rajasthan	Chlorpyriphos
11.	Uttar Pradesh	Aldrin, Chlorpyriphos, Dichlorvos, Phorate, Chlordane, Chlorfenvinfos, DDT-T, Heptachlor, Monocrotophos, Ethion, Endosulfan-T, HCH-A & HCH-B
12.	Andhra Pradesh	Cyfluthrin-Beta, Chlorpyriphos, HCH-A, HCH-D & Endosulfan-T
13.	Delhi & NCR	Cypermethrin-alpha, Cypermethrin, Chlorpyriphos, Alachlor, Parathion-ME & Pendimethalin

(*Source*: AINP on PR, ICAR, Govt. of India)

Safe use of insecticide to minimise its residue in fruits

Today, India has roughly 16% of the world's population and 2% of world's geographic area. Only 56% of our total geographic area is cropped. With fixed and limited cultivable land on one hand and increasing population in the other, India cannot afford to have low crop productivity and hence what we produce needs to be protected from pests and diseases. The burgeoning problems of insect infestation must be mitigated and that too by the application of insecticides as we do not have sufficient number of non-chemical alternatives in our arsenal. Only thing, we have to shift ourselves from the conventional approach to the safer synthetic alternatives so that the hazardous insecticide residue problems would be minimized in fruits for our own consumption and for export as well. And we are in the transition of this shifting process.

Low toxic insecticides shifting from higher toxicity

We need more number of safer insecticides that will create minimum risk towards non-targets. Today, modern insecticides registered in our country, such as Chlorantraniliprole, Hexythiazox, Lufenuron, Pyridalyl etc. are of low of mammalian toxicity. They also don't persist for a long time in the environment, particularly in the environment of tropical countries. Hopefully, these safer insecticides will replace the all of the conventional insecticides in near future giving complete protection to crop without affecting biodiversity.

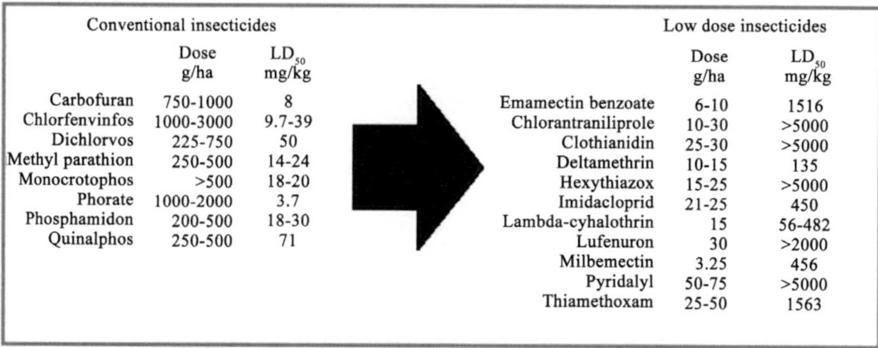

Conventional insecticides

	Dose g/ha	LD_{50} mg/kg
Carbofuran	750-1000	8
Chlorfenvinfos	1000-3000	9.7-39
Dichlorvos	225-750	50
Methyl parathion	250-500	14-24
Monocrotophos	>500	18-20
Phorate	1000-2000	3.7
Phosphamidon	200-500	18-30
Quinalphos	250-500	71

Low dose insecticides

	Dose g/ha	LD_{50} mg/kg
Emamectin benzoate	6-10	1516
Chlorantraniliprole	10-30	>5000
Clothianidin	25-30	>5000
Deltamethrin	10-15	135
Hexythiazox	15-25	>5000
Imidacloprid	21-25	450
Lambda-cyhalothrin	15	56-482
Lufenuron	30	>2000
Milbemectin	3.25	456
Pyridalyl	50-75	>5000
Thiamethoxam	25-50	1563

Fig. 5: Shifting towards safer synthetic alternatives

Newer mode of action combating resistance

A major threat toward insecticide use is rapid resistance development in target insects. The resistant insects are able to metabolize the insecticide, thereby reducing its concentration within the insect to an ineffective level. To avoid this problem, farmers need to change insecticides with different modes of action. Now, we have insecticides with varieties of modes of action in our hand. It is possible now to target different sites in insect bio-chemical system.

Table 6: Some target sites for insecticides.

Insecticides	Target Site
Acetamiprid, Clothianidin, Dinotefuran, Imidacloprid, Nitenpyram, Thiacloprid, Thiamethoxam,	Nicotinic acetylcholine receptor (nAChR) agonists (Nerve action)
Spinosad	Nicotinic acetylcholine receptor (nAChR) allosteric activators (Nerve action)
Abamectin, Emamectin benzoate, Lepimectin, Milbemectin	Chloride channel activators (Nerve and muscle action)
Hydroprene, Kinoprene, Methoprene, Pyriproxyfen, Fenoxycarb	Juvenile hormone mimics Growth regulation
Pymetrozine	Modulators of Chordotonal Organs (Nerve action)
Hexythiazox	Mite growth inhibitors (Growth regulation)
Bacillus thuringiensis subsp. israelensis *Bacillus thuringiensis* subsp. Aizawai, *Bacillus thuringiensis* subsp. Kurstaki, *Bacillus thuringiensis* subsp. tenebrionis, *Bacillus* sphaericus	Microbial disruptors of insect midgut membranes

Insecticides	Target Site
Diafenthiuron, Propargite	Inhibitors of mitochondrial ATP synthase
(Energy metabolism)	
Diflubenzuron, Flufenoxuron, Lufenuron, Novaluron, Buprofezin	Inhibitors of chitin Biosynthesis
Amitraz	Octopamine receptor agonists (Nerve action)
Chlorantraniliprole, Flubendiamide	Ryanodine receptor modulators (Nerve and muscle action)

Natural products with newer mode of action and leads

Throughout the history, plant products have been successfully exploited as plant protection chemicals. The chemistry of organocarbamate insecticides started from the isolation and identification of physostigmin, also known as esserine, from the calabar beans. The insecticidal properties of the several *Chrysanthemum* species were known for centuries in Asia and Africa. The pyrethrins, six naturally occurring terpenoids esters in chrysanthemum flowers, possess very good knock down effect against many insects, but they are highly photo labile under sunlight. The chemical manipulation in the structures of natural pyrethrins helped in developing a large number of photostable pyrethrin-like compounds collectively known as 'synthetic pyrethroids'. Azadirachtin, a strong antifeedant isolated from neem seed kernel is being exploited in plant protection, though it has limitation in its stability. The sun screen based formulation of azadirachtin is an integral part of the integrated pest management programme. The metabolites of animal origin have also been utilized as insecticides directly or indirectly. For an instance, the useful nereis toxin analogues were designed and developed from the lead molecule nereis toxin, which was isolated from a marine annelid, *Lumbrineris heteropoda*. The mammalian toxicities of all the analogues are much lower than that of the parent molecule nereis toxin. In future too, the investigation of bioactive natural products of plant, animal and microbial origin would lead to generate low mammalian toxic, low dose and environment benign molecules, which will be able to curb the insecticide residue problem.

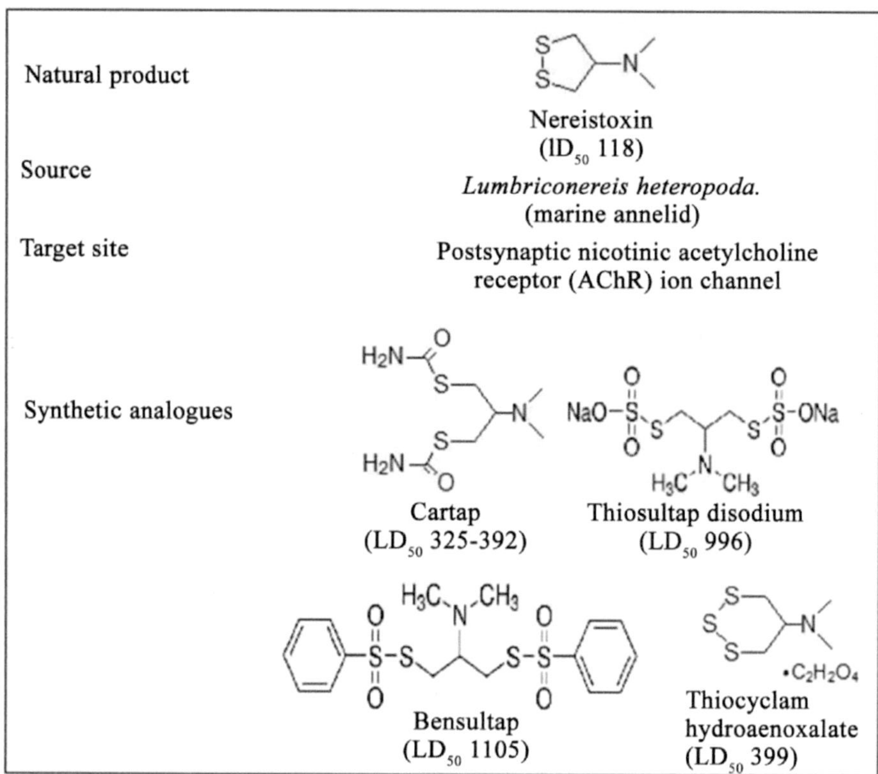

Fig. 6: Genesis of synthetic nereis toxin analogues targeting postsynaptic AChR ion channel

Biological formulations utilizing insects' own pests and diseases

Biological control of insects by using living insect pathogens is an eco-friendly self-sustaining insect management strategy. Once a component of this strategy gets established in the crop field, a continuous pressure builds upon the target insect to keep its population below economic threshold. Albeit these formulations of living organisms have some limitations, we have some formulations of *Bacillus thuringiensis* bacteria, Nuclear polyhedrosis virus (NPV) etc., which are important components in the integrated pest management programme.

Insecticide based management practices

The problem of insecticide residues in fruits can largely be avoided through good agriculture practice (GAP). A set of best management practices (BMP) is needed to be followed during insecticide application.

Selection of insecticide, rate of application and application method

Farmers need to select the insecticide and the appropriate method of application by scouting the area to evaluate the extent of damage by the insects. Application rate and label claims must strictly be followed to avoid unnecessary hazards. We should carefully consider all aspects of the pest problem, level of infestation and locational importance before recommending and/or applying any pesticide.

Information on the label available in/on the container

Pesticide labels contain important information about applicator and environmental safety. We should always follow label directions.

Mixing and loading of insecticides

Insecticides can reach groundwater and surface water as a result of spills that occur during mixing and loading operations. Mixing and loading should be done at the site of application and as far as from wells, lakes, streams, rivers and drains. There should not be any leakage in the tank and containers. Back siphoning may occur during filling the solution in the tank. Care should be taken to prevent back siphoning.

Calibration and equipment maintenance

Frequently check and maintenance of spray nozzles, hoses, gauges and tanks is needed. Proper calibration is the key to applying accurate rates of insecticides. Inaccurate tank volumes and pressure gauges or worn nozzles also may cause improper application. Under-application usually results in poor control, which may require retreatment. Over-application of insecticide may cause crop damage and may risk environment.

Avoidance of insecticide application in case of heavy rain forecast

Insecticides are most susceptible to be washed off from target sites by heavy rains during the first several hours after application.

Care to be taken to minimize overspray and drift

Wind speed, temperature and humidity all affect pesticide spray drift. Drift can be reduced by lowering boom heights and using nozzles that produce large droplet sizes.

Safe storage of insecticides

Insecticides need to be stored in a secure place and should be stored in their original containers with the labels clearly visible.

Proper disposal of insecticide containers

Containers should be pressured-rinsed thoroughly after use, punctured and buried in a pit.

Maintenance of records

Records of application details (product trade name, application rate, water volume used, insects and crops, date of application, etc.).

Education of farmers translating innovation

The greatest challenge, however, is neither technological nor scientific. Rather, by far the greatest challenge remains in convincing farmers. The success of the entire endeavor of developing safer technology lies in the adoption of that by users. Each and every stake holder of this system should be responsible for this translation. Strong networks between research organizations, extension groups and farmers are to be built up.

Reference

Abian, J., Durand, G. and Barcelo, D. 2002. Analysis of chlorotriazines and their degradation products in environmental samples by selecting various operating modes in thermospray HPLC/MS/MS. Journal of Agricultural and Food Chemistry, 41(8):1264-1273.

Aktar, W., Sengupta, D. and Chowdhury, A. 2009. Impact of pesticides use in agriculture: Their benefits and hazards. Interdisciplinary Toxicology, 2(1), 1–12. https://doi.org/10.2478/V10102-009-0001-7

Boskovic, N., Brandstatter-Scherr, K., Sedlacek, P., Bilkova, Z., Bielska, L. and Hofman, J. 2020. Adsorption of epoxiconazole and tebuconazole in twenty different agricultural soils in relation to their properties. Chemosphere, 261, 127637. https://doi.org/10.1016/J.Chemosphere.2020.127637

Carriger, J. F., Rand, G. M., Gardinali, P. R., Perry, W. B., Tompkins, M. S., & Fernandez, A. M. (2007). Pesticides of Potential Ecological Concern in Sediment from South Florida Canals: An Ecological Risk Prioritization for Aquatic Arthropods. Http://Dx.Doi.Org/10.1080/15320380500363095, 15(1), 21–45. https://doi.org/10.1080/15320380500363095

Carvalho, F. P. 2017. Pesticides, environment, and food safety. *Food and Energy Security*, 6(2), 48–60. https://doi.org/10.1002/FES3.108

Chakraborty, P., Zhang, G., Li, J., Sivakumar, A. and Jones, K. C. 2015. Occurrence and sources of selected organochlorine pesticides in the soil of seven major Indian cities: Assessment of air–soil exchange. Environmental Pollution, 204:74-80.

Cooper, J. and Dobson, H. 2007. The benefits of pesticides to mankind and the environment. Crop Protection, 26(9):1337-1348.

FAO, 2018. Pesticide Use Data-FAOSTAT. Retrieved from http://www.fao.org/faostat/en/#data/RP

Farhan, M., Wajid, A., Hussain, T., Jabeen, F., Ishaque, U., Iftikhar, M., Daim, M. A. and Noureen, A. 2021. Investigation of oxidative stress enzymes and histological alterations in tilapia exposed to chlorpyrifos. Environmental Science and Pollution Research, 28(11):13105-13111.

Food and Agriculture Organization of the United Nations (2002). International Code of conduct onthe Distributionand Use of pesticides.25thOctober, 2002. http://www.fao.org/WAICENT/FAOINFO/AGRICULT/code.pdf

Fontana, A. R., Lana, N. B., Martinez, L. D., and Altamirano, J. C. 2010. Ultrasound-assisted leaching-dispersive solid-phase extraction followed by liquid–liquid microextraction for the determination of polybrominated diphenyl ethers in sediment samples by gas chromatography–tandem mass spectrometry. Talanta, 82(1):359-366.

Fresh Fruits and Vegetables. (N.D.). Retrieved February 19, 2022, from https://apeda.gov.in/apedawebsite/six_head_product/FFV.htm

Gao, C., Yang, S., Wang, W. and Gao, L. (N.D.). Influence of Humic Acid Colloid on Adsorption of DDT in the Riverbed Sediments.

Geng, Y., Ma, J., Zhou, R., Jia, R., Li, C. and Ma, X. 2017. Assessment of insecticide risk to human health in groundwater in Northern China by using the China-PEARL model. Pest Management Science, 73(10), 2063–2070. https://doi.org/10.1002/PS.4572

Gill, H. K. and Garg, H. 2014. Pesticides: Environmental Impacts and Management Strategies. In Pesticides - Toxic Aspects. Intech Open. https://doi.org/10.5772/57399

GOI, 2020. Statistical Database. Directorate of Plant Protection, Quarantine & Storage, GOI. Retrieved from http://ppqs.gov.in/statistical-database

Kim, K. H., Kabir, E. and Jahan, S. A. 2017. Exposure to pesticides and the associated human health effects. Science of The Total Environment, 575, 525–535. https://doi.org/10.1016/J.SCITOTENV.2016.09.009

Kiritani K, Yamamura K. Exotic insects and their pathways for invasion. In: Invasive Species: Vectors and Management Strategies. Washington: Island Press; 2003. pp. 44-67 - Search Results - PubMed. (n.d.). Retrieved February 21, 2022, from https://pubmed.ncbi.nlm.nih.gov/?term=Kiritani K, Yamamura K. Exotic insects and their pathways for invasion. In: Invasive Species: Vectors and Management Strategies. Washington: Island Press; 2003. pp. 44-67

Krattiger, A. F. 1997. Insect resistance in crops: a case study of Bacillus thuringiensis (Bt) and its transfer to developing countries. ISAAA Briefs No. 2, International service for the acquisition of agri-biotech applications: Ithaca, p 42

Lamichhane, J. R. 2017. Pesticide use and risk reduction in European farming systems with IPM: An introduction to the special issue. *Crop Protection*, 97:1-6.

Luo, C., Huang, Y., Huang, D., Liu, M., Xiong, W., Guo, Q. and Yang, T. 2018. Migration and Transformation Characteristics of Niclosamide in a Soil-Plant System. *ACS Omega*, 3(2): 2312-2321.

Miller, S. F. 2016. The effects of weed control technological change on rural communities: Http://Dx.Doi.Org/10.1177/003072708201100405, 11(4):172-178.

OECD 2003. Descriptions of selected key generic terms used in chemical hazard/risk

Other Fresh Fruits. (n.d.). Retrieved February 19, 2022, from https://apeda.gov.in/apedawebsite/SubHead_Products/Other_Fresh_Fruits.htm

Pesticide Management Bill 2020, must address important concerns. (n.d.). Retrieved February 20, 2022, from https://www.downtoearth.org.in/blog/agriculture/pesticide-management-bill-2020-must-address-important-concerns-69303

Qin, F., Gao, Y. X., Guo, B. Y., Xu, P., Li, J. Z. and Wang, H. L. 2014. Environmental behavior of benalaxyl and furalaxyl enantiomers in agricultural soils. Http://Dx.Doi.Org/10.1080/03601234.2014.929482, 49(10):738-746.

Quan, G., Yin, C., Chen, T. and Yan, J. 2015. Degradation of Herbicide Mesotrione in Three Soils with Differing Physicochemical Properties from China. Journal of Environmental Quality, 44(5):1631-1637.

Robinson, D. E., Mansingh, A. and Dasgupta, T. P. 1999. Fate and transport of ethoprophos in the Jamaican environment. Science of The Total Environment, 237-238, 373-378.

Sethunathan, N., Adhya, T. K. and Raghu, K. 1982. Microbial degradation of pesticides in tropical soils. In: Matsumura F., and Krishna Murti, C. R. (eds), Biodegradation of Pesticides, pp. 91-115. Plenum Press, New York.

Singh, D. K. 2012. Pesticides and Environment. Pestic. Chem. Toxicol, 1, pp-114–122.

Slavin, J. L. and Lloyd, B. 2012. Health Benefits of Fruits and Vegetables. *Advances in Nutrition*, *3*(4), 506. https://doi.org/10.3945/AN.112.002154

TAAS, (Trust for Advancement of Agricultural Sciences), SPS, (Society of Pesticide Science), IPS, (Indian Phytopathological Society), ESI, (Entomological Society of India), 2020. Stakeholders Dialogue on Current Challenges and Way Forward for Pesticides Management - A Road Map. Indian Phytopathological Society.

Tariq, S. R. and Nisar, L. 2018. Reductive transformation of profenofos with nanoscale Fe/Ni particles. Environmental Monitoring and Assessment, 190(3):1-10.

Tcaciuc, A. P., Borrelli, R., Zaninetta, L. M. and Gschwend, P. M. 2018. Passive sampling of DDT, DDE and DDD in sediments: accounting for degradation processes with reaction–diffusion modeling. Environmental Science: Processes & Impacts, 20(1):220-231.

The World Health Report 2002: Reducing Risks, Promoting Healthy Life - World Health Organization - Google Books. (n.d.). Retrieved February 19, 2022, from https://books.google.co.in/books?hl=en&lr=&id=epuQi1PtY_cC&oi=fnd& pg = PR9 &ots=N4I39XBh Nn&sig=qejhiSmREGjZ vi1jQNNUWh5E9U&redir_esc=y #v= onepage&q&f=false

Walker, A. and Welch, S. J. 1992. Further studies of the enhanced biodegradation of soil-applied herbicides. Weed Research, 23:19–27.

Wickramaarachchi, W. A. R. T., Chaudhary, M., and Patil, J. (Eds.) 2017. Facilitating Microbial Pesticide Use in Agriculture in South Asia. SAARC Agriculture Centre.

Zhao, Y., Wendling, L. A., Wang, C. and Pei, Y. 2016. Behavior of chlorpyrifos and its major metabolite TCP (3,5,6-trichloro-2-pyridinol) in agricultural soils amended with drinking water treatment residuals. Journal of Soils and Sediments, 17(4):889-900.

Colour Plates

Plate 1: Pests of Mango

Mango hopper on tree trunk during off-season

Leaf cutting weevil and its damage

Mango leaf webber and its damage

Pupation of *Cricula* on mango leaf

Vegetative malformation due to mite and fungus

Association between mango mealy bug and ant

Plate-2: Pests of Mango

Damage symptom of leaf cutting weevil in mango sapling

Damage symptom of mango shoot gall psylla

Sooty mold due to hopper

Red banded mango fruit borer entering inside the fruit

Adult trunk borer

Fruit borer larva entering into the mango trunk for pupation

Plate 3: Pests of Citrus

Adult of citrus butterfly

Larva of citrus butterfly

Damage by citrus butterfly

Citrus shoot borer

Damage by citrus shoot borer

Citrus fruit sucking moth

Plate 4: Pests of Citrus, Banana, Guava.

Adults of citrus trunk borer

Grubs of trunk borer causing damage

Live mine of citrus leaf miner

Mealybug on banana leaf

Rugose spiralling whitefly infested guava leaf

Rugose spiralling whitefly on banana leaf

Plate 5: Pests of Papaya, Banana, Litchi and Pineapple.

Papaya mealybug infested leaf

Banana leaf and fruit scarring beetle infesting fruits

Pseudostem weevil grub infesting pseudostem

Mite infested litchi leaf, erinium

Litchi trunk infested by bark eating caterpillar

Mealybug infested pinapple plant

Plate 6: Pests of Temperate Fruit Crops

Codling moth infested apple fruit

Codling moth infested apple fruit

Flat headed borer larvae

Shot hole borer infestation

Pear psyllid on tender twig

Pear psyllid infested pear fruit

Plate 7: Pests of Temperate Fruit Crops and Pomegranate

Peach leaf curl aphid on almond

San jose scale infestation on pear

Infested twig with large peach aphid colony

Small walnut aphid colony on walnut leaf

Anar butterfly infestation on pomegranate

Stem borer infestation on trunk

Annexures

Annexure-I: Chemical pesticides and their trade names

Technical	Brand Name	Dose/Acre
1. Abamectin 1.8% EC	Vertimec (Syngenta), Tagmec 1.9 EC (Tropical), ABC 1.85% EC (KR)	50-100 ml/ acre, 5-10ml/ pump
2. Acephate 75%SP	Starthene(Swal),Orthene(Arysta), Missile (Devidayal), Megastar(MIL), Lancer (UPL), Oval (PI Ind.), Rasayan Phate (KR), Acefex (Excel), Kingmax (Vimax), Asataf (TATA), Accent 787 (Sumil), Miltaf (IIL), Bheem (Kilpest), Tagace (Tropical), Lucid (Cheminova), Lion (SuperCSL), Sritaf (Crystal), Ortain (Coromandel), Hilphate (HIL), Ample (Advance), Rythane (Ramcides), Corohamp (CAPL), Topsis (Atul), Molphate (GP), Top'O'Top (CGI), King Phate (KCS), Acesul (Sulphur Mills), Vega (PCCPL), Pace (Nagarjuna), Topsis (Atul), Tremor (BioStadt), Ace (Canary), Willace (Willowood), Chettak (GSP), Archa (Amber), Bhoochal 75 (AOL)	300-400 gm/ acre 20-25 gm/pump
3. Acephate 95%SG	Hunk (TATA)	
4. Acetamiprid 20%SP	Manik (TATA), Ekka (KR), Rapid (Crystal), Rekord (DuPont), Active (Devidayal), Award(MIL), KingPrid(KCS), Acetacel(Excel), Aceta(UPL), Echo 797 (Sumil), Stona (Vimax), Lift (Indofil), Dhan Preet (Dhanuka), Crop Pride (NACL), Sharp (IIL), Proud (Kilpest), Tagride (Tropical), Su- Pride (SuperCSL), Scuba (Coromandel), Hilprid (HIL), Armour (Advance), Acelon (CAPL), Albis (Atul), Quick (CLSL), Pounce (FMC), Molprid (GP), Tada(CGI), Glide(Safex), Prima(SulphurMills), Patron(PCCPL), Ennova (Nagarjuna), Albis (Atul), Wapkil& Rider (BioStadt), Antenna (Willowood), Nize (GSP), Shourya(Amber)	40-80 gm/ acre
5. Alpha-Cypermethrin 10% WP	Alfastar (Swal), Guru (UPL), Karfu (Kilpest), Safari (SuperCSL), Dash 5% WP (MIL), Dolphin (GSP)	200-300 ml/ acre

Technical	Brand Name	Dose/Acre
6. Alphamethrin 10%EC	Mig10 (Devidayal), Legend (MIL), Grand 301 (Sumil) Tata Alpha (TATA), Gem (Indofil), Nayak (IIL), Thrill (Tropical), Numethrin (Cheminova), Alpha (KR), Vasuki (CGI), Sixer (Sulphur Mills), Stop 10 EC (BioStadt), Alpha (Canary), Alfagold (CCIL)	
7. Bifenthrin 10% EC	Hectastar (Swal), Rock (KR), Metastar (Sumil), Markar (Dhanuka), Super Star (IIL), Barnerr (Tropical), Conister (Coromandel), Centrix (Sulphur Mills), Rockstar (Shivalik), Klintop (BioStadt), Wilthrin (Willowood), Viklap(Amber)	35-40 ml/pump
8. Bifenthrin 2.5%EC	Swal Home (Swal)	
9. Buprofezin 25%SC	Jawaa (DuPont), Flotis (Bayer), Trust (Swal), Irvy (Dow), Devifezin (Devidayal), Buprostar (MIL), Hillblaze (HIL), PI Bupro (PI Ind.), Applaud (TATA), Braun 111 (Sumil), Apple (Dhanuka), Phentom (IIL), Tagvoltage (Tropical), Tribune (Crystal), Ninja (Coromandel), KriMarch (KR), Cordon (Advance), Bravo (Sulphur Mills), Benj (Nagarjuna), Banzo (BioStadt), Jantar (Canary), Deligent (Willowood), Awaksh(Amber)	50 ml/pump
10. Carbaryl 50% WP	Devicarb (Devidayal), Kavin (KR)	
11. Carbofuran 3% GR	Furadan (FMC), Starfuron (Swal), Sumo 3G (IIL), Prachand 3G (Kilpest), Furon G (Crystal), Fury (Nagarjuna)	
12. Carbosulfan 25% EC	Marshal (FMC), Aaatank (Dhanuka), Aayudh (Coromandel)	
13. Cartap Hydrochloride 4% GR	Kitap 4G (KR), Kaardon 4G (UPL), Karvan (Devidayal), Caldan 4G (Dhanuka), Cartox (TATA), Corona GR (Sumil), Beacon GR (Indofil), Indian 4G (IIL), Kildon 4G (Kilpest), Fast 4G (Tropical), Sudan 4G (SuperCSL), Nidan 4G (Crystal), Parry/Ratna (Coromandel), Hilcarpet 4GR (HIL), Trixx (Advance), Grip (SulphurMills), Captor 4G (PCCPL), Nagarjuna 4G (Nagarjuna), Capsi GR (Atul), Dartriz 4G (BioStadt), Cartap 4G (Canary), Wiltap 4G (Willowood), Prithvi 4G (Amber), Eldan 50SP(CCIL), Paras 4G (AOL)	8-10 kg/acre

Technical	Brand Name	Dose/Acre
14. Cartap Hydrochloride 50% SP	Kitap 50SP (KR), Startop-50 (Swal), King Dan (KCS), Kaardon 50SP (UPL), Karvan Plus (Devidayal), Caldan 50SP (Dhanuka), Corona 50 (Sumil), Beacon SP (Indofil), Indian 50SP (IIL), Kildon (Kilpest), Fast 50 (Tropical), Sudan 50 (SuperCSL), Nidan 50SP (Crystal), Josh SP (Coromandel), Hilcartap 50SP (HIL), Grip 50 (SulphurMills), Captor 50SP (PCCPL), CapsiSP(Atul), Dartriz 50SP & Woktap 50SP (BioStadt), Mantar 50SP(Canary), Wiltap50SP(Willowood), Prithvi50SP(Amber), Eldan4G (CCIL), Paras 50SP (AOL)	200-400 gm/acre & 20-30 gm/acre
15. Chlorantraniliprole 0.4% GR	Ferterra (DuPont)	
16. Chlorantraniliprole 18.5% SC	Coragen (DuPont)	
17. Chlorfenapyr 10% SC	Intreprid (BASF), Lepido (PI Ind.), Record (Crystal)	
18. Chlorfluazuron 5.45% SC	Atabron (UPL)	600-800 ml/ acre
19. Chlorpyrifos 1.5% DP	Deviban 1.5 DP (Devidayal), Kilzex D &Kilzex 0.25% DP (Kilpest)	
20. Chlorpyrifos 10% GR	Deviban 10G (Devidayal), Suldrin GR (SuperCSL), Hilban 10 GR (HIL)	
21. Chlorpyrifos 20% EC	Krishan20(KR), Starban(Swal),Dursban(Dow), Megaban(MIL), Deviban 20 (Devidayal), Chloroban (UPL), ChloroKing (KCS), Banest 20 (Arysta), Tricel 20 (Excel), Uno 501 (Sumil), Lethal 20 EC (IIL), Sacban (Shivalik), Veermet (Vimax), Tafaban (TATA), Pyrifex (Safex), Gilphos (GIL), Crop Chlor(NACL), Mig-20(Kilpest), Tagban(Tropical), ShreebanTC(SKSCC), Classic-20(Cheminova), Suldrin(SuperCSL), Kaman/EldrinTC(Crystal), Trishul (Coromandel), Hilban 20 (HIL), Pyrifex (Safex), Pyriban (Aimco), Kattar 20 (GKR), Hyban (HCL), All-Drin (Ambuja), Sulban (SulphurMills), Lynch (PCCPL), Strike 20EC (BioStadt), Dangal (Canary), Chlorocil (GSP), Kartoos (Amber), Gold Ban (CCIL), Potta(AOL)	50 ml/pump

Technical	Brand Name	Dose/Acre
22. Chlorpyrifos 50% EC	Krishan 50 (KR), Predator (Dow), Tricel 48 EC (Excel), Megaban Super (MIL), DevibanTC(Devidayal), Sacban(Shivalik), Chowkidar(GIL), Lethal 50 (IIL), Tagban-TC (Tropical), Integer (Coromandel), Hilban 50 (HIL), Leafan-50 (Advance), Oshchlor (Oshnic), Kartoos 50 (Amber), Gold Ban TC (CCIL), Hitlor(AOL)	40-50 ml/pump
23. Chromafenozide 80% WP	Dodger (PI Ind.)	
24. Clothianidin 50% WDG	Dentotsu (Sumitomo)	50 gm/kg of seed
25. Cyantraniliprole 10OD	Benevia (DuPont)	2 ml/litre of water
26. Cypermethrin 10% EC	Kricyp-10 (KR), Starcip-10 (Swal), Megatrin (MIL), Cyper King (KCS), Ustaad (UPL), Devicyper 10 (Devidayal), SuperKiller 10 (Dhanuka), Cypermil(IIL), Cypercot10E(Kilpest), Challenger10(Tropical),Shakti-10 (Cheminova), Super Jet (SuperCSL), Cypercil (GSP), Action-10(CCIL),Shera 10 (AOL)	350-500 ml/acre
27. Cypermethrin 25% EC	Cymbush (Syngenta), Kricyp-25 (KR), Starcip-25 (Swal), Megacyper (MIL), Cyrux (UPL), Colt (PI Ind.), Devicyprin 25 (Devidayal), MaxKill (Vimax),Pyro401(Sumil), SuperKiller25(Dhanuka), CropCyper(NACL), Super Fighter (IIL), Cypercot 25E (Kilpest), Chellenger 25(Tropical), Shakti-25 (Cheminova), Super Jet+ (SuperCSL), Jackpot-25 (Crystal), Cyperkill (Coromandel), Hirange 25EC (Yesell), Cypersul 25 (Sulphur Mills), Auzar 25EC (BioStadt), Cyper (Canary), Cobra (GSP), Anth (Amber), Action-25 (CCIL), Shera 25 (AOL	220-300 ml / acre & 30-40 ml/pump
28. Cypermethrin 5% EC	Smash 5 EC (BioStadt)	
29. Deltamethrin 11% EC	Decis 100 (Bayer)	
30. Deltamethrin 2.8% EC	Decis 2.8 (Bayer), Shastra (Devidayal), Tagcis 100 – 11% EC (Tropical), Tagcis&Tagcis Flow 2.5% SC (Tropical)	

Technical	Brand Name	Dose/Acre
31. Diafenthiuron 50% SP	Polo/Pegasus (Syngenta), Krijet/Krijet Super (KR), Alero (Sumil), Logo/Gama (IIL), Ferotia (Coromandel), Derby (BioStadt), Ruby/Declare (GSP)	25 gm/pump
32. Dichlorovos (DDVP) 76% EC	Doom (UPL), Fume (KR), Starchlor (Swal), Divine (MIL), Devikol (Devidayal), DDVP King (KCS), Divap (PI Ind.), Crywap 75 EC (Vimax), Kush 760 (Sumil), Nuvan (IIL), Kilvos 76 E (Kilpest), Divipan (Tropical), Suvan (SuperCSL), Marvex Super (Coromandel), Hilvos (HIL), Avon (Advance), Hyvap (HCL), Demand (Sulphur Mills), Drone (PCCPL), Steamer (Canary), Haq (Amber), Badal 76 (Bharat), Nuvos (CCIL)	250-350 ml/acre
33. Dicofol 18.5% EC	Klin (KR), Starkel (Swal), Tiktok (UPL), Might (Devidayal), Sumfol (Sumil), Colonel-S (Indofil), Tagfol (Tropical), Hilfol 18.5 & 46 EC (HIL), Difol 18.5 SC (Sulphur Mills)	
34. Diflubenzuron 2% GR	Barcelo GR (Bayer)	1.25-3.0 kg/ha in clear water
35. Diflubenzuron 2% TB	Barcelo TB (Bayer)	0.5-1.0 ppm /40 lit. of water
36. Dimethoate 30% EC	Rogor (Cheminova), Tara-909 (Swal), Nugor (UPL), Progor (Pioneer), Devigon 30 (Devidayal), Tafgor (TATA), Crop Diamond (NACL), Rogorin (IIL), Kilgore 30 E (Kilpest), Tagore (Tropical), Kigor(SKACC)	
37. Dinotefuran 20% SG	Osheen (PI Ind.), Token (Indofil), Oshin 20SG (BioStadt)	
38. Emamectin Benzoate 5% SG	Proclaim(Syngenta), Missile(Crystal), Emstar5(KR), Starclaim(Swal), Robot (Excel), Spolit (UPL), Empower (Sumil), Tatkal (Gharda), Wegon (Vimax), EM-1 (Dhanuka), Xplode (IIL), Benzer (Coromandel), Emzet (Sulphur Mills), Trust (Nagarjuna), Bioclaim&Inclaim (BioStadt), Claim (Canary), Emacto (Willowood), Proceed (Amber)	80-100 gm/acre
39. Endosulfan 35% EC	Endostar (Swal), Thiokill (UPL), King Sulfan (KCS), Crop Endo (NACL), Kildofan (Kilpest), SuSulfan (SuperCSL), Hildan (HIL), Dhira (PCCPL), Endon (CCIL)	
40. Endosulfan 4% DP	Kildofan 4DP (Kilpest)	

Technical	Brand Name	Dose/Acre
41. Ethion 50% EC	Tafethion (TATA), Krithion (KR), Mit-505 (Swal), Mitkill (UPL), King Mite (KCS), Fosmite (PI Ind.), Deviastra (Devidayal), Vithion (Vimax), Fighter (Kilpest), Sumite (SuperCSL), Hilmite (HIL), Shakti (IPL)	
42. Etofenprox 10% EC	Trebon Excel 10EC (BioStadt)	
43. Etofenprox 30% EC	Dash 30EC (BioStadt)	
44. Fenobucarb (BPMC) 50% EC	Knock (KR), Hulchul (Devidayal), Blast (Tropical), Hopcide 50EC (BioStadt), Break (Willowood), Aandhi (Bharat)	
45. Fenpropathrin 10% EC	Danitol (Sumitomo)	
46. Fenpropathrin 30% EC	Meothrin (Sumitomo)	
47. Fenpyroximate 5% EC	Sedna (TATA), K-Mit (KR), Pyromite (Excel), Dynamite Plus (IIL), Wilgate (Willowood)	20-30 ml/pump
48. Fenvalerate 0.4% DP	Devifen 0.4 DP (Devidayal), Fencot 0.4 D (Kilpest)	
49. Fenvalerate 20% EC	Krifen (KR), Fenkill (UPL), King Pen (KCS), Fenval 20 EC (Isagro Asia), Devifen20(Devidayal), Tatafen10EC(TATA), Fencot20E(Kilpest), Final 20(AOL)	
50. Fipronil 0.3% GR	Regent GR (Bayer), Ruler 0.3GR (KR), Stargazette GR (Swal), Devigent 0.3GR(Devidayal), Sonic(TATA), FlorinaGR(Sumil), FaxGR(Dhanuka), Sargent GR (IIL), Tagagent GR (Tropical), Su-Nil GR (SuperCSL), Salvo GR (Coromandel), Fiprox G (Sulphur Mills), Avtar (PCCPL), Monil (Atul), Janbaaz GR (BioStadt), Fipro 0.3 (Canary), Vycm(Amber)	10 kg/acre
51. Fipronil 2.92% EC	Agenda 25 EC (Bayer)	100 ml/1 Lit. of water

Technical	Brand Name	Dose/Acre
52. Fipronil 5% SC	Regent SC (Bayer), Ruler (KR), Stargazette (Swal), Sonic Flo (TATA), Devigent Plus (Devidayal), Fipro King (KCS), Vizent (Vimax), Sonic Flo (TATA), Rabid (CLSL), Florina 444 (Sumil), Mahaveer (Gharda), Fax SC (Dhanuka), Sargent SC (IIL), Tagagent SC (Tropical), Su-Nil (SuperCSL), Crigent SC (Crystal), Salvo SC (Coromandel), Frazer (Advance), Fiprosh EC (Oshnic), Fiprox (Sulphur Mills), Monil SC (Atul), Janbaaz SC(BioStadt), Getter (Canary), Refree (GSP)	25-30 ml/pump
53. Fipronil 80% WG	Jump (Bayer), Ruler 80 (KR)	6-9 gm/pump
54. Flonicamid 50 WG	Ullala (UPL)	60-80 gm/ acre
55. Flubendiamide 20% WG	Fluton (PI Ind.), Takumi (TATA), Suraksha (Nagarjuna)	
56. Flubendiamide 39.35% SC	Fame (Bayer)	25-50 gm/ acre
57. Hexythiazox 5.45% EC	Endurer (Coromandel), Dimite (Nagarjuna), Xmite& Maiden (BioStadt)	
58. Imidachloprid 17.8% SL	Confidor (Bayer), Seamer (DuPont), Josh (KR), Imidastar (Swal), Imiden (Dow), Courage (MIL), Midas 2000 (Devidayal), King Dor (KCS), Jumbo (PIInd.),Imidacel17.8&20(Excel), ImidaGold(UPL), Expert100(Sumil), Sacdor (Shivalik), V-Mida (Vimax), Tatamida (TATA), Imigrow (CLSL), Expert 100 (Sumil), Atom (Indofil), Media (Dhanuka), Crop Mida (NACL), Victor (IIL), Kildor (Kilpest), Tropical Magic (Tropical), Chemida (Cheminova), Sumida (SuperCSL), Parrymida (Coromandel), Hilmida (HIL), Hotshot (Sulphur Mills), Radiant (PCCPL), Mida (Nagarjuna), Ribo(Atul), Ultimo 200SL (BioStadt), Farrata (Canary), Leopard 17.8 SL (Willowood), Imidore (CCIL), Midas (AOL)	70-100 ml/acre 15 ml/pump
59. Imidachloprid 30.5% SC	ConfidorSuper(Bayer), JoshPlus(KR), Protect(MIL), SuperKing(KCS), Jumbo(PIInd.), MaxMida(Vimax), Termax(TATA), ExpertSuper(Sumil), Victor Super (IIL), Tropical Magic Super (Tropical), Sumida 30.5 (SuperCSL), Confidence Super (Crystal), Total Control (Oshnic), Hotshot Plus(SulphurMills), Leopard30.5SC(Willowood),PunjSuper(Amber)	5-6 ml/pump

Technical	Brand Name	Dose/Acre
60. Imidachloprid 48% SL/FS	Gaucho 48 FS (Bayer), Imigo 600 FS (UPL), Tatamida 600 FS (TATA), Tropical Magic 600 FS (Tropical)	5-9 gm/kg of seed for seed treatment
61. Imidachloprid 70% WG / WS	Admire (Bayer), Tatamida 70WS (TATA), Josh 70 (KR), Dzire (Excel), Global 777 (Sumil), Ad-Fyre (Dhanuka), Victor Plus (IIL), Tagmyre& Tropical Magic 70WS (Tropical), Pactus (Cheminova), Sumida 70 WG (SuperCSL), Looper & Confidence 555 WS (Crystal), Hilmida WS (HIL), Pronto (Sulphur Mills), Eka (PCCPL), Mida-G (Nagarjuna), Edan(BioStadt), Hamster(Canary), Leopard70WG (Willowood), Jatan(Amber)	3-5 gm/kg seed for seed treatment
62. Indoxacarb 14.5% SC	Avaunt (DuPont), Care (KR), Fego (UPL), Index (Devidayal), Vindoxa (Vimax), Indimax (Sumil), Dhawa Gold (Dhanuka), Avval (IIL), Kaal (Kilpest), Tagpower (Tropical), Super Indoxa (SuperCSL), Vantage (Sulphur Mills), Amsac (Atul), Indica (Willowood), Arav (Amber), SuperDoxa (AOL)	160-200 ml/acre & 10-20 ml/pump
63. Lambda Cyhalothrin 2.5% EC	Lambdastar (Swal), Ninja (Devidayal), Reeva 2.5 (TATA), Flex 801 (Sumil), Agent Plus (Indofil), Bravo 2500 (IIL), Sumo (Kilpest), Demand (Tropical), Spider (SuperCSL), Judo (Crystal), Hillambda (HIL), Jaguar (Sulphur Mills), Xylo (Atul), Kozuka (BioStadt), Lambada2.5 (Canary), Samurai (GSP)	20-25 ml/pump
64. Lambda Cyhalothrin 10% WP	Icon (Syngenta)	
65. Lambda Cyhalothrin 4.9% CS	Matador (Syngenta), LOC++ (KR), Novell (Sumil), Metro (IIL), Challanger (Sulphur Mills), Farata (Canary), Spanner (GSP), Sip (Amber)	20-25 ml/pump
66. Lambda Cyhalothrin 5% EC	Karate(Syngenta),LOC-5(KR),Judo(Devidayal),Richo(Vimax),Reeva 5 (TATA), Dragon 701 (Sumil), Bravo 5000 (IIL), Sumo Plus (Kilpest), Tag Command (Tropical), Spider Plus (SuperCSL), Judo Plus (Crystal), Hillambda (HIL), Lakshya (Advance), Judho 5 EC (Yesell), Mustang (Sulphur Mills), Ray (PCCPL), Xylo-5 (Atul), Lambada5 (Canary), Sunny (Willowood), Santri(GSP), Remand(Amber), Kataar(CCIL), Thiachi(AOL)	20-25 ml/pump
67. Lufenuron 5% EC	Match (Syngenta), Modern Guard (MIL), Lunox (Crystal)	1 ml/L water

Technical	Brand Name	Dose/Acre
68. Malathion 5% DP	Devimalt 5 DP (Devidayal), Kropmal D (Kilpest)	
69. Malathion 50% EC	Kthion (KR), Devimalt 50 (Devidayal), Milthion (IIL), Lakshya (Kilpest), Tagthion (Tropical), Cythion (Coromandel), Hilmala / Hilthion (HIL), Sulmathion (Sulphur Mills), Malathion 57EC (BioStadt)	
70. Metaflumizone 22% SC	Tagline (Tropical)	
71. Metaldehyde 2.5% DP	Snailkil (PI Ind.)	
72. Methomil 40% SP	Kinet (KR), Lannate (DuPont), Dunet (Dhanuka), Dash (Indofil), Draagon (Tropical), Scorpio (Sulphur Mills)	300-500 ml
73. Methyl Parathion 2% DP	Gramexin (Pioneer), Devithion 2DP (Devidayal), Kildot 2DP (Kilpest), Tagpar (Tropical)	
74. Methyl Parathion 50% EC	Devithion 50 (Devidayal), Kildot 50 E (Kilpest)	
75. Monocrotophos 36% SL	Phoskill (UPL), Rasayanphos (KR), Monostar (Swal), Megamono (MIL), King Phos (KCS), Crotocel (Excel), Devimono (Devidayal), Monocip (Vimax), Monogil (GIL), Kevin 360 (Sumil), Crop Mono (NACL), Monocil (IIL), Kilphex 36% WSC (Kilpest), Macrophos (Tropical), Sumo (SuperCSL), Luphos (Crystal), Parryfos/Monophos (Coromandel), Hilcron (HIL), Admono (Advance), Monofos 36 (Safex), Monosul (Sulphur Mills), Shura (PCCPL)	350-500 ml/acre
76. Niclosamide 70% WP	Deadbol 70WP (BioStadt)	
77. Novaluron 10% EC	Rimon (Indofil), Noval (KR), Remostar (Swal), Novamax (Vimax), DNA (Willowood)	
78. Permethrin 25% EC	Perkill (UPL), Agniban (Devidayal), Hawk (IIL), Tag Bush (Tropical), Permasect (Coromandel)	
79. Phenthoate 50% EC	Phendal (Coromandel), Salvo 50EC (BioStadt)	

Technical	Brand Name	Dose/Acre
80. Phorate 10% SG/CG	Foratox (PI Ind.), Kaymet (KR), Starphor 10G (Swal), Umet (UPL), Thimet10CG (IIL), Tuskar 10 G (Kilpest), Helmet (Tropical), Sriphort (Crystal), Hilforate CG (HIL), Hilmet 10G (Canary)	5-10 kg/acre
81. Phosphamidon 40% SL	Starmidon (Swal), Kinadon Plus (UPL), Don 400 (Devidayal)	350-500 ml/acre
82. Profenophos 50% EC	Curacron (Syngenta), Celcron (Excel), Jashn (TATA), Kriphos (KR), Proven (MIL), King Cron (KCS), Carina (PI Ind.), Devi-soldier (Devidayal), Profenofos 50 EC (GAICL), Banjo (IIL), Maxcron (Vimax), Jashn (TATA), Bahadur 201- 40% (Sumil), Tagpro (Tropical), Aurifos (Cheminova), Sucron (SuperCSL), Kilcron (Crystal), Ajanta (Coromandel), Hilfos (HIL), Laser (Advance), Profos (Sulphur Mills), Kombat (PCCPL), Profex (Nagarjuna), Orax (Atul), Prahar & Prudent (BioStadt), Simcron (GSP), Profax (AOL)	50 ml/pump
83. Propergite 57% EC	Omite (Dhanuka), Propastar (Swal), Simbaa (PI Ind.), Mastamite (Chemtura), Teeka (Nagarjuna)	400-600 ml/acre
84. Pymetrozine 50% WG	Chess (Syngenta)	80-240 gm/acre
85. Quinalphos 1.5% DP	Deviquin 1.5 DP (Devidayal), Keterphos 1.5 D (Kilpest)	
86. Quinalphos 25% EC	Ekalux (Syngenta), Krilux (KR), Starlux 25 EC (Swal), Deviquin 25 (Devidayal), Virsta (Vimax), Flash (Indofil), Keterphos 25 E (Kilpest), Vazra-25 (Cheminova), Hilquin (HIL), Battle (GKR), Krush (BioStadt)	
87. Spinetoram 11.7% SC	Delrgate (Dow)	1 ml/L water
88. Spinosad 2.5% EC	Success (Dow)	240-280 ml/ acre
89. Spinosad 45% SC	Spintor (Bayer), Tracer (Dow), One-Up (Dhanuka), Taffin (TATA), Conserve (Nagarjuna),	60 ml/acre 6 ml/pump
90. Spiromesifen 22.9% SC	Oberon (Bayer), Voltage (PI Ind.)	250ml/acre 25ml/pump
91. Spirotetramat 150 OD	Movento (Bayer)	2ml/litre of water

Technical	Brand Name	Dose/Acre
92. Temephos 50% EC	Terminate (Devidayal)	
93. Thiachloprid 21.7% SC	Alanto (Bayer), Splendour (Cheminova)	100ml/acre 10ml/pump
94. Thiodicarb 75% WP	Larvin (Bayer)	0.5 gm/L of water
95. Thiamethoxam 25% WG	Actara (Syngenta), Kri-Oxm (KR), Battalion (Swal), Theme (MIL), King Tara (KCS), Maxima (PI Ind.), Wonderex (Excel), Renova (UPL), Devitara (Devidayal), Maestro 707 (Sumil), Maxtara (Vimax), Giltara (GIL), Click (Indofil), Areva(Dhanuka), Arrow(IIL), Actor(Kilpest), Tagxone(Tropical), SuperTara(SuperCSL), ExtraSuper(Crystal), Optra(Coromandel), Spike (Sulphur Mills), Dxtar (Nagarjuna), Spora (Atul), Evident &Tiomax (BioStadt), Willoxam (Willowood), Abhedh (Amber), Agrostar (AOL)	40-80 gm/acre &8-10 gm/pump
96. Thiamethoxam 30% FS	Slayer Pro (GSP), Abhedh Super (Amber)	
97. Thiamethoxam 70% WS	Slayer (GSP)	0.5 ml/L of water
98. Thiamethoxam 75% SG	Capcadis (Syngenta)	50-65 gm/ acre as a soil application with fertilizeror sand
99. Tolfenpyrad 15 EC	Keefun (PI. Ind.)	1 L/ha
100. Triazophos 20% EC	Jane 20% (KR)	
101. Triazophos 40% EC	Josh(UPL), Tarzan(KR), Triazostar(Swal), Truzo(MIL), Trizocel(Excel), Current 440 (Devidayal), Traz (Sumil), Ghatak (Dhanuka), Crop Tryzo (NACL), Titan (IIL), Kilthion (Kilpest), Kargil (Tropical), Trifos-40 (Cheminova), Suthion (SuperCSL), Hilazofos (HIL), Trizo (Advance), Triaceo (Sulphur Mills), Myza (Atul), Fulstop 40EC (BioStadt), Wilphos (Willowood), Trizo (GSP), Devak (Amber), Junoon 40 (AOL)	350-500 ml/acre & 40-50 ml/pump
Combined Insecticides		
1. Acephate 50% + Imidachloprid 1.8% SP	Lancer Gold (UPL), Stargold (Swal)	

Technical	Brand Name	Dose/Acre
2. Chlorantraniliprole 9.3% + Lambda Cyhalothrin 4.6% ZC	Ampligo (Syngenta)	80-100 ml/acre, 8-10 ml/ pump
3. Chlorpyrifos 16% + Alphamethrin 1% EC	Anth Super (KR), Legend Plus (MIL), Chlorthrin (Devidayal), Pincer 115 (Sumil), Aflatoon (IIL), Alert (Tropical), Logan (SuperCSL), Zoro (Crystal),Twins (Sulphur Mills), Aaghaat (Bharat)	30-35 ml/pump
4. Chlorpyrifos 21% + Fenobucarb (BPMC) 10.5% EC	Perfek 31.5 EC (BioStadt)	
5. Chlorpyrifos 50% + Cypermethrin 5% EC	Double Star (Swal), Anth (KR), Sinergy (MIL), Combi King (KCS), Koranda 505 (TATA), Sac 505 (Shivalik), Hamla 550 (Gharda), Super Strong 505 (Sumil), Lethal Super 505 (IIL), Pradhan (Kilpest), Action 505 (Tropical), Nurocomb (Cheminova), Super 505 (SuperCSL), Catchh (Coromandel), Hilhunter (HIL), Lynch + Combi (PCCPL), Cannon (Nagarjuna), Balio (Atul), Ulka 505 (BioStadt), Cyklon (GSP), Panther (Amber), Nagraj 505 (CCIL)	350-400 ml/acre 35-40 ml/pump
6. Cypermethrin 3% + Quinalphos 20% EC	Viraat (UPL), Alert (Devidayal)	350-500 ml/acre
7. Deltamethrin 0.27% + Buprofezin 5.65% EC	Devikey (Devidayal), Tusker (Crystal)	
8. Deltamethrin 1% + Triazophos 35% EC	Tiger (KR), Truzo Super (MIL), Tricada (Sumil), Combi DT (Devidayal), Shark (IIL), Shock (Kilpest), Tridelta (Tropical), Delfosse (Advance), Fusion(Sulphur Mills), Scoop (Willowood), Anaconda Plus (CCIL)	35-40 ml/pump
9. Ethion 40% + Cypermethrin 5% EC	Nagata (TATA), Colfos (PI Ind.), Jashn (KR), Mitplus (Swal), Mit Plus (Swal), Cyperton(MIL), Eagle405(Devidayal), RimJhim(IIL), Sumite-405 (SuperCSL)	350-400 ml/acre 35-40ml/pump
10. Imidachloprid 19.81% + Beta-cyfluthrin8.49% OD	Solomon (Bayer)	0.75-1 ml/L of water
11. Imidachloprid 40% + Ethiprole 40% WG	Glamore (Bayer)	50-0 gm acre

Technical	Brand Name	Dose/Acre
12. Imidachloprid 40% + Fipronil 40% WG	Lesenta (Bayer)	100-200 gm/ acre
13. Flubendiamide 19.92% + Thiachloprid 19.92% SC	Belt Expert (Bayer)	0.3-0.5 ml/L of water
14. Indoxacarb 14.5% + Acetamiprid 7.7% SC	Caesar (DuPont), Indoprid (Willowood)	
15. Profenophos 40% + Cypermethrin 4% EC	Polytrin C (Syngenta), Kriphos Super (KR), Correct (MIL), Prosper-44 (Devidayal), Profeno King (KCS), Roket (PI Ind.), Hitcel (Excel), Banjo Super (IIL), Maxcron Super (Vimax) Protrin (Sumil), Razor (Kilpest), Pataka (Tropical), Sumit 99 (SuperCSL), Kilcron Plus (Crystal), Impact (Advance), Cypro (Sulphur Mills), Profex Super (Nagarjuna), Orax Super (Atul), Genesis (Willowood), Ajanta Super (Coromandel), Legend Super (CCIL)	400-600 ml/acre & 35-40 ml/pump
16. Thiamethoxam 1% + Chlorantraniliprole 0.5% GR	Virtako (Syngenta)	2.5-4 kg/ acre
17. Thiamethoxam 12.6% + Lambda Cyhalothrin 9.5% ZC	Alika (Syngenta)	15 ml/pump
18. Thiamethoxam17.5% + Chlorantraniliprole 8.8% GR	Voliam Flexi (Syngenta)	0.5-1 ml/L of water
19. Flubendiamide 8.33 + Deltamethrin 5.56 SC	Fenos Quick (Bayer)	1 ml/L water
20. Spirotetramat 11.01% + Imidacloprid 11.01% w/w SC	Movento Energy (Bayer)	0.5-1 ml/ litre of water

Annexure II: Pesticide manufacturers in India

AIMCO, Akhand Jyoti, 8th Road, Santa Cruz (East), Mumbai - 400005, India. Website: www.aimco.com

ADAMA Ltd., IKP Knowledge Park, Plot No: DS -13, IKP Knowledge Park, Sy. No. 542/2, Genome Valley, Turkapally, Shameerpet Medchal-Malkajgiri district, Hyderabad, Telangana, 500101. Website: www.adama.com

Advance Pesticides, Somnath Park, Near Dndori Naka, Panchavati, Nasik-422003, Maharashtra, India. Website: www.advancepesticides.com

Amber Crop Science Pvt. Ltd., CSC-GH-4/FF-5, DDA Market, Meera Apartments, Outer Ring Road, Paschim Vihar, New Delhi- 110063, India. Website: www.ambercrops.com

Atul Ltd, Atul House Plot Number A-68, Road Number 21 Wagle Estate Thane 400 604, Maharashtra, India. Website: www.atul.co.in

BASF Group Companies in India, 1st Floor, VIBGYOR Towers, Plot No. C - 62, 'G' Block, Bandra-Kurla Complex, Mumbai - 400 051, Maharashtra. Website : www.india.basf.com.

Bayer Crop Science Limited, Bayer House, Central Avenue, Hiranandani Gardens, Powai, Mumbai – 400 076, Maharashtra, India. Website : www.bayer.co.in

Bhaskar Agro Chemicals Limited, No. 6 - 3 - 347 / 9, Dwaraka Puri Colony - 503, Riviera Apartment - 500 082, Hyderabad, Telangana.

Biostadt India Limited, Poonam Chambers, 'A' Wing, 6th Floor, Dr. A. B. Road, Worli, Mumbai - 400 018, India. Website: www.biostadt.com

Cheminova India Ltd., "Keshava" 7th Floor, Bandra - Kurla Complex, Bandra (E), Mumbai 400051 Website : www.cheminovaindia.in

Coromandal Indag Products Pvt Ltd., 62, S T Road, Chetpet, Chennai – 600031, Tamil Nadu. Website : http://indiankanoon.org

Coromandel International Limited, Coromandel House, Sardar Patel Road, Secunderabad 500003, Telangana, India. Website: www.coromandel.biz

Crop Chemicals India Ltd. (CCIL), C-63-64-65, Focal Point, Industrial Area, Kot Kapura. Distt. Faridkot, Punjab Pin-151204, India. Website: www.cropchemicals.co.in

Crop Life Science Ltd. (CCSL), 209, "PRIMATE", Nr.Judges Bunglow Cross Road, Bodakdev,Ahmedabad -380 015, Gujarat, India. Website: www.croplifescience.com

Crystal Crop Protection, B-95, Wazirpur Industrial Area, Wazirpur, Delhi-110052, India. Website: www.crystalcropprotection.com

Cyanamid Agro Ltd., Mahindra Tower, RBC 4th Floor, A & B Wing, Dr. G.M. Bhosle Road, Maharashtra, India. Website : www.bseindia.com

Devidayal Agro Chemicals, No. 123/124, Mittal Chamber,228 Narimon Point, Mumbai, Maharashtra, 400021, India.

Dhanuka Agritech Limited, 14th Floor, Building 5A, DLF Cyber Terrace, Cyber City, DLF Phase III Gurgaon- 122002, Haryana, India. Website : www.dhanuka.com

Dow Agro Sciences India Pvt. Ltd., Corporate Park, Unit No.1, V.N.Purav Marg, Chembur, Mumbai– 400 071, Maharastra. Website : www.dowagro.com/india

DuPont Crop Protection, E. I. DuPont India Private Limited, 6th Floor, Tower C, DLF Cyber Greens, Sector-25A, DLF City, Phase III, Gurgaon - 122002, Haryana, India. Website:www.dupont.co.in

E I D Parry (India) Limited, Dare House, New No.2, Old 234, NSC Bose Road, Chennai – 600 001. Website : www.eidparry.com

Excel Crop Care Limited, 13/14 Aradhana Industrial Development Corporation, Near Virwani Industrial Estate, Goregaon East, Mumbai - 400063. Website : www.excelcropcare.com

FMC India Ltd., Embassy Star (1st Floor), No 8 Palace Road, Bangalore – 560052, India Website : www.fmc.in

Gharda Chemicals Ltd., Avashi, Maharashtra, 415722, India Website : www.gharda.com

Gujarat Agro Industries Corporation Ltd., Khet - Udyog Bhavan, Opp. Old High Court, Navrangpura, Ahmedabad - 380 014. Website : www.gujagro.org

GSP Crop Science Private Limited, 403, Lalita Complex, 353/3 Rasala Road, Nr. Jain Temple, Navrangpura, Ahmedabad, 380009, Gujarat, India

Hindustan Insecticides Limited, 2nd Floor, Core - 6, SCOPE Complex, 7, Lodhi Road, New Delhi - 110 003, India. Website : www.hil.gov.in

Hyderabad Chemical Limited, A-24/25 APIE, Balanagar, Hyderabad - 500037, Andhra Pradesh, India. Website : www.hyderabadchemicals.com

ICI Zeneca Ltd., Block N1, 12th FloorManyata Embassy Business Park, Rachenahalli, Outer Ring Road, 560045 Bangalore, India. Website : www.astrazeneca.com

Indian Oil Corporation Ltd., Corporate Office, 3079/3, J B Tito Marg, Sadiq Nagar, New Delhi – 110049. Website : www.iocl.com

Indofil Industries Limited, Kalpataru Square – 4th floor, Kondivita Rd., Off Andheri Kurla Rd., Andheri – East, Mumbai- 400 059, Maharashtra, India. Website : https://indofilcc.com

Insecticides (India) Ltd., 401-402, Lusa Tower, Azadpur Commercial Complex, Delhi – 110033 Website : www.insecticidesindia.com

Kilpest India limited, 7-C, Industrial Area, Govindpura, Bhopal, India. Website : www.kilpest.com

Krishirasayan (KR), 29, Elgin Rd, Gaza Park, Sreepally, Bhowanipore, Kolkata, West Bengal 700020, India. Website: www.krishirasayan.com

Makhteshim-Agan India Pvt. Ltd., Plot No: DS -13, IKP Knowledge Park, Sy. No. 542/2, Genome Valley, Turkapally, Shameerpet, Ranga Reddy District, Hyderabad, AP 500078, India. Website : www.ma-india.com

Meghamani Organics Ltd., 183-184, Meghmani House, Shree Niwas Society, Paldi, Ahmedabad, Gujarat 380007, India. Website : www.meghmani.com

Modern Insecticides Limited (MIL), Corp. Office : 2nd Floor, Kesar Complex, Near Malhar Cinema, Gurdev Nagar, Ludhiana, India. Website: www.milworld.co

Nagarjuna Agro Chemicals Pvt. Ltd., G-01, D.No. 6-3-1218/6/2, Street No.6, Spring Heaven, Umanagar, Begumpet, Hyderabad-500016, A. P, India. Website : www.nagarjunaagrochemicals.com

Nagarjuna Fertilizers and Chemicals Limited, Nagarjuna Hills, Hyderabad - 500 082, Andhra Pradesh, India. Website : www.nagarjunafertilizers.com

Paushak Limited, Alembic Road, Vadodara – 390 003, India. Website : www. paushak.com

PI Industries Limited (PI), Udaisagar Road, Udaipur – 313001, Rajasthan, India. Website: www.piindustries.com

Pest Control (India) Pvt. Ltd., 2, 3, 4 Floor, 'Narayani' Ambabai Temple Compound, Near Bank of Maharashtra, Aarey Road Goregaon (West), Mumbai - 400 062, Maharashtra, India Website: www.pestcontrolindia.com

Punjab Chemicals and Crop Protection Limited, Milestone-18, Ambala-Kalka Road, Village & P.O. Bhankharpur, Derabassi, Dist. SAS Nagar, (Mohali), Punjab-140201, India. Website: www.punjabchemicals.com

Rallis India Limited, 156/157, 15th Floor, Nariman Bhavan, 227, Nariman Point, Mumbai – 400 021 Website: www.rallis.co.in

Sabero Organics Gujarat Limited, (A subsidiary of Coromandel International Ltd.) *A Murugappa Group Company,* Bezzola Commercial Complex, 3rd floor, A - Wing, Suman Nagar, Sion Trombay Road, Chembur, Mumbai – 400 071, India. Website : www.sabero.com

Safex Chemicals, 4th & 5th Floor, NDM-1, Netaji Subhash Place, New Delhi-110034, India. Website: www.safexchemicals.com

Shaw Wallace and Company, "Wallace House," 4, Bankshall Street, Kolkata, West Bengal, 700001. Website : www.shawwallace.lk

Shivalik Crop Science Pvt. Ltd., Neelam Cinema Complex, Sector 17-E, Chandigarh Pin: 160017, India. Website: www.shivalikcropsciences.com

Southern pesticides, 10-5-3/2/2, Masab Tank, Hyderabad – 500028. Website : www.southernag.com

Spec India Ltd., "SPEC House", Parth Complex, Near Swastik Cross Roads, Navarangpura, Ahmedabad 380 009, INDIA. Website : www.spec-india.com

Sulphur Mills Limited, 349 – Business Point, Western Express Highway, Andheri (E), Mumbai – 400069. India. Website: www.sulphurmills.com

Sumil Chemical Industries Pvt. Ltd., 349- Business Point, Western Express Highway,Andheri(E), Mumbai – 400069, India. Website: www.sumilchem.com

SWAL Corporation Limited, UPL House, 4th Floor, CTS No. 610B/2, Bandra Village, Off Western Express Highway, Behind Teachers Colony, Bandra (East), Mumbai 400 051. Website: www.swal.in

Syngenta India Ltd., Amar Paradigm, S. No. 110/ 11/ 3, Baner Road, Baner, Pune 411 045 Website : www.syngenta.com

Tropical Agrosystem India (P) Ltd., Jhaver Centre, IV th Floor, 72, Marshalls Road, Chennai, Tamilnadu, India 600008. Website: www.tropicalagro.in

UPL, UPL House, 610 B/2, Bandra Village, Off Western Express Highway, Bandra (East), Mumbai 400 051, India. Website: www.upl-ltd.com

Vimax Crop Science Limited, D-87, Kuvadva GIDC, Nr. National Highway 8-B, Rajkot - 360 003. Gujarat,India. Website: www.vimaxcropscience.com

Vital Crop Science Pvt. Ltd., 101, Vaibhav Chamber, 7/1, Ushaganj, Chhawani, Indore - 1, India Website : www.indianyellowpages.com/vitalcropscience

Willowood, The Lounge, 4/F, 23 Chetla Central Rd, Kolkata – 700027 West Bengal, India. Website: www.willowood.com

Annexure-III: Insecticides registered for pest management in fruit crops in India

Pesticides	Crops	Pests	Dosage (per ha)			Waiting period (days)
			a.i. (gm)	Formulation (gm/ml per l.)	Dilution in water (Litre)	
Abamectin1.9%EC	Grapes	Mites	0.014/l.	0.75ml/l.	500-1000	03
Bifenthrin 8%SC	Apple	Mites	60 gm	750 ml/tree	10 litre/tree	11
Buprofezin 25%SC	Mango	Hoppers	0.025-0.05	1-2ml/l.	5-10 litre/tree	20
	Grapes	Mealy bugs	250-375	1000-1500	500-1000	07
Carbofuran 3 % CG	Apple	Wooly aphid	5/tree	166/tree	. . .	. . .
	Citrus	Nematode	360	12000	. . .	. . .
		Leaf miner	1500	50000	. . .	. . .
	Banana	Rhizome weevil	1g/sucker	33gm/sucker	. . .	. . .
		Aphid	50gm/sucker	166gm/sucker	. . .	. . .
		Nematode	1.5gm/sucker	50gm/sucker	. . .	. . .
	Peach	Leaf curl aphid	1000	33300	. . .	. . .
	Mandarins	Soft green scale	0.40gm/plant	13.30gm/plant	. . .	. . .
Chlorpyrifos 20% EC	Apple	Aphid	0.05%	3750-5000	1500-2000	. . .
	Ber	Leaf hopper	0.03%	2250-3000	1500-2000	. . .
	Citrus	Citrus black aphid	0.02%	1500-2000	1500-2000	. . .
Cyantraniliprole 10.26%OD	Grapes	Thrips	70	700	1000	. . .
	Pomegranate	Thrips	75	750	1000	. . .
		Whitefly	90	900	1000	. . .
Cyenopyrafen 30 % SC	Apple	Mite	60-90	200-300	1000	. . .
Deltamethrin 02.80 % EC	Mango	Mango hopper	0.03-0.05%	0.33-0.5ml/litre	As per spray field requirement	01
Dicofol18.50 % EC	Litchi	Red spider mite	0.05%	2700 – 4050	1000 – 5000	15-20

Pesticides	Crops	Pests	Dosage (per ha)			Waiting period (days)
			a.i. (gm)	Formulation (gm/ml per l.)	Dilution in water (Litre)	
Diafenthiuron 50.00% WP	Citrus	Mites	1.0 g/l	2.0 g/l	2-3 liter/ha.	30
Dimethoate 30.00% EC	Apricot	Aphid	0.03%	1485-1980	1500-2000	-
	Banana	Aphid, Lace wing bug	0.03%	1485-1980	1500-2000	-
	Citrus	Black aphid	0.03%	1485-1980	1500-2000	-
	Fig	Fig Jassid	0.03%	1485-1980	1500-2000	-
		Mealy bug	0.03%	2475-3300	1500-2000	-
	Mango	Hopper	0.05%	2475-3300	1500-2000	-
Emamectin benzoate 5.00% SG	Grapes	Thrips	11.00	220.0	500-1000	05
Fenazaquin 10.00% EC	Apple	Red spider mite, Two spotted mite	40.0	400	1000	30
Fipronil 80.00% WG	Grapes	Thrips	40.0-50.0	50-62.5	750-1000	10
Hexythiazox 05.45% w/w EC	Apple	European RedMite	0.002%	0.04%	10ltr./tree	15
Imidacloprid 17.80% SL	Mango	Hopper	0.40-0.80 g/tree	2.0-4.0 ml/tree	10 litre	45
	Citrus	Leaf miner, Psylla	10.0	50.0	Depending on size of tree & Protection equipment used	15
	Grapes	Flea beetle	0.06-0.08	300.0-400.0	1000	32
Lambda-cyhalothrin 04.90% CS	Grapes	Thrips & Flea beetle	12.50	250.0	500-1000	07

Pesticides	Crops	Pests	Dosage (per ha)			Waiting period (days)
			a.i. (gm)	Formulation (gm/ml per l.)	Dilution in water (Litre)	
Lambda-cyhalothrin 05.00% EC	Mango	Hoppers	0.0025-0.005%	0.5-1.0 ml/l of water	-	07
Malathion 50.00% EC	Apple	Sanjose scale, Wooly aphid	0.05%	1500-2000	1500-2000	-
	Mango	Mealy scale, Mango hopper	0.075%	2250-3000	1500-2000	-
	Grape	Beetle	500.0	1000	1500-2000	-
	Grapes	Mealy bug	500.0	1250	500-1000	10
Methyl Bromide 98.00% w/w	Dry Fruits	Rust Red Flour Beetle	Air tight cover	24 -32 gm/m3	24 hrs waiting Period 72 hrs	As when residues not to exceed 25 ppm
Oxydemeton-methyl 25.00% EC	Apple	Sanjose scale	0.07%	4200-5600	1500-2000	-
		Wooly Aphid	0.025%	1500-2000	1500-2000	-
	Banana	Tingid bug	0.025%	1500-2000	1500-2000	-
		Aphids	0.05%	3000-4000	1500-2000	
	Mango	Hoppers	0.025%	1500-2000	1500-2000	-
	Peaches	Leaf curl aphids	0.025%	1500-2000	1500-2000	-
Propargite 57.00% EC	Apple	European red mite, Two spotted mite	2.85-5.7/tree	5-10 ml/tree	10 lit/tree	09

Pesticides	Crops	Pests	Dosage (per ha)			Waiting period (days)
			a.i. (gm)	Formulation (gm/ml per l.)	Dilution in water (Litre)	
Quinalphos 25.00% EC	Apple	Wooly Aphid	0.05%	3000-4000	500-1000	-
	Banana	Tingid bug	0.05%	3000-4000	500-1000	-
	Citrus	Scale	0.07%	4200-5600	500-1000	-
		Citrus butterfly	0.025%	1500-2000	500-1000	-
	Pomegranate	Scales	0.08%	4800-6400	500-1000	-
Spinosad 45.00% SC	Grapes	Thrips	25 ml/100 lit	250	1000	15
Spiromesifen 22.90% SC	Apple	European Red Mite & Red spider mite	72 (0.03%)	300	1000	30
Thiacloprid 21.70% SC	Apple	Thrips	0.01-0.012%	0.04-0.05%	As per size of tree	30
Thiamethoxam 25.00% WG	Mango	Hoppers	25.0	100	1000	30
	Citrus	Psylla	25.0	100	1000	20
*Verticillium lecanii*01.15% WP	Citrus	Mealy bug	02.50 kg	500-550 L	-	-
Metaldehyde	Citrus	Snails, Slugs	Available in ready to use 2.5% Dust			

(*Source*: Central Insecticide Board & Registration Committee, Directorate of Plant Protection, Quarantine & Storage, Department of Agriculture & Farmers Welfare, Government of India)

Annexure - IV: Insecticides/Pesticides banned, refused registration and restricted use in India

(As on 01.04.2022)

1. Insecticides/Formulations banned in India

Pesticides banned for manufacture, import and use	
Sl. No.	Insecticides/Acaricides/Nematicides/Rodenticide
1.	Aldicarb (vide S.O. 682 (E) dated 17th July 2001)
2.	Aldrin
3.	Benzene Hexachloride
4.	Carbaryl (vide S.O 3951(E) dated 8th August, 2018)
5.	Chlorbenzilate (vide S.O. 682 (E) dated 17th July 2001)
6.	Chlordane
7.	Chlorofenvinphos
8.	Copper Acetoarsenite
9.	Diazinon (vide S.O 3951(E) dated 8th August, 2018)
10.	Dibromochloropropane (DBCP) (vide S.O. 569 (E) dated 25th July 1989)
11.	Dichlorovos (Vide S.O. 3951 (E), dated 08.08.2018)
12.	Dieldrin (vide S.O. 682 (E) dated 17th July 2001)
13.	Endosulfron (vide ad-Interim order of the Supreme Court of India in the Writ Petition (Civil) No. 213 of 2011 dated 13th May, 2011 and finally disposed of dated 10th January, 2017)
14.	Endrin
15.	Ethyl Parathion
16.	Ethylene Dibromide (EDB) (vide S.O. 682 (E) dated 17th July 2001)
17.	Fenthion (vide S.O 3951(E) dated 8th August, 2018)
18.	Heptachlor
19.	Lindane (Gamma-HCH)
20.	Menazon
21.	Methyl Parathion (vide S.O 3951(E) dated 8th August, 2018)
22.	Phorate (Vide S.O. 3951 (E), dated 08.08.2018)
23.	Phosphamidon (Vide S.O. 3951 (E), dated 08.08.2018)
24.	Sodium Cyanide (banned for Insecticidal purpose only vide S.O 3951(E) dated 8th August, 2018) *
25.	Tetradifon
26.	Thiometon (vide S.O 3951(E) dated 8th August, 2018)
27.	Toxaphene (Camphechlor) (vide S.O. 569 (E) dated 25th July 1989)
28.	Triazophos (Vide S.O. 3951 (E), dated 08.08.2018)
29.	Trichlorfon (Vide S.O. 3951 (E), dated 08.08.2018)

2. Pesticide formulations banned for import, manufacture and use

1.	Carbofuron 50% SP (vide S.O. 678 (E) dated 17th July 2001)
2.	Methomyl 12.5% L
3.	Methomyl 24% formulation
4.	Phosphamidon 85% SL

3. Pesticides Withdrawn

(Withdrawal may become inoperative as soon as required complete data as per the guidelines is generated and submitted by the Pesticides Industry to the Government and accepted by the Registration Committee. (S.O 915(E) dated 15th Jun,2006)

1.	Formothion
2.	Paradichlorobenzene (PDCB)
3.	Warfarin (vide S.O. 915 (E) dated 15th June 2006)

* Regulation to be continued in the extant manner for non-insecticidal uses.

4. Pesticides refused registration

Sl. No.	Insecticides/Acaricides/Nematicides/Rodenticide
1.	Azinphos Ethyl
2.	Azinphos Methyl
3.	Calcium Arsenate
4.	Carbophenothion
5.	Dicrotophos
6.	EPN
7.	Lead Arsenate
8.	Leptophos (Phosvel)
9.	Mephosfolan
10.	Mevinphos (Phosdrin)
11.	Thiodemeton / Disulfoton
12.	Vamidothion

5. Pesticides restricted for use in the country

Sl. No.	Insecticides	Details of Restrictions
1.	Aluminium Phosphide	The Pest Control Operations with Aluminium Phosphide may be undertaken only by Govt./Govt. undertakings / Govt. Organizations / pest control operators under the strict supervision of Govt. Experts or experts whose expertise is approved by the Plant Protection Advisor to Govt. of India except [1]Aluminium Phosphide 15 % 12 g tablet and [2]Aluminum Phosphide 6 % tablet. [RC decision circular F No. 14-11(2)-CIR-II (Vol. II) dated 21- 09-1984 and G.S.R. 371(E) dated 20th may 1999]. [1]Decision of 282nd RC held on 02-11-2007 and, [2]Decision of 326th RC held on 15-02-2012. The production, marketing and use of Aluminium Phosphide tube packs with a capacity of 10 and 20 tablets of 3 g each of Aluminium Phosphide are banned completely. (S.O.677 (E) dated 17thJuly, 2001)
2.	Cypermethrin	Cypermethrin 3 % Smoke Generator is to be used only through Pest Control Operators and not allowed to be used by the General Public. [Order of Honorable High Court of Delhi in WP(C) 10052 of 2009 dated 1407-2009 and LPA-429/2009 dated 08-09-2009]
3.	Dazomet	The use of Dazomet is not permitted on Tea. (S.O.3006 (E) dated 31st Dec, 2008)
4.	Dichloro Diphenyl Trichloroethane (DDT)	The use of DDT for the domestic Public Health Programme is restricted up to 10,000 Metric Tonnes per annum, except in case of any major outbreak of epidemic. M/s Hindustan Insecticides Ltd., the sole manufacturer of DDT in the country may manufacture DDT for export to other countries for use in vector control for public health purpose. The export of DDT to Parties and State non-Parties shall be strictly in accordance with the paragraph 2(b) article 3 of the Stockholm Convention on Persistent Organic Pollutants (POPs). (S.O.295 (E) dated 8th March, 2006) Use of DDT in Agriculture is withdrawn. In very special circumstances warranting the use of DDT for plant protection work, the state or central Govt. may purchase it directly from M/s Hindustan Insecticides Ltd. to be used under expert Governmental supervision. (S.O.378 (E) dated 26thMay, 1989)
5.	Fenitrothion	The use of Fenitrothion is banned in Agriculture except for locust control in scheduled desert area and public health. (S.O.706 (E) dated 03rdMay, 2007)

Sl. No.	Insecticides	Details of Restrictions
6.	Methyl Bromide	Methyl Bromide may be used only by Govt./Govt. undertakings/Govt. Organizations / Pest control operators under the strict supervision of Govt. Experts or Experts whose expertise is approved by the Plant Protection Advisor to Govt. of India. [G.S.R.371 (E) dated 20thMay, 1999 and earlier RC decision]
7.	Monocrotophos	Monocrotophos is banned for use on vegetables. (S.O.1482 (E) dated 10thOct, 2005)

Source: Directorate of Plant Protection Quarantine & Storage, Faridabad, 2022.